Using Aspen Plus® in Thermodynamics Instruction

Using Aspen Plus® in Thermodynamics Instruction

A Step-by-Step Guide

Stanley I. Sandler

Department of Chemical and Biomolecular Engineering
University of Delaware
Newark, DE

WILEY

A Joint Publication of the American Institute of Chemical Engineers and John Wiley & Sons, Inc.

Published by John Wiley & Sons, Inc., Hoboken, New Jersey.
Published simultaneously in Canada.

Library of Congress Cataloging-in-Publication Data:

Sandler, Stanley I., 1940- author.
Using Aspen plus in thermodynamics instructions : a step by step guide / by Stanley I. Sandler, Department of Chemical and Biomolecular Engineering, University of Delaware.
pages cm
Includes index.
ISBN 978-1-118-99691-1 (paperback)
1. Thermodynamics–Data processing. 2. Aspen plus. I. Title.
QD504.S26 2015
536'.7028553–dc23

2014029629

Printed in the United States of America.

10 9 8 7 6 5 4 3 2 1

Contents

Preface

Aspen Plus® is a very powerful process simulator—that is, a tool for modeling chemical processes, including complete chemical and pharmaceutical plants and petroleum refineries. As such, it requires accurate models of thermodynamic properties and phase behavior.

The purpose of this book is to introduce the reader to the use of Aspen Plus in courses in thermodynamics; consequently, very few of the process simulation capabilities are considered here. In undergraduate chemical engineering degree programs, process simulation is heavily used in the capstone design course, and this is where its details and intricacies are generally taught. This book serves as a prelude to instruction in the more complex process simulation and provides a coherent approach to introducing the Aspen Plus simulator in undergraduate thermodynamics courses. I hope it will make such courses more interesting and relevant by allowing calculations of real processes that would otherwise be very tedious. One advantage of doing such calculations by computer is that repetitive calculations with varying parameters are quickly achieved, so that the student gains experience into the ways in which different input parameters affect the output. Such calculations develop engineering insight. Any instructor knows that asking students to do repetitive calculations by hand is met by moans and groans. Doing a calculation for one case is an important learning activity, while doing many cases by hand has much less pedagogical return for the student's time investment.

This book provides the reader with a self-study, step-by-step guide to doing thermodynamic calculations in Aspen Plus. It provides actual screenshots of the Aspen Plus interface to solve example problems of specific types, including vapor–liquid, liquid–liquid, vapor–liquid–liquid and chemical reaction equilibria, and simple applications to liquefaction, distillation, and liquid–liquid extraction. One important feature is that learning occurs by means of illustration. It is not a book of rules but of specific examples, encouraging readers to generalize from those examples and apply what they have observed to a specific problem. Designed for self-study, this book is not meant as an in-class textbook but for out-of-class use. However, there are places in this book where it is useful to refer to fundamental thermodynamic principles. In such cases, for convenience, I provide explicit references to my thermodynamics textbook, *Chemical, Biochemical, and Engineering Thermodynamics*, 4th ed., published by John Wiley & Sons, Inc., in 2006. However, the same material can be found in any standard thermodynamics textbook, so this book can be used with any thermodynamics textbook.

Let me reiterate that although the Aspen Plus program is designed to do process simulation, the purpose of this book is not to emphasize simulation. Examples of using the program for simulation are included, however, because some thermodynamics calculations can only be done in Aspen Plus by using simulation. These include vapor–liquid and vapor–liquid–liquid equilibrium flash calculations, especially an adiabatic flash (i.e., Joule–Thomson expansion) and chemical reaction equilibrium calculations. Please keep in mind that the Aspen Plus program has far greater capabilities than are demonstrated here.

This book is meant to be a step-by-step guide for individual study. As such, the book contains many screen images produced using Aspen Plus®. These screen images of Aspen

All suggestions for improvement would be greatly appreciated. Please communicate those to me at sandler@udel.edu.

Finally, I want to acknowledge the assistance provided by Aspen Technology, Inc., that provided an individual license to use Aspen Plus® so that I could work on this manuscript at home while attending to my wife during her final illness. I especially want to thank Chau-Chyun Chen, a former Aspen Technology employee, who made this possible and also provided many helpful comments on an early version of the manuscript. Suphat Watanasiri of Aspen Technology was also helpful in the process.

STANLEY I. SANDLER

January 2015

An Introduction for Students

As you are beginning to see in your courses, thermodynamic calculations for other than ideal gases can be quite time consuming. Also, in class you may have considered a single device, such as a Joule–Thomson valve, or in the case of liquefaction, just a few devices (e.g., a compressor, a heat exchanger, a Joule-Thomson valve and a separator.) Such calculations can require many iterations, both for each unit operation (e.g., equation of state calculations for the compressor) and, if there are recycle streams as in the Linde process, for the overall process as well. You can imagine how difficult and time-consuming such calculations would be for a whole chemical plant or petroleum refinery, with very many different pieces of process equipment and a complex web of many recycle streams. So how are such calculations done in industry, or how will you do them efficiently in your design course, especially if you want to consider many different design options? The answer is by using a complicated computer program known as a process simulator. Such a computer program allows the user to put together a flow sheet of the equipment in the process being considered and to connect all the equipment by the flows of mass (and in the case of heat exchangers and some other equipment by the flows of heat). Then, after the user specifies the components, the feed composition, the conditions, the constraints, and the thermodynamic models to be used, he or she is able to compute the amounts and compositions of all the streams in the process. After seeing the results, the user can easily make changes to the inlet stream and the conditions (e.g., temperature and/or pressure at various points of the process), and rerun the simulation. In this way the engineer acquires an understanding of how the process responds to changes in conditions, allowing the engineer to optimize the process for various metrics: profitability, minimum carbon dioxide or other waste emissions, minimum energy use, etc.

Why introduce process simulation in a course on thermodynamics? Several compelling reasons exist. First, as you progress through your thermodynamics classes you will see that the calculations involved become increasingly more complicated. This alone justifies the use of some type of computer software. Second, a calculation for a single set of process specifications can be tiresome, though doing only one allows the student to understand the basis of the calculation. But this understanding, while important, does not provide the student or engineer with any insight into the way the process will respond to changes in the variables or whether the current set of specifications is optimum. Such knowledge comes only from calculations involving a collection of other operational specifications, and these calculations can be done rapidly with a process simulator, allowing the user to better understand the behavior of the process. In this way he or she can develop engineering intuition that would not come from doing only a single calculation. Third, choosing the correct thermodynamic model(s) is critical to obtaining meaningful results in process simulation, so that thermodynamics and process simulation are linked together in a very important way.

I want to emphasize this last point since it is so important. As an example, suppose the process we want to model contains liquids, but we tell the process simulator to use the ideal gas law to describe the system. What will happen? The process simulator will provide answers, but the results will be nonsensical. A process simulator will do computations exactly as instructed, but it is unable to determine whether the result obtained is meaningful or not. That is the job of the engineer. In computer lingo there is an acronym, GIGO, which means "garbage in, garbage out." Here it translates to bad thermodynamics, bad results. Unfortunately, in my consulting experience, I have observed another application of GIGO, meaning "garbage in, gospel out." That is, the user of a process simulator accepts the results obtained without critical evaluation. This is often due to the fact that the engineer has had difficulty getting a completely unrelated process simulation calculation to converge (or converge to a reasonable answer) and has tried different thermodynamic models until one yielded a reasonable answer. From then on, the engineer has tended to use that model for all other processes, including the ones for which it is completely inappropriate. This is a serious error of engineering judgment that could prove expensive to rectify or dangerous if a chemical plant were built according to those faulty specifications.

The central point, therefore, is that while using a process simulator can eliminate the tedium of process calculations, the results will only be meaningful if the user has a sufficient understanding of thermodynamics and thermodynamic models to make informed choices. Anyone using a process simulator (or, for that matter, any computer tool for calculations) should carefully check the results against his or her engineering intuition, experience, and knowledge of thermodynamics. For example, in the chapter on chemical reactions, the principle of Le Chatelier and Braun can provide guidance as to how the equilibrium state of a chemically reacting system might shift in response to a pressure change. If the result of a process simulation indicates otherwise, one should make sure that the input and process choices are correct. Similarly, if a change of temperature in a process produces a result that is counterintuitive, further analysis is required. The point is that one should not blindly accept any result derived from a computer. Rather, one must analyze all the factors to see whether they make sense.

While most chemical and petroleum companies initially developed their own in-house process simulators, the expense of maintaining them, of introducing new thermodynamic and equipment models as they became available, and of servicing users became untenable. Consequently, the field of process simulation is now dominated by software from a few commercial vendors and some freeware through the CAPE-OPEN project. A web search will yield a number of available process simulators. Here we will consider only one, Aspen Plus—arguably, the one possessing the largest user base and, in addition, made available to universities at a very affordable price.

The main use of Aspen Plus is for process simulation, and various books and courses are devoted to instructing students in how to employ it. That is not our purpose here; we are interested solely in the way Aspen Plus can be used in undergraduate courses in thermodynamics. Therefore, while Aspen Plus has a wide range of capabilities, we will consider only the following:

1. Basic process simulation
2. Phase equilibria (vapor–liquid, liquid–liquid, and vapor–liquid–liquid)
3. The thermodynamic data regression capabilities
4. **Property analysis** (pure fluids and mixtures)
5. The **NIST TDE** (thermodynamics data engine)
6. The **Property Method Selection Assistant**
7. Simple distillation and extraction

Note: Throughout this text a word or words in **bold** font refer to a specific aspect or function of the Aspen Plus simulator identified as such. Readers of this book when using it to follow an example while using the Aspen Plus simulator, should select, by clicking on it, the relevant text indicated in **bold**.

Chapter 1

Getting Started With Aspen Plus®

Aspen Plus® is a process simulation program that can also be used for many types of thermodynamic calculations, or to retrieve and/or correlate thermodynamic and transport data. In this book it will largely be used for thermodynamic calculations, such as computing phase equilibria and regressing parameters in thermodynamic models, and also for some very simple process simulations, merely to introduce the concept.

To start, open the **Aspen Plus V8.x**, which you may have to locate depending on the setup of your computer. [It may be on your desktop or you may have to follow the path **All Programs>Aspen Tech>Process Modeling V8.x>Aspen Plus>Aspen Plus V8.x**. In doing this you will also see paths to the large collection of specialized Aspen Plus modules that will not be considered here.] The interface is somewhat different for Aspen Plus V8.0 and Aspen Plus V8.2 and higher. For Aspen Plus V8.0 continue here, while for Aspen V8.2 (or higher) go to Fig. 1-2a.

[The screen images shown below and throughout this book were produced using Aspen Plus®. These screen images of Aspen Plus® are reprinted with permission of Aspen Technology, Inc. AspenTech®, aspenONE®, Aspen Plus®, and the AspenTech leaf logo are trademarks of Aspen Technology, Inc. All rights reserved.]

Figure 1-1a Aspen Plus V8.0 Start-up

When you open **Aspen Plus V8.0**, you will briefly see the Aspen logo in Fig. 1-1a. There is then a slight delay while the program connects to the server, and then the **Getting Started** page shown in Fig. 1-1b appears. There you will see a list of **Product News** items (that changes as it updates on a regular basis). From this window you will be able to start a new **Simulation** or open one of your previous simulations that will appear (in the future) in the list under **Recent Cases**.

Using Aspen Plus® in Thermodynamics Instruction: A Step-by-Step Guide, First Edition. Stanley I. Sandler.

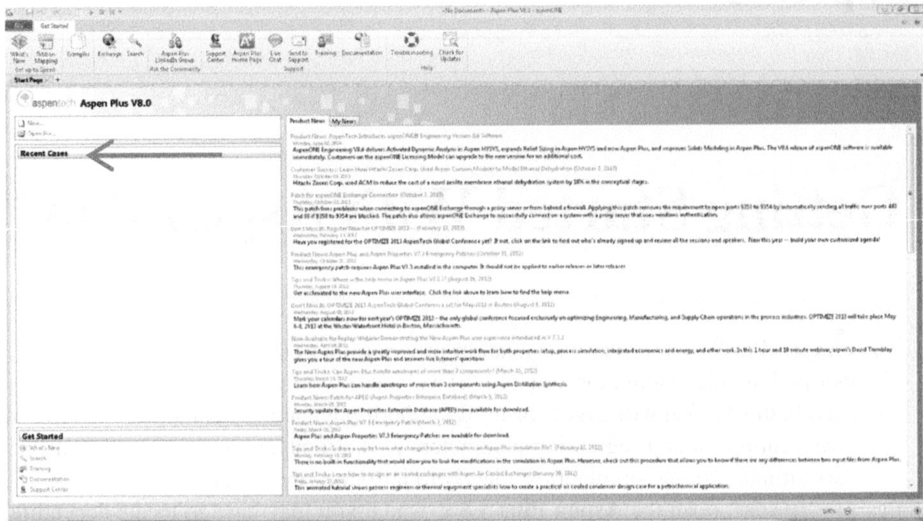

Figure 1-1b Aspen Plus V8.0 Start-up

To proceed, click on **New..**, which brings up the window in Fig. 1-3.

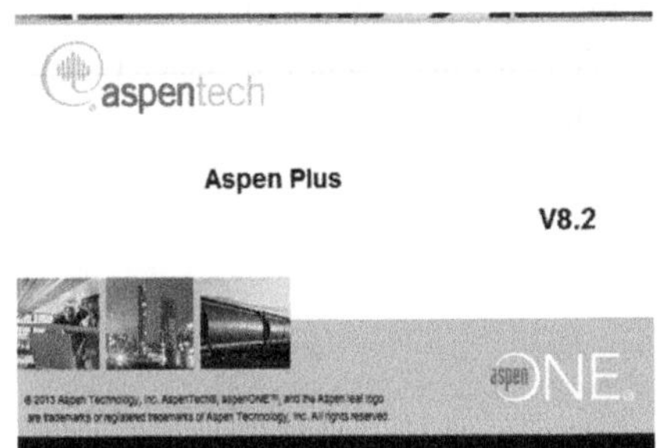

Figure 1-2a Aspen Plus V8.2 Start-up

When you open **Aspen Plus V8.2**, you will briefly see the Aspen logo in Fig. 1-2a. There is then a slight delay while the program connects to the server, and then the **Exchange** window shown in Fig. 1-2b appears.

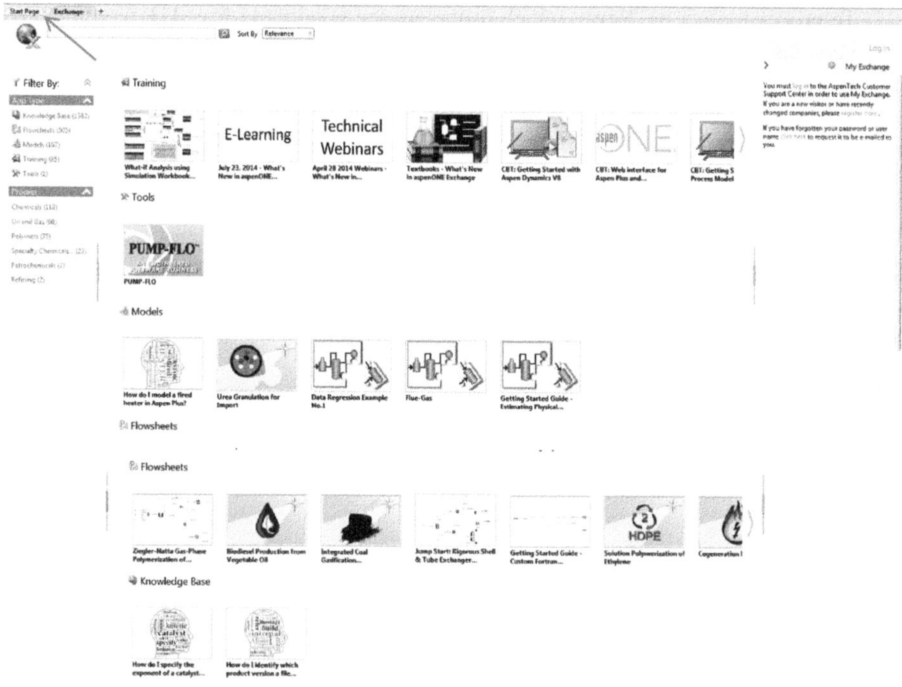

Figure 1-2b

This window contains flow sheets and information about a number of processes, training information, pre-prepared models for specialized unit operations, and other items. These will be ignored here as the emphasis is on thermodynamic modeling. Click on the **Start Page** tab as shown by the arrow in Fig. 1-2b, which will bring up the **Start Page** shown in Fig. 1-2c. There you will see a list of **Product News** items (that changes as it updates on a regular basis). From this window you will be able to start a new **Simulation** or open one of your previous simulations that will appear (in the future) in the list under **Recent Cases**.

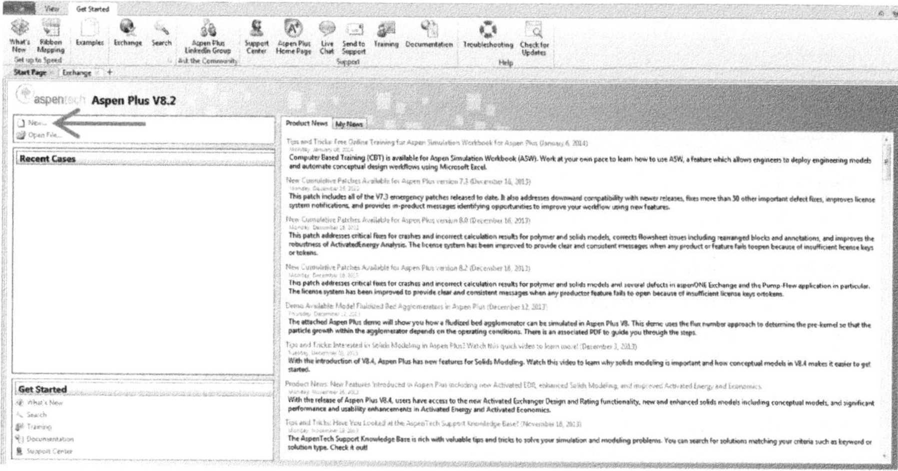

Figure 1-2c

To proceed, click on **New..**, which brings up the window in Fig. 1-3. Continuation for all versions of Aspen Plus V8.0 and higher.

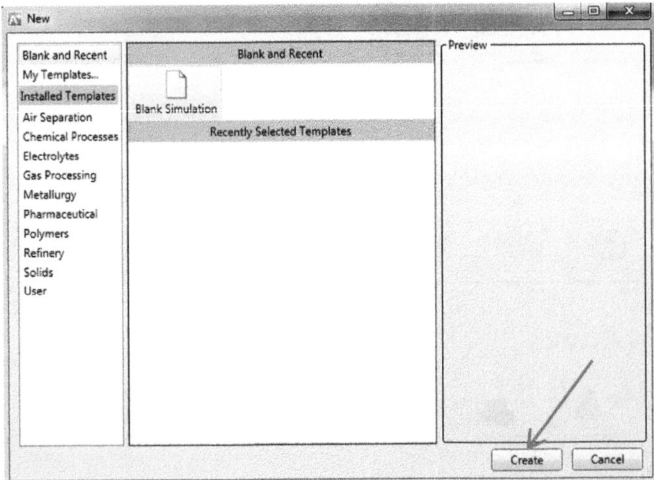

Figure 1-3

Click on **Blank Simulation** and then **Create**. This will bring up Fig. 1-4.

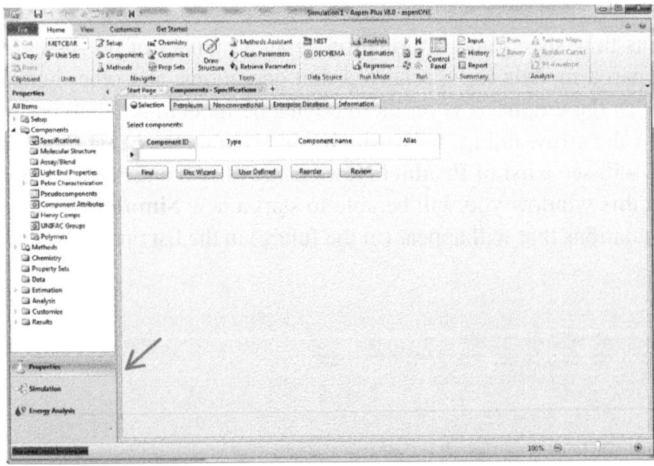

Figure 1-4

On the lower-left-hand corner of this window, there are three choices. The first, which Aspen Plus opens with, is **Properties**; the drop-down menu under **Components>Specifications** is used to specify the component or components for the calculation, and the drop-down menu under **Methods** is used to specify the thermodynamic models and parameters that will be used in the calculation. The second general area is **Simulation** that will take you to a flow sheet window, to be discussed later, and the third is **Energy Analysis** that will not be considered here. The default is to start with **Properties**.

We will proceed by entering the component propane. There are two ways to enter component names. The simplest and most reliable to ensure that you will get the correct component and its properties from the Aspen Plus database is to click on the **Find** box that brings up the **Find Compounds** window and then enter the component name by typing in propane and then clicking on **Find Now**, which produces the window in Fig. 1-5.

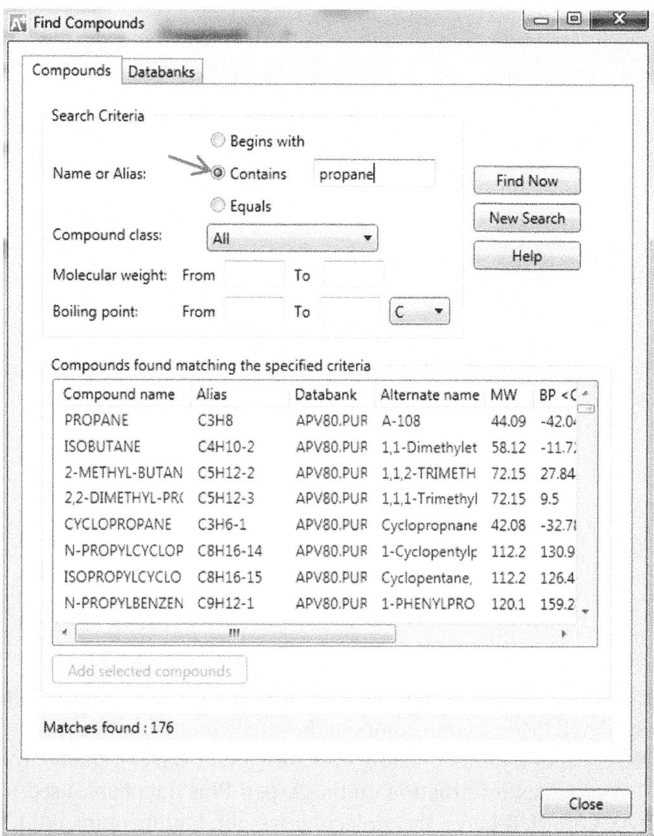

Figure 1-5

A long list of 176 compounds in Fig. 1-5 is generated because the default **Contains** was used in the **Find Compounds** window; as a result every compound in the database that contains propane either in its compound name (e.g., propane, but also cyclopropane) or in its alternate name (e.g., isobutane is also known as 2-methyl propane) appears in the list. The compound we are interested in happens to be first on the list here, but that will not always be the case. Therefore, a better way to proceed in the **Find Compounds** window is to click **Equals** instead of the default **Contains**, and then click **Find Now**, which produces instead a list containing only propane (Fig. 1-6).

[Note that Aspen Plus has a large number of data banks of pure component and mixture thermodynamic and transport properties data. Generally these are called up automatically by the program. Here the pure component names and properties were obtained as shown from the data bank APV80.PURE.x, where 80 indicates the version number of Aspen Plus (here 8.0) and PURE.x is version x of the pure component properties data bank. Other data banks will be encountered later in this book.]

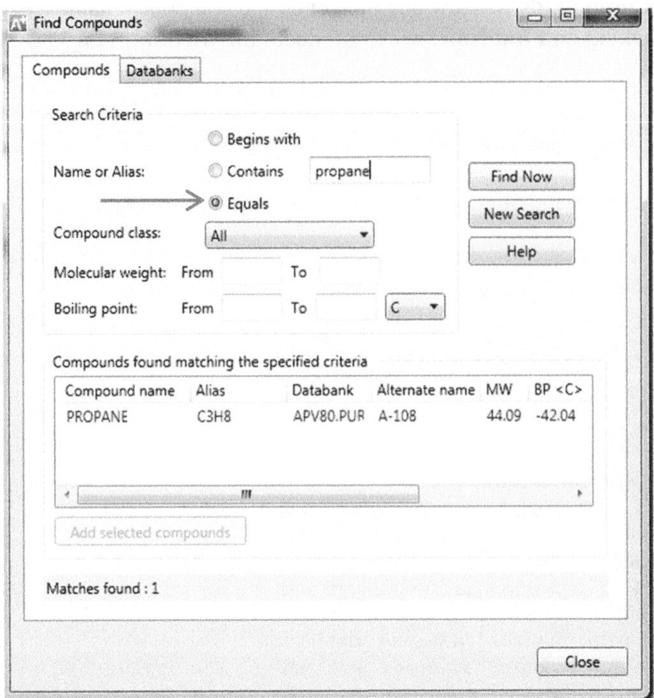

Figure 1-6

Click on **PROPANE** and then **Add selected compounds**, and for this example, then click on **Close**. In cases considered later, several compounds will be sequentially added following this procedure with the exception of not clicking on **Close** after each compound. [Note that the window of Fig. 1-6 provides information on the Aspen Plus data bank used to obtain the data for propane (APV80.PUR here), the molecular weight, boiling point, etc.] You will then see the following (Fig. 1-7):

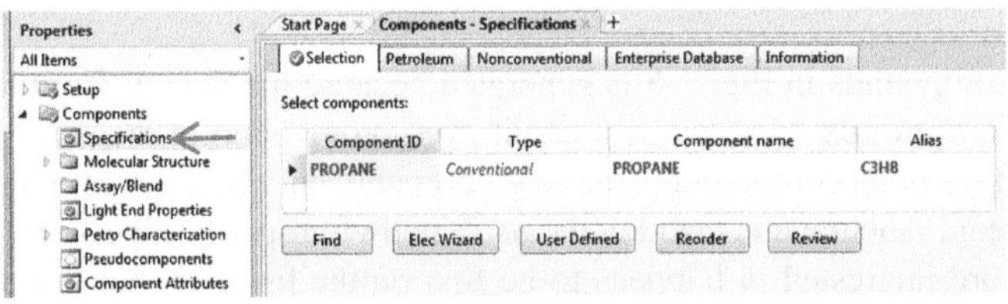

Figure 1-7

[Another alternative is to type in all or part of the name directly in the **Components-Specification** window and see whether Aspen Plus finds the correct name.] Note that propane has now been added to the **Select components** list and that both **Components** and **Specifications** now have check marks indicating that sufficient information has been provided to proceed to the next step. However, this may not be sufficient information for the problem of interest to the user. If the problem to be solved involves a mixture, one or more additional components may be added following the procedure described above except that the **Close** button in the **Find Compounds** window is used only after all the components have been added.

The next step is to go to **Methods** by clicking on it. The window in Fig. 1-8 appears and here a number of thermodynamic models can be used. For a simple hydrocarbon system at the pressures here, a simple cubic equation of state can be used, for example, the Peng–Robinson or Soave–Redlich–Kwong equation. Here, and frequently throughout this book, the Peng–Robinson (indicated by **PENG-ROB** in Aspen Plus) will be used, though any other equation of state for which parameters are available can be used. Generally, simple equations of state, such as the Peng–Robinson and other cubic equations of state produce results that are not of great accuracy, but they provide adequate descriptions of both the vapor and liquid states, and are adequate for thermodynamic calculations for pure fluids and mixtures that are nonpolar and do not hydrogen bond. The Peng–Robinson equation has been chosen from the drop-down **Base method** menu to calculate the thermodynamic properties of this nonpolar alkane. [Note that if you need help in choosing a thermodynamic model, you can click on **Methods Assistant**… for help. The **Methods Assistant** will be discussed in Chapter 8.] After accepting the Peng–Robinson equation by pressing **Enter**, **Methods** on the left-hand side of Fig. 1-8 will also have a check.

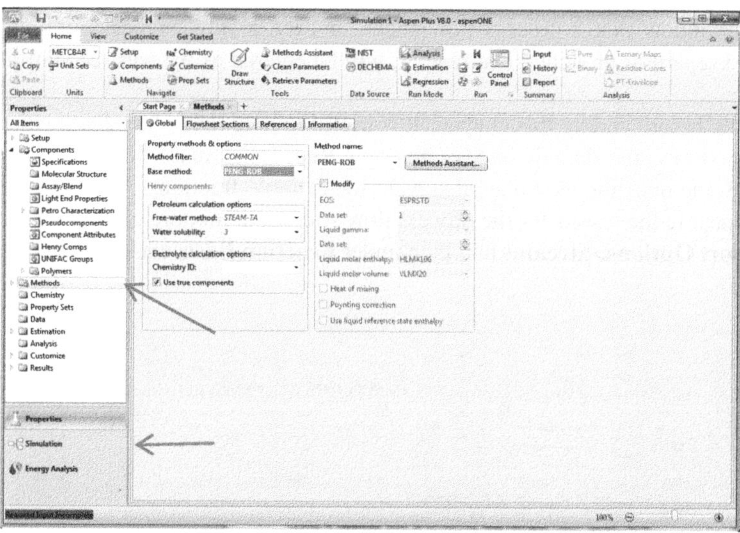

Figure 1-8

Clicking on **Simulation** brings up the **Main Flowsheet** window of Fig. 1-9 together with the **Model Palette** at the bottom of the window.

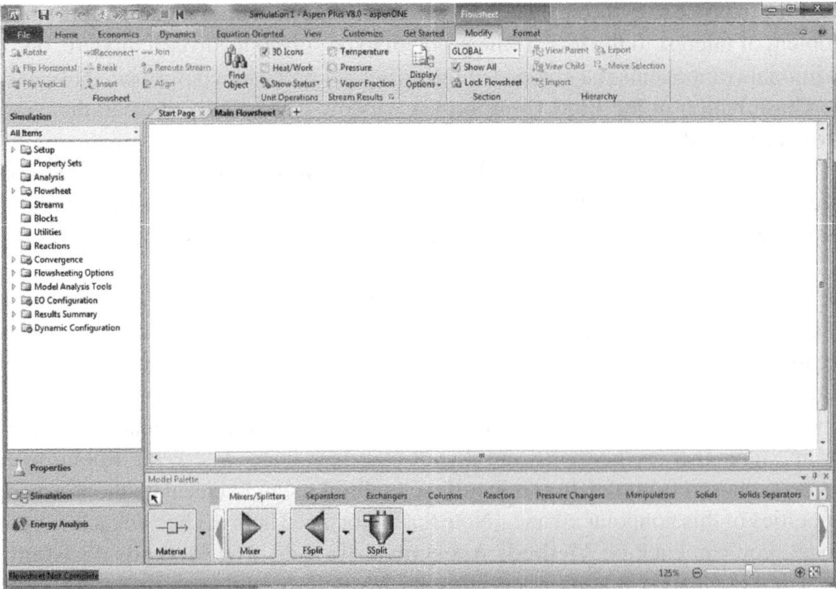

Figure 1-9

It is in this window that a process flow sheet, or even a single process unit such as a vapor–liquid separator or a chemical reactor, will be entered as we will see starting in the next chapter.

However, before ending this chapter, it is useful to note that when doing calculations involving mixtures, the default in **Streams** and **Reports** is mass flow rates and molar flow rates. Sometimes the user may also want to see mole fractions. To ensure that mole fractions appear in the results for the process flow streams in mixture calculations, click on **Setup>Report Options>Streams** and then under **Fraction Basis** click on **Mole** as shown in Fig. 1-10.

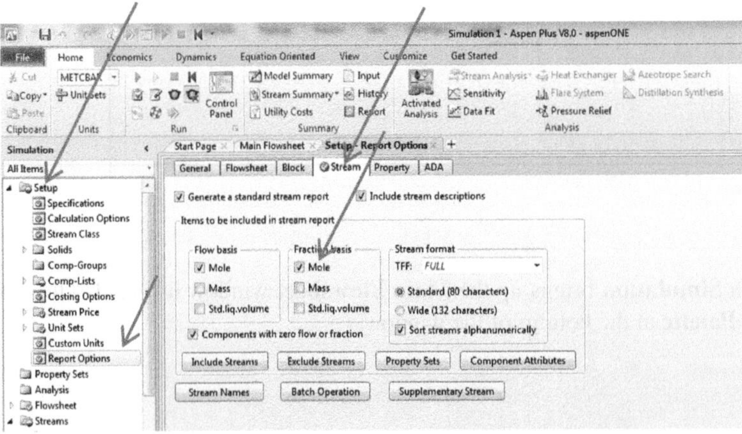

Figure 1-10

What has been described in this chapter needs to be done when starting any new Aspen Plus calculation. In the following chapters it will be assumed that you can successfully navigate through the **Setup**, **Components, and Methods** windows. Therefore, this material is not generally repeated in the chapters that follow.

There is one point to note here when using Aspen Plus. This program is designed for (largely) steady-state process flow calculations, not static calculations. That is, while we may be interested in a thermodynamic calculation for 1 mole of a species, Aspen Plus expects a flow rate. So instead of using the program to solve a problem for a change of state for a fixed amount of material, say 1 mole, in the **Streams** setup one will have to specify a flow rate such as 1 mol/min, 1 mol/hr, 1 kmol/hr. That choice will not affect the equilibrium state or compositions of the streams, but will affect the heat and work flows that are computed in Aspen Plus for the process equipment (referred to as blocks in Aspen Plus) as they depend linearly on the flow rates.

PROBLEMS

1.1. In the Aspen setup choose *n*-pentane as the component and the Soave–Redlich–Kwong equation of state as the base method.

1.2. Set up Aspen to use *n*-pentane and *n*-hexane as the two components in a binary mixture, and the Soave–Redlich–Kwong equation of state as the base method.

1.3. Set up Aspen to use ethanol and water as the two components in a binary mixture, and the NRTL model as the base method.

Chapter 2

Two Simple Simulations

Aspen Plus® was developed as a process simulator. This will be illustrated here for two very simple cases involving pure fluids, merely to provide the student simple examples of the main use of a process simulator. You will use the process simulation capabilities extensively in later courses and especially in the design course. However, in thermodynamics courses it is mostly the thermodynamic and phase equilibrium calculational abilities and the databases that will be used.

As motivation and as an introduction to the use of Aspen Plus as a process simulator, we consider here, as the first of two examples, a simple process for the liquefaction of propane. This process starts with propane vapor at ambient conditions (298 K and 1 bar), which is compressed to 15 bar, cooled back down to 298 K, expanded through an adiabatic (i.e., Joule–Thomson) valve to 1 bar, and then the resulting gaseous and liquid streams are separated. Since propane is not an ideal gas throughout the process because of the conditions (high pressure and the presence of both a gas and a liquid), the Peng–Robinson equation of state will be used as the thermodynamic model, though other equations of state could have been chosen.

Here we start by opening the **Aspen Plus User Interface** as described in the previous chapter. Also, as described in the previous chapter, *n*-propane will be chosen as the single component, and the Peng–Robinson equation of state as the thermodynamic model. Next click on **Simulation**, which brings up a blank **Main Flowsheet** window with the **Model Palette** at the bottom as shown in Fig. 2-1.

We then start constructing the flow sheet. The first piece of process equipment in this liquefaction example is the compressor to adiabatically and reversibly compress the feed propane from 1 bar and 298 K to 15 bar. We place the compressor on the flow sheet by first choosing the **Pressure Changers** tab in the **Model Palette** at the bottom of the screen, which lists six different types of pressure changing devices, choosing the compressor, **Compr**, by first clicking on it, then moving the cursor (now a cross) to the **Main Flowsheet** window, choosing where you want to put the compressor, and then left clicking to produce the window in Fig. 2-1.

Using Aspen Plus® in Thermodynamics Instruction: A Step-by-Step Guide, First Edition. Stanley I. Sandler.

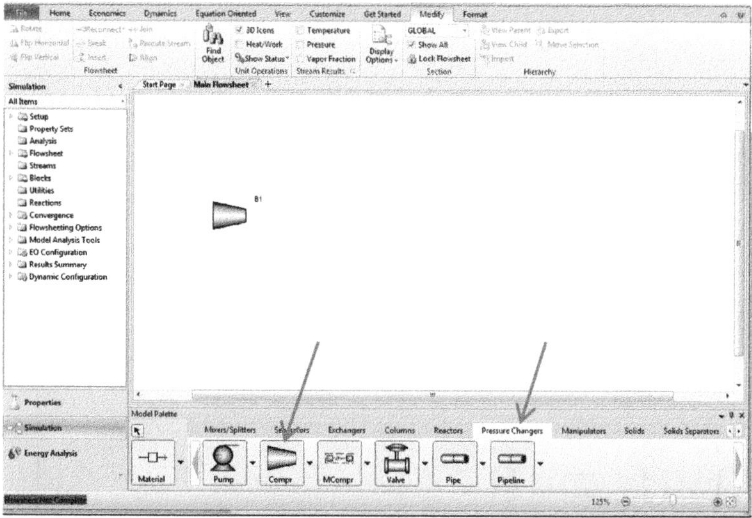

Figure 2-1

Next go to the **Exchangers** tab in the **Model Palette**. Four types of heat exchangers appear; choose the first one, **Heater,** and add it to the flow sheet as shown in Fig. 2-2.

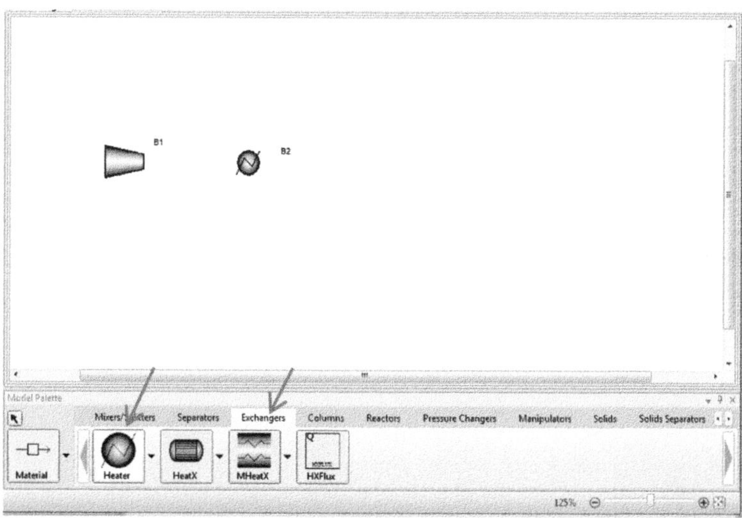

Figure 2-2

Next, go back to **Pressure Changers** and add the **Valve**. Finally, go to the **Separators** tab and add the **Flash2** (two-phase) separator to your flow sheet, which should now look like the window in Fig. 2-3.

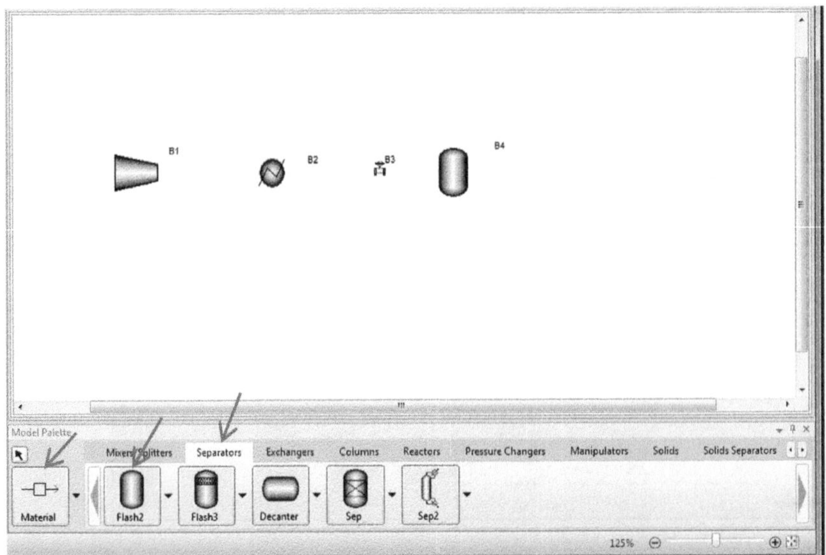

Figure 2-3

Note that in the lower-right-hand corner of the screen, in red, is the statement “Flowsheet Not Complete.” Next we have to connect the material flow streams to the process equipment. This is done by clicking on the **Material** icon at the lower-left corner of the **Model Palette**. Start with the feed stream by clicking on the compressor. The window should then be as in Fig. 2-4.

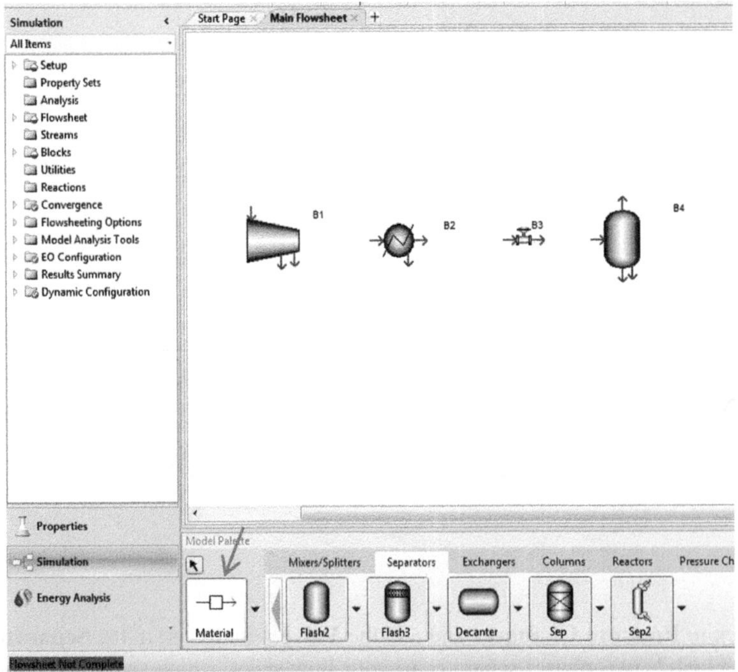

Figure 2-4

Each red arrow indicates a required input or outlet stream for each piece of equipment, while a blue arrow is an optional stream. For the compressor, the red arrow pointing into it is the required feed stream, and there is a required outlet stream (red) and an optional outlet stream (blue), for example, if the outlet contained both gas (red stream) and liquid (blue stream). In this example, all blue streams will be ignored. As the first step in completing the flow sheet, the feed stream will be added to the compressor and the red outlet stream will be connected to the heat exchanger. This is done by clicking and holding on an arrow, moving the cursor to the desired location, and releasing it. Doing this for the inlet (feed) arrow to the compressor, and connecting the outlet from the compressor to the inlet to the heat exchanger produces Fig. 2-5.

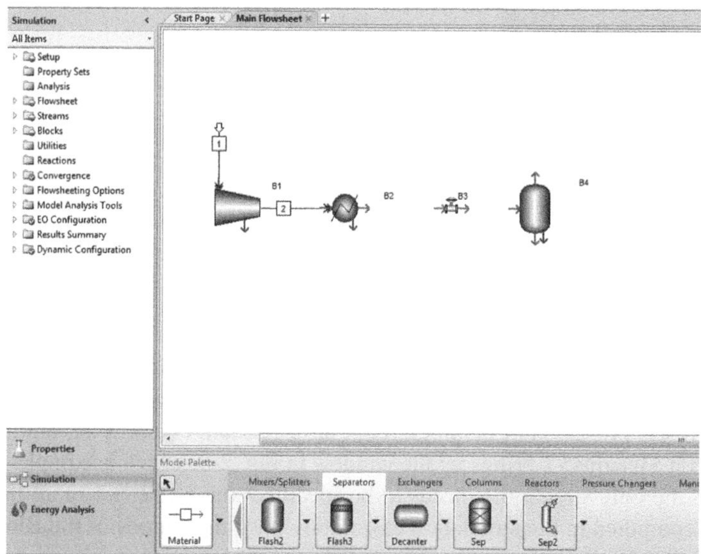

Figure 2-5

The bigger open arrow (above stream 1) pointing inward identifies it as an external inlet stream. A similar arrow pointing outward indicates a stream exiting the process.

Now completing the flow sheet by connecting the remaining streams one obtains Fig. 2-6.

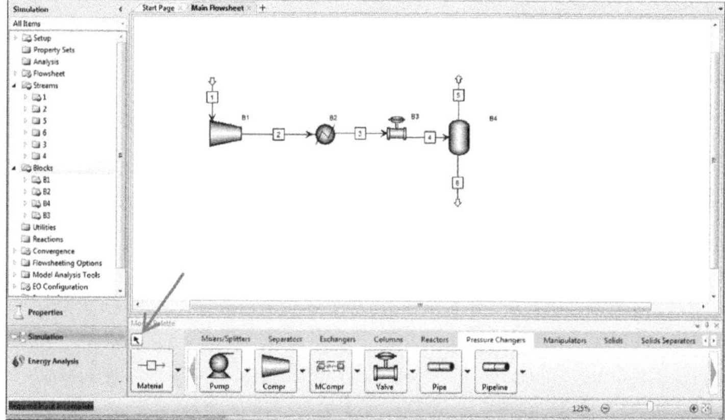

Figure 2-6

To get out of the mode of adding **Material** (streams), click on the arrow above that icon.

Some suggestions for putting together a nice flow sheet:

1. Unless you are very steady with the mouse, the streams are likely to appear jagged. To prettify your flow sheet, you can click on a stream and then on **Align blocks**. I suggest starting at the end of the flow sheet and continuing stepwise backward to the feed; otherwise you may introduce jagged feed streams upstream or downstream.
2. I like to rename the blocks (process units) and streams with names easier to recognize in the output. However, I retain the initial numbers so that the order in the output will be the same as in the flow sheet.

The revised flow sheet based on these suggestions appears in Fig. 2-7.

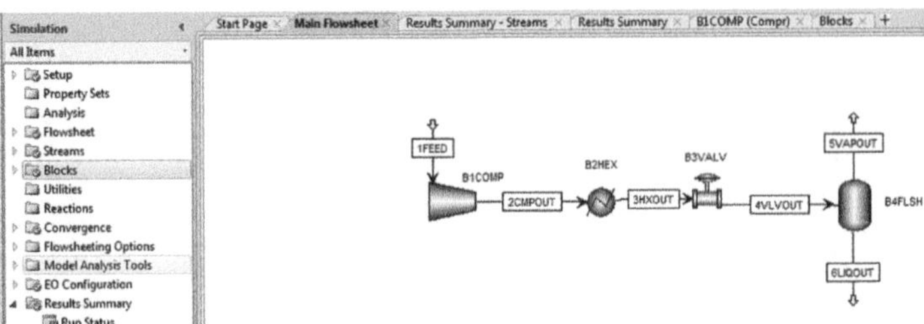

Figure 2-7

To proceed, the user must now provide the specifications for the process, which includes the inlet stream component(s) and conditions, pressures and temperatures where needed, and the type of each process unit or block (e.g., an isentropic compressor, an adiabatic valve) Clicking on **Streams>1** brings up the window in Fig. 2-8, which is completed as in Fig. 2-9. This is the only stream that needs to be specified; the properties of all the other streams will be computed in the simulation calculation once the actions of the **Blocks** are chosen.

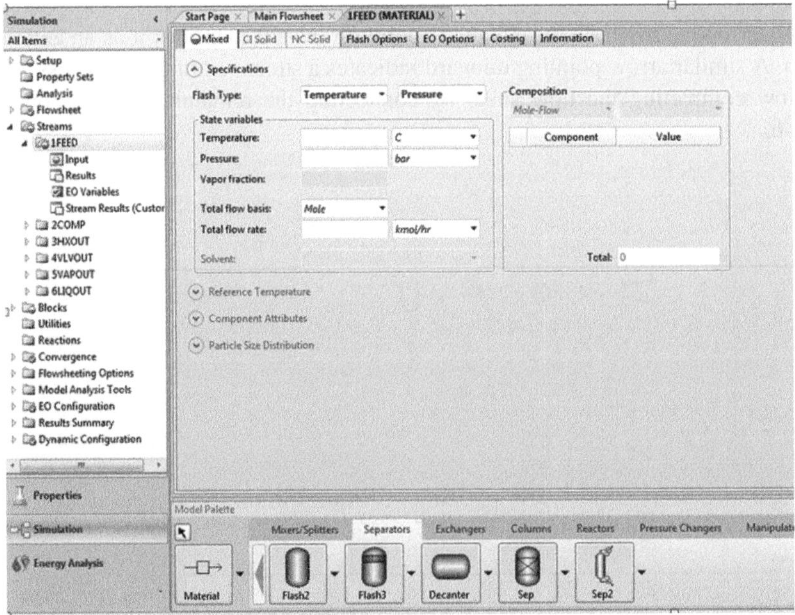

Figure 2-8

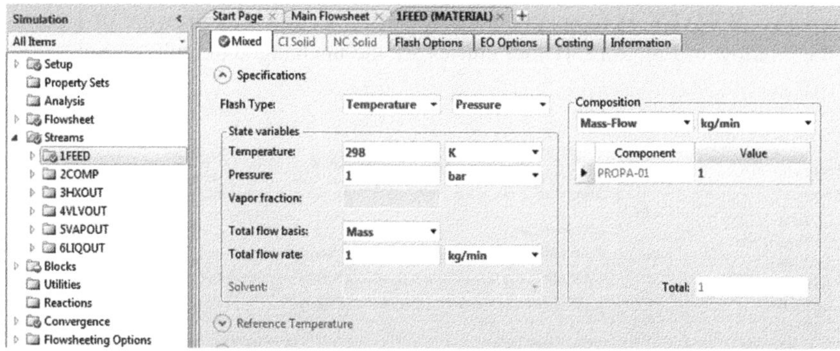

Figure 2-9

We then move on to the specifications for the **Blocks** (process units). Starting at **B1COMP**, the compressor, we specify that it is to be a perfect (all efficiencies equal to 1) isentropic compressor with a discharge pressure of 15 bar; this is shown in Fig. 2-10.

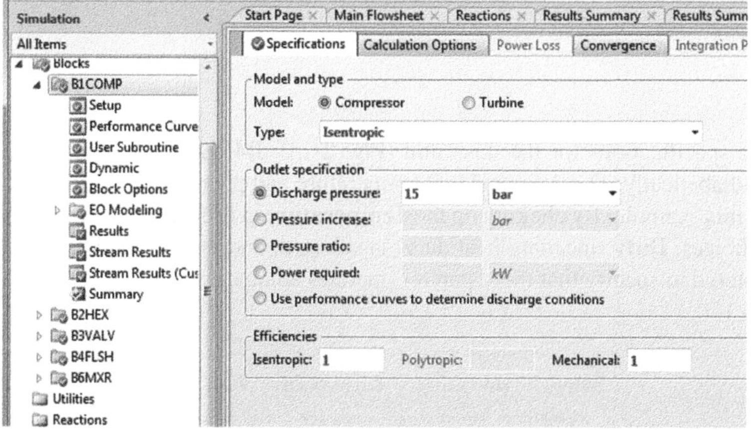

Figure 2-10

Next the heat exchanger, **B2HEX**, in this example is to operate with no pressure loss (15 bar) and an exit temperature of 298 K (see Fig. 2-11.)

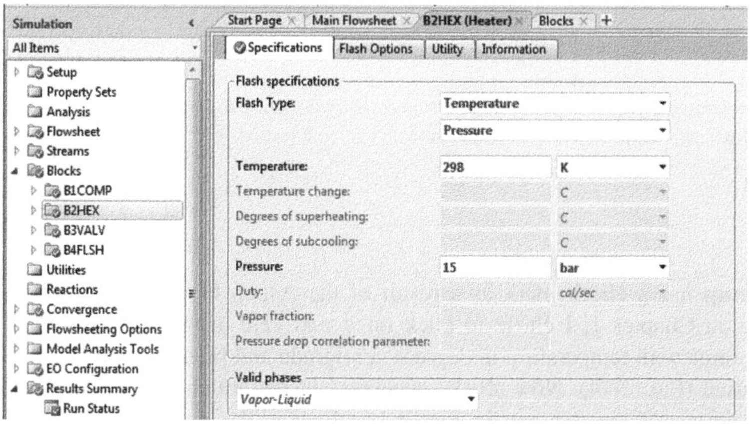

Figure 2-11

The valve, **B3VALV**, is to operate adiabatically with an outlet pressure of 1 bar as shown in the window of Fig. 2-12. [Other entries are the unchanged defaults.]

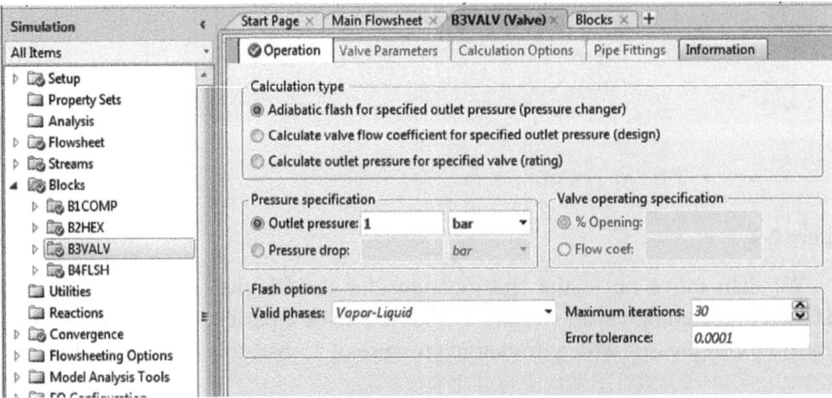

Figure 2-12

Finally, the specifications for the flash unit (Fig. 2-13), **B4FLSH**, are that it operates at 1 bar and adiabatically. The default is that temperature and pressure are fixed, which is not the case in this example. By clicking on the **Temperature**, a drop-down menu appears, and from the choices, **Duty** (meaning heat duty) is selected, and then in the **Duty** drop-down box 0 is entered to specify that the separator operates adiabatically.

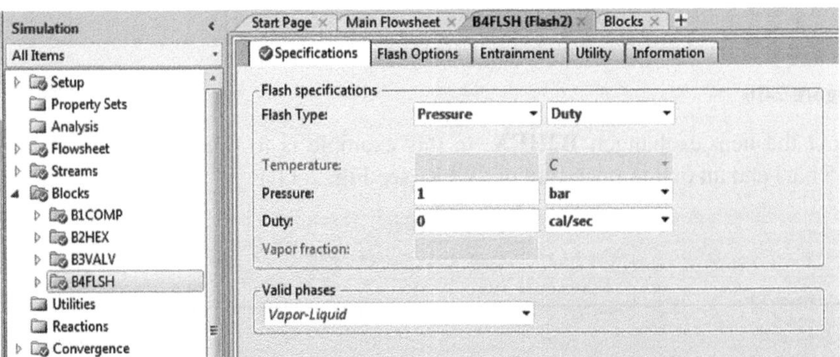

Figure 2-13

Though **Setup** has a check mark as a result of the Aspen Plus defaults, following the instructions in Chapter 1, I chose to click on it and give a title to the simulation and that metric units with temperature in degrees centigrade and pressure in bar (METCBAR) are being used (Fig. 2-14). Now all the necessary boxes are checked, and in the lower-right-hand corner of the window in Fig. 2-14, there is the message "**Required Input Complete**."

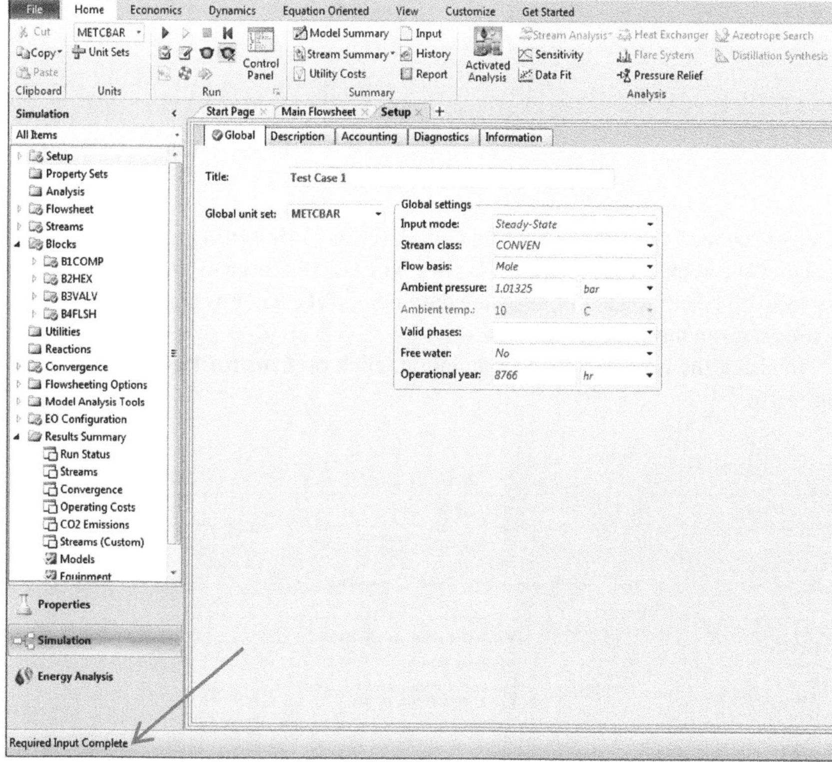

Figure 2-14

We are now ready to run the simulation. There are five ways to start the simulation. The first way is to press the **F5** key (not function F5, just the F5 key). The second and third ways are to press one of the two forward arrow keys on the **Main Toolbar** (Fig. 2-15a) and click on the forward arrow above **Run** (which will gray if not all the information for the simulation has been entered, but turn dark blue when all necessary data have been entered). The final two ways are to click on one of the Aspen Plus **Next** keys, N▸, on the **Main Toolbar** that will bring up the message of Fig. 2-15b. Clicking OK will then run the simulation.

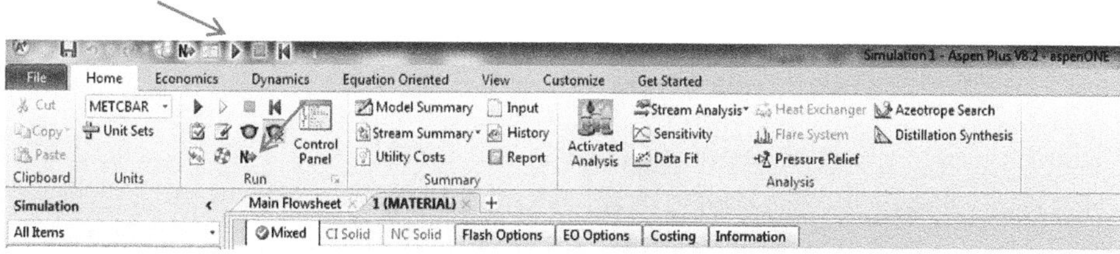

Figure 2-15a

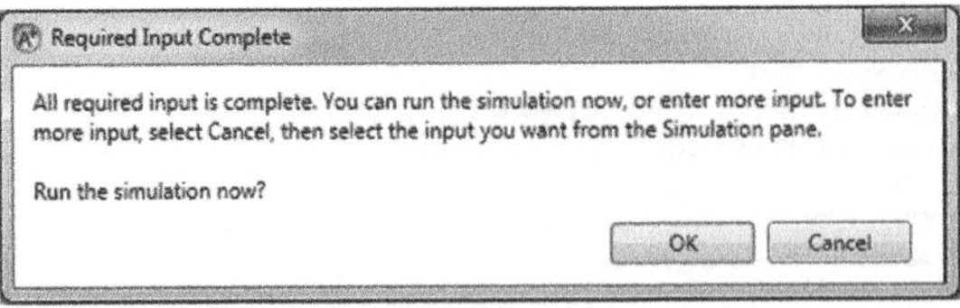

Figure 2-15b

As a digression, it is useful to note that wherever one is in setting up a simulation, clicking on one of the Aspen Plus **Next** keys, **N▸**, will take the user to the next part of the setup that needs to be completed or, as is the case here, if the setup is complete by clicking **OK**, the user can run the simulation.

To check the progress of the simulation, click on **Control Panel** on the main toolbar (Fig. 2-16).

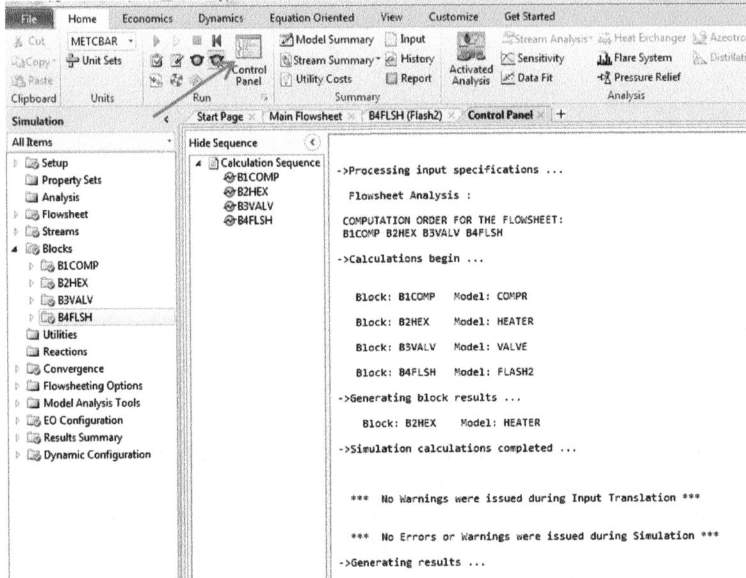

Figure 2-16

After running the simulation, the popup message in Fig. 2-17 appears.

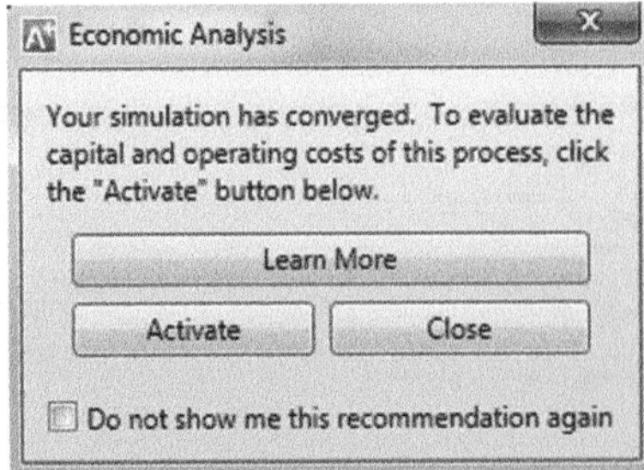

Figure 2-17

Click on **Close** since we are not interested in an **Economic Analysis**. Also, I suggest clicking the **Do not show me this recommendation again** box. Now going to **Results Summary** and then **Streams** brings up a window (Fig. 2-18) containing the table of stream results.

	1FEED	2CMPOUT	3HXOUT	4VLVOUT	5VAPOUT	6LIQOUT
Substream: MIXED						
Mole Flow kmol/hr						
PROPA-01	1.36065	1.36065	1.36065	1.36065	0.532814	0.827838
Mole Frac						
PROPA-01	1	1	1	1	1	1
Total Flow kmol/hr	1.36065	1.36065	1.36065	1.36065	0.532814	0.827838
Total Flow kg/hr	60	60	60	60	23.4952	36.5048
Total Flow l/min	552.553	44.6261	2.03024	165.683	164.638	1.04474
Temperature C	24.85	126.217	24.85	-42.5381	-42.5381	-42.5381
Pressure bar	1	15	15	1	1	1
Vapor Frac	1	1	0	0.391588	1	0
Liquid Frac	0	0	1	0.608413	0	1
Solid Frac	0	0	0	0	0	0
Enthalpy cal/mol	-25031.8	-23262.6	-28862.8	-28862.8	-26144.6	-30612.3
Enthalpy cal/gm	-567.659	-527.537	-654.536	-654.536	-592.894	-694.211
Enthalpy cal/sec	-9460.98	-8792.29	-10908.9	-10908.9	-3869.5	-7039.44
Entropy cal/mol-K	-64.3595	-64.3595	-81.426	-80.3676	-68.5807	-87.9539
Entropy cal/gm-K	-1.45951	-1.45951	-1.84654	-1.82254	-1.55524	-1.99458
Density mol/cc	4.10413e-05	0.000508168	0.0111698	0.000136873	5.39379e-05	0.0132064
Density gm/cc	0.00180978	0.0224084	0.492552	0.00603563	0.00237847	0.582359
Average MW	44.0965	44.0965	44.0965	44.0965	44.0965	44.0965

Figure 2-18

Clicking on **Stream Table** adds a reformatted version of this table to the flow sheet as shown in Fig. 2-19. You should only do this for the final case for a simulation if you are varying parameters. Doing this for every intermediate case would lead to a very cluttered diagram.

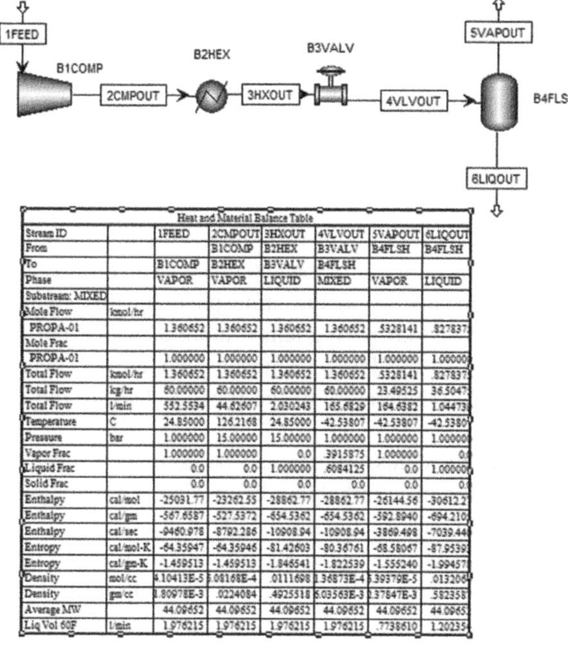

Heat and Material Balance Table

Stream ID		1FEED	2CMPOUT	3HXOUT	4VLVOUT	5VAPOUT	6LIQOUT
From			B1COMP	B2HEX	B3VALV	B4FLSH	B4FLSH
To		B1COMP	B2HEX	B3VALV	B4FLSH		
Phase		VAPOR	VAPOR	LIQUID	MIXED	VAPOR	LIQUID
Substream: MIXED							
Mole Flow	kmol/hr						
PROPA-01		1.360652	1.360652	1.360652	1.360652	.5328141	.827837
Mole Frac							
PROPA-01		1.000000	1.000000	1.000000	1.000000	1.000000	1.00000
Total Flow	kmol/hr	1.360652	1.360652	1.360652	1.360652	.5328141	.827837
Total Flow	kg/hr	60.00000	60.00000	60.00000	60.00000	23.49525	36.5047
Total Flow	l/min	552.5534	44.62607	2.030243	165.6829	164.6382	1.04473
Temperature	C	24.85000	126.2168	24.85000	-42.53807	-42.53807	-42.5380
Pressure	bar	1.000000	15.00000	15.00000	1.000000	1.000000	1.00000
Vapor Frac		1.000000	1.000000	0.0	.3915875	1.000000	0.
Liquid Frac		0.0	0.0	1.000000	.6084125	0.0	1.00000
Solid Frac		0.0	0.0	0.0	0.0	0.0	0.
Enthalpy	cal/mol	-25031.77	-23262.55	-28862.77	-28862.77	-26144.56	-30612.2
Enthalpy	cal/gm	-567.6587	-527.5372	-654.5362	-654.5362	-592.8940	-694.210
Enthalpy	cal/sec	-9460.978	-8792.286	-10908.94	-10908.94	-3869.498	-7039.44
Entropy	cal/mol-K	-64.35947	-64.35946	-81.42603	-80.36761	-68.58067	-87.9539
Entropy	cal/gm-K	-1.459513	-1.459513	-1.846541	-1.822539	-1.555240	-1.99457
Density	mol/cc	4.10413E-5	5.08168E-4	.0111698	1.36873E-4	5.39379E-5	.013206
Density	gm/cc	1.80978E-3	.0224084	.4925518	6.03563E-3	2.37847E-3	.582358
Average MW		44.09652	44.09652	44.09652	44.09652	44.09652	44.0965
Liq Vol 60F	l/min	1.976215	1.976215	1.976215	1.976215	.7738610	1.20235

Figure 2-19

It is sometimes useful to copy the data in the output table and put it into Excel or other spreadsheet software for further manipulation (i.e., eliminating some rows that are not of interest, computing differences, etc.) This is easily done by clicking on **Copy All** and then pasting the information in Excel, producing a table as shown in Fig. 2-20.

	1FEED	2CMPOUT	3HXOUT	4VLVOUT	5VAPOUT	6LIQOUT
	B1COMP	B2HEX	B3VALV	B4FLSH		
		B1COMP	B2HEX	B3VALV	B4FLSH	B4FLSH
	VAPOR	VAPOR	LIQUID	MIXED	VAPOR	LIQUID
Substream: MIXED						
Mole Flow kmol/hr						
PROPA-01	1.360652	1.360652	1.360652	1.360652	0.532814	0.827838
Mole Frac						
PROPA-01	1	1	1	1	1	1
Total Flow kmol/hr	1.360652	1.360652	1.360652	1.360652	0.532814	0.827838
Total Flow kg/hr	60	60	60	60	23.49525	36.50475
Total Flow l/min	552.5534	44.62607	2.030243	165.6829	164.6382	1.044739
Temperature C	24.85	126.2168	24.85	-42.5381	-42.5381	-42.5381
Pressure bar	1	15	15	1	1	1
Vapor Frac	1	1	0	0.391588	1	0
Liquid Frac	0	0	1	0.608413	0	1
Solid Frac	0	0	0	0	0	0
Enthalpy cal/mol	-25031.8	-23262.6	-28862.8	-28862.8	-26144.6	-30612.3
Enthalpy cal/gm	-567.659	-527.537	-654.536	-654.536	-592.894	-694.211
Enthalpy cal/sec	-9460.98	-8792.29	-10908.9	-10908.9	-3869.5	-7039.44
Entropy cal/mol-K	-64.3595	-64.3595	-81.426	-80.3676	-68.5807	-87.9539
Entropy cal/gm-K	-1.45951	-1.45951	-1.84654	-1.82254	-1.55524	-1.99458
Density mol/cc	4.10E-05	5.08E-04	0.01117	1.37E-04	5.39E-05	0.013206
Density gm/cc	1.81E-03	0.022408	0.492552	6.04E-03	2.38E-03	0.582359
Average MW	44.09652	44.09652	44.09652	44.09652	44.09652	44.09652
Liq Vol 60F l/min	1.976215	1.976215	1.976215	1.976215	0.773861	1.202354

Figure 2-20

In the results table we see that for each 60 kg/hr (1 kg/min) of entering propane vapor feed, the process produces 36.5 kg/hr of liquid propane (stream 6) and releases 23.5 kg/hr of gaseous propane (stream 5), presumably to the atmosphere in this simple process. The temperature of these streams is 230.6 K (the temperature at which the vapor pressure of propane is 1 bar according to the Peng–Robinson equation of state). Also, by looking at stream 2, we see that the temperature of the propane leaving the compressor is 399.4 K.

The compressor work required for this process can be computed using the energy balance and the difference between the enthalpies of its inlet stream (stream 1 of −9460.978 cal/sec) and outlet stream (stream 2 of −8792.286 cal/sec). However, a simpler way is to scroll down to and click on **Models** under **Results Summary**, and then click on the tab **Compr** at the bottom of the screen. Looking at **Net Work Required**, we see that 2.8 kW is needed for this system to produce 36.5 kg/hr of liquid propane from a feed of 60 kg/hr (1 kg/min) at 298 K and 1 bar as seen in Fig. 2-21.

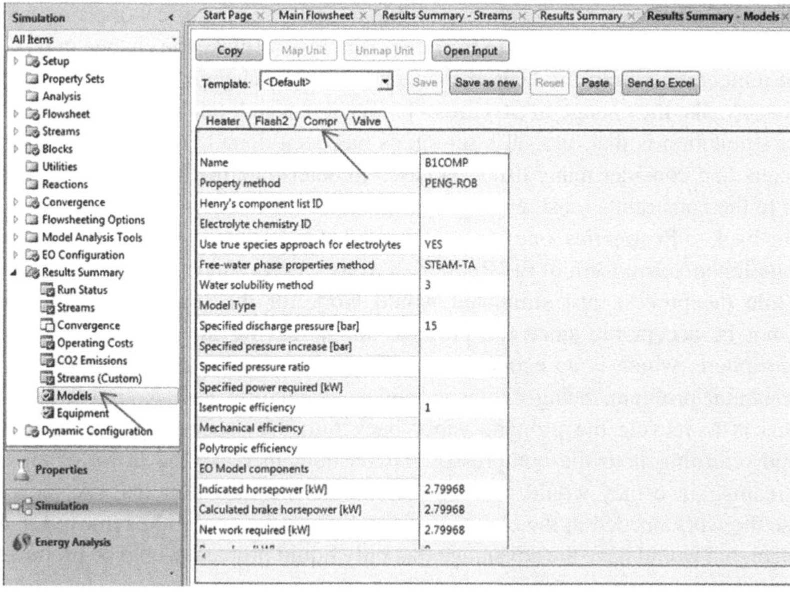

Figure 2-21

Similarly, to determine the amount of cooling needed in the heat exchanger, click on the **Heater** tab to see in Fig. 2-22 that 2116.65 cal/sec must be removed from the compressed propane at 399.4 K to cool it to 298 K.

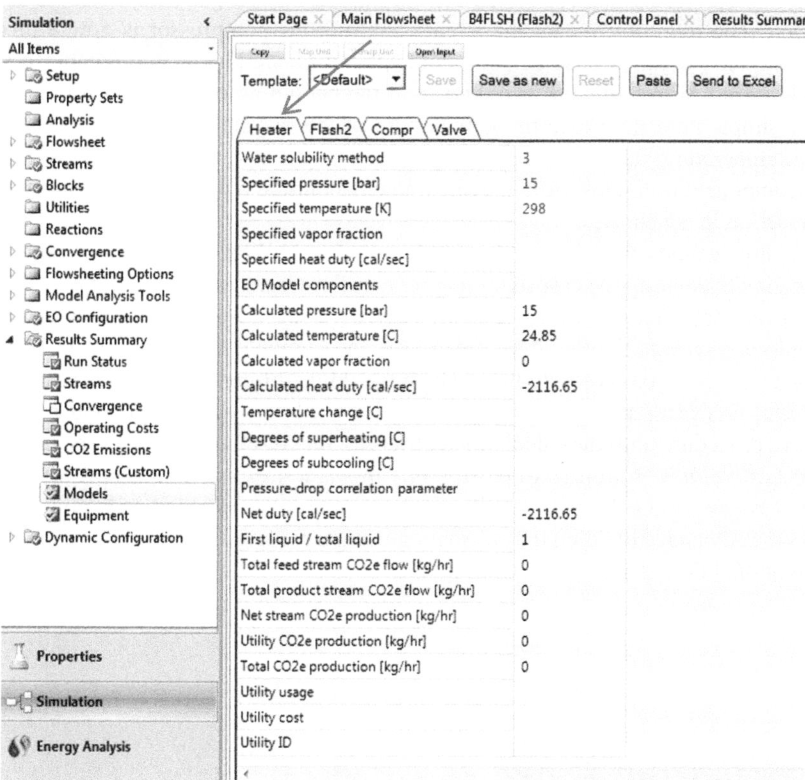

Figure 2-22

With the simulation working, it is now possible to easily see the effect of changing process parameters, such as the exit pressure from the compressor (make the change in **B1COMP**), the exit temperature from the heat exchanger (make the change in **B2HEX**), or the feed conditions (make the change in **Streams>1**). Try it! Indeed, one of the real advantages of process simulation is that once it is set up, as has been done here, it is very easy to vary parameters and consider many different cases to determine the best operating conditions subject to the constraints (cost, energy use, minimum loss of gaseous propane, etc.) Also, by going back to **Properties**, one can examine the effect of using another equation of state, for example the Soave form of the Redlich–Kwong (SRK) equation.

While the process just simulated would work for the liquefaction of propane, it would not be acceptable since the propane vapor leaving the separator is released to the atmosphere, which is an economic problem, since it is a valuable resource, and an environmental problem, being explosive and a greenhouse gas. A simple way to deal with this is to recycle the propane vapor back into the process by mixing it with the feed and returning it to the compressor. This would increase the flows of all the process streams, since they would now be the sum of the feed flow and the recycle flow, increase the work needed in the compressor, and affect the heat to be removed in the heat exchanger, but would have the advantage that only liquid propane would be produced in the process.

The previous calculation without recycle was a once-through calculation. That is, whether doing the calculation by hand or with Aspen Plus, the initial feed flow and conditions were known at the start of the calculation, and the action of each block (or unit operation) on the stream would be calculated sequentially through the process until the amounts and state of the final streams were obtained. However, by recycling the gaseous propane, the calculation has become more complicated and generally must be solved by iteration. The difficulty is that the stream entering the compressor is now a mixture of the initial feed stream whose amount and temperature are known, but combined with the recycle stream whose amount is unknown at the beginning of the calculation. [Note that in this simple pure fluid case the temperature of the recycle stream is known because it is the temperature at which the vapor pressure of propane is 1 bar. For more complex multicomponent problems, little may be known about the recycle stream temperature and composition in advance.]

So how would one solve such a problem by hand? What one would do is a first-pass calculation by solving the case assuming no recycle stream, and then calculate the amount of propane that is recycled for that case. Then do a second-pass calculation by including the first-pass recycle flow with the fresh feed and compute the new flows in every block to obtain a second-pass recycle flow, which would be different from the first-pass recycle flow. Use this second-pass recycle flow to do a third-pass calculation, and continue this iterative procedure until the calculation converges, that is, there is only a very small change in the recycle flow from pass to pass. While this is a tiresome calculation when done by hand, it is easily done here by modifying the flow sheet we have already considered, and shows the advantage of doing the calculation using a process simulator such as Aspen Plus. The modified flow sheet is shown in Fig. 2-23.

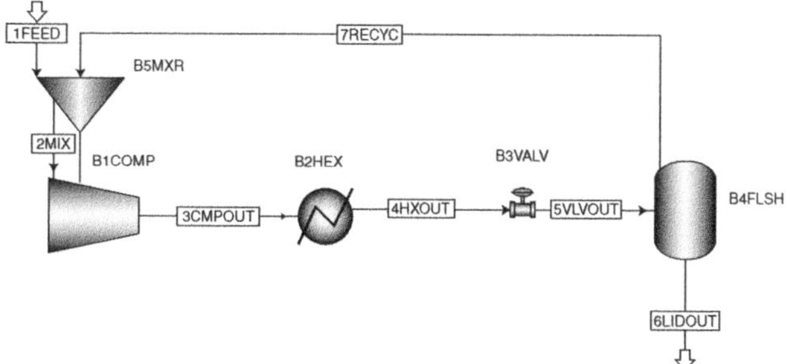

Figure 2-23

The change that has been made from the previous example is to add a **Mixer**, Block **B5MIX**, and route the vapor stream from the separator and the feed stream to the mixer. Also, for convenience, the streams have been renumbered. Everything else in the simulation is unchanged. This simulation is then run.

We see by clicking on the **Control panel**, which produces the results in Fig. 2-24, that the calculation converged on the third iteration.

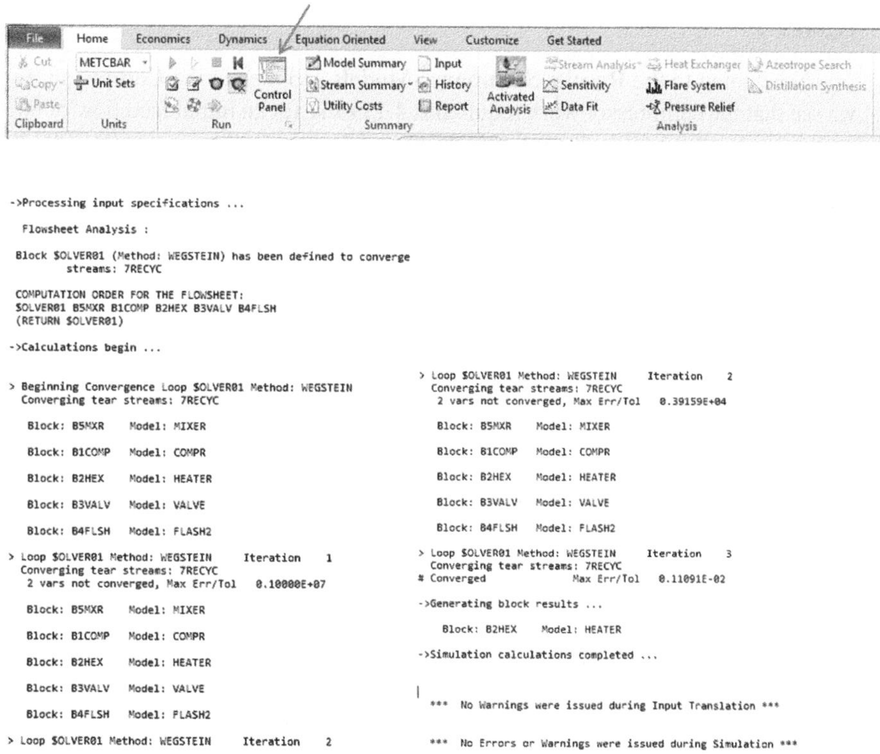

Figure 2-24

The stream results seen in **Results Summary>Streams** are given in the window of Fig. 2-25. This shows that the recycle flow is 38.62 kg/hr of propane for each 60 kg/hr of propane initially fed into the process. [Note that the first-pass value of the vapor exiting the separator from the previous calculation was 23.5 kg/hr. The correct value here was obtained by iteration.] Also, we see that since the cold recycle stream (−42.5°C) is mixed with the warmer feed (24.8°C), the inlet stream to the compressor (**Stream 2**) is now −0.45°C.

Material | Heat | Load | Work | Vol.% Curves | Wt. % Curves | Petro. Curves | Poly. Curves | Solids

Display: Streams Format: FULL Stream Table Copy All

	1FEED	2MIX	3CMPOUT	4HXOUT	5VLVOUT	6LIQOUT	7RECYC
Substream: MIXED							
Mole Flow kmol/hr							
PROPA-01	1.36065	2.2364	2.2364	2.2364	2.2364	1.36065	0.875744
Mole Frac							
PROPA-01	1	1	1	1	1	1	1
Total Flow kmol/hr	1.36065	2.2364	2.2364	2.2364	2.2364	1.36065	0.875744
Total Flow kg/hr	60	98.6173	98.6173	98.6173	98.6173	60	38.6173
Total Flow l/min	552.553	827.232	75.4048	3.33695	272.32	1.71716	270.603
Temperature C	24.85	-0.451647	134.133	24.85	-42.5381	-42.5381	-42.5381
Pressure bar	1	1	15	15	1	1	1
Vapor Frac	1	1	1	0	0.391588	0	1
Liquid Frac	0	0	0	1	0.608413	1	0
Solid Frac	0	0	0	0	0	0	0
Enthalpy cal/mol	-25031.8	-25467.5	-23074.1	-28862.8	-28862.8	-30612.3	-26144.6
Enthalpy cal/gm	-567.659	-577.541	-523.263	-654.536	-654.536	-694.211	-592.894
Enthalpy cal/sec	-9460.98	-15821	-14334.1	-17930.2	-17930.2	-11570.2	-6359.99
Entropy cal/mol-K	-64.3595	-65.8869	-63.8921	-81.426	-80.3676	-87.9539	-68.5807
Entropy cal/gm-K	-1.45951	-1.49415	-1.44892	-1.84654	-1.82254	-1.99458	-1.55524
Density mol/cc	4.10413e-05	4.50578e-05	0.000494309	0.0111698	0.000136873	0.0132064	5.39379e-05

Figure 2-25

Now going to the **Results Summary>Models** and then to the **Compr** tab (Fig. 2-26) we see that the compressor work required is 4.199 kW for an initial feed flow of 60 kg/hr of propane to produce 60 kg/hr of liquid propane. Thus the energy cost is 0.0700 kW per kg

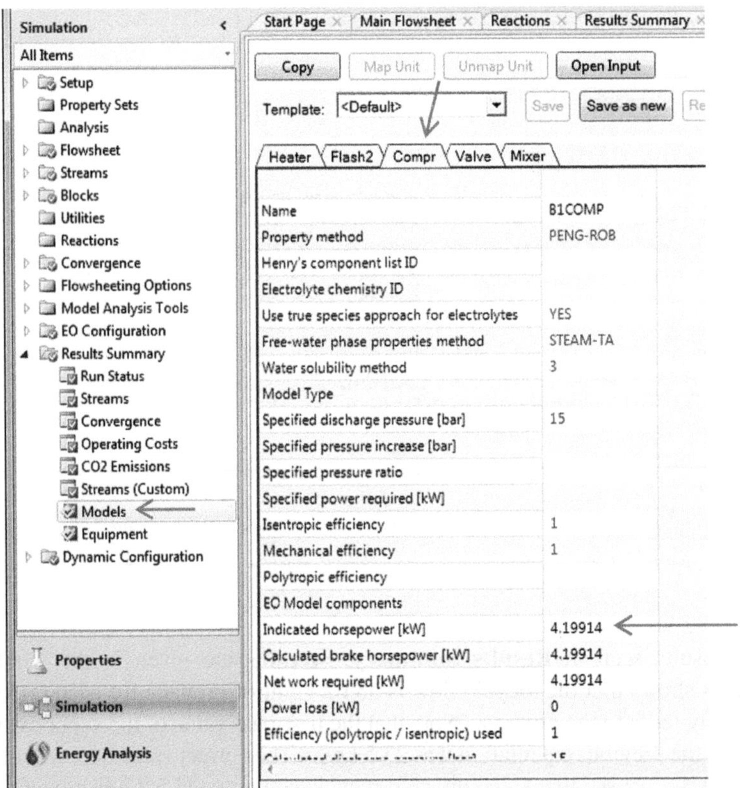

Figure 2-26

liquid propane per hr or 70.0 W per kg liquid propane per hr. This is to be compared with the non-recycle case of 2.8 kW needed to produce 36.5 kg/hr of liquid propane or 76.7 W per kg of liquid propane per hr. So, not only is the recycle process more efficient in terms of the energy required per kg of liquid propane produced, but it also successfully recovers as liquid all the propane fed into the process, and not discarding 23.5 kg/hr or 23.5 × 100/60 = 39.1% of it as in the once through, non-recycle process.

Clicking on **Stream Table** in **Results Summary>Streams** results in the addition to the flow sheet shown in Fig. 2-27.

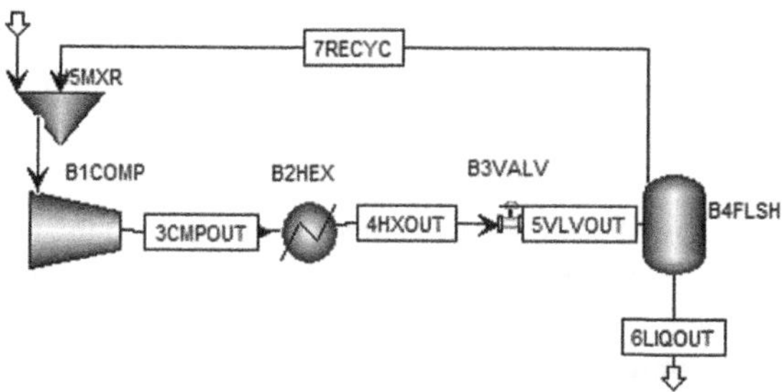

Test case 1 with recycle								
Stream ID		1FEED	2MIX	3CMPOUT	4HXOUT	5VLVOUT	6LIQOUT	7RECYC
From			B5MXR	B1COMP	B2HEX	B3VALV	B4FLSH	B4FLSH
To		B5MXR	B1COMP	B2HEX	B3VALV	B4FLSH		B5MXR
Phase		VAPOR	VAPOR	VAPOR	LIQUID	MIXED	LIQUID	VAPOR
Substream: MIXED								
Mole Flow	kmol/hr							
C3H8		1.360652	2.236411	2.236411	2.236411	2.236411	1.360660	.8757590
Total Flow	kmol/hr	1.360652	2.236411	2.236411	2.236411	2.236411	1.360660	.8757590
Total Flow	kg/hr	60.00000	98.61793	98.61793	98.61793	98.61793	60.00039	38.61792
Total Flow	l/min	552.5534	827.2363	66.52637	3.336973	272.3168	1.717166	270.6072
Temperature	K	298.0000	272.6981	373.9317	298.0000	230.6134	230.6120	230.6120
Pressure	atm	.9869233	.9869233	14.80385	14.80385	.9869233	.9869233	.9869233
Vapor Frac		1.000000	1.000000	1.000000	0.0	.3915776	0.0	1.000000
Liquid Frac		0.0	0.0	0.0	1.000000	.6084224	1.000000	0.0
Solid Frac		0.0	0.0	0.0	0.0	0.0	0.0	0.0
Enthalpy	cal/mol	-25031.77	-25467.53	-23853.06	-28862.77	-28862.77	-30612.27	-26144.56
Enthalpy	cal/gm	-567.6587	-577.5406	-540.9284	-654.5362	-654.5362	-694.2105	-592.8940
Enthalpy	cal/sec	-9460.978	-15821.07	-14818.12	-17930.28	-17930.28	-11570.25	-6360.093
Entropy	cal/mol-K	-64.35947	-65.88694	-65.88694	-81.42603	-80.36767	-87.95393	-68.58067
Entropy	cal/gm-K	-1.459513	-1.494153	-1.494153	-1.846541	-1.822540	-1.994578	-1.555240
Density	mol/cc	4.10413E-5	4.50579E-5	5.60282E-4	.0111698	1.36876E-4	.0132064	5.39379E-5
Density	gm/cc	1.80978E-3	1.98690E-3	.0247064	.4925518	6.03573E-3	.5823586	2.37847E-3
Average MW		44.09652	44.09652	44.09652	44.09652	44.09652	44.09652	44.09652
Liq Vol 60F	l/min	1.976215	3.248170	3.248170	3.248170	3.248170	1.976228	1.271955

Figure 2-27

This simple example was meant to illustrate two concepts. First is that recycle streams, which are frequently used in the chemical industry, lead to more efficient processes and greater product recovery. Second, that recycle streams increase the calculational complexity over that of a once-through process in that quantity and state (here temperature) of the inlet stream to the compressor are not known at the start of the analysis. Therefore, iterative calculations are required. That is, one can start by doing the once-through calculation to obtain the recycle stream amount, add the recycle stream back to feed, redo the once-through calculation with the new feed to compute the amount of the new recycle stream, etc. This is what the process simulator will do very efficiently. In fact, commercial processes, for example, a petroleum refinery, have a large number of recycle streams that would require months of hand calculations, but can be done in minutes or less using a process simulator such as Aspen Plus.

As the second example of a simple simulation, we consider computing the coefficient of performance of an automobile air conditioner using a vapor-compression refrigeration cycle with HFC-134a (which is the DuPont name, the chemical name is 1,1,1,2 tetrafluoroethane) as the working fluid. This is Illustration 5.2-2 in *Chemical, Biochemical and Engineering Thermodynamics*, 4th ed., S. I. Sandler (John Wiley & Sons, Inc., 2006). In this illustration a thermodynamic chart was used to obtain the properties of HFC-134a (see p. 161 of that text). However, it is inconvenient to have to manually use charts of thermodynamic data in computer-based process simulation. So what has to be done is to describe the properties of HFC-134a using the Peng–Robinson equation of state, which is introduced in Section 6.4 of the textbook mentioned above. As this equation of state provides a good, but approximate, description of HFC-134a, and as there were also uncertainties in reading the graph in the textbook, the results of using Aspen Plus will be in reasonable, but not perfect, agreement with those in the textbook.

We start by opening the **Aspen Plus User Interface** as described in the previous chapter. Also, using the procedures described in the previous chapter, 1,1,1,2 tetrafluoroethane is chosen as the single component, and the Peng–Robinson equation of state as the thermodynamic model. Use the **Find Compounds**, enter 1,1,1,2 in **Begins with** in the first window in Fig. 2-28a, and then scroll down the list to find and click on 1,1,1,2 tetrafluoroethane, and then click **Add selected compounds**.

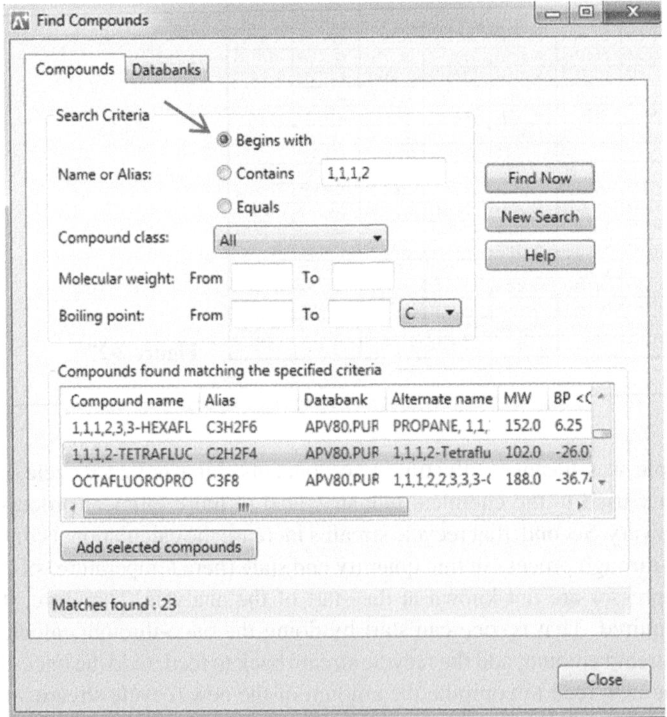

Figure 2-28a

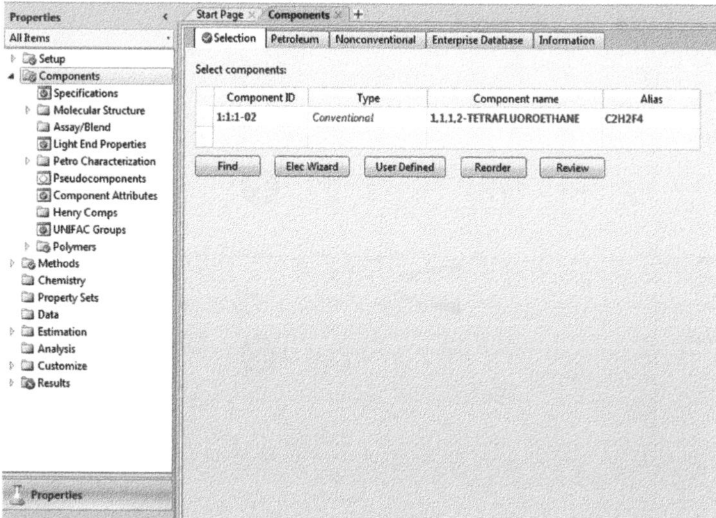

Figure 2-28b

Next click on **Methods** and populate it as shown in Fig. 2-29.

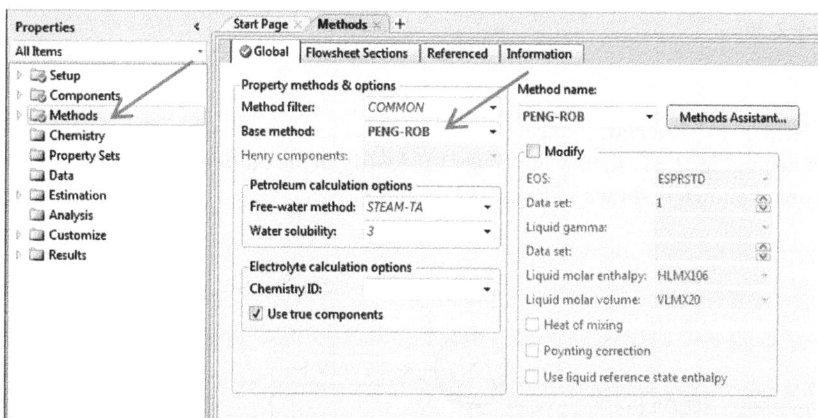

Figure 2-29

Then click on **Simulation**, which brings up a blank **Main Flowsheet** window with the **Model Palette** at the bottom, and populate it as shown in Fig. 2-30. Next, click on each of the block names sequentially (B1, B2, etc.) and rename them so that they can easily be identified in the results in the output. I then clicked on the **B2VALV** icon and rotated it 90° to the left, and on the **B3BLR** icon and rotated it 180°, so that the final flow sheet would look nicer.

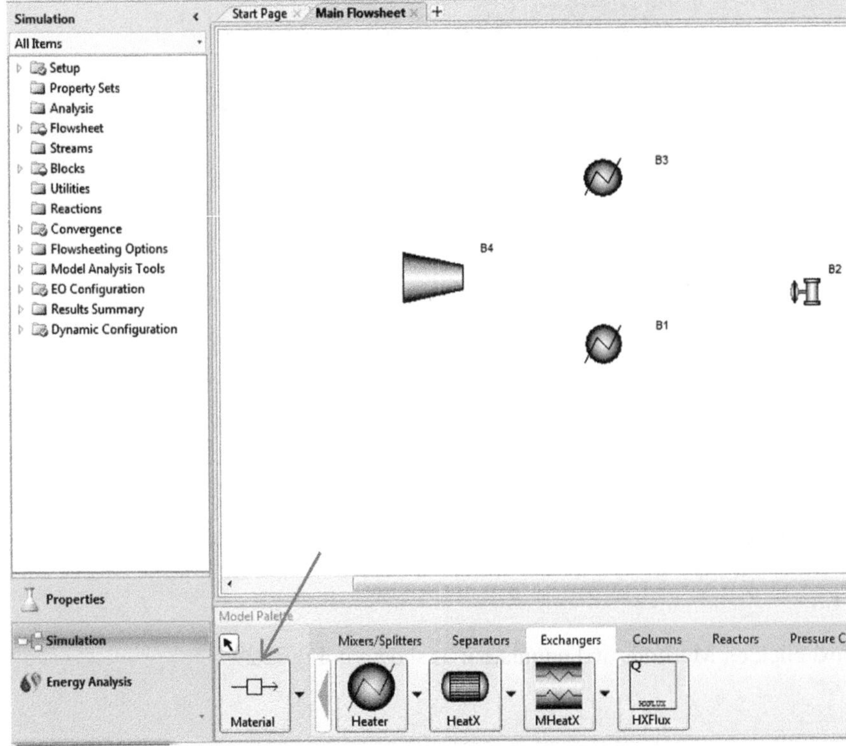

Figure 2-30

Next go to **Material,** connect all the flow sheet blocks, and then relabel the flow streams as shown in Fig. 2-31, so they are easier to identify in the output. The completed flow sheet window is shown is shown in Fig. 2-32.

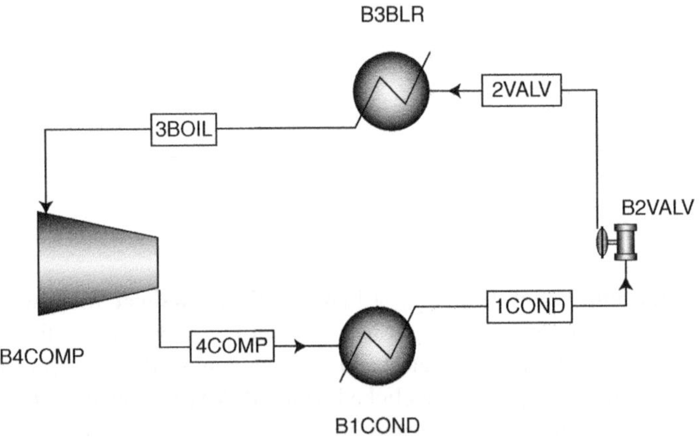

Figure 2-31

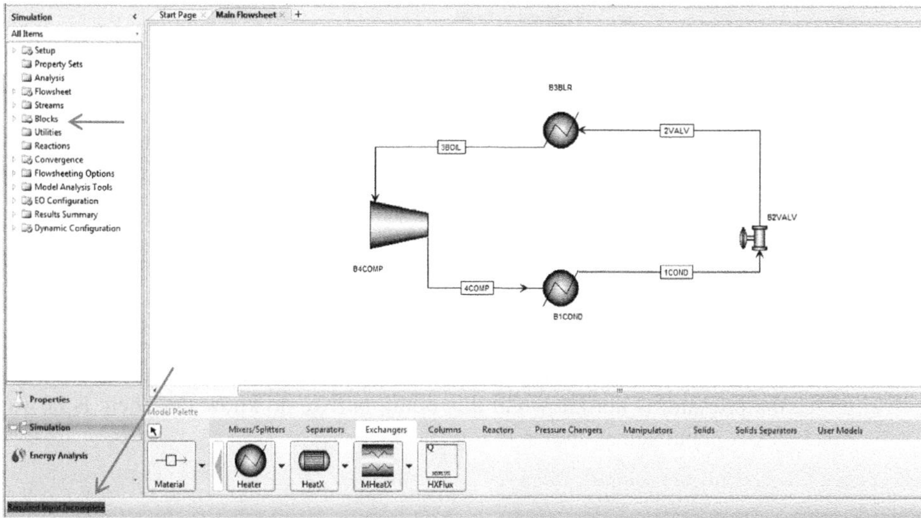

Figure 2-32

Even though this refrigeration cycle is operating as a closed system, we need to specify one flow rate through the system. We will do this by specifying a flow rate of 1 kmol/sec leaving the condenser, that is, of stream **1COND,** at an inlet **Temperature** of 278.15 K and zero **Vapor fraction** as shown in Fig. 2-33.

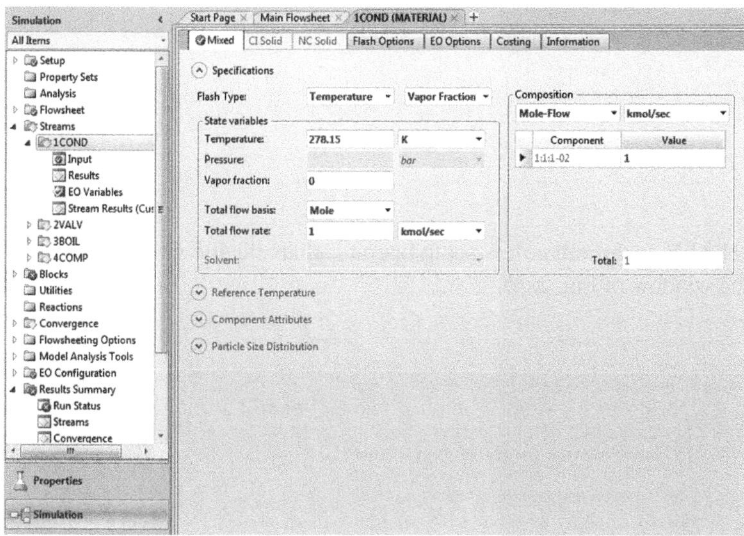

Figure 2-33

Note that in the lower left-hand corner of Fig. 2-32 is the message "**Required Input Incomplete**." Looking at the list of items in the simulation, we see there is a red flag at **Blocks**. Next go to **Blocks** (see Fig. 2-34).

Figure 2-34

Starting with the condenser, **B1COND**, we know the temperature leaving the condenser is 55°C; the pressure is not known at this point, but we do know that there is no vapor. So, this block is populated as shown in Fig. 2-35.

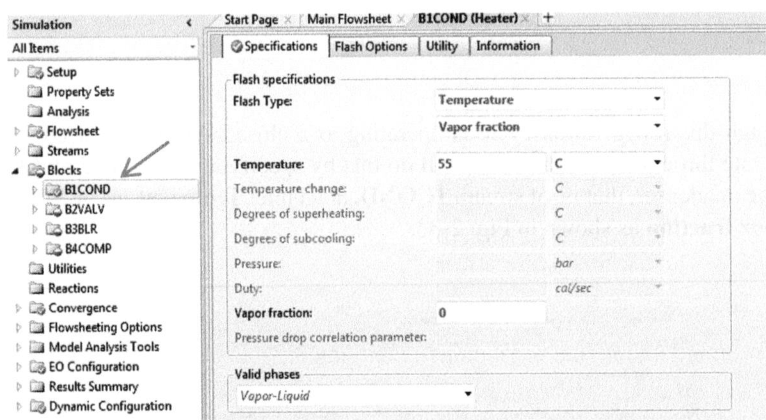

Figure 2-35

Block 2, **B2VALV**, is the valve. It is set to operate adiabatically with an outlet pressure of 3.5 bar in the window of Fig. 2-36.

Figure 2-36

Block 3, **B3BLR**, is the boiler that is set to operate at 55°C and 3.5 bar in Fig. 2-37.

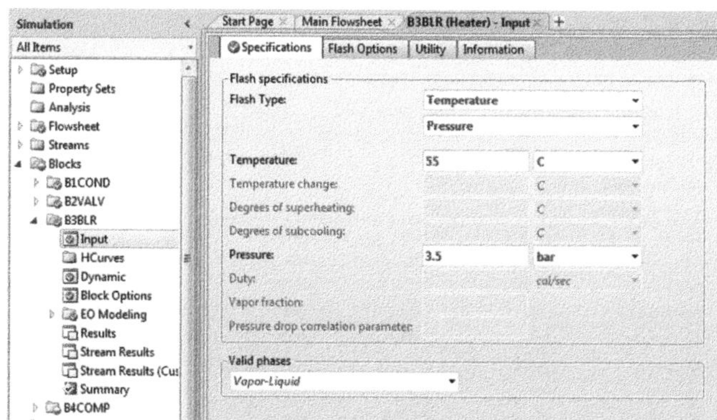

Figure 2-37

The final block is the compressor, **B4COMP**. To be consistent with Illustration 5.2-2 in the textbook, we will assume the compressor operates isentropically, even though in a real design we would use the isentropic efficiency reported by the manufacturer, and we will use an outlet pressure of 14.93 bar; these are set in the window of Fig. 2-38.

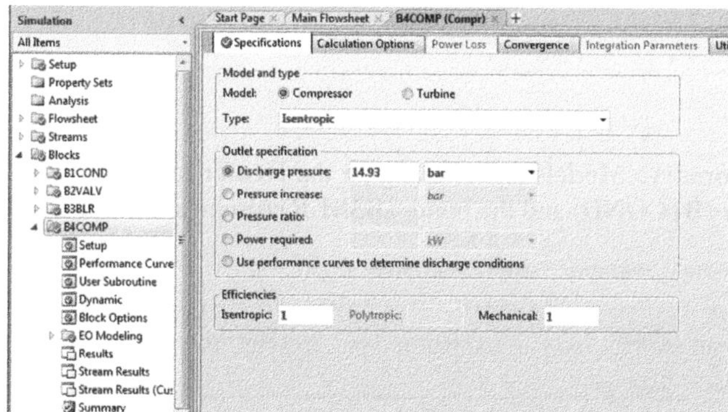

Figure 2-38

We now see in Fig. 2-39 that all the blocks have been specified and are checked.

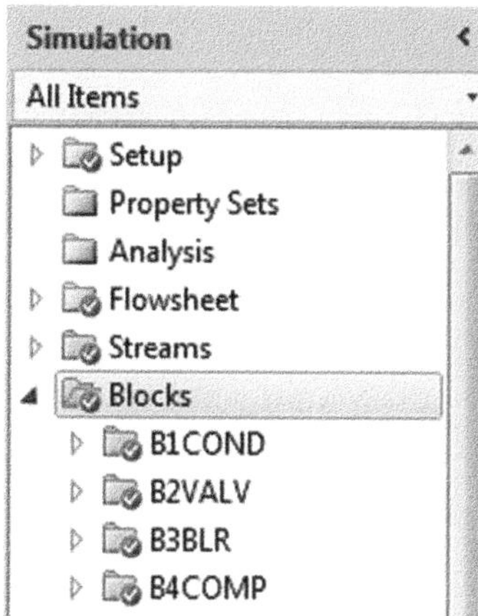

Figure 2-39

Next, run the simulation using **N>** on the main toolbar, then click on **Results Summary>Streams** to see the table in Fig. 2-40.

	1COND	2VALV	3BOIL	4COMP
Substream: MIXED				
Mole Flow kmol/hr				
1:1:1-02	3600	3600	3600	3600
Total Flow kmol/hr	3600	3600	3600	3600
Total Flow kg/hr	367313	367313	367313	367313
Total Flow l/min	5695.27	141301	443877	109050
Temperature C	55	5.13053	55	107.188
Pressure bar	14.958	3.5	3.5	14.93
Vapor Frac	0	0.381373	1	1
Liquid Frac	1	0.618628	0	0
Solid Frac	0	0	0	0
Enthalpy cal/mol	-217470	-217470	-213430	-212500
Enthalpy cal/gm	-2131.42	-2131.42	-2091.77	-2082.7
Enthalpy cal/sec	-2.1747e+08	-2.1747e+08	-2.1343e+08	-2.125e+08
Entropy cal/mol-K	-70.6991	-70.0696	-55.8433	-55.8433
Entropy cal/gm-K	-0.692914	-0.686745	-0.547314	-0.547314
Density mol/cc	0.010535	0.000424625	0.000135173	0.000550205
Density gm/cc	1.07491	0.0433251	0.0137918	0.0561382
Average MW	102.032	102.032	102.032	102.032
Liq Vol 60F l/min	5078.26	5078.26	5078.26	5078.26

Figure 2-40

Going to **Results Summary>Models** and clicking on the **Heater** tab gives the heat duty for both the condenser (**B1COND**) and the boiler (**B3BLR**) results as shown in Fig. 2-41.

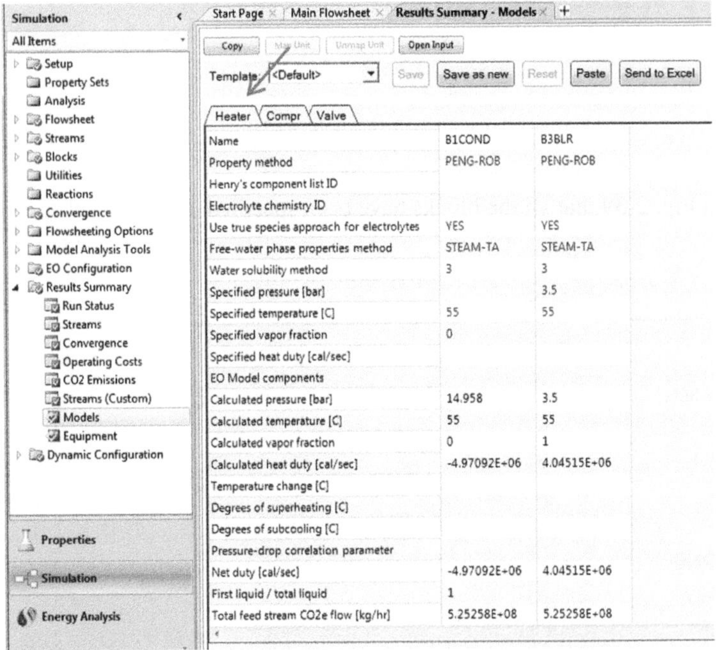

Name	B1COND	B3BLR
Property method	PENG-ROB	PENG-ROB
Henry's component list ID		
Electrolyte chemistry ID		
Use true species approach for electrolytes	YES	YES
Free-water phase properties method	STEAM-TA	STEAM-TA
Water solubility method	3	3
Specified pressure [bar]		3.5
Specified temperature [C]	55	55
Specified vapor fraction	0	
Specified heat duty [cal/sec]		
EO Model components		
Calculated pressure [bar]	14.958	3.5
Calculated temperature [C]	55	55
Calculated vapor fraction	0	1
Calculated heat duty [cal/sec]	-4.97092E+06	4.04515E+06
Temperature change [C]		
Degrees of superheating [C]		
Degrees of subcooling [C]		
Pressure-drop correlation parameter		
Net duty [cal/sec]	-4.97092E+06	4.04515E+06
First liquid / total liquid	1	
Total feed stream CO2e flow [kg/hr]	5.25258E+08	5.25258E+08

Figure 2-41

Finally, clicking on the **Compr** tab gives work required for the compressor, block **B4COMP**, in Fig. 2-42.

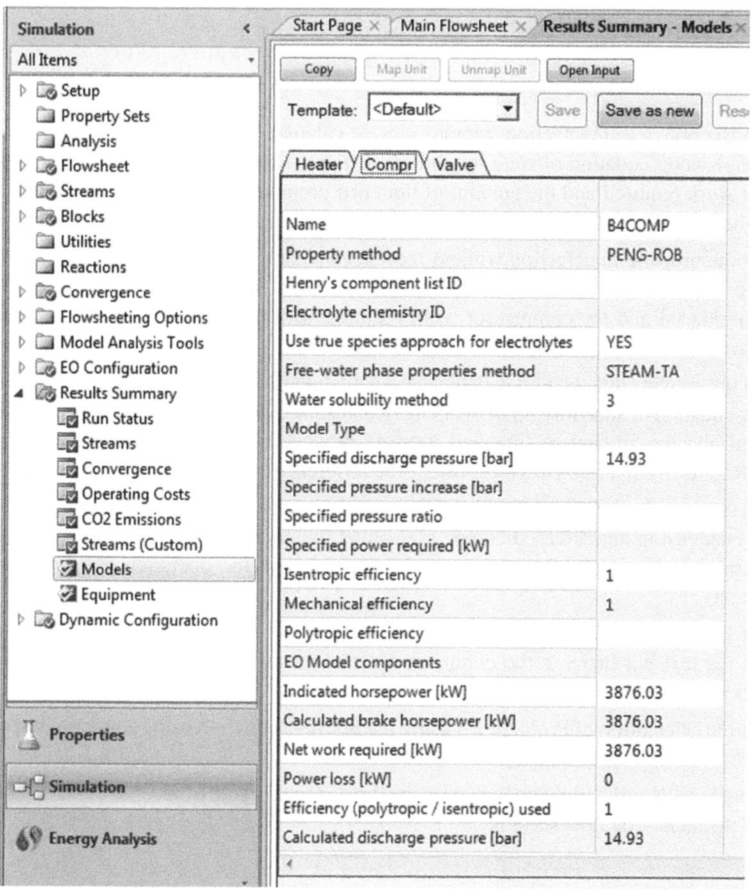

Figure 2-42

Looking at these results, we see that the work in the compressor is 3876 kW and the heat removal in the boiler is 4045 kcal/sec, both based on a flow of 1 kmol/sec. As 1 kW = 0.239 kcal/sec, 3876 kW = 926.4 kcal/sec and the coefficient of performance is

$$\text{C.O.P.} = \frac{Q_B}{W} = \frac{4045}{926.4} = 4.366$$

which is somewhat higher than the value of 4.07 reported in the textbook. The difference is due to the uncertainty of reading the chart in the textbook, and that the Peng–Robinson equation of state is an imperfect representation of the properties of HFC-134a (or any other fluid).

It is now easy to change any of the operating variables, and see the effect on the coefficient of performance. A very real advantage of using process simulation software is that once a simulation is set up, it is very easy to change parameters and determine the effect. In this way many variations can be studied and a process optimized.

The following chapters deal with the use of Aspen Plus for thermodynamic and phase equilibrium properties, predictions, and regression. Later chapters will deal with simple simulations of distillation, extraction, and chemical reactions.

PROBLEMS

2.1. Repeat the propane liquefaction without recycle calculation in the text but using the Soave–Redlich–Kwong equation of state instead of the Peng–Robinson equation. Compare the compressor work required and the amount of liquefied propane produced with the values obtained using the Peng–Robinson equation. Do you consider the differences significant?

2.2. Repeat the propane liquefaction without recycle calculation with the Peng–Robinson equation of state using an increased compressor outlet pressure of 20 bar. Compute the amount of liquid produced and the compressor work required with the operation at 15 bar discussed in the text.

2.3. Repeat the propane liquefaction with recycle calculation in the text but using the Soave–Redlich–Kwong equation of state instead of the Peng–Robinson equation. Compare the compressor work required and the amount of liquefied propane produced with the values obtained using the Peng–Robinson equation. Do you consider the differences significant?

2.4. Repeat the propane liquefaction with recycle calculation using the Soave–Redlich–Kwong equation of state using an increased compressor outlet pressure of 20 bar. Compare the amount of liquid produced and the compressor work required with the operation at 15 bar discussed in the text.

2.5. Repeat the calculation of the coefficient of performance for the automobile air conditioning unit in the text but increase the compressor outlet pressure to 20 bar. Has the coefficient of performance increased?

2.6. Repeat the calculation of Problem 2.5 using the Soave–Redlich–Kwong equation of state rather than the Peng–Robinson equation.

2.7. As a more challenging problem, consider the three-stage recycle Linde process for the liquefaction of methane. The flow sheet is

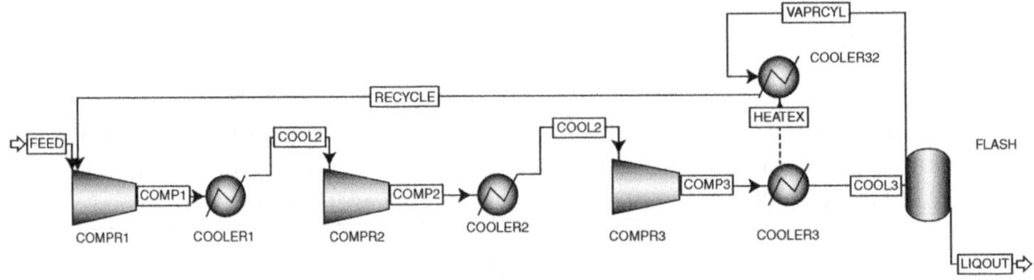

The operating conditions, based on Illustration 5.1-1 in *Chemical, Biochemical and Engineering Thermodynamics*, 4th ed., S. I. Sandler (John Wiley & Sons, Inc., 2006) are as follows: compression stage 1 from 1 to 5 bar, compression stage 2 from 5 to 25 bar, and compression stage 3 from 25 to 100 bar. Intercooling between stages to 280 K, and the feed is at 1 bar and 280 K.

This process is more complicated than the previous ones by the number of units involved, but especially that the hot methane leaving compressor 3 is cooled by the cold vapor leaving the separator and being recycled. Thus, COOLER3 and COOLER32 are actually two parts of a single piece of equipment, a countercurrent heat exchanger. The two exchangers are connected by a heat stream. The heat stream is placed on the flow sheet by going to the lowerleft-hand corner of the **Model Palette** and clicking on the down arrow by the **Material** to bring up a drop-down menu and choose **Heat**, and then connect the two heat exchangers.

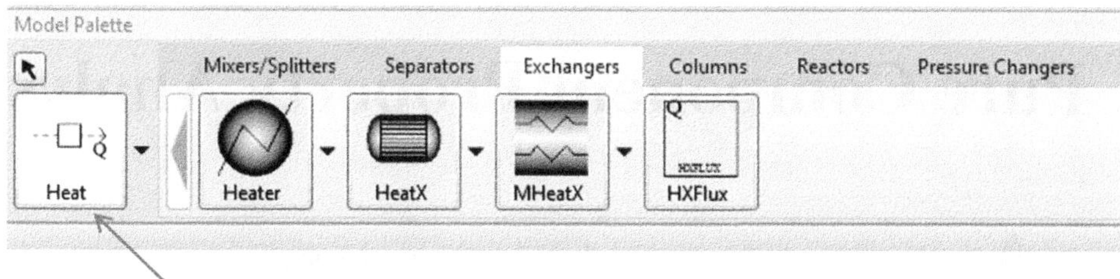

Determine the work required in each compressor and the total work needed for each kilogram of liquid methane produced. Use the Peng–Robinson equation of state in your calculations.

Chapter 3

Pure Component Property Analysis

Aspen Plus® provides rather simple and easy ways to compute a wide variety of thermodynamic properties and phase behavior of pure components and, as will be shown in later chapters, for mixtures using either activity coefficient models (low pressure) or equations of state. In this chapter we consider only pure component properties and phase behavior. Mixture property analyses are considered in Chapters 5 and 6.

The starting point is to start the Aspen Plus program that will open up at **Properties**, go to **Simulation**, and choose **Blank Simulation** as in Fig. 3-1, go to **Setup** to choose the title and the units you wish to use.

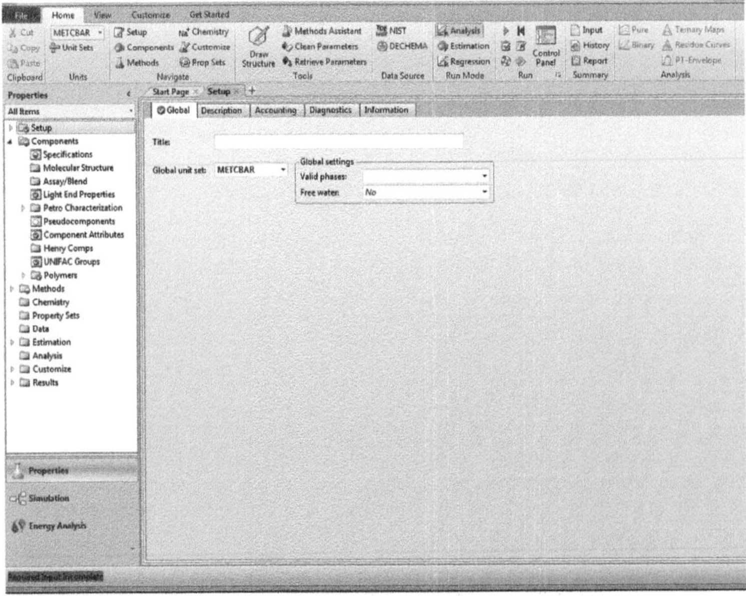

Figure 3-1

Then go to **Components>Specifications** and add one or more components following the method described in Chapter 1. Here ethanol will be used as an example as shown in Fig. 3-2.

Using Aspen Plus® in Thermodynamics Instruction: A Step-by-Step Guide, First Edition. Stanley I. Sandler.

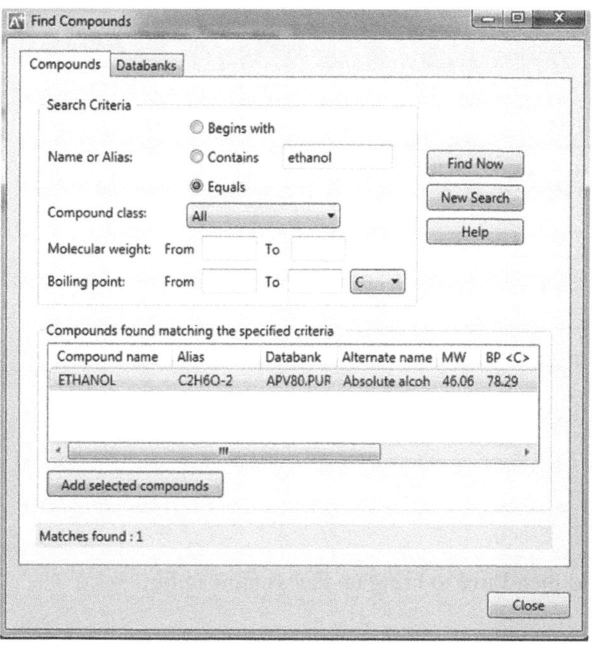

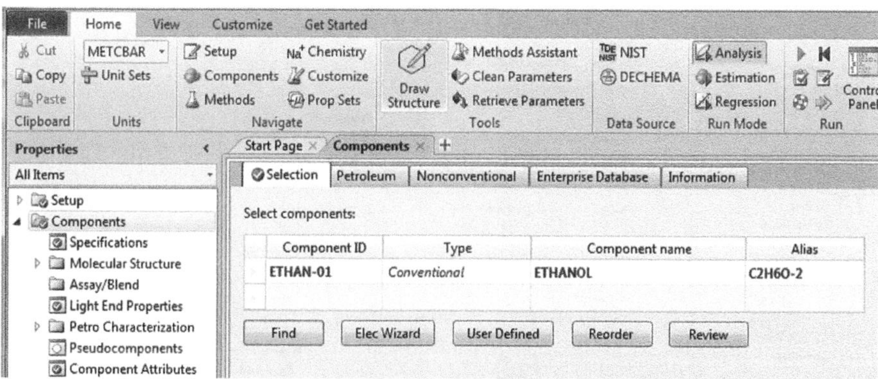

Figure 3-2

Next go to **Methods** (Fig. 3-3) and choose a **Base method**. The Peng–Robinson equation of state is chosen here. As a general rule, one should not use the generalized form of the Peng–Robinson equation, that is, the form in which the equation of state parameters are obtained from only the critical temperature, critical pressure, and the acentric factor, for a strongly polar compound such as ethanol. However, in Aspen Plus the equation of state parameters have also been fitted to vapor pressure data, and therefore, provide a good description of this property. In fact, the predicted and measured normal boiling points (the temperature at which the vapor pressure is 1 atm) are both 351.5°C. Note that this choice of method will affect the values of the pressure/density dependent properties (i.e., enthalpies, entropies, etc.), but not the values of the ideal gas properties.

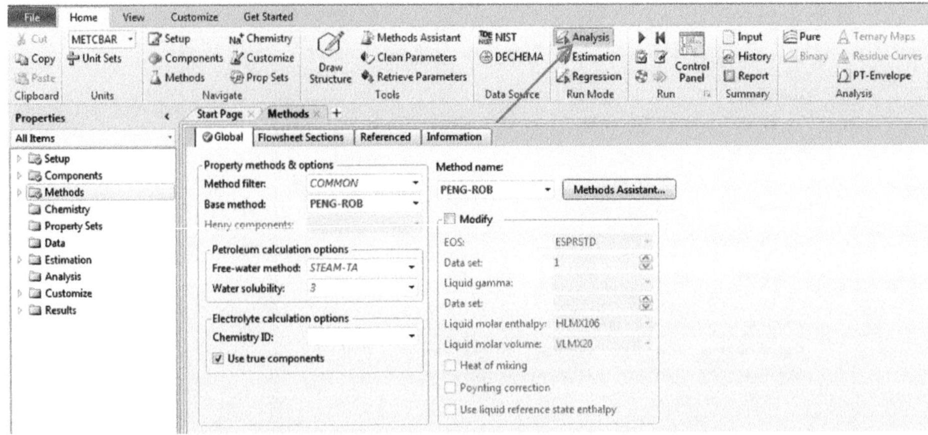

Figure 3-3

Then click on **Analysis** and then **Pure** to bring up the window in Fig. 3-4.

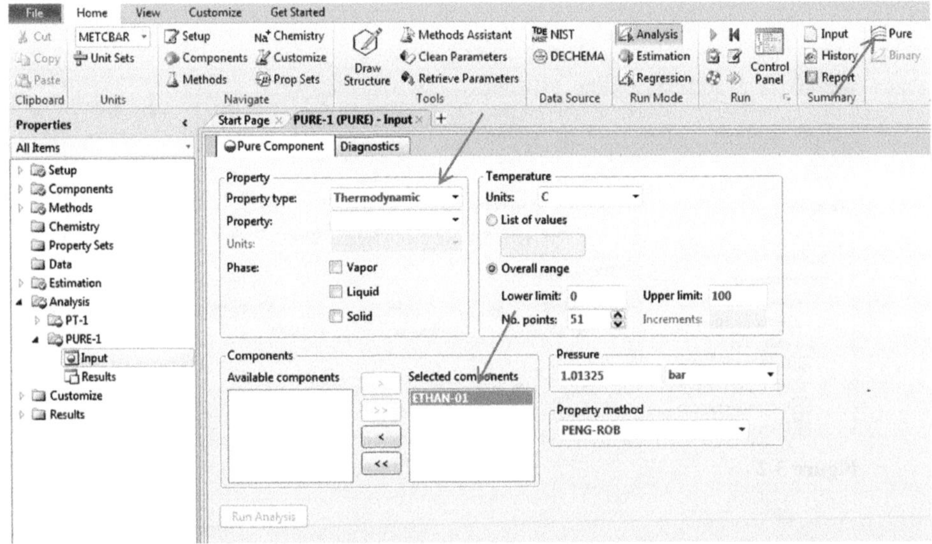

Figure 3-4

The **Property types** available are Thermodynamic, Transport, and All. **Thermodynamic** will be used here. Clicking on **Property** brings a drop-down menu with a long list of abbreviations. When the cursor is on an item, a description of it appears. Also, from the list of **Available components** (that may contain one or several depending on how many you added earlier), click on the component of interest and then on the > to move it to **Selected components**. A completed window displaying the vapor pressure (denoted as PL) of ethanol from 250 to 600 K is shown in Fig. 3-5.

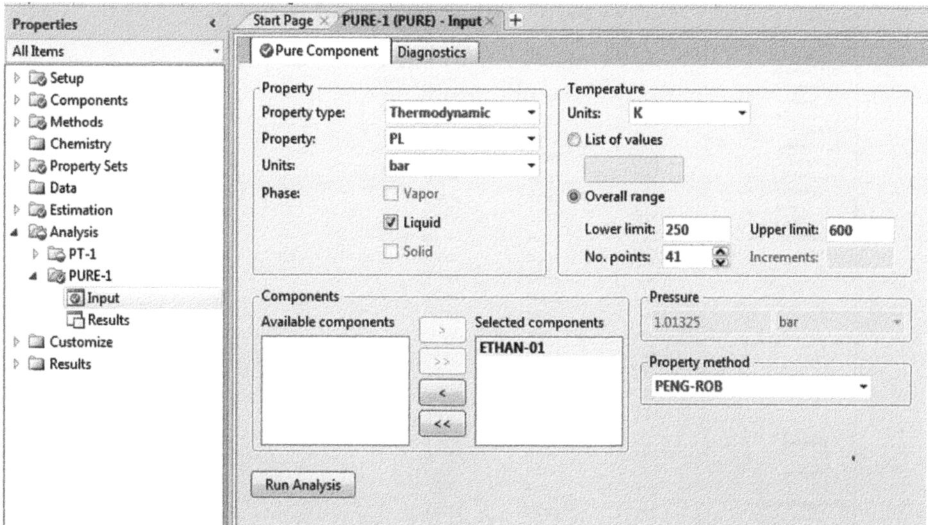

Figure 3-5

Now by clicking on **Run Analysis** Fig. 3-6 is generated.

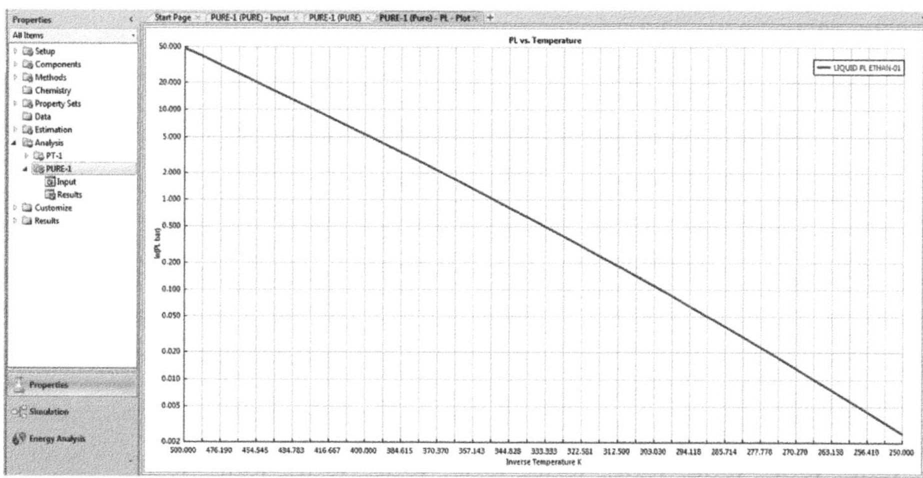

Figure 3-6

Then click on **Analysis>Pure-1>Results** to see the table of results in Fig. 3-7. [Note that to determine the normal boiling point from the results in the table, one would have to interpolate the liquid–vapor pressures to obtain a vapor pressure of 1.01325 bar. However, since the table result of 1.00803 is very close to 1.01325 bar, the predicted normal boiling point will be very close to 350 K.]

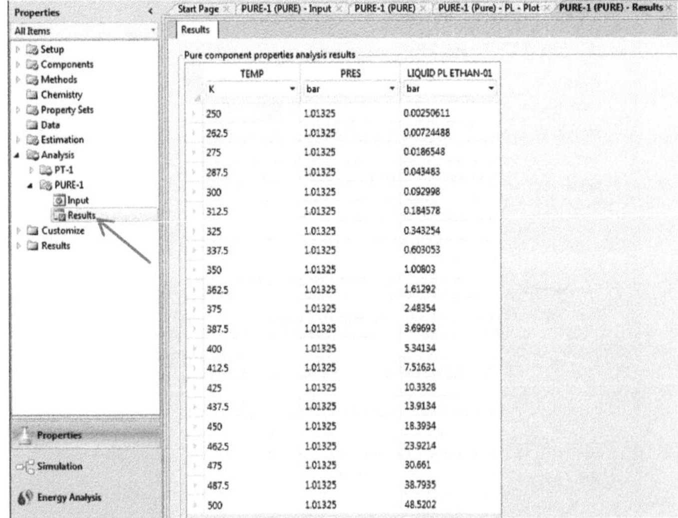

Pure component properties analysis results

TEMP	PRES	LIQUID PL ETHAN-01
K	bar	bar
250	1.01325	0.00250611
262.5	1.01325	0.00724488
275	1.01325	0.0186548
287.5	1.01325	0.043483
300	1.01325	0.092998
312.5	1.01325	0.184578
325	1.01325	0.343254
337.5	1.01325	0.603053
350	1.01325	1.00803
362.5	1.01325	1.61292
375	1.01325	2.48354
387.5	1.01325	3.69693
400	1.01325	5.34134
412.5	1.01325	7.51631
425	1.01325	10.3328
437.5	1.01325	13.9134
450	1.01325	18.3934
462.5	1.01325	23.9214
475	1.01325	30.661
487.5	1.01325	38.7935
500	1.01325	48.5202

Figure 3-7

A similar procedure can be used to calculate the ideal gas heat capacity, **CPIG**, of ethanol as shown in Fig. 3-8.

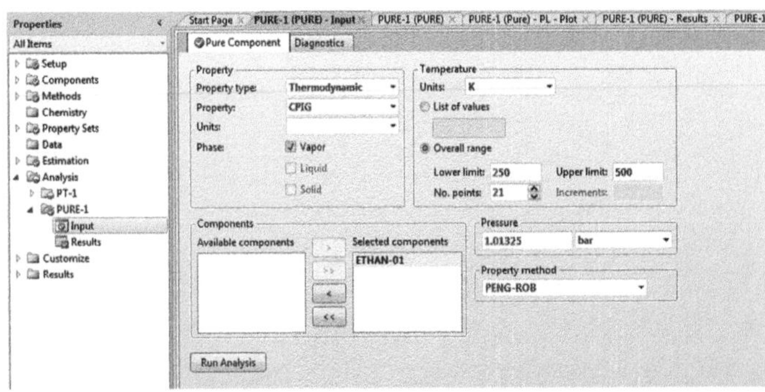

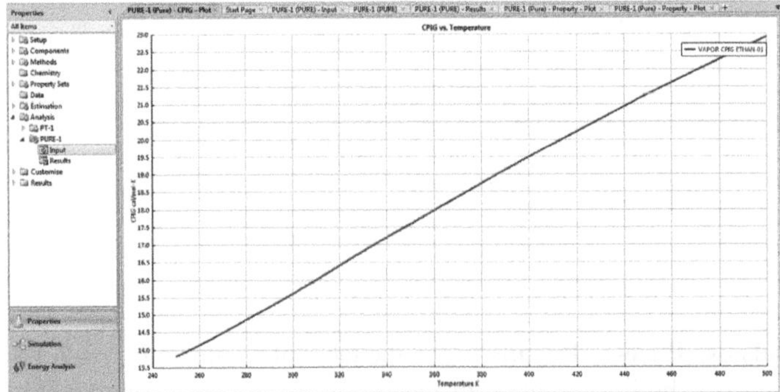

Figure 3-8

Similarly, to generate the enthalpies at 20 bar as a function of temperature, but now for both the vapor and liquid phases, follow the procedure in Fig. 3-9.

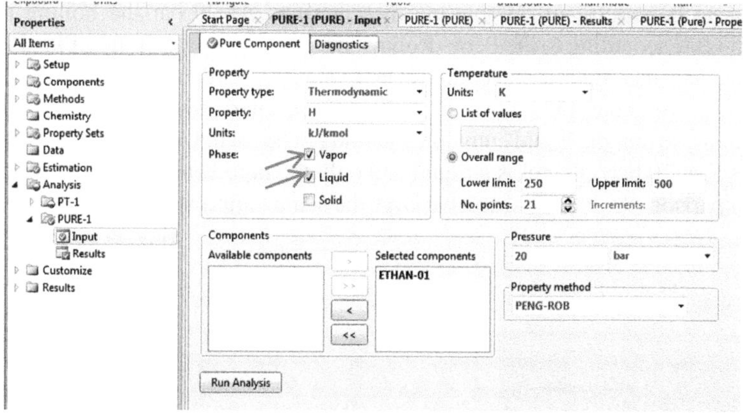

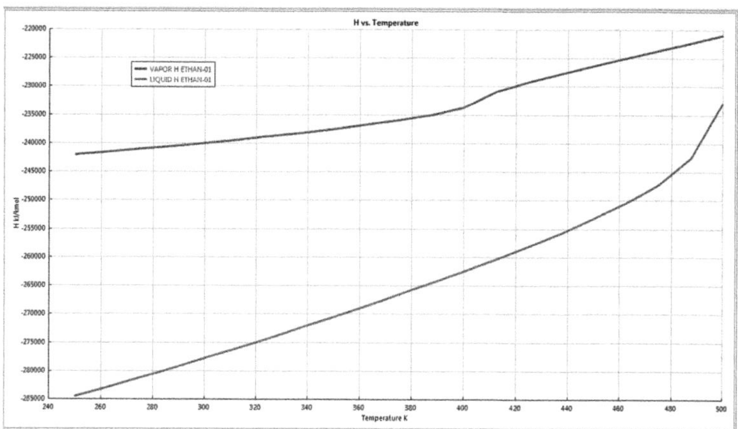

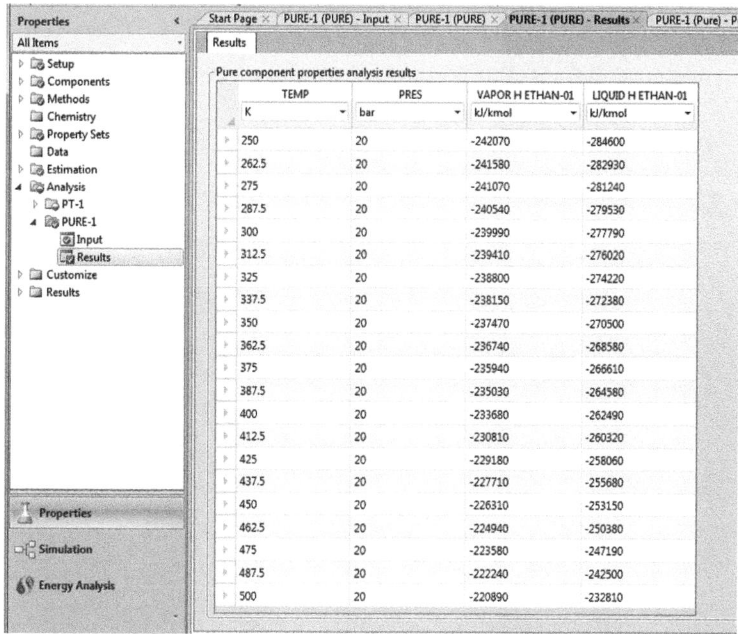

TEMP	PRES	VAPOR H ETHAN-01	LIQUID H ETHAN-01
K	bar	kJ/kmol	kJ/kmol
250	20	-242070	-284600
262.5	20	-241580	-282930
275	20	-241070	-281240
287.5	20	-240540	-279530
300	20	-239990	-277790
312.5	20	-239410	-276020
325	20	-238800	-274220
337.5	20	-238150	-272380
350	20	-237470	-270500
362.5	20	-236740	-268580
375	20	-235940	-266610
387.5	20	-235030	-264580
400	20	-233680	-262490
412.5	20	-230810	-260320
425	20	-229180	-258060
437.5	20	-227710	-255680
450	20	-226310	-253150
462.5	20	-224940	-250380
475	20	-223580	-247190
487.5	20	-222240	-242500
500	20	-220890	-232810

Figure 3-9

One has to be careful in interpreting these results. At 20 bar the boiling point of ethanol is 453.8 K according to the Peng–Robinson equation of state. **Aspen Plus Property Analysis** does not take this into account. Consequently, the reported liquid enthalpies at temperatures above about 453.8 K at 20 bar are for a liquid, while ethanol is a vapor at these conditions, so that the liquid enthalpies should not be used. Likewise, for temperatures below 453.8 K at 20 bar, ethanol is a liquid and only the liquid enthalpies are relevant.

For comparison, the enthalpy at 2 bar over the same temperature range is as shown in Fig. 3-10.

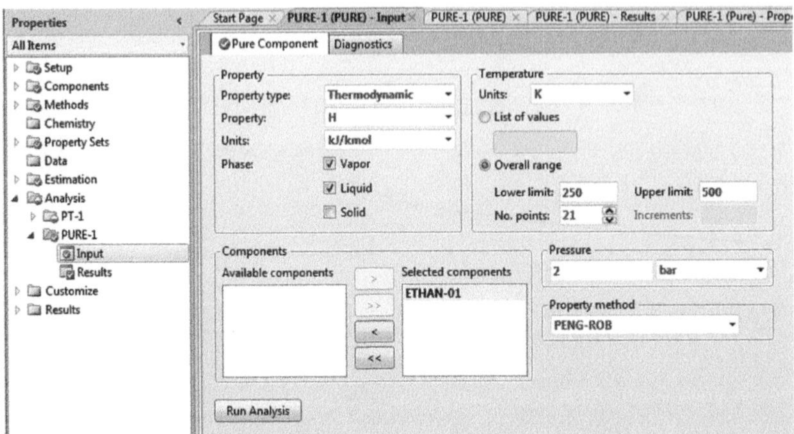

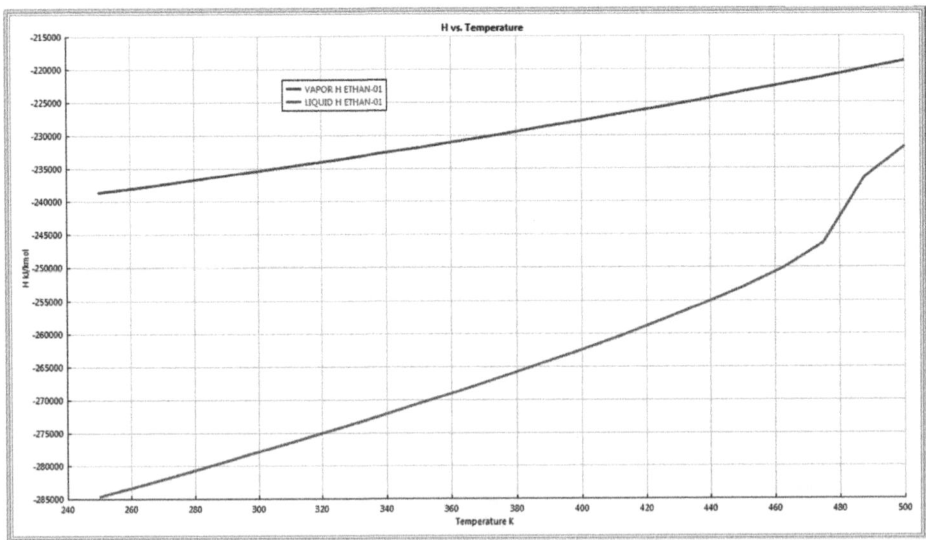

Figure 3-10

Results

Pure component properties analysis results

TEMP	PRES	VAPOR H ETHAN-01	LIQUID H ETHAN-01
K	bar	kJ/kmol	kJ/kmol
250	2	-238640	-284680
262.5	2	-237840	-283010
275	2	-237030	-281320
287.5	2	-236200	-279610
300	2	-235350	-277860
312.5	2	-234490	-276090
325	2	-233600	-274290
337.5	2	-232690	-272440
350	2	-231750	-270560
362.5	2	-230800	-268630
375	2	-229820	-266640
387.5	2	-228810	-264600
400	2	-227780	-262490
412.5	2	-226730	-260300
425	2	-225660	-258000
437.5	2	-224570	-255570
450	2	-223450	-252950
462.5	2	-222320	-250010
475	2	-221160	-246310
487.5	2	-219980	-236540
500	2	-218780	-231640

Figure 3-10 *(Continued)*

At 2 bar the predicted boiling point (using the Peng–Robinson equation of state) of ethanol is 369.8 K. So that at 2 bar pressure one should use liquid enthalpies at temperatures below about 370 K, and vapor enthalpies at higher temperatures.

The Gibbs energies can be calculated at 2 bar in a similar manner as shown in Fig. 3-11.

At 2 bar

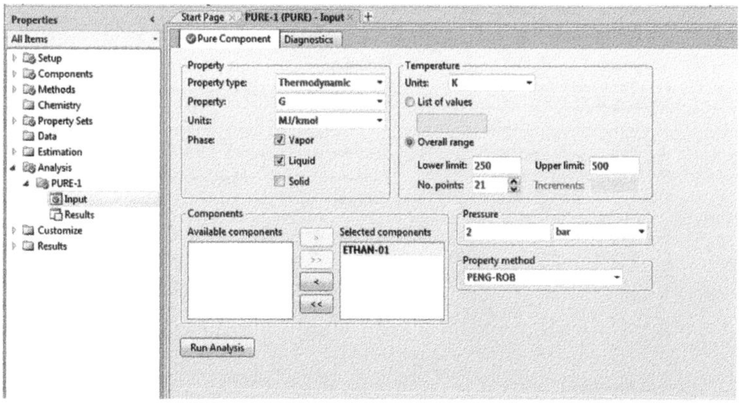

Figure 3-11

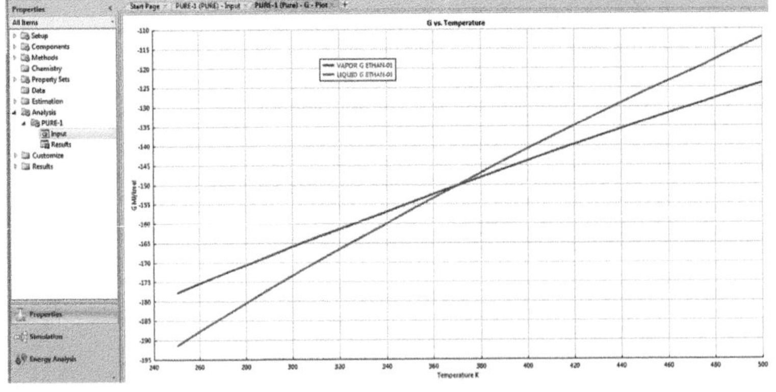

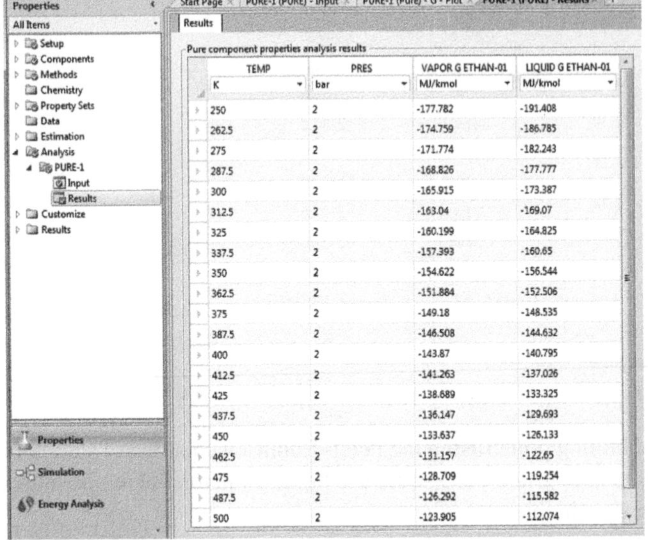

Pure component properties analysis results

TEMP K	PRES bar	VAPOR G ETHAN-01 MJ/kmol	LIQUID G ETHAN-01 MJ/kmol
250	2	-177.782	-191.408
262.5	2	-174.759	-186.785
275	2	-171.774	-182.243
287.5	2	-168.826	-177.777
300	2	-165.915	-173.387
312.5	2	-163.04	-169.07
325	2	-160.199	-164.825
337.5	2	-157.393	-160.65
350	2	-154.622	-156.544
362.5	2	-151.884	-152.506
375	2	-149.18	-148.535
387.5	2	-146.508	-144.632
400	2	-143.87	-140.795
412.5	2	-141.263	-137.026
425	2	-138.689	-133.325
437.5	2	-136.147	-129.693
450	2	-133.637	-126.133
462.5	2	-131.157	-122.65
475	2	-128.709	-119.254
487.5	2	-126.292	-115.582
500	2	-123.905	-112.074

Figure 3-11 (*Continued*)

and at 20 bar

Properties | All Items | Setup | Components | Methods | Chemistry | Property Sets | Data | Estimation | Analysis | PURE-1 | Input | Results | Customize | Results

Start Page | PURE-1 (PURE) - Input | PURE-1 (Pure) - G - Plot | PURE-1 (PURE) - Results

Pure Component | Diagnostics

Property
Property type: Thermodynamic
Property: G
Units: MJ/kmol
Phase: Vapor, Liquid, Solid

Temperature
Units: K
List of values
Overall range
Lower limit: 250 Upper limit: 500
No. points: 21 Increments:

Components
Available components
Selected components: ETHAN-01

Pressure
20 bar

Property method
PENG-ROB

Run Analysis

Figure 3-12

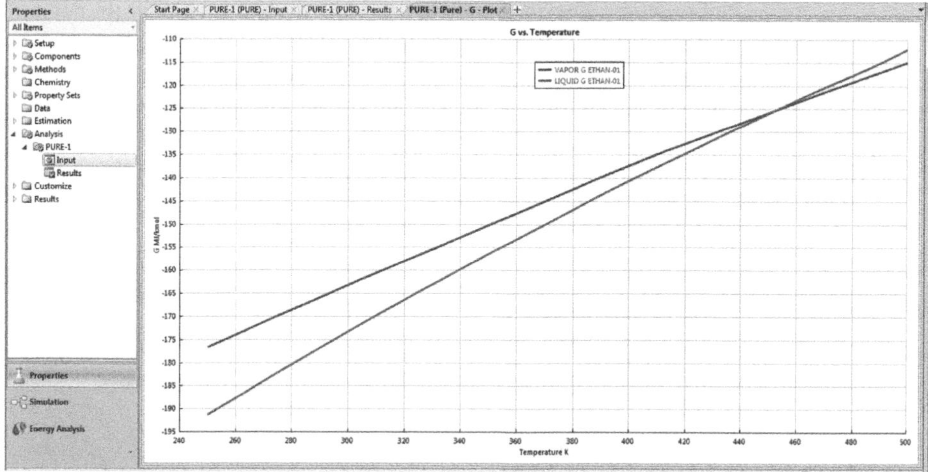

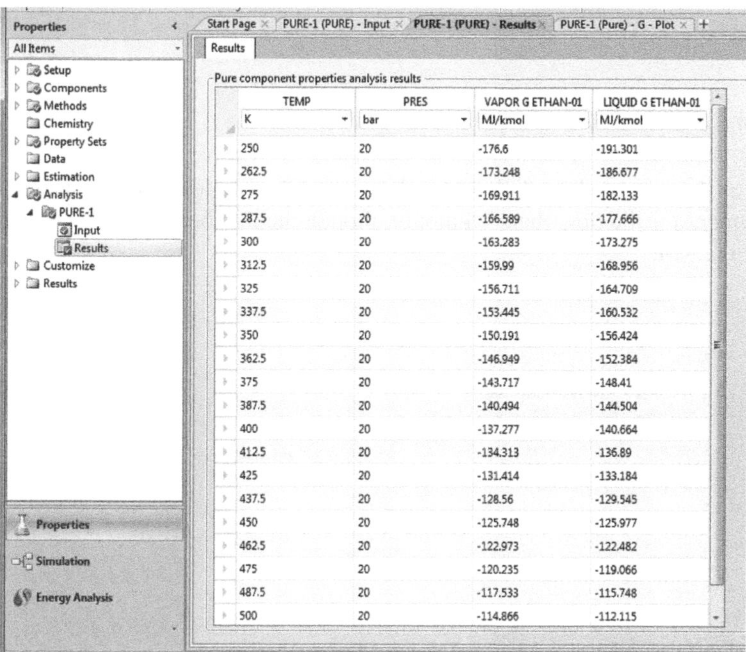

Pure component properties analysis results

TEMP	PRES	VAPOR G ETHAN-01	LIQUID G ETHAN-01
K	bar	MJ/kmol	MJ/kmol
250	20	-176.6	-191.301
262.5	20	-173.248	-186.677
275	20	-169.911	-182.133
287.5	20	-166.589	-177.666
300	20	-163.283	-173.275
312.5	20	-159.99	-168.956
325	20	-156.711	-164.709
337.5	20	-153.445	-160.532
350	20	-150.191	-156.424
362.5	20	-146.949	-152.384
375	20	-143.717	-148.41
387.5	20	-140.494	-144.504
400	20	-137.277	-140.664
412.5	20	-134.313	-136.89
425	20	-131.414	-133.184
437.5	20	-128.56	-129.545
450	20	-125.748	-125.977
462.5	20	-122.973	-122.482
475	20	-120.235	-119.066
487.5	20	-117.533	-115.748
500	20	-114.866	-112.115

Figure 3-12 *(Continued)*

In these figures of the Gibbs energies at the two different pressures we see that there is a temperature at which the vapor and liquid Gibbs energies lines cross, so that at each temperature there is a single point at which Gibbs energies of the vapor and liquid are the same. These are the temperatures of the vapor–liquid transition, that is, the temperatures at which the liquid will boil at each of the pressures. Or, said another way, these are temperatures at which the vapor pressure computed for ethanol is 2 bar (Fig. 3-11) and 20 bar (Fig. 3-12). [Remember from thermodynamics that the Gibbs energy of an equilibrium state is a minimum at a fixed temperature and pressure. So that at each temperature, the phase of the lowest Gibbs energy in the figure is the equilibrium state.]

Aspen Plus provides some thermodynamic properties information in the default **Results** output. However, many other thermodynamic properties are also calculated internally, and many of these can be made part of the output by using a user-defined **Prop-Sets** option. This is illustrated here by calculating and outputting some of the properties of carbon dioxide.

Start a new simulation, then go to **Components** to enter the component of interest, here carbon dioxide, and then go to **Methods** and choose the Peng–Robinson equation of state, **PENG-ROB**, as the property method, as shown in Fig. 3-13.

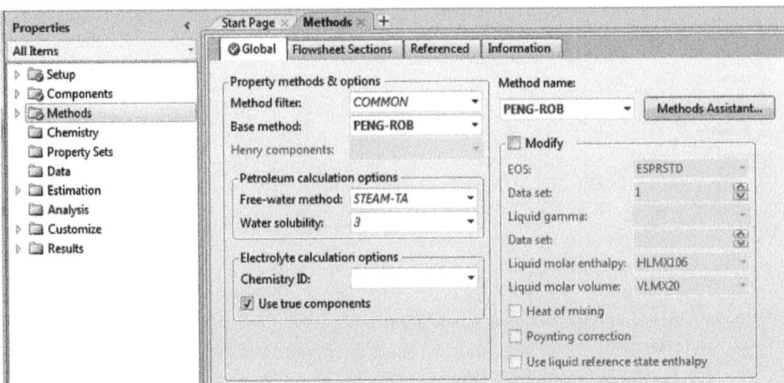

Figure 3-13

Next, on the main toolbar click on **Analysis,** then **Pure** followed by **PT-Envelope**, and put in a number for the flow (here 1 kmol/hr, though the number and unit is irrelevant) as shown in Fig. 3-14.

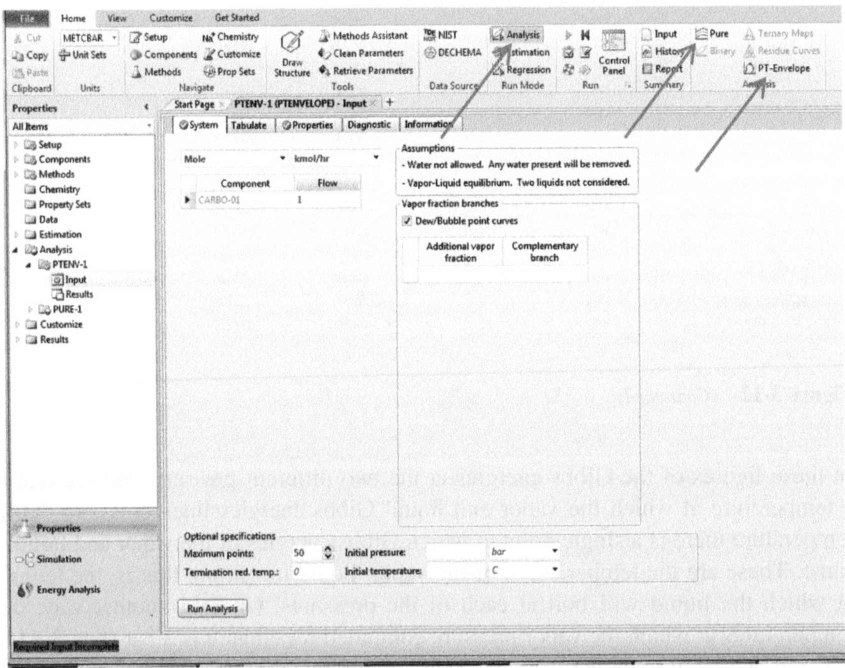

Figure 3-14

Finally click on **Run Analysis**, which produces the graph in Fig. 3-15.

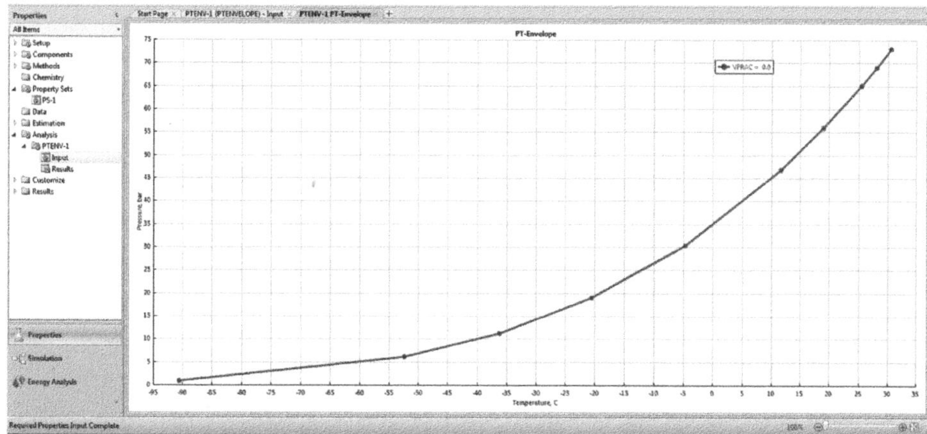

Figure 3-15

Then, going to **Analysis>PTENV-1>Results** brings up the table in Fig. 3-16.

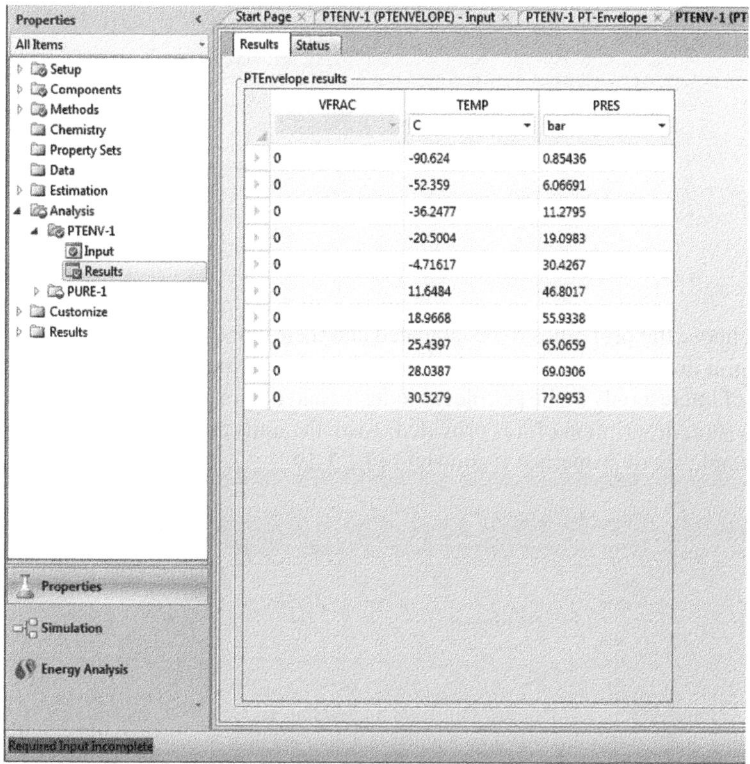

VFRAC	TEMP	PRES
	C	bar
0	-90.624	0.85436
0	-52.359	6.06691
0	-36.2477	11.2795
0	-20.5004	19.0983
0	-4.71617	30.4267
0	11.6484	46.8017
0	18.9668	55.9338
0	25.4397	65.0659
0	28.0387	69.0306
0	30.5279	72.9953

Figure 3-16

We can get the results for other thermodynamic properties by adding **Property Sets** as follows. Click on **Property Sets** and then **New** as shown in Fig. 3-17.

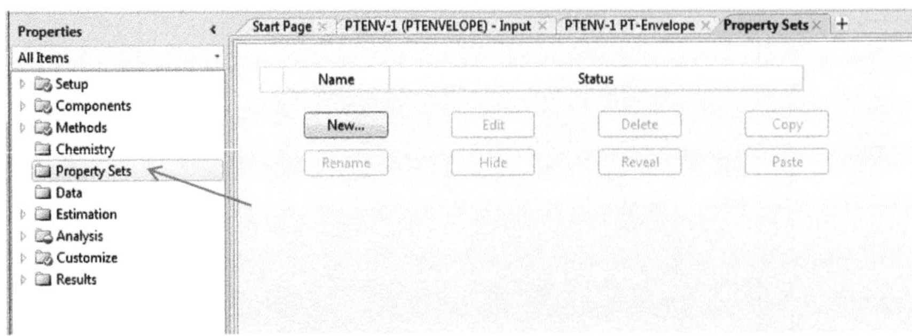

Figure 3-17

Then either accept the default ID **PS-1** or choose a name of your own as indicated in Fig. 3-18, then click **OK**.

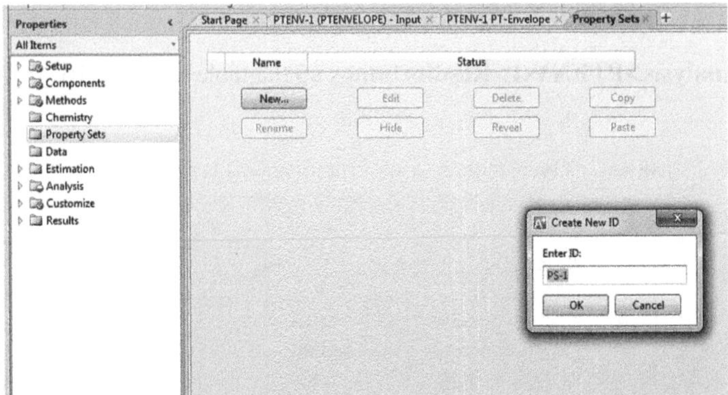

Figure 3-18

Next, choose the properties to be tabulated and their units. Note that the properties are chosen from a drop-down menu that has many entries, including many with specialized definitions of interest only in the petroleum industry and others. As the cursor is moved over any entry, a short description of it is provided. Also, the units are chosen from a drop-down menu. A sample set of properties is shown in Fig. 3-19.

Figure 3-19

Here the choices are temperature (to make sure it appears when this **Property Set** is used in other than the **PT envelope** calculations), the fugacity coefficient **PHI**, and the enthalpy (**DH**) and entropy (**DS**) departures from ideal gas behavior. The units also have to be chosen, and both the properties and units are selected from drop-down menus. Next, as shown in Fig. 3-20, click on the **Qualifiers** tab and enter the phases for which you want data displayed.

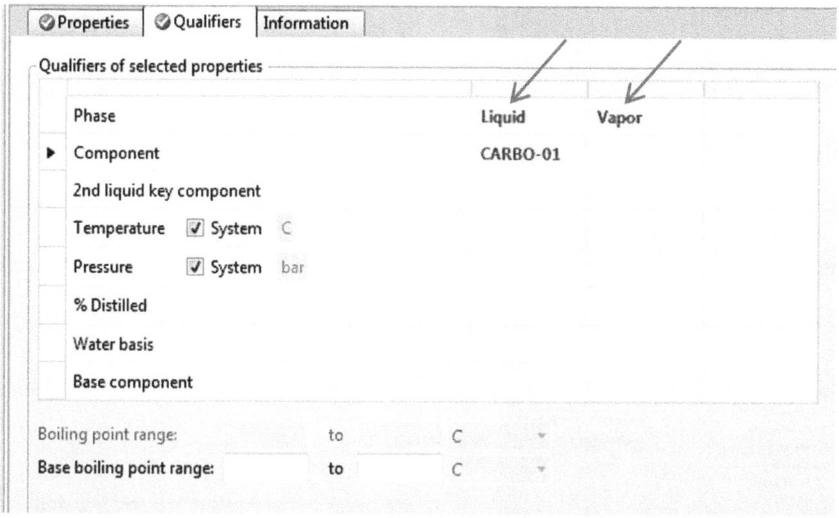

Figure 3-20

The PS-1 **Property Set** is now complete. This property set is added to the **PT-1** analysis as follows. Go to **Properties>Analysis>PTENV-1>Input** and click on the **Tabulate** tab and choose the **Prop-Set PS-1** that has just been set up. This is shown in Fig. 3-21.

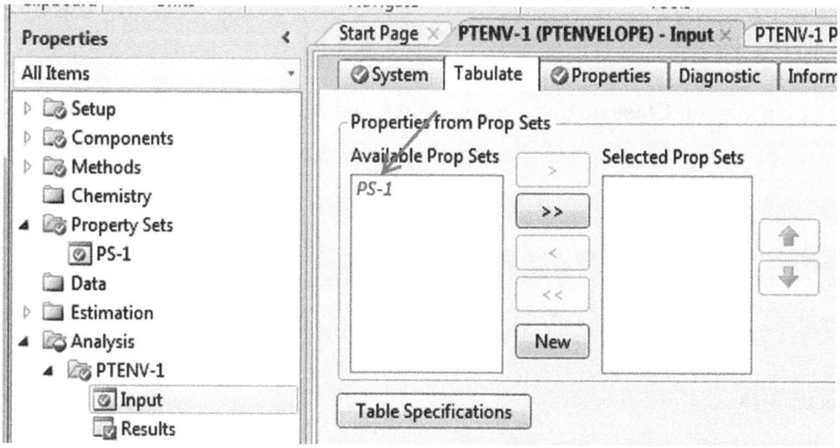

Figure 3-21

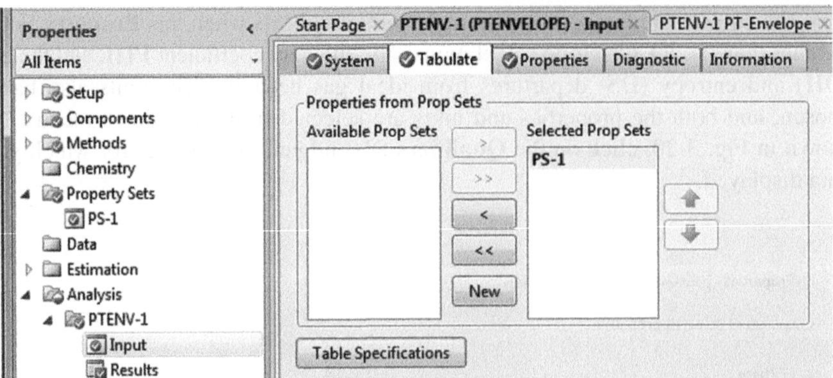

Figure 3-21 (*Continued*)

Next click on **Table Specifications** and populate the pop-up window as shown in Fig. 3-22.

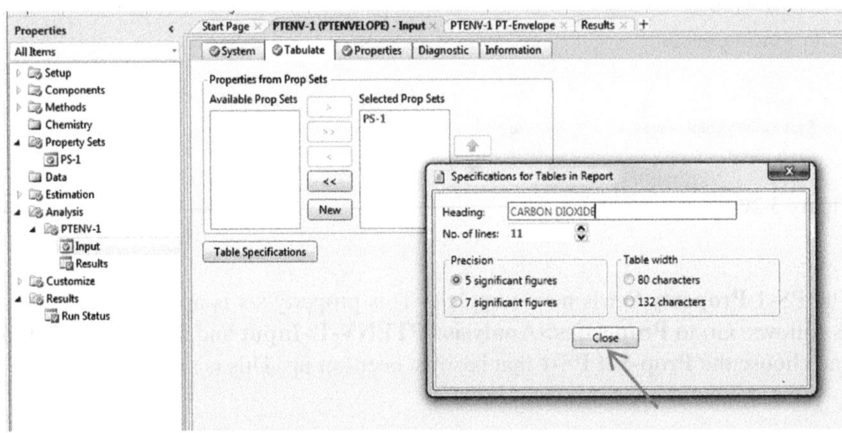

Figure 3-22

After clicking on **Close** go to the main toolbar and click on the **Run** icon (Fig. 3-23)

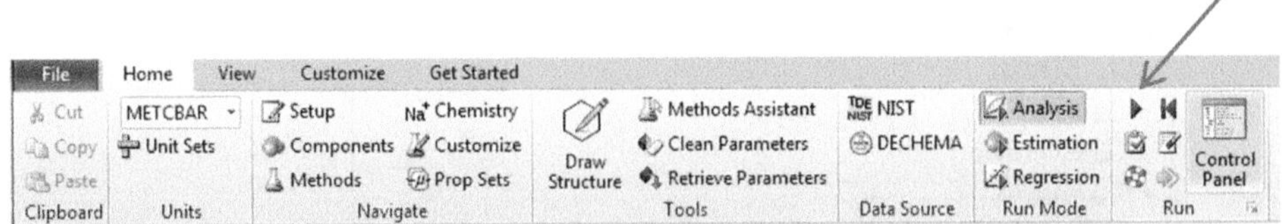

Figure 3-23

and then on **Results** to see the table in Fig. 3-24.

PTEnvelope results

VFRAC	TEMP C	PRES bar	TOTAL TEMP K	LIQUID PHI CARBO-01	VAPOR PHI CARBO-01	LIQUID DH CARBO-01 kJ/kmol	VAPOR DH CARBO-01 kJ/kmol	LIQUID DS CARBO-01 kJ/kmol-K	VAPOR DS CARBO-01 kJ/kmol-K
0	-90.624	0.85436	182.526	0.981693	0.981693	-17270.2	-73.5248	-93.046	1.16894
0	-52.359	6.06691	220.791	0.922195	0.922195	-18680.3	-422.704	-84.9988	-16.1211
0	-36.2477	11.2795	236.902	0.881544	0.881544	-14827.6	-741.003	-81.5771	-22.1156
0	-20.5004	19.0983	252.65	0.833903	0.833903	-13929.8	-1209.36	-78.0392	-27.691
0	-4.71617	30.4267	268.434	0.779786	0.779786	-12850.7	-1907.29	-74.0916	-33.3239
0	11.6484	46.8017	284.798	0.718748	0.718748	-11379.9	-3040.14	-69.0787	-39.7958
0	18.9668	55.9338	292.117	0.690317	0.690317	-10486	-3603.99	-66.1606	-43.2897
0	25.4397	65.0659	298.59	0.66477	0.66477	-9394.48	-4788.42	-62.6743	-47.2816
0	28.0387	69.0906	301.189	0.654431	0.654431	-8761.41	-5385.22	-60.6624	-49.4462
0	30.5279	72.9953	303.678	0.644495	0.644495	-7727.98	-6380.29	-57.5558	-52.9207

Figure 3-24

Note that the additional thermodynamic properties we chose in property set **PS-1** are now displayed. Next we will create a second and different property set **PS-2** following the same procedure. The properties now are the pressure, Gibbs energy, enthalpy, entropy, and constant pressure heat capacity with the same qualifiers as in the previous case as shown in Fig. 3-25.

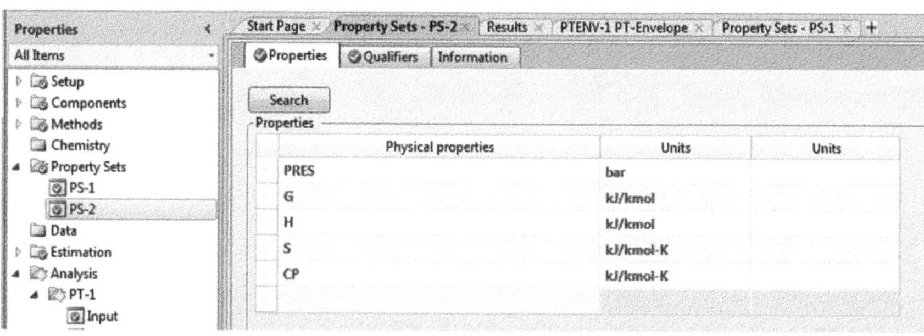

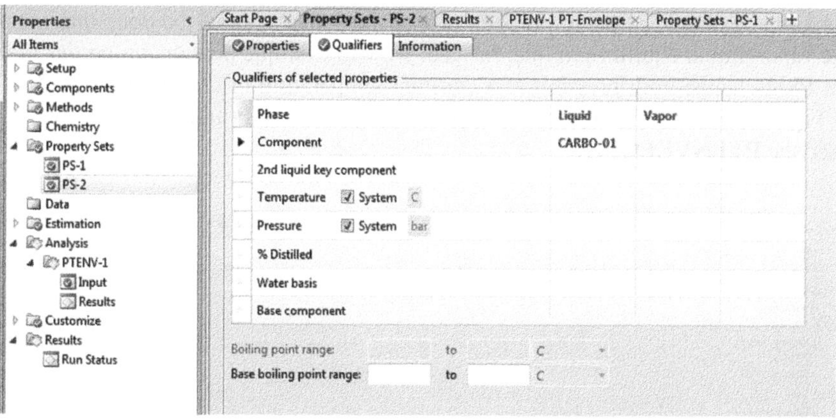

Figure 3-25

Even though we could use both **Property Sets** in the **PTENV-1** analysis, for simplicity and to keep the output visible on these pages, we will instead use only **Prop Set PS-2** here. This is accomplished by going to **Properties>Analysis>PTENV-1>Input>Tabulate** and adding only **PS-2** in the **Selected Prop Sets** as shown in Fig. 3-26.

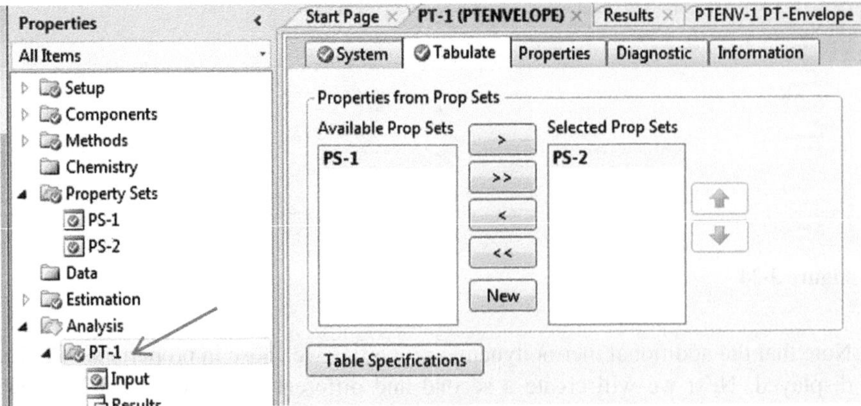

Figure 3-26

Then, clicking on the **Run icon** on the main toolbar, and going to **Results** gives the table in Fig. 3-27.

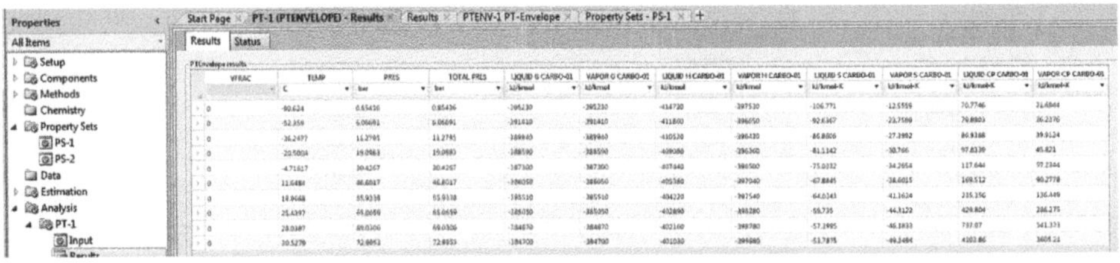

Figure 3-27

As a final example of using property sets, a third analysis will be done that is not along the vapor–liquid equilibrium line. Instead, we will examine how the properties in property set PS-2 change with pressure at a constant temperature of 290 K. Therefore, as shown in Fig. 3-28, the **GENERIC** type property analysis is used (unlike the previous cases that used the **PTENVELOPE**).

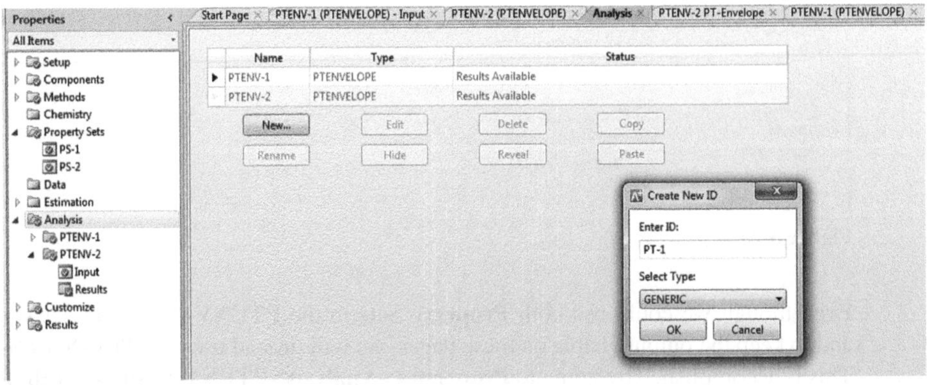

Figure 3-28

Going to **PT-1 Input**, the following window appears. Here it is important to click on **Point(s) without flash** as shown in Fig. 3-29.

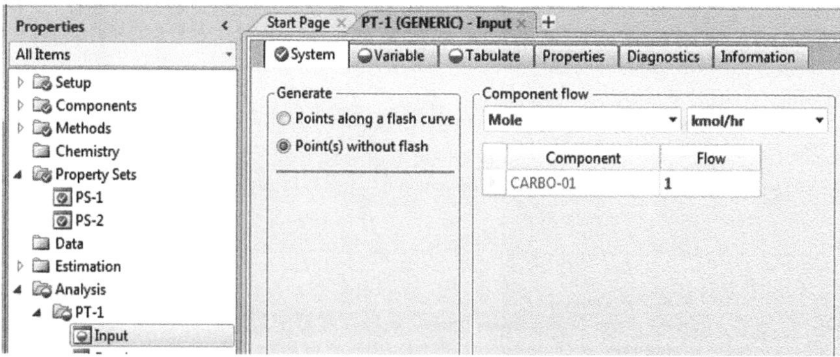

Figure 3-29

Then click on the **Variable** tab and populate the window as shown in Fig. 3-30.

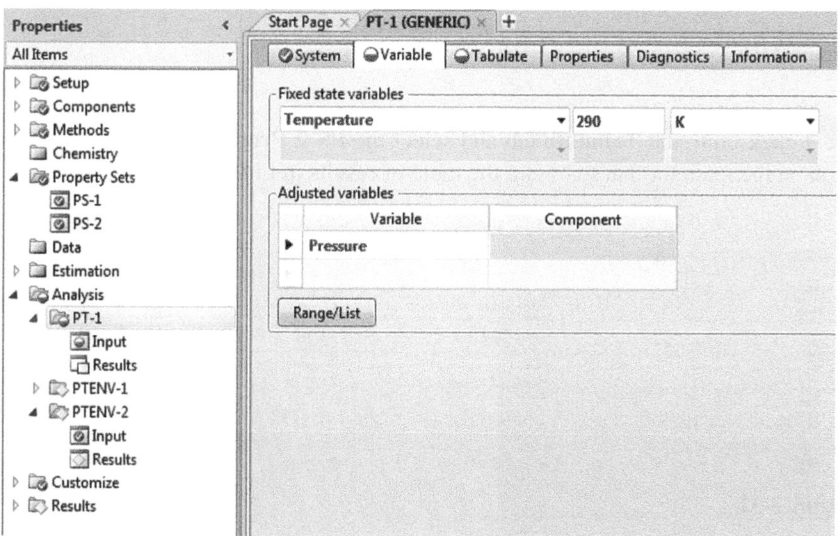

Figure 3-30

Here the temperature will be fixed at 290 K and the pressure varied from 1 to 100 atm. To do this we choose **Pressure** as the adjusted variable, and then click on the **Range/List** and add the list of pressures to be studied, here a collection of pressures from 1 to 100 bar as shown in Fig. 3-31.

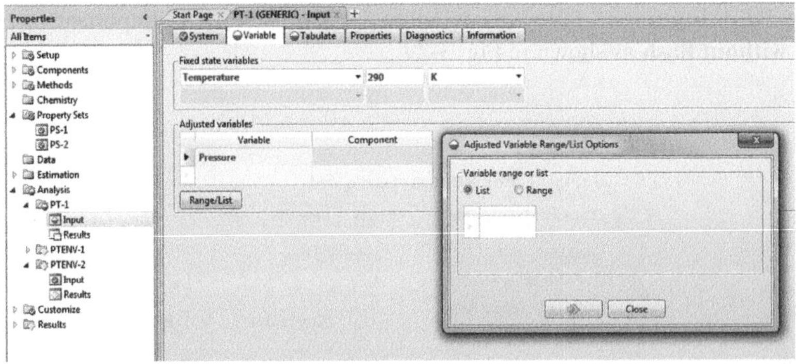

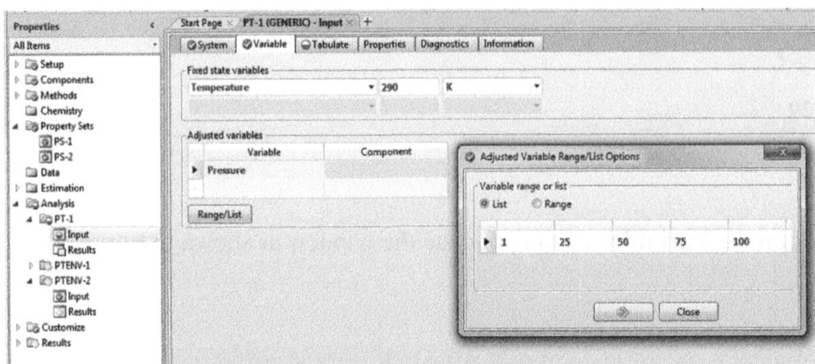

Figure 3-31

Next click on to the **Tabulate** tab and select the **PS-2 Prop Set.** Finally, click on the run icon on the main toolbar to obtain the table of results in Fig. 3-32.

PRES	TOTAL PRES	LIQUID G CARBO-01	VAPOR G CARBO-01	LIQUID H CARBO-01	VAPOR H CARBO-01	LIQUID S CARBO-01	VAPOR S CARBO-01	LIQUID CP CARBO-01	VAPOR CP CARBO-01
bar	bar	kJ/kmol	kJ/kmol	kJ/kmol	kJ/kmol	kJ/kmol-K	kJ/kmol-K	kJ/kmol-K	kJ/kmol-K
1	1	-385610	-394400	-400440	-393860	-51.1497	1.86806	173.012	37.0528
25	25	-385680	-387000	-401570	-395040	-54.7815	-27.729	363.047	46.0983
50	50	-385690	-385750	-404460	-396950	-64.7421	-38.6231	232.209	91.0944
75	75	-385340	-385560	-405090	-399470	-67.4001	-47.9533	147.383	9.46841
100	100	-385410	-385570	-405410	-399670	-68.9595	-48.6041	123.48	10.2544

Figure 3-32

Note that the Peng–Robinson prediction for the vapor pressure of CO_2 at 290 K is 52.61 atm, so that below this pressure carbon dioxide is a vapor and it is a liquid above it.

Finally note that in setting up the **Property Sets**, there is a very large choice of properties that could be tabulated. Some are transport properties, not of interest here, and others are only used in specialized applications and will also not be considered here. The procedure to list any of these properties is the same as has just been described, and will not be illustrated here.

PROBLEMS

3.1. Repeat the property analysis done for ethanol in the text using the Soave–Redlich–Kwong equation of state, and comment on the differences in the results from using the Peng–Robinson equation of state.

3.2. Do a property analysis as in the text for *n*-hexane using an equation of state of your choice.

3.3. Do a property analysis as in the text for *n*-hexadecane using an equation of state of your choice.

3.4. Do a property analysis as in the text for water and steam using the Peng–Robinson equation of state.

3.5. Repeat Problem 3.4 using the IAPWS-95 (International Association for the Properties of Water and Steam) equation of state.

3.6. Discuss the differences in the results for water and steam based on using the Peng–Robinson and IAPWS-95 equations of state. Which set of results do you believe to be more accurate?

Chapter 4

The NIST ThermoData Engine (TDE)

One of the important features of the Aspen Plus® system versions 7.3 and later is the availability of the NIST ThermoData Engine (TDE). It is a huge database of thermodynamic and transport properties and phase behavior, and also a thermodynamic data correlation, evaluation, and prediction tool. It is contained in Aspen Plus and Aspen Properties as a result of collaboration with the United States National Institute of Standards and Technology (NIST). The TDE is an excellent source of thermodynamic data, which is especially important for designing separations and purifications processes.

The ThermoData Engine contains critically evaluated, published experimental thermodynamic and transport property data for pure components and mixtures comprising the following elements: C, H, N, O, F, Cl, Br, I, S, and P that are stored in the program database that contains measured property data for over 17,000 compounds. The results of the TDE evaluations include model parameters and numerical values of properties that can be used in process calculations. The user's own experimental data can also be correlated with the **Aspen Plus Data Regression System**, discussed in later chapters, and then used in simulation.

As an example of using the TDE, a blank simulation is started and, following the procedures discussed in Chapter 1, benzene and *n*-hexane have been entered as the two chemicals of interest in the **Components>Specifications>Select components** drop-down menu resulting in Fig. 4-1.

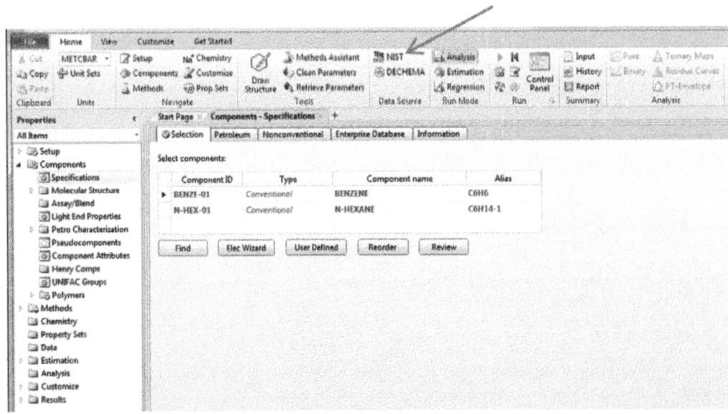

Figure 4-1

Using Aspen Plus® in Thermodynamics Instruction: A Step-by-Step Guide, First Edition. Stanley I. Sandler.

Next, click on **NIST** on the main toolbar. [Note using **DECHEMA** also provides access to a large amount of experimental data, but that database is available only by paid subscription. The use of the **NIST TDE** is free.] This brings up the window in Fig. 4-2 in which the **Property data type** was chosen to be **Pure** (binary mixture data will be considered later), and benzene (**BENZE-01**) has been chosen from the drop-down menu (the other choice being hexane).

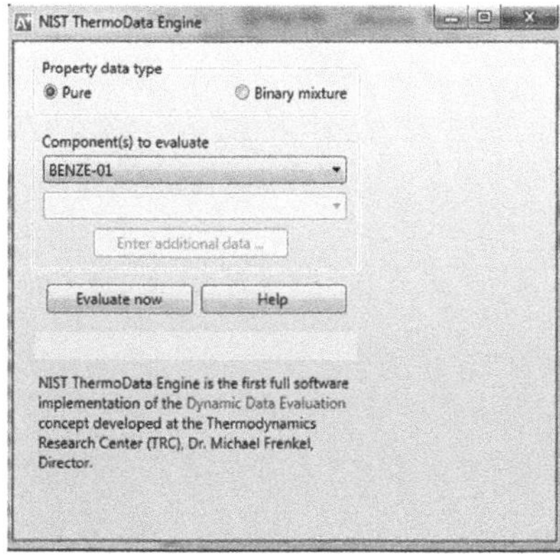

Figure 4-2

Clicking on **Evaluate now** produces the window in Fig. 4-3 with a list of properties available for benzene, and also includes numerical values for fixed point properties, such as the normal boiling point, melting point, critical properties.

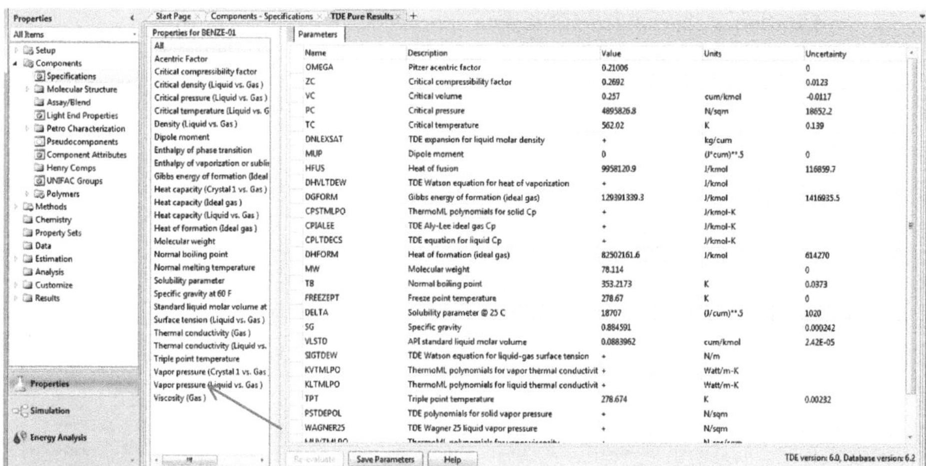

Figure 4-3

Now, as an example, clicking on **Vapor pressure (Liquid vs. Gas)** in the window on the left produces results such as those in Fig. 4-4, depending on which tab is selected. The default tab choice is the parameters in the Wagner vapor pressure equation, which is a very good equation for correlation when sufficient data have been available to fit its parameters. The Wagner 25 equation, using the NIST notation, is

$$\ln p_{ri}^{*,l} = (C_{1i}(1 - T_{ri}) + C_{2i}(1 - T_{ri})^{1.5} + C_{3i}(1 - T_{ri})^{3} + C_{4i}(1 - T_{ri})^{6})/T_{ri}$$

$$\text{for } C_{5i} \leq T \leq C_{6i} \text{ with } T_{ri} = T/T_{ci} \text{ and } p_{ri}^{*,l} = p_{i}^{*,l}/p_{ci}$$

where T_{ci} and p_{ci} are the critical temperature and pressure of component i, respectively.

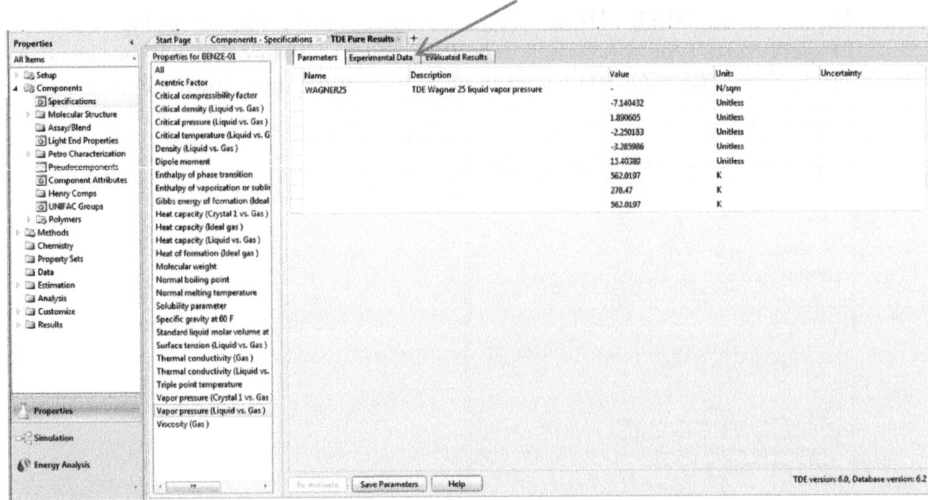

Figure 4-4

Clicking on the **Experimental Data** tab results in the list of Fig. 4-5 of the available experimental vapor pressure data for benzene, only some of which can be shown without scrolling. Note that some of the data have been classified as rejected for various reasons, including being considered inaccurate based on the other data available. Also, by clicking on **Citation**…, the source of the data appears.

Experimental data for TDE Wagner 25 liquid vapor pressure

No.	Temperature (K)	Vapor pressure (Liquid vs. Gas) (N/sqm)	Uncertainty	Accept/Reject	Citation
1	352.42	98530	4926.5	Accept	Citation...
2	352.42	98525.2	2463.1	Accept	Citation...
3	353.22	101791.3	5089.6	Accept	Citation...
4	352.92	101325	5066.3	Accept	Citation...
5	272.55	666.6	73.3	Reject	Citation...
6	274.45	1333.2	113.3	Reject	Citation...
7	276.45	1999.8	124	Reject	Citation...
8	278.25	2666.5	153.3	Reject	Citation...
9	279.55	3333.1	180	Reject	Citation...
10	286.94	6666.1	244	Reject	Citation...
11	293.24	9999.18	257.3	Accept	Citation...
12	313.23	23320	1410	Reject	Citation...
13	314.43	25050	1510	Accept	Citation...
14	326.33	40460	2430	Accept	Citation...
15	327.83	41930	2520	Accept	Citation...
16	337.22	60010	3610	Accept	Citation...
17	337.32	60540	3640	Accept	Citation...
18	346.02	80870	4860	Accept	Citation...
19	346.52	81870	4920	Accept	Citation...
20	352.32	99620	5990	Accept	Citation...
21	353.02	101380	6090	Accept	Citation...

Citation for TDE Wagner 25 liquid vapor pressure at 352.42 (K)

Bruhl, J. W. Justus Liebigs Ann. Chem., 1880, 200, 139-231 The Chemical Constitution of Organic Compounds in Relation to Their Density and Their Light Propagating Ability

TDE version: 6.0, Database version: 6.2

Figure 4-5

Finally, clicking on the **Evaluated Results** tab produces the list of recommended vapor pressure values as a function of temperature, shown in Fig. 4-6, based on the correlation of the accepted data using the Wagner 25 equation mentioned earlier.

Parameters | Experimental Data | Evaluated Results

Evaluated results for TDE Wagner 25 liquid vapor pressure

No.	Temperature (K)	Vapor pressure (Liquid vs. Gas) (N/sqm)
1	278.47	4735
2	281.31	5514.3
3	284.14	6398.5
4	286.98	7398.3
5	289.81	8525.2
6	292.65	9791.4
7	295.48	11209.8
8	298.32	12794.2
9	301.15	14558.8
10	303.99	16519
11	306.82	18690.8
12	309.66	21090.7
13	312.5	23736.5
14	315.33	26646.2
15	318.17	29838.9
16	321	33334.4
17	323.84	37153.1
18	326.67	41316.4
19	329.51	45846.2
20	332.34	50765.2
21	335.18	56096.8
22	338.02	61865
23	340.85	68094.7
24	343.69	74811.2
25	346.52	82040.5

Figure 4-6

Next, moving on to mixtures, once again we start by clicking on **TDE NIST** on the main toolbar that results in the small pop-up menu below. This time we proceed by choosing **Binary mixture** as the **Property data type**, choosing the two components of interest from the drop-down menu that lists the two (or more) components entered previously, as shown in Fig. 4-7, and then clicking on **Retrieve data** produces the window in Fig. 4-8, only some of which can be seen without scrolling. When this table appears, close the **NIST ThermoData Engine** pop-up window of Fig. 4-7.

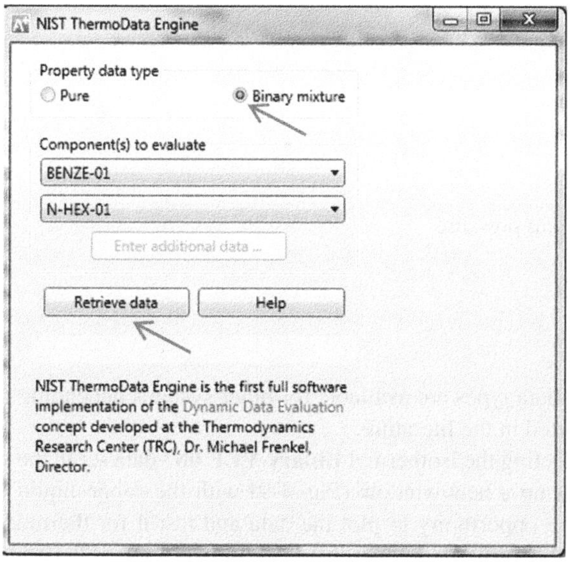

NIST ThermoData Engine

Property data type

Pure Binary mixture

Component(s) to evaluate

BENZE-01

N-HEX-01

Enter additional data ...

Retrieve data Help

NIST ThermoData Engine is the first full software implementation of the Dynamic Data Evaluation concept developed at the Thermodynamics Research Center (TRC), Dr. Michael Frenkel, Director.

Figure 4-7

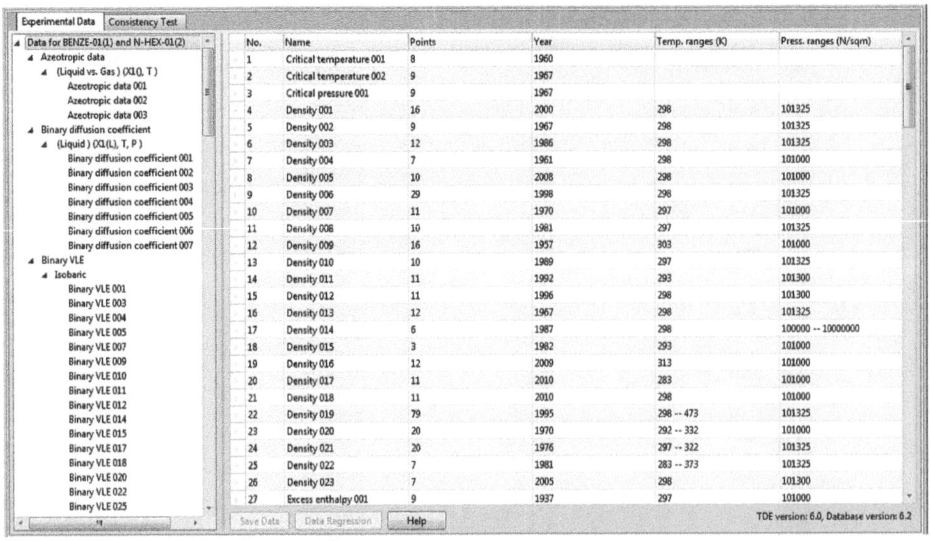

Experimental Data | Consistency Test

Data for BENZE-01(1) and N-HEX-01(2)
- Azeotropic data
  - (Liquid vs. Gas) (X1(), T)
    - Azeotropic data 001
    - Azeotropic data 002
    - Azeotropic data 003
- Binary diffusion coefficient
  - (Liquid) (X1(L), T, P)
    - Binary diffusion coefficient 001
    - Binary diffusion coefficient 002
    - Binary diffusion coefficient 003
    - Binary diffusion coefficient 004
    - Binary diffusion coefficient 005
    - Binary diffusion coefficient 006
    - Binary diffusion coefficient 007
- Binary VLE
  - Isobaric
    - Binary VLE 001
    - Binary VLE 003
    - Binary VLE 004
    - Binary VLE 005
    - Binary VLE 007
    - Binary VLE 009
    - Binary VLE 010
    - Binary VLE 011
    - Binary VLE 012
    - Binary VLE 014
    - Binary VLE 015
    - Binary VLE 017
    - Binary VLE 018
    - Binary VLE 020
    - Binary VLE 022
    - Binary VLE 025

No.	Name	Points	Year	Temp. ranges (K)	Press. ranges (N/sqm)
1	Critical temperature 001	8	1960		
2	Critical temperature 002	9	1967		
3	Critical pressure 001	9	1967		
4	Density 001	16	2000	298	101325
5	Density 002	9	1967	298	101325
6	Density 003	12	1986	298	101325
7	Density 004	7	1961	298	101000
8	Density 005	10	2008	298	101000
9	Density 006	29	1998	298	101325
10	Density 007	11	1970	297	101000
11	Density 008	10	1981	297	101325
12	Density 009	16	1957	303	101000
13	Density 010	10	1989	297	101325
14	Density 011	11	1992	293	101300
15	Density 012	11	1996	298	101300
16	Density 013	12	1967	298	101325
17	Density 014	6	1987	298	100000 -- 10000000
18	Density 015	3	1982	293	101000
19	Density 016	12	2009	313	101000
20	Density 017	11	2010	283	101000
21	Density 018	11	2010	298	101000
22	Density 019	79	1995	298 -- 473	101325
23	Density 020	20	1970	292 -- 332	101000
24	Density 021	20	1976	297 -- 322	101325
25	Density 022	7	1981	283 -- 373	101325
26	Density 023	7	2005	298	101300
27	Excess enthalpy 001	9	1937	297	101000

Save Data | Data Regression | Help

TDE version: 6.0, Database version: 6.2

Figure 4-8

Note that more data types are available than are shown in Fig. 4-8, and these can be seen by scrolling down in the window above during simulation. For this well-studied system, the following types of experimental data are available:

- Azeotropic pressure
- Binary diffusion coefficient
- Binary liquid–liquid equilibrium compositions (Binary LLE)
- Binary vapor–liquid equilibrium compositions (Binary VLE)
  - Isobaric
  - Isothermal
  - Other
- Critical pressure
- Critical temperature
- Density
- Excess enthalpy
- Heat capacity at constant pressure
- Surface tension
- Thermal conductivity
- Viscosity

More or fewer data and data types are available for other systems depending on what has been measured and reported in the literature.

Now, as an example, selecting the isothermal **Binary VLE 087** data set in the window on the left in Fig. 4-8 brings up a new window (Fig. 4-9) with the vapor–liquid equilibrium data and its source. The opportunity to plot the data and test it for thermodynamic consistency will be discussed shortly. [Note that the databases in Aspen Plus are updated regularly, and, as a result, the identifying number for data set may change. Consequently the data identified here as **Binary VLE 087** may have a different identifying number in your version of Aspen Plus.]

Start Page | Components | Components - Specifications | TDE Binary Results

Experimental Data | Consistency Test

Binary VLE 045
Binary VLE 047
Binary VLE 057
Binary VLE 091
Isothermal
Binary VLE 031
Binary VLE 035
Binary VLE 037
Binary VLE 039
Binary VLE 050
Binary VLE 051
Binary VLE 052
Binary VLE 053
Binary VLE 055
Binary VLE 060
Binary VLE 061
Binary VLE 062
Binary VLE 063
Binary VLE 069
Binary VLE 071
Binary VLE 073
Binary VLE 075
Binary VLE 076
Binary VLE 077
Binary VLE 078
Binary VLE 079
Binary VLE 080
Binary VLE 082
Binary VLE 085
Binary VLE 087
Binary VLE 089
Binary VLE 093
Binary VLE 095
Others

No.	Liquid mole fraction BENZE-01	Temperature (K)	Vapor mole fraction BENZE-01	Total pressure (N/sqm)
1	0.0277	333.13	0.0271	76546
2	0.1094	333.13	0.106	76430
3	0.1513	333.13	0.144	76326
4	0.1963	333.13	0.1826	76056
5	0.2301	333.13	0.2107	75831
6	0.3224	333.13	0.2854	74950
7	0.3505	333.13	0.307	74626
8	0.377	333.13	0.3269	74309
9	0.3914	333.13	0.3369	74102
10	0.4032	333.13	0.347	73910
11	0.4486	333.13	0.3812	73227
12	0.4952	333.13	0.4163	72370
13	0.5154	333.13	0.4322	71987
14	0.5296	333.13	0.4418	71714
15	0.5792	333.13	0.4812	70622
16	0.6088	333.13	0.5038	69952
17	0.6314	333.13	0.5215	69421
18	0.6502	333.13	0.5375	68864
19	0.6868	333.13	0.5689	67866
20	0.7177	333.13	0.5955	66912
21	0.7503	333.13	0.6266	65828
22	0.7698	333.13	0.6455	65157
23	0.7932	333.13	0.6696	64229
24	0.8254	333.13	0.7054	62845

Seez, C.; Compostizo, A.; Rubio, R. G.; Crespo Colin, A.; Diaz Pena, M. Fluid Phase Equilib., 1985, 24, 241 p, T, x, y, Data of Benzene + n-Hexane and Cyclohexane + n-Heptane Systems

Display Uncertainty

Save Data | Data Regression | Help

TDE version: 7.1, Database version: 7.3

Figure 4-9

With any experimental data set open, under the **Home** tab of the main toolbar, the **Plot** group displays buttons for ways you can plot that data (see Fig. 4-10).

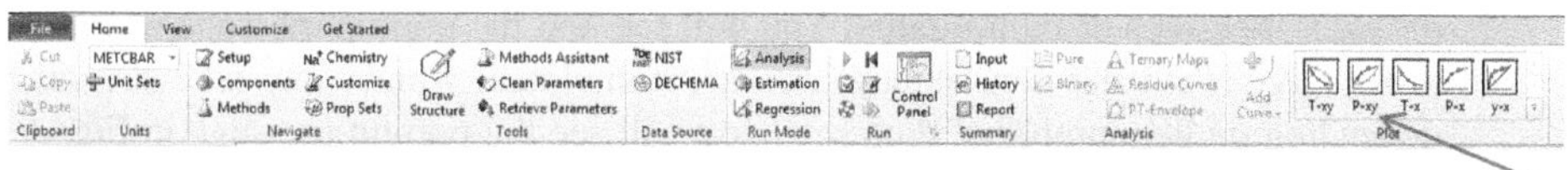

Figure 4-10

Since this data set is P-xy data, that plot has been selected resulting in Fig. 4-11.

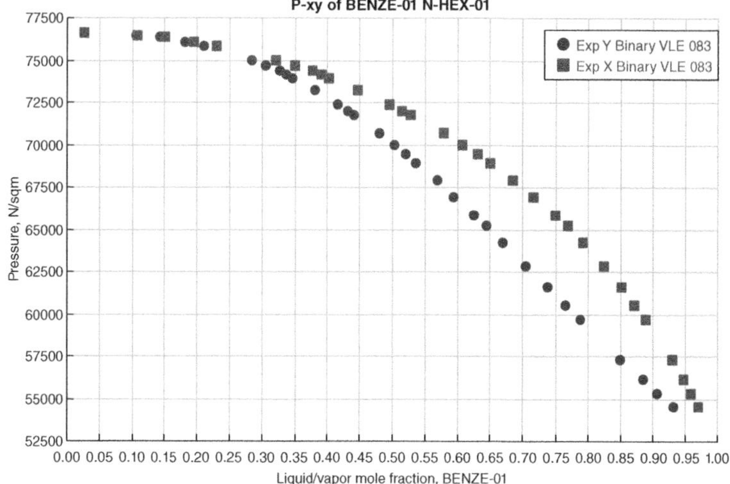

Figure 4-11

Clicking on the y-x plot results in Fig. 4-12.

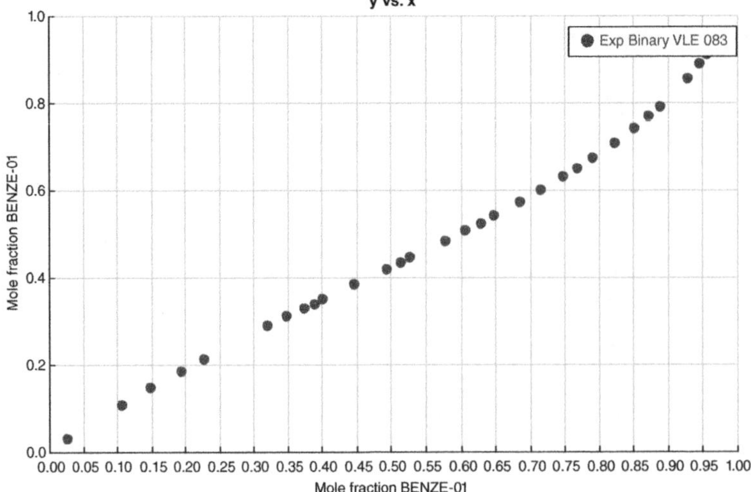

Figure 4-12

Clicking on **Consistency Test** tab brings up Fig. 4-13. Click on **Run Consistency Tests**, which initiates thermodynamic consistency tests on all the relevant data, and thereby provides measures of their quality. You will see first the pop-up message of Fig. 4-14 and press **OK**. When the consistency tests are completed, which may take several minutes depending on the number of data sets, you will see the pop-up message in Fig. 4-15 over the results in Figs. 4-16a and 4-16b.

Start Page | Components | Components - Specifications | TDE Binary Results | +

Experimental Data | Consistency Test

- Data for BENZE-01(1) and N-HEX-01(2)
  - Binary VLE
    - Isobaric
      - Binary VLE 001
      - Binary VLE 003
      - Binary VLE 004
      - Binary VLE 005
      - Binary VLE 008
      - Binary VLE 010
      - Binary VLE 011
      - Binary VLE 012
      - Binary VLE 013
      - Binary VLE 015
      - Binary VLE 016
      - Binary VLE 018
      - Binary VLE 019
      - Binary VLE 021
      - Binary VLE 022
      - Binary VLE 024
      - Binary VLE 026
      - Binary VLE 027
      - Binary VLE 028
      - Binary VLE 029
      - Binary VLE 033
      - Binary VLE 041
      - Binary VLE 043
      - Binary VLE 045
      - Binary VLE 047
      - Binary VLE 057
      - Binary VLE 091
    - Isothermal
      - Binary VLE 031
      - Binary VLE 035
      - Binary VLE 037

No.	Name	Points	Year	Isotherm (K)	Isobaric (N/sqm)	Overall data quality	Herington test	Van Ness test	Piont test	Infinite dilution test	EOS test	Endpoint test
1	Binary VLE 001	11	1911	---	102000	---	---	---	---	---	---	---
2	Binary VLE 003	15	1933	---	101000	---	---	---	---	---	---	---
3	Binary VLE 004	11	1933	---	97900	---	---	---	---	---	---	---
4	Binary VLE 005	15	1933	---	96500	---	---	---	---	---	---	---
5	Binary VLE 008	38	1955	---	101000	---	---	---	---	---	---	---
6	Binary VLE 010	13	1957	---	405000	---	---	---	---	---	---	---
7	Binary VLE 011	13	1957	---	610000	---	---	---	---	---	---	---
8	Binary VLE 012	13	1957	---	810800	---	---	---	---	---	---	---
9	Binary VLE 013	10	1957	---	1013000	---	---	---	---	---	---	---
10	Binary VLE 015	6	1960	---	101320	---	---	---	---	---	---	---
11	Binary VLE 016	29	1960	---	101320	---	---	---	---	---	---	---
12	Binary VLE 018	6	1960	---	101320	---	---	---	---	---	---	---
13	Binary VLE 019	29	1960	---	101320	---	---	---	---	---	---	---
14	Binary VLE 021	4	1961	---	101320	---	---	---	---	---	---	---
15	Binary VLE 022	17	1961	---	101320	---	---	---	---	---	---	---
16	Binary VLE 024	7	1961	---	101325	---	---	---	---	---	---	---
17	Binary VLE 026	16	1963	---	26660	---	---	---	---	---	---	---
18	Binary VLE 027	6	1963	---	40000	---	---	---	---	---	---	---
19	Binary VLE 028	12	1963	---	40000	---	---	---	---	---	---	---
20	Binary VLE 029	14	1963	---	53330	---	---	---	---	---	---	---
21	Binary VLE 033	12	1963	---	101000	---	---	---	---	---	---	---
22	Binary VLE 041	24	1967	---	101000	---	---	---	---	---	---	---
23	Binary VLE 043	26	1967	---	101320	---	---	---	---	---	---	---
24	Binary VLE 045	24	1967	---	101000	---	---	---	---	---	---	---
25	Binary VLE 047	10	1969	---	101320	---	---	---	---	---	---	---
26	Binary VLE 057	10	1971	---	101300	---	---	---	---	---	---	---
27	Binary VLE 091	16	1992	---	101325	---	---	---	---	---	---	---

Run Consistency Tests | Help

TDE version: 7.1, Database version: 7.3

Figure 4-13

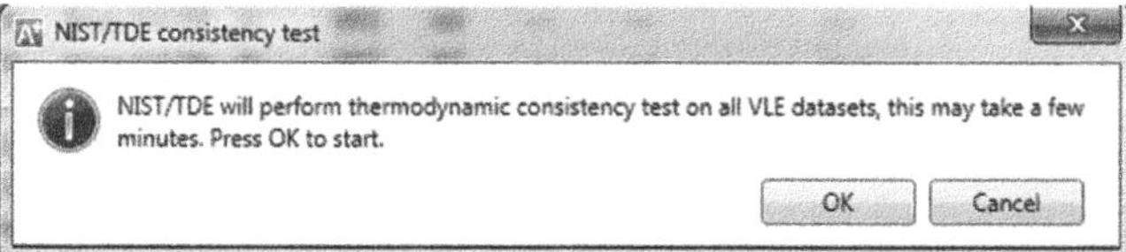

Figure 4-14

When the consistency tests are completed, which may take several minutes depending on the number of data sets, you will see the pop-up message in Fig. 4-15 over the results in Figs. 4-16a and 4-16b.

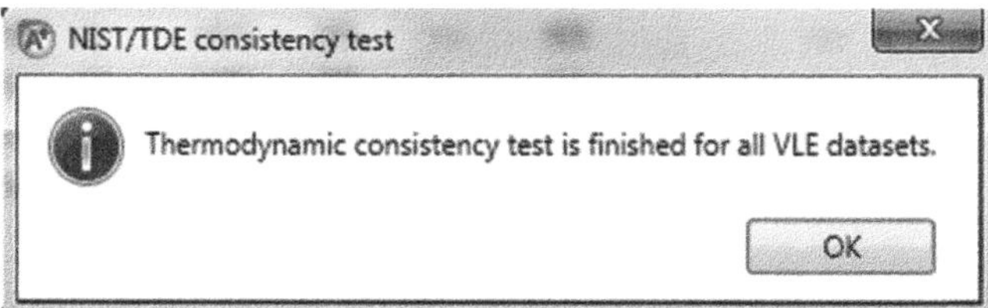

Figure 4-15

Some of the results are shown in Figs. 4-16a and 4-16b for data sets for which there are sufficient data (generally at least 10 data points) to do a consistency test. As examples of the results of the consistency tests, Fig. 4-16a shows the results for the first half of the Binary VLE data sets, and Fig. 4-16b shows the results for the remaining data sets. The Herington test is the equal area test as described, for example, in Section 10.2 of *Chemical, Biochemical and Engineering Thermodynamics*, 4th ed., S. I. Sandler (John Wiley & Sons, Inc., 2006). The other tests will not be described here. Note that the **Binary VLE 087** data set discussed above passes all but the infinite dilution test, as there are no infinite dilution data in the data set. Based on the results of the consistency tests, this data set can be considered to be reliable.

Experimental Data | Consistency Test

Data for BENZE-01(1) and N-HEX-01(2)
- Binary VLE
  - Isobaric
    - Binary VLE 001
    - Binary VLE 003
    - Binary VLE 004
    - Binary VLE 005
    - Binary VLE 008
    - Binary VLE 010
    - Binary VLE 011
    - Binary VLE 012
    - Binary VLE 013
    - Binary VLE 015
    - Binary VLE 016
    - Binary VLE 018
    - Binary VLE 019
    - Binary VLE 021
    - Binary VLE 022
    - Binary VLE 024
    - Binary VLE 026
    - Binary VLE 027
    - Binary VLE 028
    - Binary VLE 029
    - Binary VLE 033
    - Binary VLE 041
    - Binary VLE 043
    - Binary VLE 045
    - Binary VLE 047
    - Binary VLE 057
    - Binary VLE 091
  - Isothermal
    - Binary VLE 031
    - Binary VLE 035
    - Binary VLE 037

No.	Name	Points	Year	Isotherm (K)	Isobaric (N/sqm)	Overall data quality	Herington test	Van Ness test	Piont test	Infinite dilution test	EOS test	Endpoint test
1	Binary VLE 001	11	1911	---	102000	0.129	---	---	---	---	---	0.259
2	Binary VLE 003	15	1933	---	101000	0.5	---	---	---	---	---	1
3	Binary VLE 004	11	1933	---	97900	0.723	Passed	Passed	---	Failed	---	---
4	Binary VLE 005	15	1933	---	96500	0.5	---	---	---	---	---	1
5	Binary VLE 008	38	1955	---	101000	0.898	Passed	Passed	---	Failed	---	1
6	Binary VLE 010	13	1957	---	405000	0.53	Passed	Passed	---	Failed	---	0.732
7	Binary VLE 011	13	1957	---	610000	0.866	Passed	Passed	---	Failed	---	1
8	Binary VLE 012	13	1957	---	810800	0.339	Failed	Failed	---	Failed	---	0.648
9	Binary VLE 013	10	1957	---	1013000	0.23	Failed	Failed	---	Failed	---	0.445
10	Binary VLE 015	6	1960	---	101320	0.25	---	---	---	---	---	---
11	Binary VLE 016	29	1960	---	101320	0.904	Passed	Passed	---	Failed	---	---
12	Binary VLE 018	6	1960	---	101320	0.25	---	---	---	---	---	---
13	Binary VLE 019	29	1960	---	101320	0.904	Passed	Passed	---	Failed	---	---
14	Binary VLE 021	4	1961	---	101320	0.5	---	---	---	---	---	1
15	Binary VLE 022	17	1961	---	101320	0.762	Passed	Passed	---	Failed	---	---
16	Binary VLE 024	7	1961	---	101325	0.333	---	---	---	---	---	0.667
17	Binary VLE 026	16	1963	---	26660	0.783	Passed	Passed	---	Failed	---	---
18	Binary VLE 027	6	1963	---	40000	0.25	---	---	---	---	---	---
19	Binary VLE 028	12	1963	---	40000	0.827	Passed	Passed	---	Failed	---	---
20	Binary VLE 029	14	1963	---	53330	0.956	Passed	Passed	---	Failed	---	---
21	Binary VLE 033	12	1963	---	101000	0.573	Failed	Passed	---	Failed	---	---
22	Binary VLE 041	24	1967	---	101000	0.889	Passed	Passed	---	Failed	---	---
23	Binary VLE 043	26	1967	---	101320	1	Passed	Passed	---	Passed	---	---
24	Binary VLE 045	24	1967	---	101000	0.889	Passed	Passed	---	Failed	---	---
25	Binary VLE 047	10	1969	---	101320	0.69	Failed	Passed	---	Failed	---	---
26	Binary VLE 057	10	1971	---	101300	0.689	Failed	Passed	---	Failed	---	---
27	Binary VLE 091	16	1992	---	101325	0.902	Passed	Passed	---	Failed	---	1

Figure 4-16a

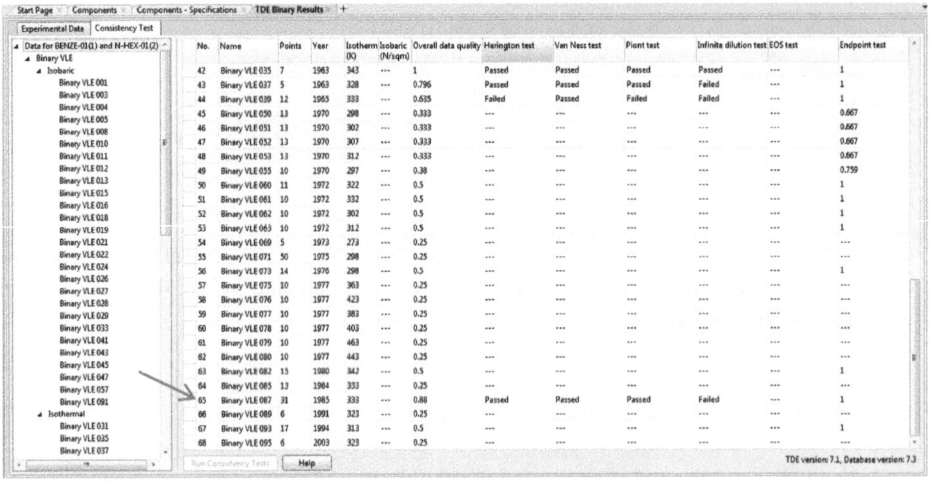

Start Page | Components | Components - Specifications | TDE Binary Results

Experimental Data | Consistency Test

Data for BENZE-01(1) and N-HEX-01(2)
Binary VLE
Isobaric
Binary VLE 001
Binary VLE 003
Binary VLE 004
Binary VLE 005
Binary VLE 008
Binary VLE 010
Binary VLE 011
Binary VLE 012
Binary VLE 013
Binary VLE 015
Binary VLE 016
Binary VLE 018
Binary VLE 019
Binary VLE 021
Binary VLE 022
Binary VLE 024
Binary VLE 026
Binary VLE 027
Binary VLE 028
Binary VLE 029
Binary VLE 033
Binary VLE 041
Binary VLE 043
Binary VLE 045
Binary VLE 047
Binary VLE 057
Binary VLE 091
Isothermal
Binary VLE 031
Binary VLE 035
Binary VLE 037

No.	Name	Points	Year	Isotherm (K)	Isobaric (N/sqm)	Overall data quality	Herington test	Van Ness test	PPiont test	Infinite dilution test	EOS test	Endpoint test
42	Binary VLE 035	7	1963	343	---	1	Passed	Passed	Passed	Passed	---	1
43	Binary VLE 037	5	1963	328	---	0.796	Passed	Passed	Passed	Failed	---	1
44	Binary VLE 039	12	1965	333	---	0.635	Failed	Passed	Failed	Failed	---	1
45	Binary VLE 050	13	1970	298	---	0.333	---	---	---	---	---	0.667
46	Binary VLE 051	13	1970	302	---	0.333	---	---	---	---	---	0.667
47	Binary VLE 052	13	1970	307	---	0.333	---	---	---	---	---	0.667
48	Binary VLE 053	13	1970	312	---	0.333	---	---	---	---	---	0.667
49	Binary VLE 055	10	1970	297	---	0.38	---	---	---	---	---	0.759
50	Binary VLE 060	11	1972	322	---	0.5	---	---	---	---	---	1
51	Binary VLE 061	10	1972	332	---	0.5	---	---	---	---	---	1
52	Binary VLE 062	10	1972	302	---	0.5	---	---	---	---	---	1
53	Binary VLE 063	10	1972	312	---	0.5	---	---	---	---	---	1
54	Binary VLE 069	5	1973	273	---	0.25	---	---	---	---	---	---
55	Binary VLE 071	50	1975	298	---	0.25	---	---	---	---	---	---
56	Binary VLE 073	14	1976	298	---	0.5	---	---	---	---	---	1
57	Binary VLE 075	10	1977	363	---	0.25	---	---	---	---	---	---
58	Binary VLE 076	10	1977	423	---	0.25	---	---	---	---	---	---
59	Binary VLE 077	10	1977	383	---	0.25	---	---	---	---	---	---
60	Binary VLE 078	10	1977	403	---	0.25	---	---	---	---	---	---
61	Binary VLE 079	10	1977	463	---	0.25	---	---	---	---	---	---
62	Binary VLE 080	10	1977	443	---	0.25	---	---	---	---	---	---
63	Binary VLE 082	15	1980	342	---	0.5	---	---	---	---	---	1
64	Binary VLE 085	13	1984	333	---	0.25	---	---	---	---	---	---
65	Binary VLE 087	31	1985	333	---	0.88	Passed	Passed	Passed	Failed	---	1
66	Binary VLE 089	6	1991	323	---	0.25	---	---	---	---	---	---
67	Binary VLE 093	17	1994	313	---	0.5	---	---	---	---	---	1
68	Binary VLE 095	6	2003	323	---	0.25	---	---	---	---	---	---

Run Consistency Tests | Help

TDE version: 7.1, Database version: 7.3

Figure 4-16b

Finally, it should be noted that the **NIST TDE** provides real experimental data for pure fluids and mixtures, unlike the **Property Analysis** methods for pure fluids (Chapter 3) and mixtures (Chapters 5 and 6) that are based on activity coefficient models (Chapter 5) and equations of state (Chapter 6). If available, experimental data are always preferred, especially for the design of a real process. However, the approximate **Property Analysis** methods can be useful for screening. For example, in mixtures, in deciding in advance whether or not an azeotrope is likely to occur that would make distillation difficult.

Also, while vapor–liquid equilibrium data were considered here, many other data types for both pure components and binary mixtures can be accessed through the **NIST TDE**.

PROBLEMS

4.1. Use the NIST TDE to obtain experimental data such as vapor pressures, heat capacities, and densities of ethanol, and compare the results with the property analysis predictions in Chapter 3. Comment on the differences between the predictions and the experimental data, and the likely cause.

4.2. Use the NIST TDE to obtain experimental data such as vapor pressures, heat capacities, and densities of n-hexane, and compare the results with the property analysis predictions in Problem 3.2. Comment on the differences between the predictions and the experimental data, and the likely cause.

4.3. Use the NIST TDE to obtain experimental data of water and steam, and compare the results with the property analysis predictions in Problem 3.4. Comment on the differences between the predictions and the experimental data, and the likely cause.

4.4. Use the NIST TDE to obtain experimental data of water and steam, and compare the results with the property analysis predictions using the IAPWS-95 equation of state (Problem 3.5.) How good is the agreement between experimental data and the IAPWS-95 equation of state?

4.5. Use the NIST TDE to obtain experimental binary vapor–liquid equilibrium data for the ethanol + water mixture at 312.91 K. Plot these data. Does this mixture have an azeotrope?

4.6. Use the NIST TDE to obtain experimental binary vapor–liquid equilibrium data for the furfural + water mixture at 298.15 K. Plot these data. Does this mixture have an azeotrope?

4.7. Use the NIST TDE to obtain experimental binary vapor–liquid equilibrium data for the *n*-heptane + toluene mixture at 348 K. Plot these data. Does this mixture have an azeotrope?

4.8. Use the NIST TDE to obtain experimental isobaric binary vapor–liquid equilibrium data for the carbon tetrachloride + water mixture at 101,000 N/m^2. Plot these data. Does this mixture have an azeotrope?

4.9. Use the NIST TDE to obtain experimental isothermal binary vapor–liquid equilibrium data for the carbon dioxide + propane mixture at 270 K. Plot these data.

4.10. Use the NIST TDE to obtain experimental isothermal binary vapor–liquid equilibrium data for the carbon dioxide + *n*-butane mixture at 310.9 K. Plot these data.

Chapter 5

Vapor–Liquid Equilibrium Calculations Using Activity Coefficient Models

This chapter begins the analysis of mixture phase equilibrium calculations using Aspen Plus®. This discussion extends over three chapters with vapor–liquid equilibrium (VLE) using activity coefficient models considered here. The following chapter deals with VLE using equations of state, and the final chapter in this sequence deals with liquid–liquid and vapor–liquid–liquid equilibria, generally with activity coefficient models.

While the NIST TDE provides measured vapor–liquid equilibrium data, it is frequently of interest to know the equilibrium state of a mixture of known overall composition at a specified temperature and pressure or at conditions within a process for which no experimental data are available. That is, one may want to know whether at the specified conditions the mixture is a liquid, a vapor, a vapor–liquid (VLE) mixture, two liquids in equilibrium (LLE) or two liquids and a vapor in equilibrium (VLLE). Further, if there are multiple phases (VLE, LLE, or VLLE), what are the compositions of the equilibrium phases. Also, one may want to know how the phase behavior will change as a result of a change in temperature (heating or cooling) or pressure (expansion or compression). Such computations are frequently referred to as flash calculations. Aspen Plus does a phase equilibrium calculation on each stream during a simulation, but in the default reports, only the fraction of the stream that is vapor, liquid, and solid may be provided, not their compositions. The objective of this and the following chapters is to do more detailed calculations.

This chapter has the following parts:

Section 5.1: The prediction of vapor–liquid equilibrium (VLE) using **Property Analysis** is presented. There VLE is computed using an activity coefficient model with the parameters in the Aspen Plus data bank, parameters supplied by the user, or complete predictions are made using the UNIFAC model. Further, a related calculation allows one to examine whether the mixture has a VLE azeotrope.

Section 5.2: In this section the calculation of VLE using activity coefficients is considered in the process **Simulation** mode with the two-phase **Flash2** or three-phase **Flash3** (for possible vapor–liquid–liquid equilibrium) blocks.

Section 5.3: The last part of this long chapter deals with regressing available experimental VLE data (perhaps obtained from the NIST TDE) to obtain values of parameters in activity coefficient models.

Using Aspen Plus® in Thermodynamics Instruction: A Step-by-Step Guide, First Edition. Stanley I. Sandler.

Before going into the details, it is useful to consider the advantages and disadvantages of each of these types of calculations so you know which to choose.

The **Property Analysis** (actually, from the main toolbar **Analysis>Binary**) method to calculate VLE has the advantage that it can easily be used to generate tables and graphs of P-xy, T-xy, and Gibbs energy of mixing as a function of composition. It can also be used to study the possibility of azeotrope formation. The disadvantages are that it can only be used for binary mixtures, and that it can only be used for calculations at a fixed temperature or pressure, but not, for example, for a Joule–Thomson expansion in which only the final pressure and enthalpy are known.

The **Simulation** method provides greater flexibility, but must be set up as a simulation using a flow sheet and a separator block. This allows computations for a broad range of possible specifications. For example, the following types of phase equilibrium calculations can be done in the **Flash2** (or **Flash3**) block by choosing the type of calculation as shown in Fig. 5.0-1a to 5.0-1e.

1. Specification of the block exit temperature and pressure (Fig. 5.0-1a).

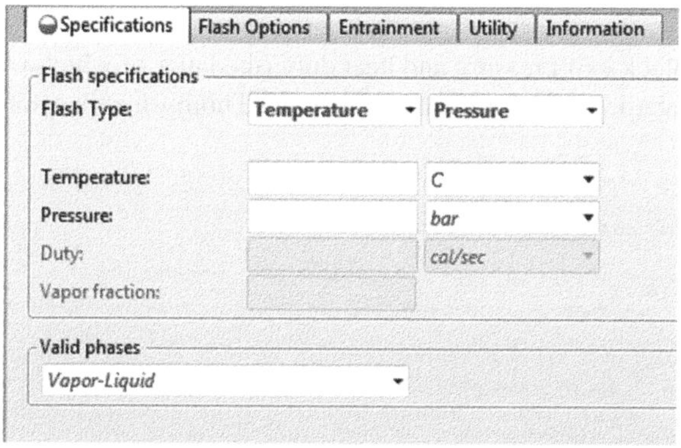

Figure 5.0-1a

2. Specification of the block exit temperature and fraction of vapor (which can be set to a very small number, for example, 0.0001, to find the bubble point pressure and composition, or to a number close to one, for example, 0.9999 to find the dew point pressure and composition), as in Fig. 5.0-1b.

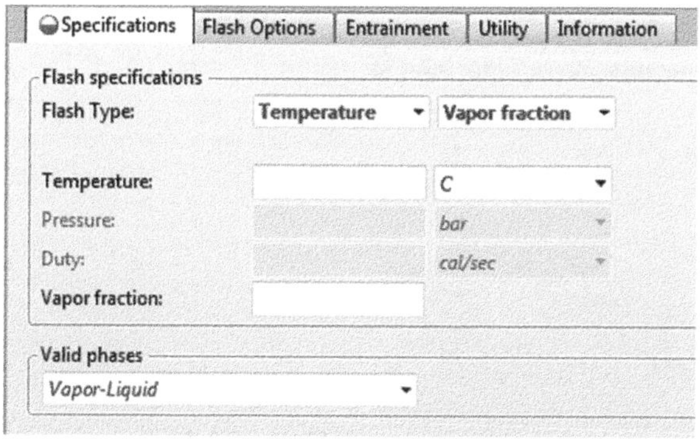

Figure 5.0-1b

3. Specification of the block exit pressure and fraction of vapor (which can be set to a very small number, for example, 0.0001, to find the bubble point temperature and composition, or to a number close to one, for example, 0.9999 to find the dew point temperature and composition), as in Fig. 5.0-1c.

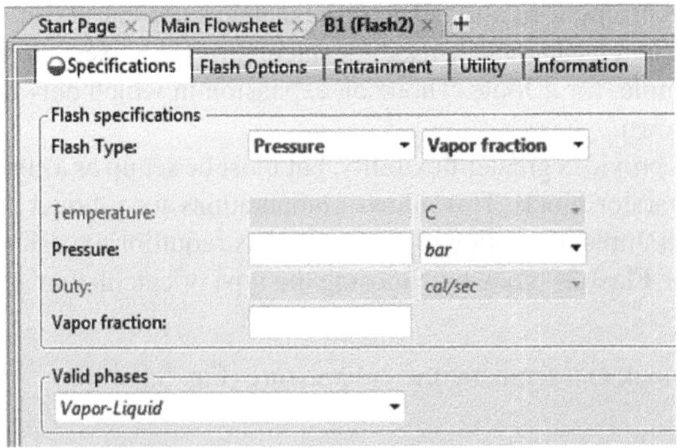

Figure 5.0-1c

4. Specification of the block exit pressure and heat duty (the latter of which would be set to zero for an adiabatic flash, for example, a Joule–Thomson expansion), as in Fig. 5.0-1d.

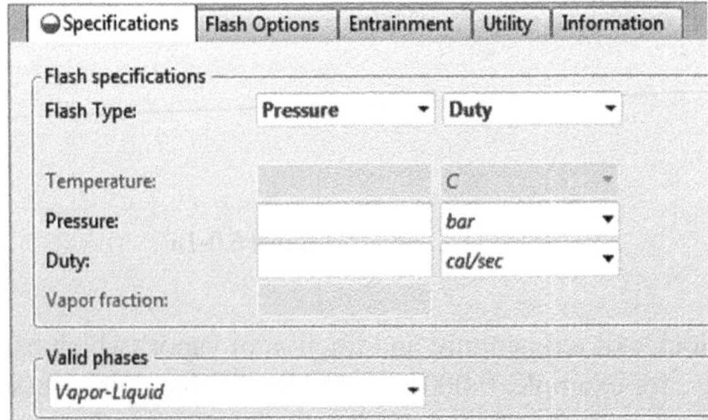

Figure 5.0-1d

5. Specification of the block exit temperature and heat duty (Fig. 5.0-1e).

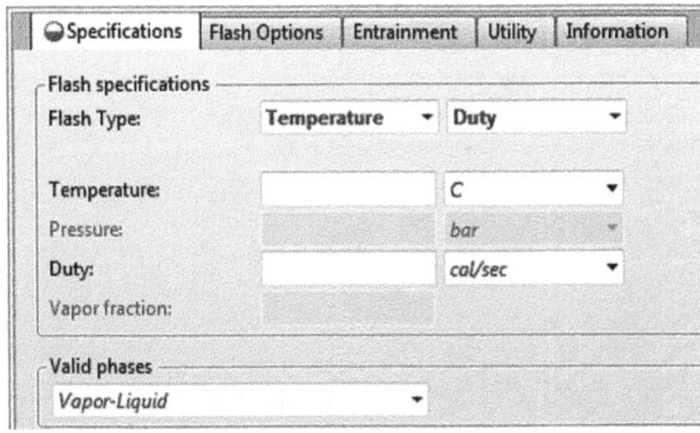

Figure 5.0-1e

Before illustrating these different methods, it is useful to make one change in the **Setup**. Since for mixtures it is likely that our interest will be in mole fractions rather than mass fractions, go to **Setup>Report Options>Stream** and in the **Fraction basis** column click on **Mole** as shown in Fig. 5.0-2.

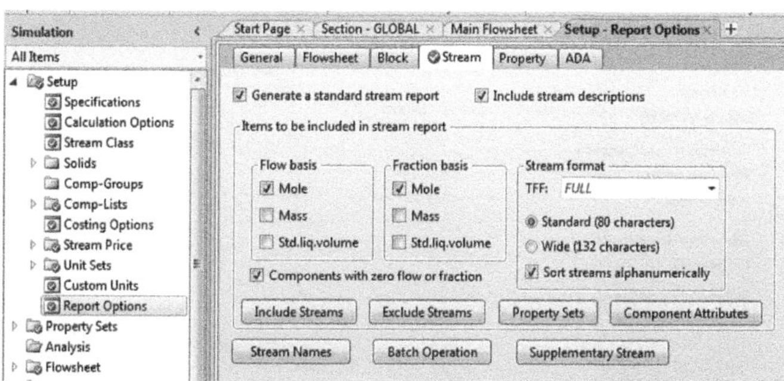

Figure 5.0-2

Throughout this chapter and elsewhere in this book, we will use activity coefficient models such as UNIQUAC, UNIFAC, NRTL, Wilson, and others without explanation. If these are unfamiliar to the reader, he or she should refer to a standard chemical engineering thermodynamics textbook.

5.1 PROPERTY ANALYSIS METHOD

As an example, the low pressure vapor–liquid equilibria of the ethanol–hexane mixture is considered. Since the mixture contains a polar component and is at low pressure, an activity coefficient model should be used, and we will consider several different ones here. We begin with the NRTL model. Start in the usual way by entering the components, here ethanol and *n*-hexane, from the **Start Page>Components>Selection** as shown in Fig. 5.1-1 and discussed in Chapter 1.

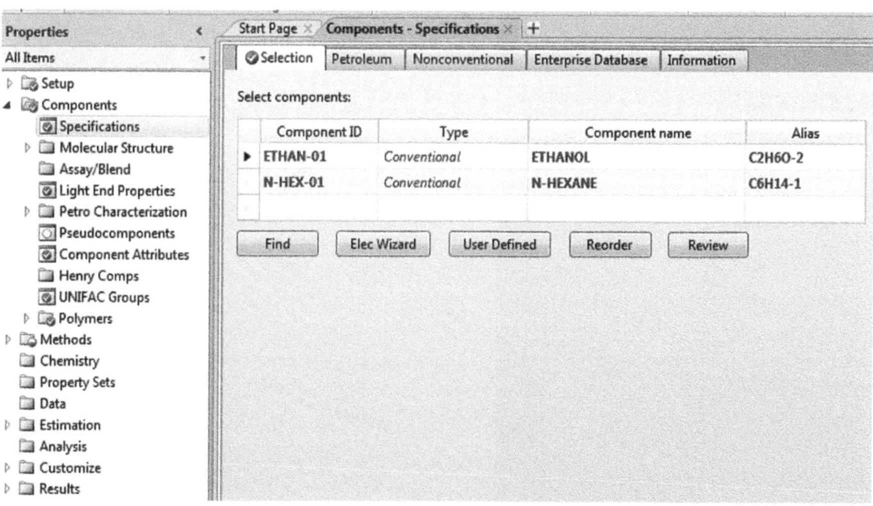

Figure 5.1-1

Then choose the **NRTL** model as the **Base method** from the drop-down menu in **Methods>Specifications** (Fig. 5.1-2).

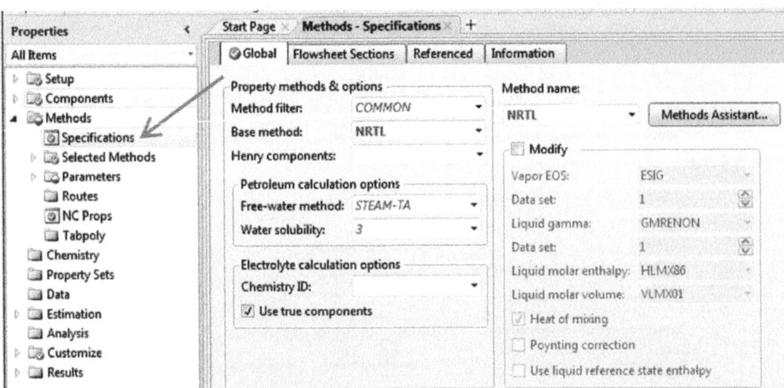

Figure 5.1-2

The Aspen default parameter values are obtained by clicking on **Parameters>Binary Interaction**. This will bring up the window in Fig. 5.1-3 showing the binary parameter sets available for this mixture and the thermodynamic model, if available, from one or more of the Aspen Plus data banks. If no binary parameters are available, this window will be blank, and in later calculations an error message may appear warning the user that the binary parameters have been set to zero.

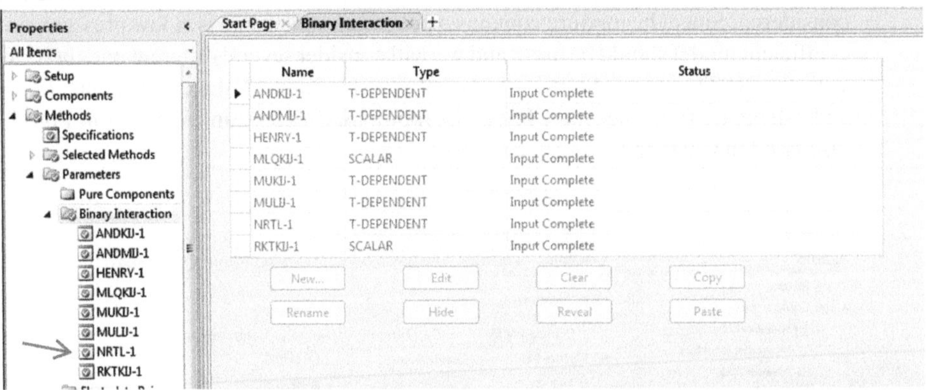

Figure 5.1-3

Clicking on **Methods>Parameters>Binary Interaction>NRTL-1** brings up the table in Fig. 5.1-4 containing the values of the model parameters.

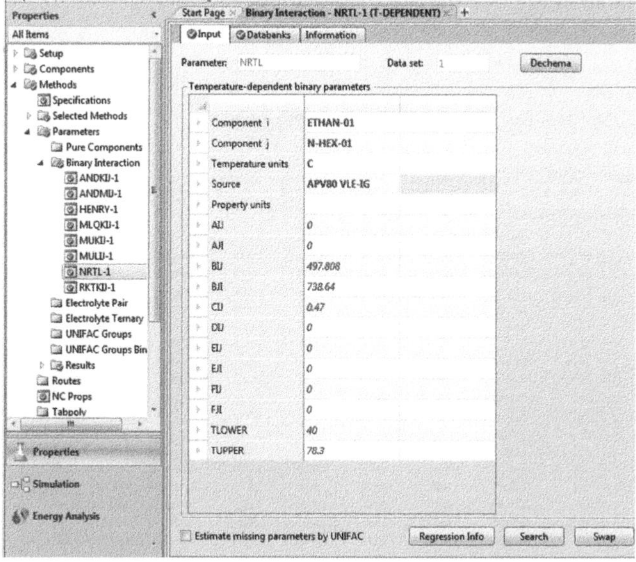

Figure 5.1-4

Note that the source of the parameters is indicated to be the Aspen Plus version 8.0 data bank (**APV80 VLE-IG**) or later depending on the version of Aspen Plus being used. Had you chosen to supply your own parameters, for example, as a result of regressing your own data or a data set obtained from **NIST TDE** as described later in this chapter, the source would then be listed as **USER**.

Now going to the **Main Toolbar** and clicking on **Binary** in the **Analysis** region brings up the following window in Fig. 5.1-5 with the default choices. Clicking on **Tools>Analysis>Property>Binary** leads to the window in Fig. 5.1-5 that has been populated with **P-xy** and **320 K** and **340 K**. [Note that a longer list of temperatures could have been entered, in which case the P-xy diagrams at all the temperatures listed would be done simultaneously, plotted on a single graph, and tabulated.]

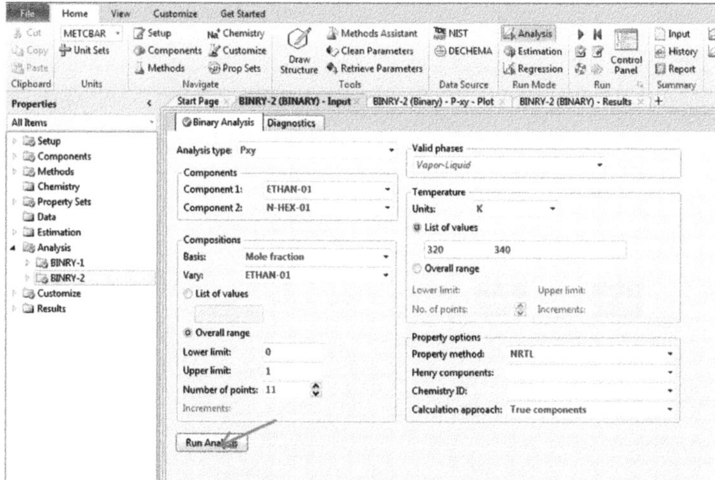

Figure 5.1-5

I have chosen, as seen in the window below, to make a **P-xy** graph at **320** and **340 K** and for simplicity of presentation at only 11 points. Clicking on **Run Analysis** and going to **Results** produces the table in Fig. 5.1-6.

Binary analysis results

TEMP	MOLEFRAC ETHAN-01	TOTAL PRES	TOTAL KVL ETHAN-01	TOTAL KVL N-HEX-01	LIQUID GAMMA ETHAN-01	LIQUID GAMMA N-HEX-01	VAPOR MOLEFRAC ETHAN-01	VAPOR MOLEFRAC N-HEX-01	LIQUID MOLEFRAC ETHAN-01	LIQUID MOLEFRAC N-HEX-01
K		bar								
320	0	0.484198	11.1364	1	21.2651	1	0	1	0	1
320	0.1	0.643286	2.86583	0.792686	7.25729	1.05313	0.286583	0.713417	0.1	0.9
320	0.2	0.651897	1.51702	0.870744	3.89306	1.17232	0.303405	0.696595	0.2	0.8
320	0.3	0.652358	1.02032	0.99129	2.62025	1.33556	0.306097	0.693903	0.3	0.7
320	0.4	0.65165	0.777486	1.14834	1.99447	1.54548	0.310995	0.689006	0.4	0.6
320	0.5	0.648157	0.638783	1.36122	1.62987	1.82216	0.319391	0.680609	0.5	0.5
320	0.6	0.640184	0.552248	1.67163	1.39174	2.21015	0.331349	0.668651	0.6	0.4
320	0.7	0.624943	0.497997	2.17134	1.22515	2.8025	0.348598	0.651402	0.7	0.3
320	0.8	0.59371	0.473576	3.1057	1.10684	3.80812	0.378861	0.621139	0.8	0.2
320	0.9	0.514488	0.508459	5.42387	1.0298	5.76317	0.457613	0.542387	0.9	0.1
320	1	0.254027	1	19.7028	1	10.3368	1	0	1	0
340	0	0.955091	12.1603	1	18.324	1	0	1	0	1
340	0.1	1.33767	3.26085	0.748795	6.88194	1.04874	0.326085	0.673915	0.1	0.9
340	0.2	1.37037	1.76233	0.809418	3.81028	1.16136	0.352466	0.647534	0.2	0.8
340	0.3	1.37417	1.19401	0.916851	2.58871	1.31915	0.358204	0.641796	0.3	0.7
340	0.4	1.37444	0.910377	1.05975	1.97415	1.52505	0.364151	0.635849	0.4	0.6
340	0.5	1.36994	0.746405	1.2536	1.61327	1.7981	0.373202	0.626799	0.5	0.5
340	0.6	1.35688	0.643716	1.53443	1.37806	2.17993	0.386229	0.613771	0.6	0.4
340	0.7	1.32868	0.579588	1.98096	1.21499	2.75583	0.405712	0.594289	0.7	0.3
340	0.8	1.26717	0.550603	2.79759	1.1008	3.71171	0.440483	0.559517	0.8	0.2
340	0.9	1.11189	0.585842	4.72742	1.02772	5.50352	0.527258	0.472742	0.9	0.1
340	1	0.633822	1	14.2496	1	9.45641	1	0	1	0

Figure 5.1-6

Clicking on the **Plot** tab brings up the somewhat ragged P-xy plots in Fig. 5.1-7 at the two temperatures.

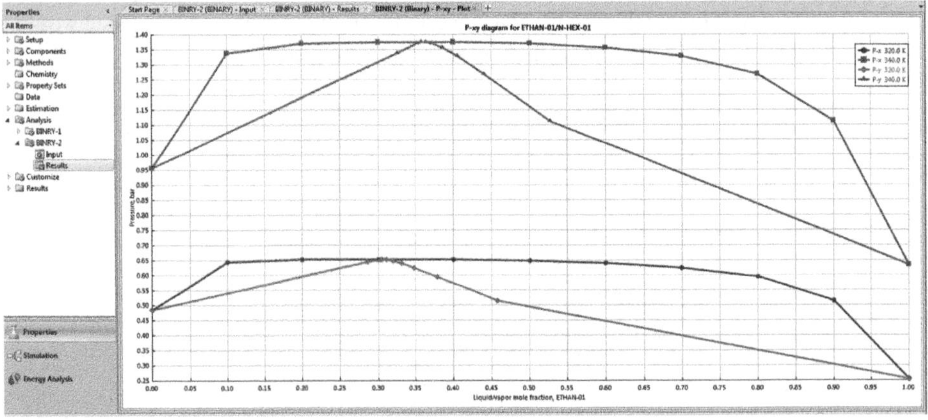

Figure 5.1-7

Had the default of 51 points been used in Fig. 5.1-5, the much smoother graph of Fig. 5.1-8 would have been obtained.

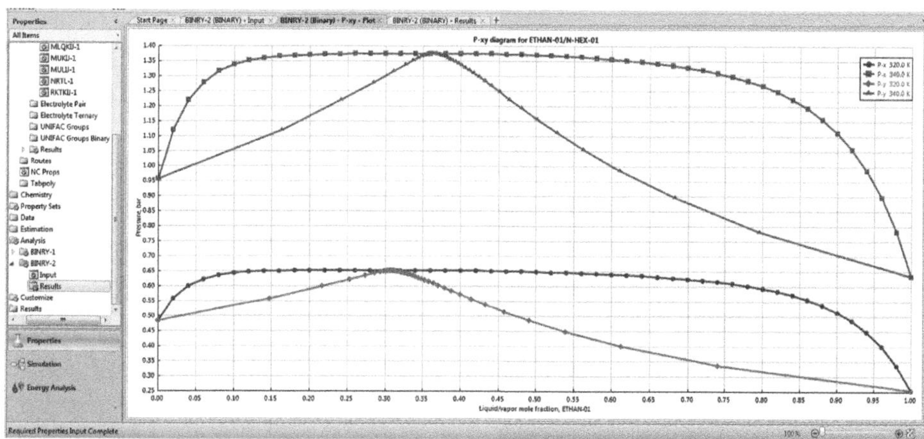

Figure 5.1-8

Following the same procedure (Fig. 5.1-9) to develop a T-xy diagram at 1 bar (using 21 points) produces the table in Fig. 5.1-10

Properties
All Items
Setup
Components
Methods
Chemistry
Property Sets
Data
Estimation
Analysis
BINRY-1
BINRY-2
Input
Results
Customize
Results

Start Page | BINRY-2 (BINARY) - Input | BINRY-2 (BINARY) - Results | BINRY-2 (Binary) - P-xy - Plot | +

Binary Analysis | Diagnostics

Analysis type: Txy

Components
Component 1: ETHAN-01
Component 2: N-HEX-01

Compositions
Basis: Mole fraction
Vary: ETHAN-01
List of values
Overall range
Lower limit: 0
Upper limit: 1
Number of points: 21
Increments:

Valid phases
Vapor-Liquid

Pressure
Units: bar
List of values
1
Overall range
Lower limit: Upper limit:
No. of points: Increments:

Property options
Property method: NRTL
Henry components:
Chemistry ID:
Calculation approach: True components

Run Analysis

Figure 5.1-9

Binary analysis results

PRES	MOLEFRAC ETHAN-01	TOTAL TEMP	TOTAL KVL ETHAN-01	TOTAL KVL N-HEX-01	LIQUID GAMMA ETHAN-01	LIQUID GAMMA N-HEX-01	VAPOR MOLEFRAC ETHAN-01	VAPOR MOLEFRAC N-HEX-01	LIQUID MOLEFRAC ETHAN-01	LIQUID MOLEFRAC N-HEX-01
bar		C								
1	0	68.3127	12.2298	1	18.1363	1	0	1	0	1
1	0.05	60.2347	5.13843	0.782188	10.8148	1.01423	0.256921	0.743079	0.05	0.95
1	0.1	58.5683	3.10231	0.76641	7.03604	1.05047	0.310231	0.689769	0.1	0.9
1	0.15	58.15	2.17352	0.792909	5.02358	1.10215	0.326028	0.673973	0.15	0.85
1	0.2	58.0411	1.65679	0.835801	3.84821	1.16603	0.331359	0.668641	0.2	0.8
1	0.25	58.0087	1.33477	0.888409	3.10481	1.24078	0.333693	0.666307	0.25	0.75
1	0.3	57.9949	1.11855	0.949195	2.60347	1.32629	0.335564	0.664436	0.3	0.7
1	0.35	57.9919	0.965418	1.01862	2.24736	1.42344	0.337896	0.662104	0.35	0.65
1	0.4	58.0056	0.852561	1.09829	1.98342	1.53407	0.341024	0.658976	0.4	0.6
1	0.45	58.0446	0.766803	1.1906	1.78076	1.66109	0.345062	0.654939	0.45	0.55
1	0.5	58.1179	0.700127	1.29987	1.62053	1.80878	0.350064	0.649936	0.5	0.5
1	0.55	58.2349	0.647486	1.43085	1.49078	1.98323	0.356117	0.643883	0.55	0.45
1	0.6	58.4073	0.605691	1.59146	1.38373	2.19312	0.363415	0.636585	0.6	0.4
1	0.65	58.6538	0.572847	1.79328	1.29422	2.45091	0.372351	0.627649	0.65	0.35
1	0.7	59.0068	0.548126	2.05437	1.21884	2.77481	0.383688	0.616312	0.7	0.3
1	0.75	59.5258	0.531851	2.40445	1.15542	3.19196	0.398888	0.601112	0.75	0.25
1	0.8	60.3254	0.526035	2.89586	1.10267	3.74369	0.420828	0.579172	0.8	0.2
1	0.85	61.6358	0.535994	3.62937	1.06012	4.49387	0.455595	0.544405	0.85	0.15
1	0.9	63.9486	0.575146	4.82368	1.028	5.54026	0.517632	0.482368	0.9	0.1
1	0.95	68.411	0.682103	7.04004	1.00735	7.01844	0.647998	0.352002	0.95	0.05
1	1	77.9763	1	12.0931	1	9.0288	1	0	1	0

Figure 5.1-10

and the graph of Fig. 5.1-11.

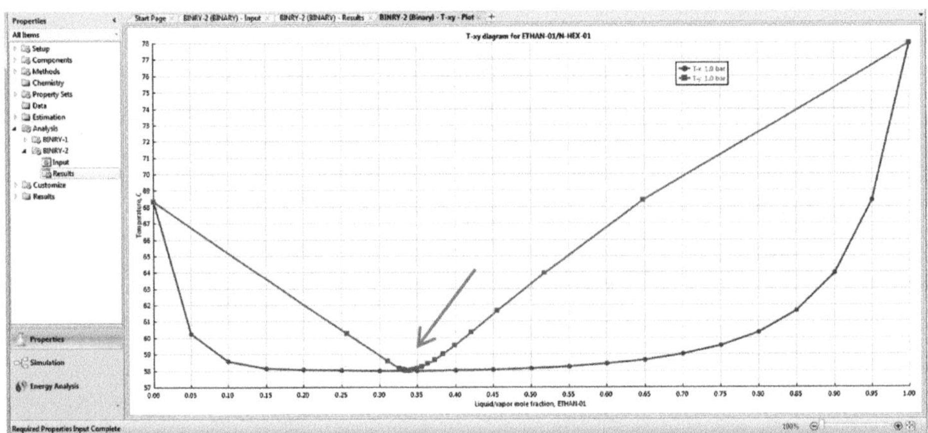

Figure 5.1-11

The plots of Figs. 5.1-8 and 5.1-11 clearly show that this system has an azeotrope (indicated by the arrow).

The remaining option on the **Binary Analysis** page is to compute the excess **Gibbs energy of mixing**; this is shown in Figs. 5.1-12 to 5.1-14

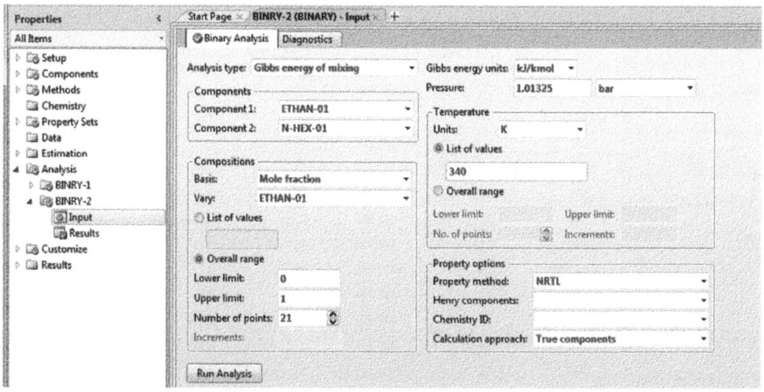

Figure 5.1-12

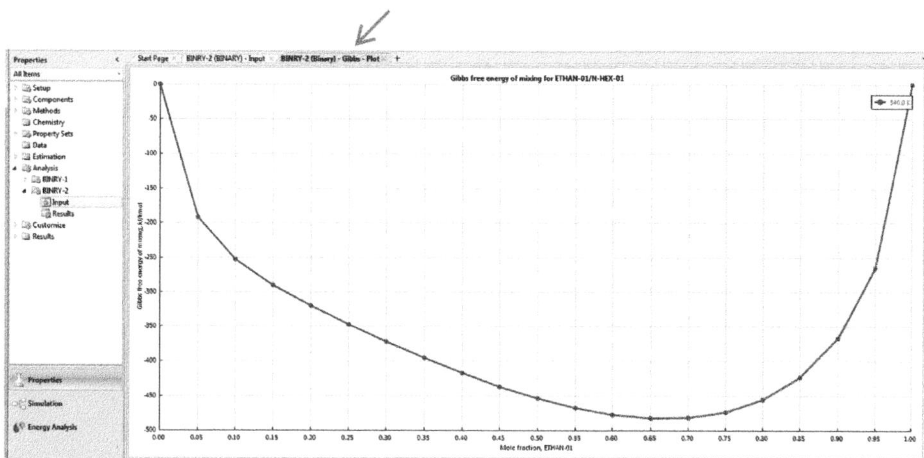

Figure 5.1-13

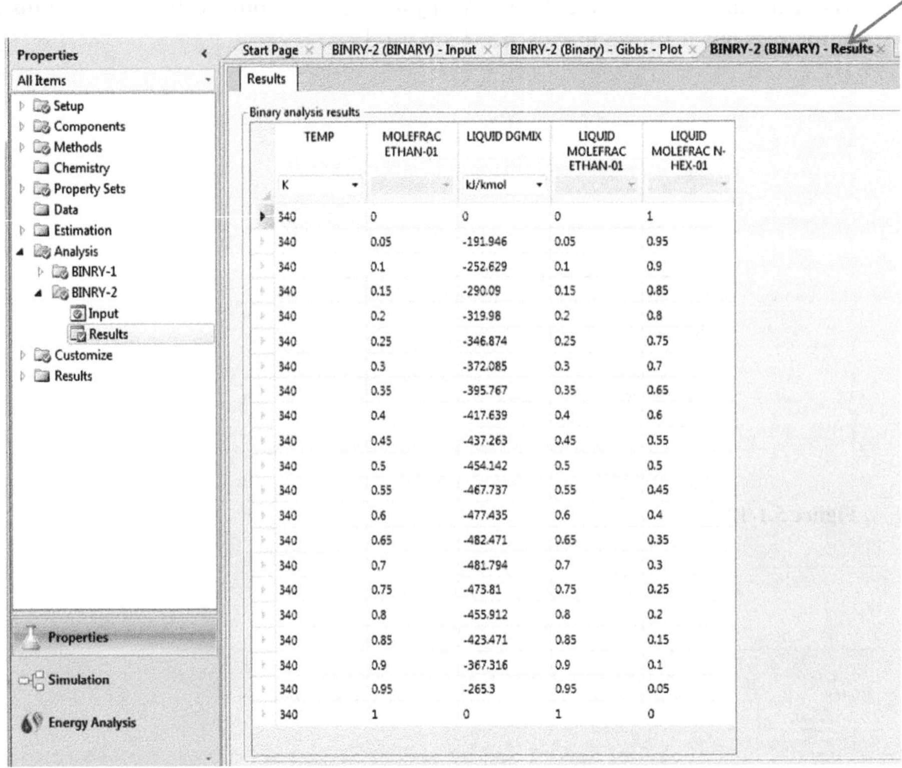

TEMP	MOLEFRAC ETHAN-01	LIQUID DGMIX	LIQUID MOLEFRAC ETHAN-01	LIQUID MOLEFRAC N-HEX-01
K		kJ/kmol		
340	0	0	0	1
340	0.05	-191.946	0.05	0.95
340	0.1	-252.629	0.1	0.9
340	0.15	-290.09	0.15	0.85
340	0.2	-319.98	0.2	0.8
340	0.25	-346.874	0.25	0.75
340	0.3	-372.085	0.3	0.7
340	0.35	-395.767	0.35	0.65
340	0.4	-417.639	0.4	0.6
340	0.45	-437.263	0.45	0.55
340	0.5	-454.142	0.5	0.5
340	0.55	-467.737	0.55	0.45
340	0.6	-477.435	0.6	0.4
340	0.65	-482.471	0.65	0.35
340	0.7	-481.794	0.7	0.3
340	0.75	-473.81	0.75	0.25
340	0.8	-455.912	0.8	0.2
340	0.85	-423.471	0.85	0.15
340	0.9	-367.316	0.9	0.1
340	0.95	-265.3	0.95	0.05
340	1	0	1	0

Figure 5.1-14

Of course, other activity coefficient models can be used. The results of computing the P-xy diagram with different models and using their Aspen Plus default parameter values are shown in Figs 5.1-15 to 5.1-18.

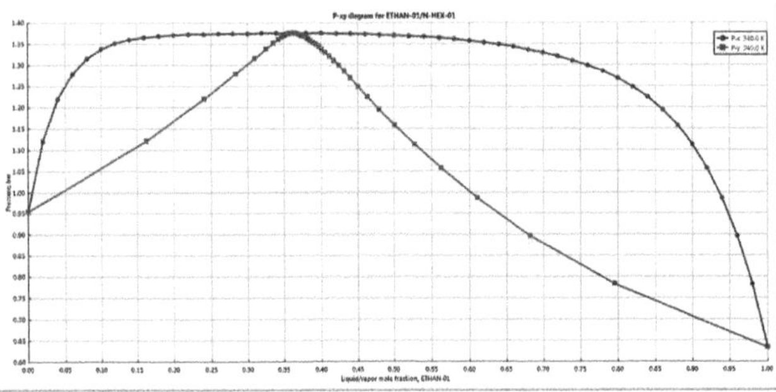

Figure 5.1-15 NRTL

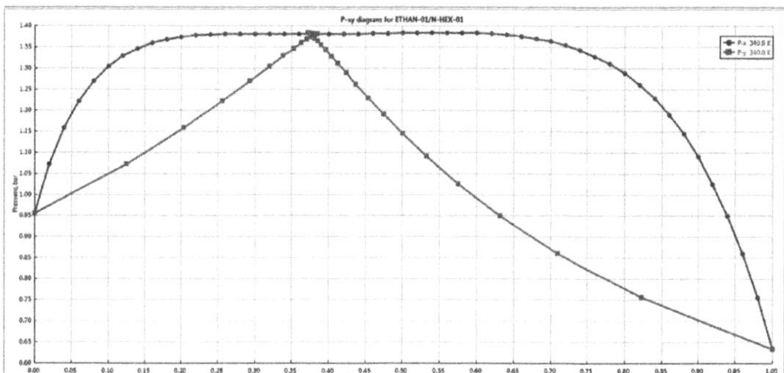

Figure 5.1-16 UNIQUAC

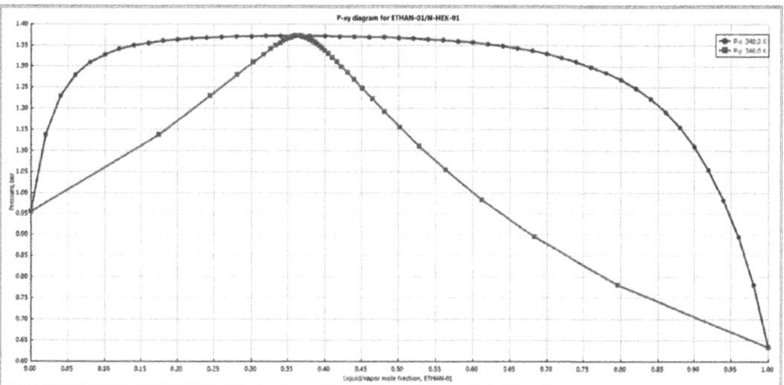

Figure 5.1-17 Wilson

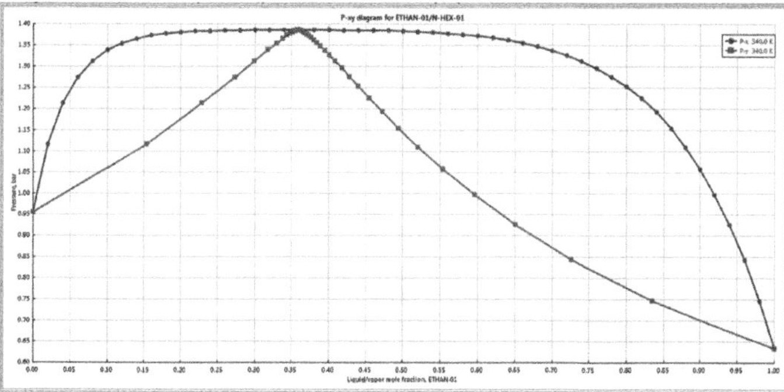

Figure 5.1-18 UNIFAC

For this mixture, all the correlative models (NRTL, UNIQUAC, and Wilson) give very similar results. Also, the predictive UNIFAC model gives similar results for this system (Fig. 5.1-8). However, since UNIFAC is completely predictive, you did not have to click on **Parameters>Binary Interaction** to get the parameters for this model.

It is not always true that the different activity coefficient models will give similar results; that is very mixture dependent. For example, while the Wilson model is generally quite good for moderately nonideal mixtures, it will not accurately predict or correlate data for very nonideal systems exhibiting liquid–liquid equilibrium, while the other models considered here do. It is for this reason that a user doing a simulation of a full process may first want to do a **Property Analysis** as was done here for each of the binary pairs in a multicomponent mixture and determine the effect of different model choices.

In Aspen Plus there is also a direct method of determining whether a mixture has an azeotrope. This is done by first changing the type of calculation from **Properties** to **Simulation**. On the main menu bar go to **Azeotrope Search** in the **Analysis** section that brings up the window in Fig. 5.1-19.

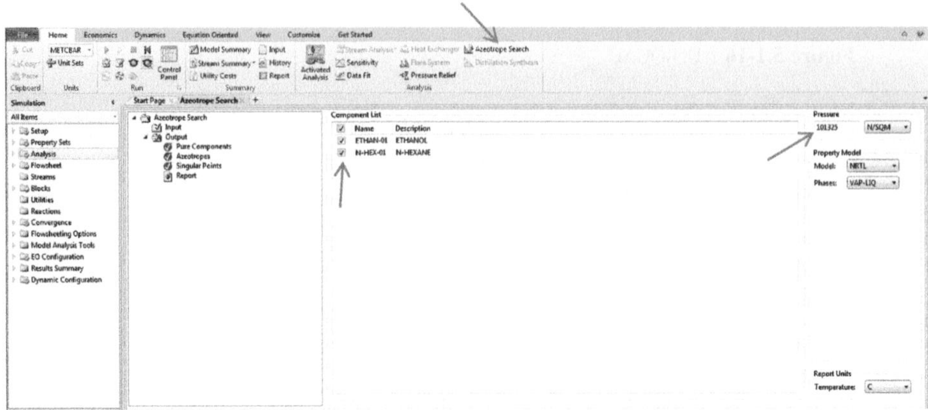

Figure 5.1-19

Choose the components, here ethanol and hexane, and set the pressure as shown in Fig. 5.1-20.

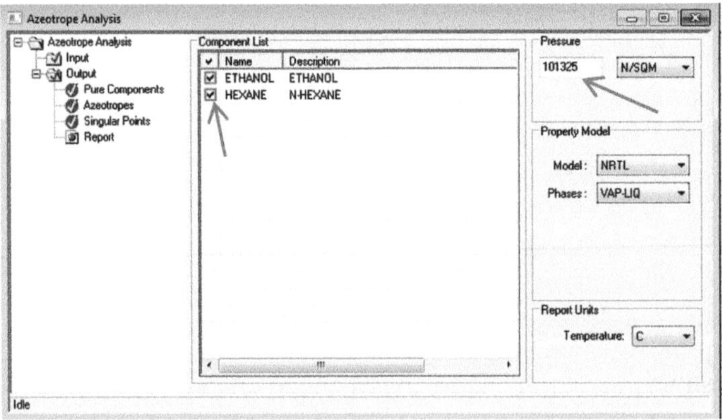

Figure 5.1-20

Then click on **Azeotropes**; we see that an azeotrope has been identified (Fig. 5.1-21)

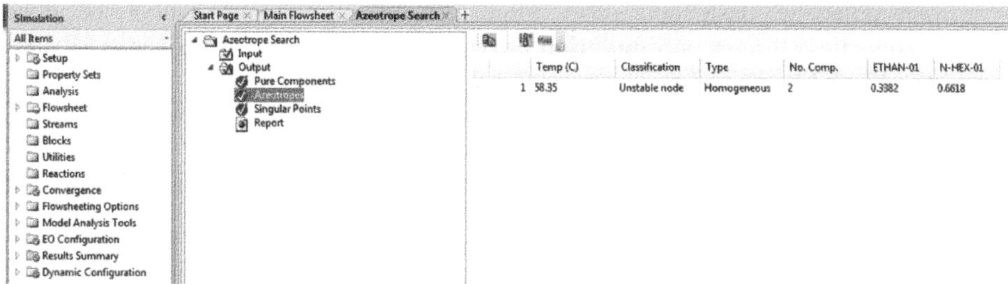

Figure 5.1-21

and clicking on **Azeotrope Search>Output>Report**, gives the results displayed in Fig. 5.1-22.

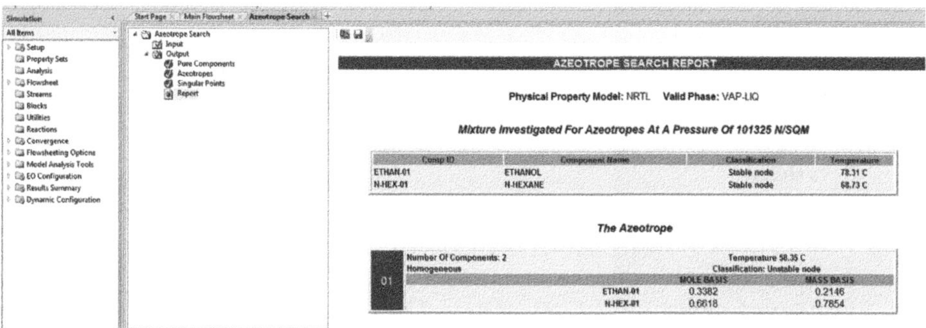

Figure 5.1-22

Repeat the calculation with the Wilson model (by going back to **Properties**, then to **Methods** and changing to Wilson, then to **Parameters>Binary Interaction** to retrieve the parameters and finally back to **Simulation** and **Azeotrope Search**) to obtain:

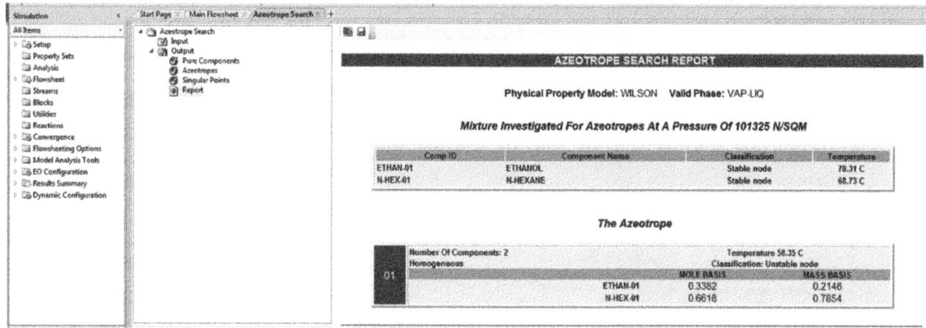

Figure 5.1-23

The predictions of the two methods agree.

5.2 THE SIMULATION METHOD

An adiabatic flash and other more general calculations cannot be done using **Properties>Binary** or **Simulation>Azeotrope Search** methods, but are easily done using the **Flash2** block. As an illustration of the simulation method, we consider an equimolar mixture of ethanol and hexane at 380 K and 10 bar, described by the NRTL model, that is to be adiabatically flashed to 1 bar using the **Flash2** separator.

The **Flowsheet**, feed input stream **Stream1** and **Block B1** (Flash2) setup is shown in Figs. 5.2-1 to 5.2-3.

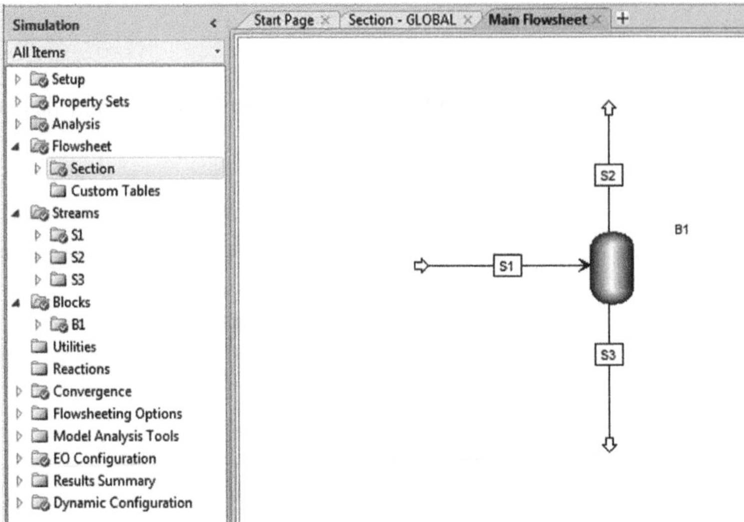

Figure 5.2-1

The input for **Stream S1** is

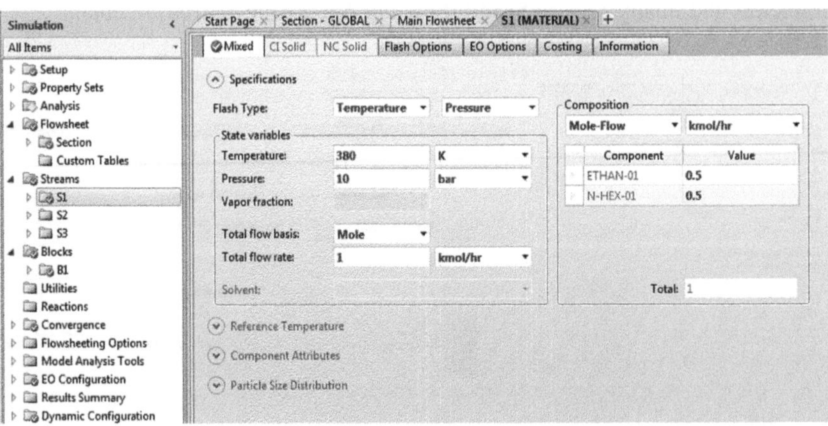

Figure 5.2-2

and the **Block B1** (**FLASH2**) setup is

Simulation | Start Page | Section - GLOBAL | Main Flowsheet | B1 (Flash2)

All Items: Setup, Property Sets, Analysis, Flowsheet, Streams, Blocks, B1, Utilities, Reactions, Convergence, Flowsheeting Options, Model Analysis Tools, EO Configuration, Results Summary, Dynamic Configuration

Specifications | Flash Options | Entrainment | Utility | Information

Flash specifications

Flash Type:	Duty	Pressure
Temperature:		C
Pressure:	1	bar
Duty:	0	cal/sec
Vapor fraction:		

Valid phases: Vapor-Liquid

Figure 5.2-3

The simulation is run by clicking on the arrow key on the main toolbar (Fig. 5.2-4)

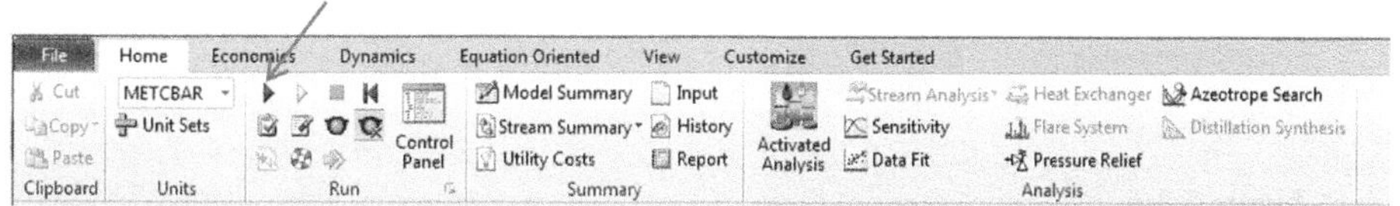

Figure 5.2-4

and the results found in **Results Summary>Streams** (Fig. 5.2-5).

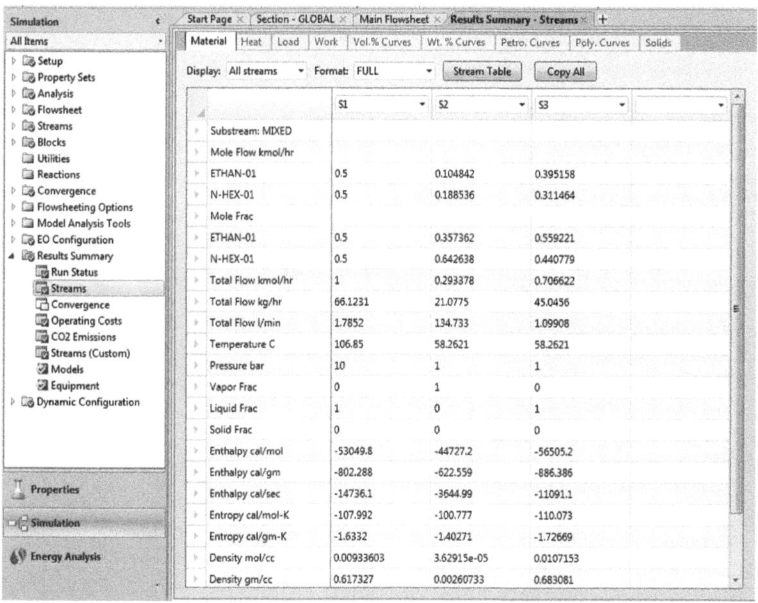

	S1	S2	S3
Substream: MIXED			
Mole Flow kmol/hr			
ETHAN-01	0.5	0.104842	0.395158
N-HEX-01	0.5	0.188536	0.311464
Mole Frac			
ETHAN-01	0.5	0.357362	0.559221
N-HEX-01	0.5	0.642638	0.440779
Total Flow kmol/hr	1	0.293378	0.706622
Total Flow kg/hr	66.1231	21.0775	45.0456
Total Flow l/min	1.7852	134.733	1.09908
Temperature C	106.85	58.2621	58.2621
Pressure bar	10	1	1
Vapor Frac	0	1	0
Liquid Frac	1	0	1
Solid Frac	0	0	0
Enthalpy cal/mol	-53049.8	-44727.2	-56505.2
Enthalpy cal/gm	-802.288	-622.559	-886.386
Enthalpy cal/sec	-14736.1	-3644.99	-11091.1
Entropy cal/mol-K	-107.992	-100.777	-110.073
Entropy cal/gm-K	-1.6332	-1.40271	-1.72669
Density mol/cc	0.00933603	3.62915e-05	0.0107153
Density gm/cc	0.617327	0.00260733	0.683081

Figure 5.2-5

Note that the temperature of the exiting streams in the adiabatic flash are almost 50°C lower than the input stream. [Question for the reader: Why aren't the vapor and liquid compositions at the azeotropic composition? Does looking at the T-xy diagram at 1 bar computed earlier using the **Properties>Binary** method help answer this question? How would the input stream compositions have to be changed to produce exit streams at the azeotropic composition? Does the mass balance provide any insight?]

As a second example, we use the **FLASH2** separator to calculate the bubble point of a benzene–decane equimolar mixture at 350 K using the NRTL model. The **FLASH2** input for this is shown in Fig. 5.2-6

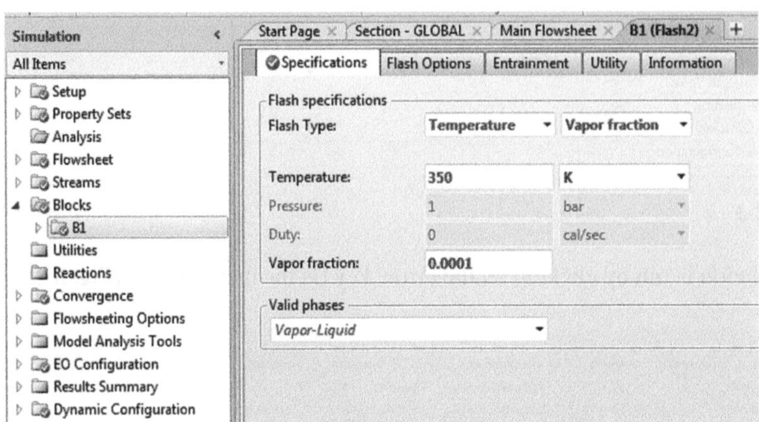

Figure 5.2-6

that leads to the results in Fig. 5.2-7.

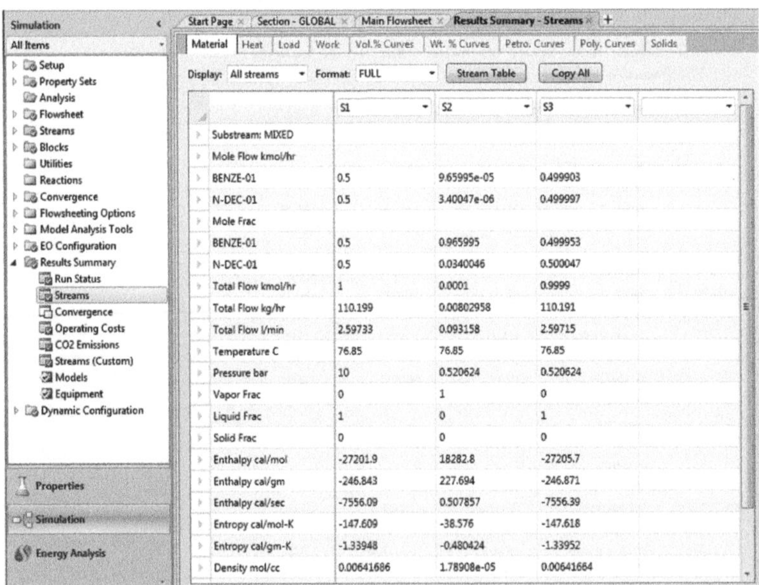

	S1	S2	S3
Substream: MIXED			
Mole Flow kmol/hr			
BENZE-01	0.5	9.65995e-05	0.499903
N-DEC-01	0.5	3.40047e-06	0.499997
Mole Frac			
BENZE-01	0.5	0.965995	0.499953
N-DEC-01	0.5	0.0340046	0.500047
Total Flow kmol/hr	1	0.0001	0.9999
Total Flow kg/hr	110.199	0.00802958	110.191
Total Flow l/min	2.59733	0.093158	2.59715
Temperature C	76.85	76.85	76.85
Pressure bar	10	0.520624	0.520624
Vapor Frac	0	1	0
Liquid Frac	1	0	1
Solid Frac	0	0	0
Enthalpy cal/mol	-27201.9	18282.8	-27205.7
Enthalpy cal/gm	-246.843	227.694	-246.871
Enthalpy cal/sec	-7556.09	0.507857	-7556.39
Entropy cal/mol-K	-147.609	-38.576	-147.618
Entropy cal/gm-K	-1.33948	-0.480424	-1.33952
Density mol/cc	0.00641686	1.78908e-05	0.00641664

Figure 5.2-7

So that at 350 K the bubble point is 0.5206 bar with a vapor mole fraction of 0.966 benzene. In a similar manner, the dew point of this mixture is calculated by starting as shown in Fig. 5.2-8,

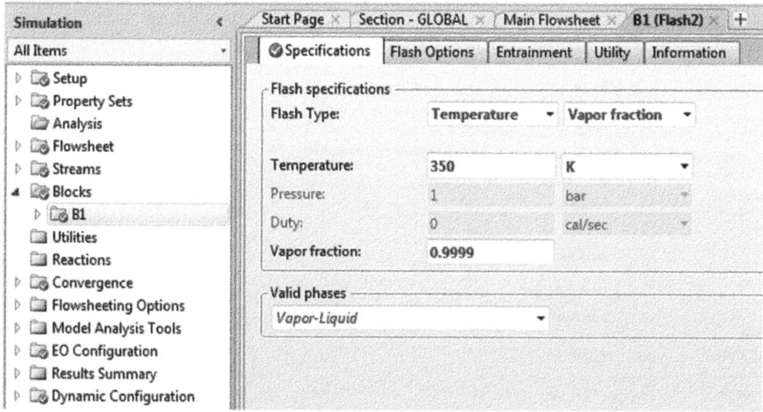

Figure 5.2-8

resulting in Fig. 5.2-9.

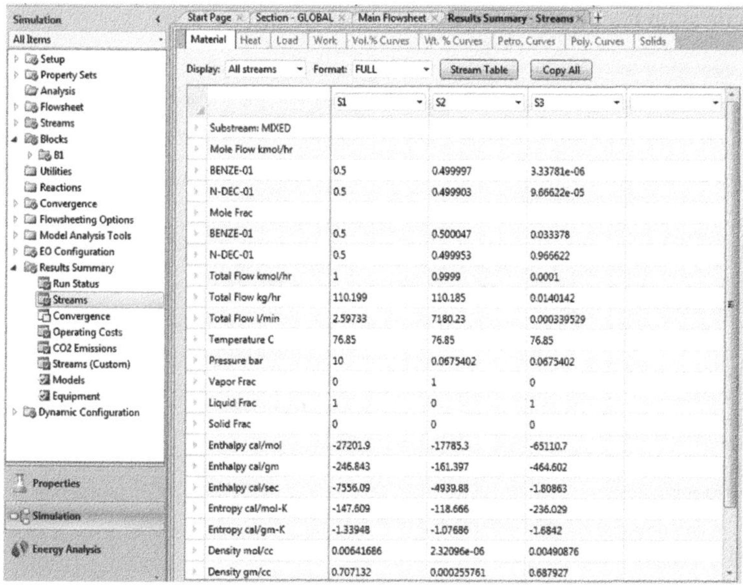

	S1	S2	S3
Substream: MIXED			
Mole Flow kmol/hr			
BENZE-01	0.5	0.499997	3.33781e-06
N-DEC-01	0.5	0.499903	9.66622e-05
Mole Frac			
BENZE-01	0.5	0.500047	0.033378
N-DEC-01	0.5	0.499953	0.966622
Total Flow kmol/hr	1	0.9999	0.0001
Total Flow kg/hr	110.199	110.185	0.0140142
Total Flow l/min	2.59733	7180.23	0.000339529
Temperature C	76.85	76.85	76.85
Pressure bar	10	0.0675402	0.0675402
Vapor Frac	0	1	0
Liquid Frac	1	0	1
Solid Frac	0	0	0
Enthalpy cal/mol	-27201.9	-17785.3	-65110.7
Enthalpy cal/gm	-246.843	-161.397	-464.602
Enthalpy cal/sec	-7556.09	-4939.88	-1.80863
Entropy cal/mol-K	-147.609	-118.666	-236.029
Entropy cal/gm-K	-1.33948	-1.07686	-1.6842
Density mol/cc	0.00641686	2.32096e-06	0.00490876
Density gm/cc	0.707132	0.000255761	0.687927

Figure 5.2-9

So the dew point pressure is 0.06754 bar with a liquid mole fraction of benzene of 0.033 and that of decane of 0.967.

Next, consider the adiabatic flash of this mixture from 400 K and 10 bar to 0.5 bar using the NRTL activity coefficient model. This set of conditions has been chosen to be used for comparison with the results of equation of state VLE calculations that will be done in the next chapter; because of the components (simple hydrocarbons), temperature, and the low pressure, this system can be described either by an activity coefficient model or an equation of state, and both should give similar (though not exactly the same) results.

The input for **Stream S1** (Fig. 5.2-10) is

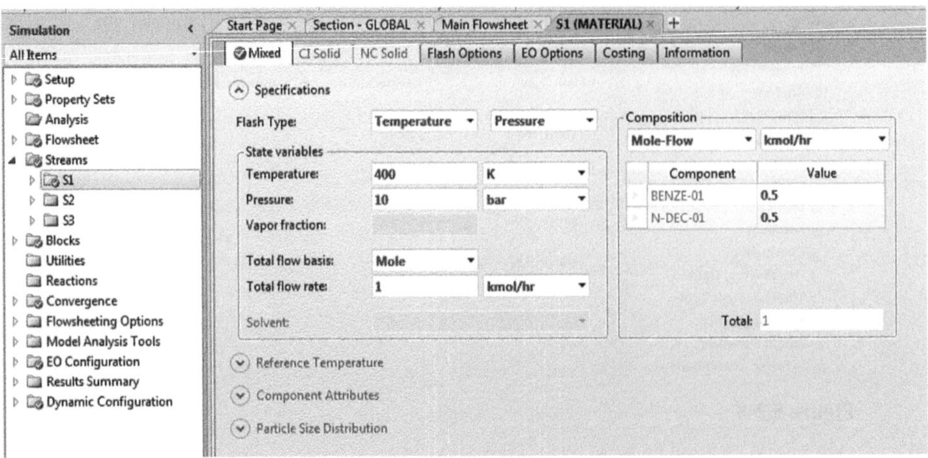

Figure 5.2-10

and the specification for **Block 1** (Fig. 5.2-11) is

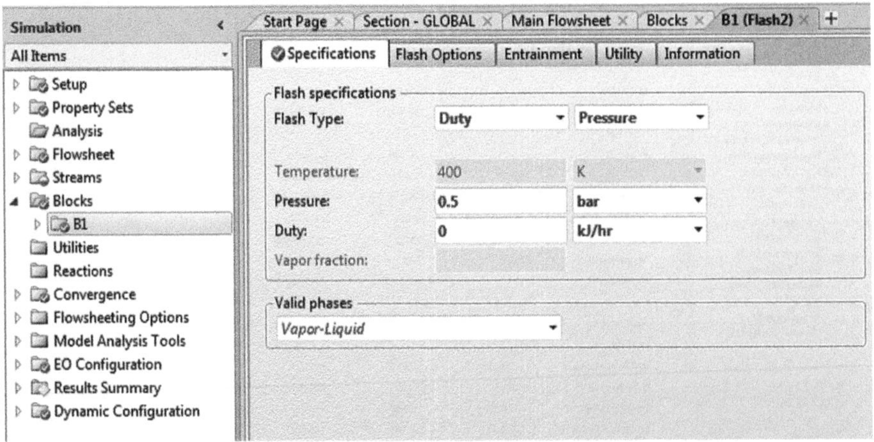

Figure 5.2-11

The results are (Fig. 5.2-12)

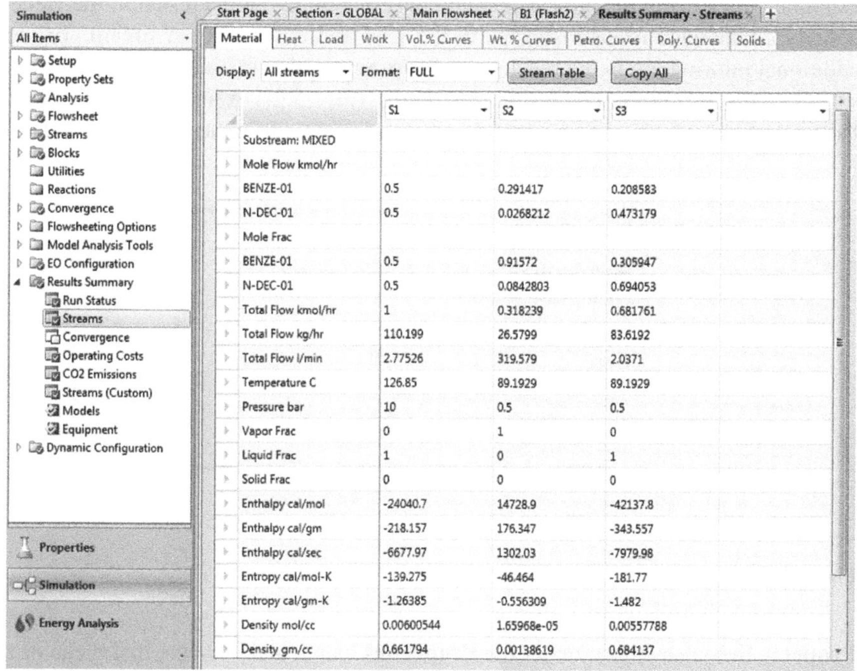

	S1	S2	S3
Substream: MIXED			
Mole Flow kmol/hr			
BENZE-01	0.5	0.291417	0.208583
N-DEC-01	0.5	0.0268212	0.473179
Mole Frac			
BENZE-01	0.5	0.91572	0.305947
N-DEC-01	0.5	0.0842803	0.694053
Total Flow kmol/hr	1	0.318239	0.681761
Total Flow kg/hr	110.199	26.5799	83.6192
Total Flow l/min	2.77526	319.579	2.0371
Temperature C	126.85	89.1929	89.1929
Pressure bar	10	0.5	0.5
Vapor Frac	0	1	0
Liquid Frac	1	0	1
Solid Frac	0	0	0
Enthalpy cal/mol	-24040.7	14728.9	-42137.8
Enthalpy cal/gm	-218.157	176.347	-343.557
Enthalpy cal/sec	-6677.97	1302.03	-7979.98
Entropy cal/mol-K	-139.275	-46.464	-181.77
Entropy cal/gm-K	-1.26385	-0.556309	-1.482
Density mol/cc	0.00600544	1.65968e-05	0.00557788
Density gm/cc	0.661794	0.00138619	0.684137

Figure 5.2-12

The results for both the NRTL and the Peng–Robinson equation of state (discussed in the next chapter) models are shown side-by-side in Fig. 5.2-13.

		PR EOS			NRTL	
		2	3		2	3
	FEED	VAPOR	LIQUID		VAPOR	LIQUID
Mole Flow kmol/hr						
BENZENE	0.5000	0.2935	0.2065		0.2914	0.2086
DECANE	0.5000	0.0285	0.4715		0.0268	0.4732
Mole Frac						
BENZENE	0.5000	0.9115	0.3045		0.9157	0.3059
DECANE	0.5000	0.0885	0.6955		0.0843	0.6941
Temperature K	400.0000	361.1887	361.1887		362.3468	362.3468
Pressure atm	9.8692	0.4935	0.4935		0.4935	0.4935
Vapor Frac	0	1	0		1	0
Liquid Frac	1	0	1		0	1

Figure 5.2-13

We see that as a result of the adiabatic flash from 10 to 0.5 bar the temperature has dropped from 400 K (126.85°C) to 89.19°C (NRTL), and that as expected the more volatile benzene is enriched in the vapor phase and the *n*-decane in the liquid phase.

There are instances where the user might want more information in the results of a completed simulation than are provided in the Aspen Plus default table of results. For example, when using the NRTL model above, one might be interested in the values of the activity coefficients of each species in the feed and exiting liquid stream. To obtain the additional information in the results, the **Prop-Sets** option is used. This is done as follows.

Go to **Setup>Report Options>Streams** (Fig. 5.2-14) and click on the option of **Property Sets**, which gives no useful options at this point in the process.

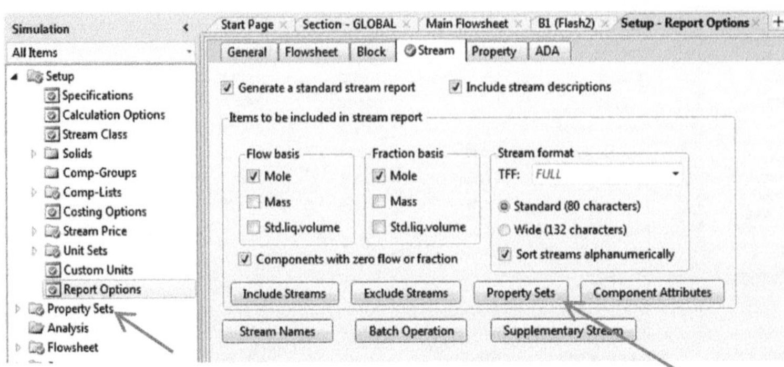

Figure 5.2-14

Property Sets also appear under the **Properties** menu, and it is here that it can be defined. Click on **Property Sets>New** and accept the default name of **PS-1** (or choose a name of your own). Next go to that **PS-1** and choose **GAMMA** under **Physical properties** from the drop-down menu that appears to see activity coefficients in the later results (Fig. 5.2-15).

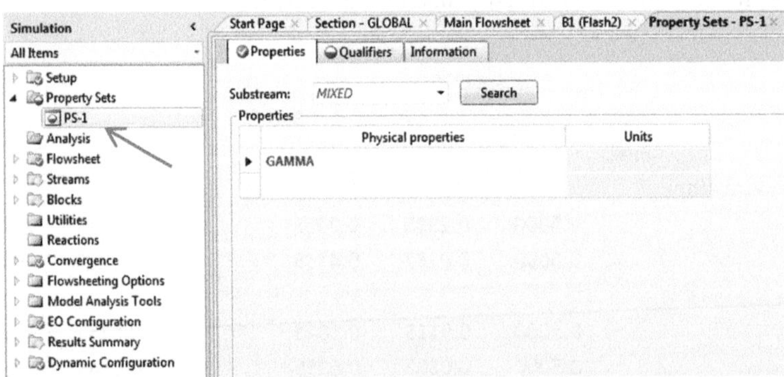

Figure 5.2-15

Then under the **Qualifiers** tab make choices for the components and conditions. Here the liquid phase, both components, at the system temperature and pressure (checked boxes) have been chosen in Fig. 5.2-16.

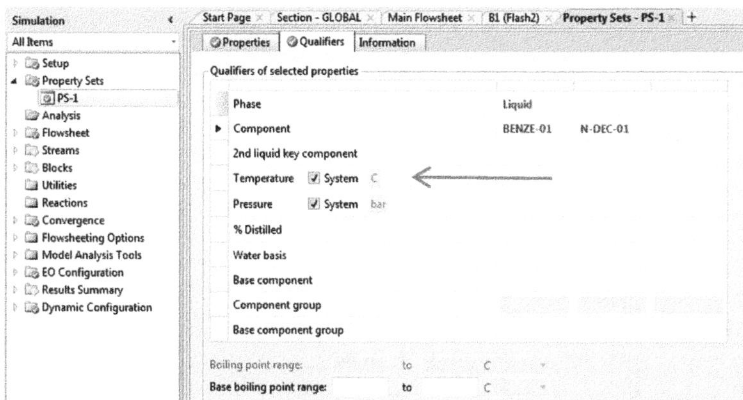

Figure 5.2-16

After completing this, go back to **Setup>Report Options>Streams** of Fig. 5.2-14 and click on **Property Sets.** Of the available **Property sets**, only **PS-1** is shown here, click on it and use the > arrow to bring it to the **Selected property sets** window as shown in Fig. 5.2-17.

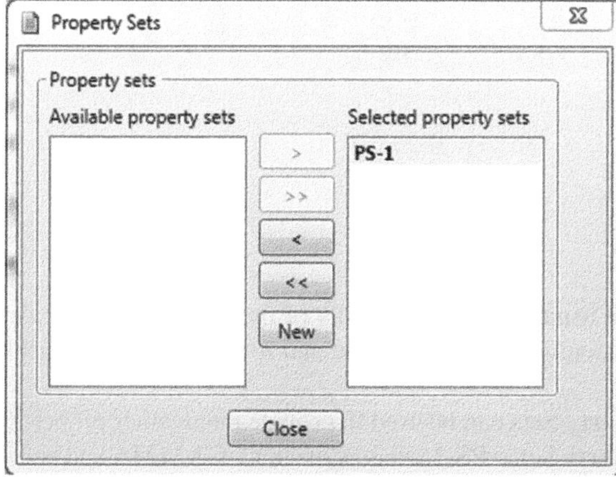

Figure 5.2-17

Now rerunning the simulation produces the output in Fig. 5.2-18.

Material | Heat | Load | Work | Vol.% Curves | Wt. % Curves | Petro. Curves | Poly. Curv

Display: All streams Format: FULL Stream Table Copy All

	S1	S2	S3	
Substream: MIXED				
Mole Flow kmol/hr				
BENZE-01	0.5	0.291417	0.208583	
N-DEC-01	0.5	0.0268212	0.473179	
Mole Frac				
BENZE-01	0.5	0.91572	0.305947	
N-DEC-01	0.5	0.0842803	0.694053	
Total Flow kmol/hr	1	0.318239	0.681761	
Total Flow kg/hr	110.199	26.5799	83.6192	
Total Flow l/min	2.77526	319.579	2.0371	
Temperature C	126.85	89.1929	89.1929	
Pressure bar	10	0.5	0.5	
Vapor Frac	0	1	0	
Liquid Frac	1	0	1	
Solid Frac	0	0	0	
Enthalpy cal/mol	-24040.7	14728.9	-42137.8	
Enthalpy cal/gm	-218.157	176.347	-343.557	
Enthalpy cal/sec	-6677.97	1302.03	-7979.98	
Entropy cal/mol-K	-139.275	-46.464	-181.77	
Entropy cal/gm-K	-1.26385	-0.556309	-1.482	
Density mol/cc	0.00600544	1.65968e-05	0.00557788	
Density gm/cc	0.661794	0.00138619	0.684137	
Average MW	110.199	83.522	122.652	
Liq Vol 60F l/min	2.3566	0.516734	1.83987	
*** LIQUID PHASE ***				
GAMMA				
BENZE-01	1.10008		1.12633	
N-DEC-01	1.02753		1.00018	

Figure 5.2-18

We see that in addition to the usual output, the results of the adiabatic flash calculation, here the liquid-phase activity coefficients for benzene and *n*-decane are now displayed at the bottom of the table.

In a similar fashion **Property Sets** can be used to provide many other properties in the output. The **Properties>Property Sets>PS-1** window in which **GAMMA** had been chosen under **Physical properties** provides a long list of other properties that can be included in the output of a simulation. However, some of the names or acronyms in the drop-down menu are not obvious, so note the short description of each option that appears when the cursor is placed over it. Many of these **Physical properties** options are for petroleum, petroleum fractions, or refinery fractions, which are mixtures of very large numbers of undefined components and characterized by empirical factors such as octane number, cetane number, boiling points of various specialized definitions. Other properties include heat capacities, both real and ideal gas, environmental factors such as chemical and biological oxygen demand, thermodynamic availability (relative to 298 K), and many others. These are not of interest in this book.

5.3 REGRESSION OF BINARY VLE DATA WITH ACTIVITY COEFFICIENT MODELS

All the calculations so far have been based on having activity coefficient model parameters available from either the Aspen Plus data bank, as supplied by the user, or by using a predictive model such as UNIFAC. This section deals with the different (inverse) problem of having experimental VLE data and determining the activity coefficient model parameters that best fit those data. In this section we consider the regression of P-xy data at constant temperature to obtain parameters in several activity coefficient models. A somewhat different situation when one only has the two infinite dilution activity coefficients in a binary mixture is considered in Chapter 8.

To use Aspen Plus to do data regression one starts as described in Chapter 1, which brings up the **Properties** window. Click on **Setup** and add a title (if you wish), and choose the units or keep the metric-centigrade-bar (**METCBAR**) unit set. **Setup** is now complete, and move on to components. Go to **Components>Specifications** and add the components. For Problem 10.2-4 in *Chemical, Biochemical and Engineering Thermodynamics*, 4th ed., S. I. Sandler (John Wiley & Sons, Inc., 2006), which is being used as an example here, the components are ethanol and benzene, and the VLE data at 45°C from I. Brown and F. Smith, *Aust. J. Chem.* **7**, 264 (1954) are

x_{eth}	y_{eth}	P(bar)	x_{eth}	y_{eth}	P(bar)
0.0000	0.0000	0.2939	0.5284	0.4101	0.4093
0.0374	0.1965	0.3619	0.6155	0.4343	0.4028
0.0972	0.2895	0.3953	0.7087	0.4751	0.3891
0.2183	0.3370	0.4088	0.8102	0.5456	0.3616
0.3141	0.3625	0.4124	0.9193	0.7078	0.3036
0.4150	0.3842	0.4128	0.9591	0.8201	0.2711
0.5199	0.4065	0.4100	1.0000	1.0000	0.2321

After both components have been added, you are then returned to the **Specifications** window that is now populated with the compounds you added. Note that **Specifications** is now checked as in Fig. 5.3-1.

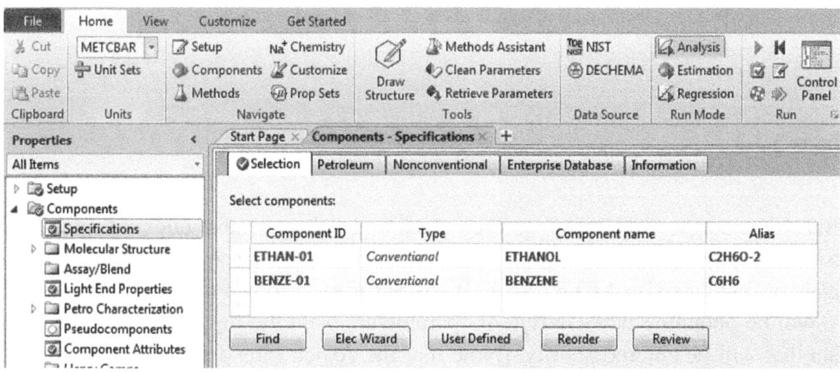

Figure 5.3-1

Next click on **Methods**, and for this example, **UNIQUAC** has been chosen as the **Base method** from the drop-down menu (Fig. 5.3-2), though any other activity coefficient model could have been used.

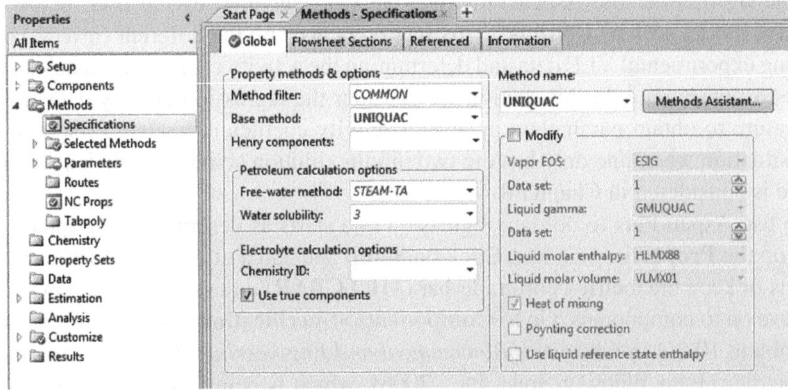

Figure 5.3-2

Next go to **Properties>Parameters>Binary Interaction**. This will bring up the list of binary interaction parameters of which some are checked including UNIQ-1 as shown in Fig. 5.3-3.

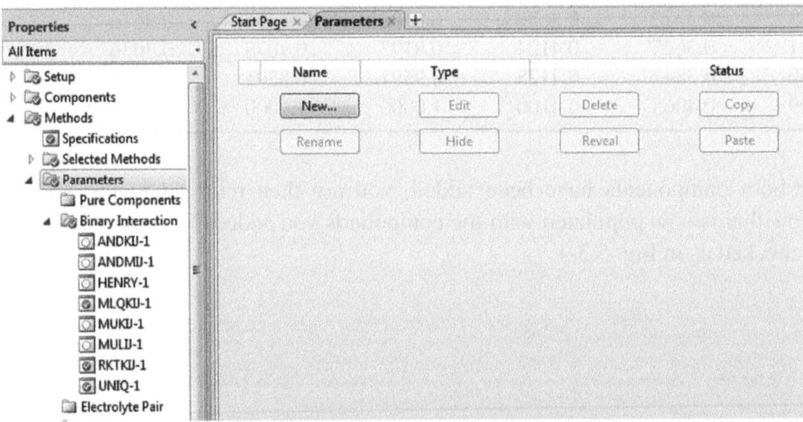

Figure 5.3-3

This will load the UNIQUAC surface areas and volumes; also clicking on **UNIQ-1** will display the UNIQUAC interaction parameters from the Aspen Plus version 8.2 or later database (indicated by **APV82 VLE-IG** as **Source**) as shown in Fig. 5.3-4. However, here we will be obtaining new UNIQUAC parameters from the regression of the experimental data that will be entered shortly. [Note that the Aspen Plus database is generally slightly changed with each upgrade of the software. So if a calculation is run from Aspen Plus 8.0, the user will see that the database used is **APV80 VLE-IG** or as here using Aspen Plus 8.2, the user will see that the database used is **APV82 VLE-IG**, etc.]

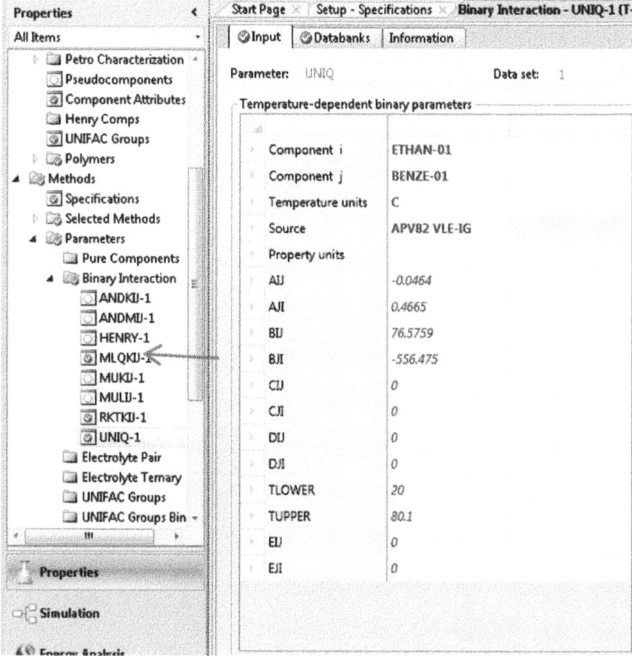

Figure 5.3-4

Next click on **Data** that brings up the window in Fig. 5.3-5 and after clicking on **New** a pop-up window appears, choose **Mixture** under **Select type**.

Figure 5.3-5

This brings you to Fig. 5.3-6,

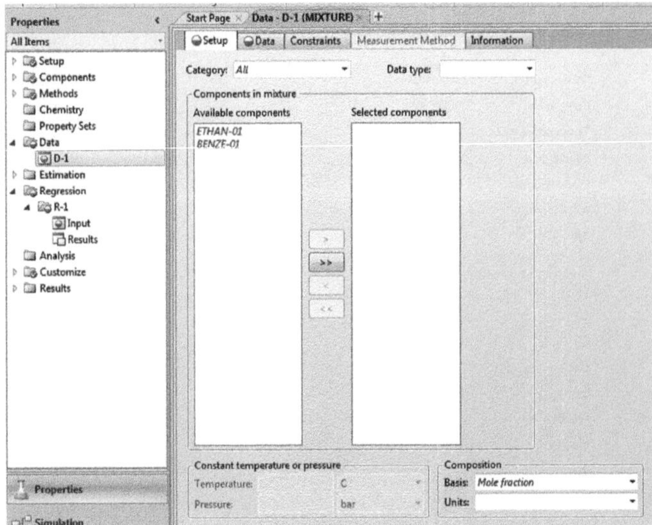

Figure 5.3-6

in which the components and the **TPZ Data type** have been chosen as shown in Fig. 5.3-7.

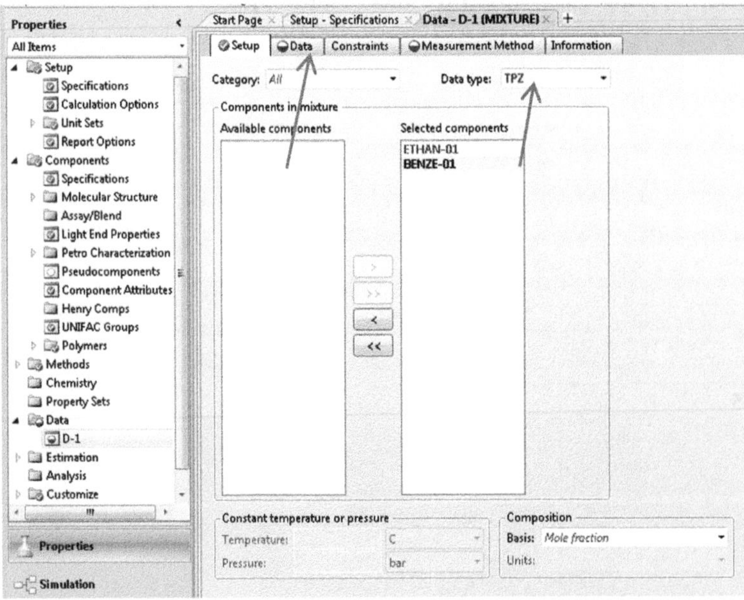

Figure 5.3-7

Next, click on the **Data** tab, and the window in Fig. 5.3-8 appears. The first data line under **Usage** has been completed with a default estimate of the uncertainty or standard deviation in the data that will be provided. The regression procedures, especially maximum likelihood methods that will be used, attempt to find the statistically most reliable estimates for the model parameters, and do this by taking into account that there are measurement errors in all variables, including temperature and pressure. These are typically reported as standard deviations, **STD-DEV**, and generally experimentalists provide only a single value for each type of data in the whole data set. The first line that appears is the Aspen Plus default values for these errors in percentage. These defaults are based on common literature estimates for high quality data of 0.1% for pressure and liquid mole fraction, and a higher error estimate of 1% for the vapor mole fraction because of the difficulties in obtaining good vapor samples. Generally the default values are sufficient, unless you have more detailed information. The completed data window for this example is shown in Fig. 5.3-8 using a single set of standard deviations for the whole data set.

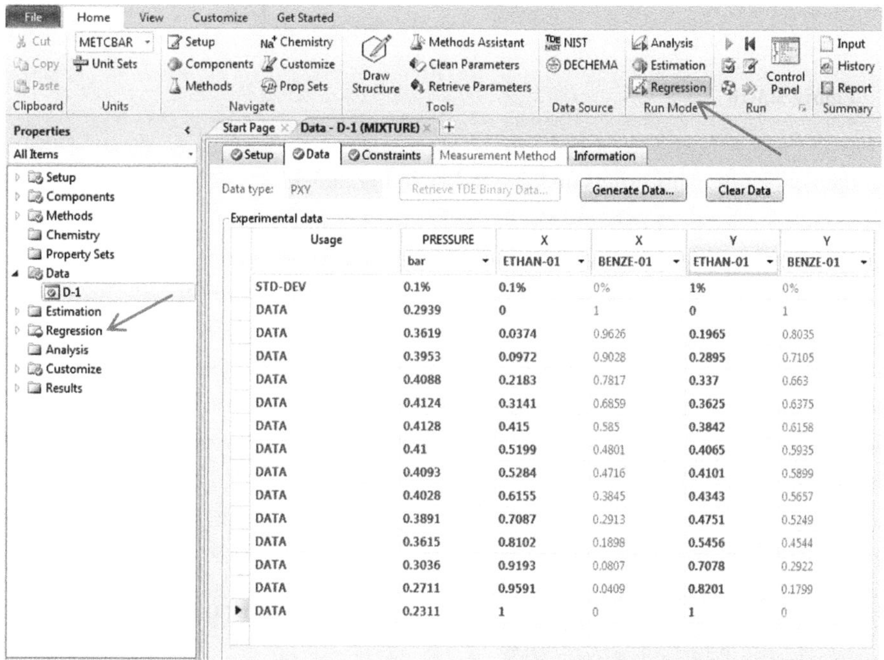

Usage	PRESSURE bar	X ETHAN-01	X BENZE-01	Y ETHAN-01	Y BENZE-01
STD-DEV	0.1%	0.1%	0%	1%	0%
DATA	0.2939	0	1	0	1
DATA	0.3619	0.0374	0.9626	0.1965	0.8035
DATA	0.3953	0.0972	0.9028	0.2895	0.7105
DATA	0.4088	0.2183	0.7817	0.337	0.663
DATA	0.4124	0.3141	0.6859	0.3625	0.6375
DATA	0.4128	0.415	0.585	0.3842	0.6158
DATA	0.41	0.5199	0.4801	0.4065	0.5935
DATA	0.4093	0.5284	0.4716	0.4101	0.5899
DATA	0.4028	0.6155	0.3845	0.4343	0.5657
DATA	0.3891	0.7087	0.2913	0.4751	0.5249
DATA	0.3615	0.8102	0.1898	0.5456	0.4544
DATA	0.3036	0.9193	0.0807	0.7078	0.2922
DATA	0.2711	0.9591	0.0409	0.8201	0.1799
DATA	0.2311	1	0	1	0

Figure 5.3-8

Next click on **Regression** in either of the two places shown in Fig. 5.3-8 and then click on **New** in the window in Fig. 5.3-9.

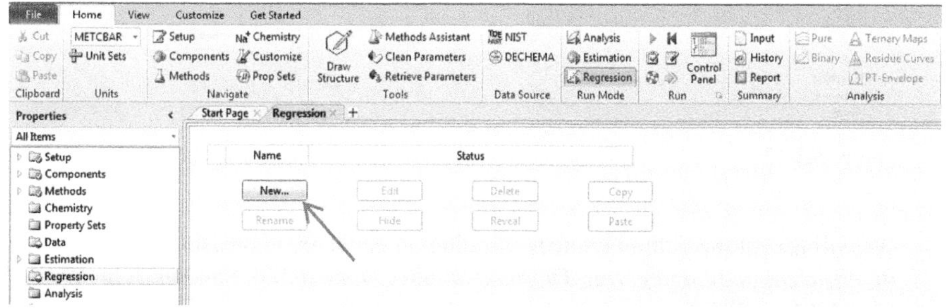

Figure 5.3-9

This brings up the small **Create new ID** window of Fig. 5.3-10; use the default ID **R-**1 or create your own,

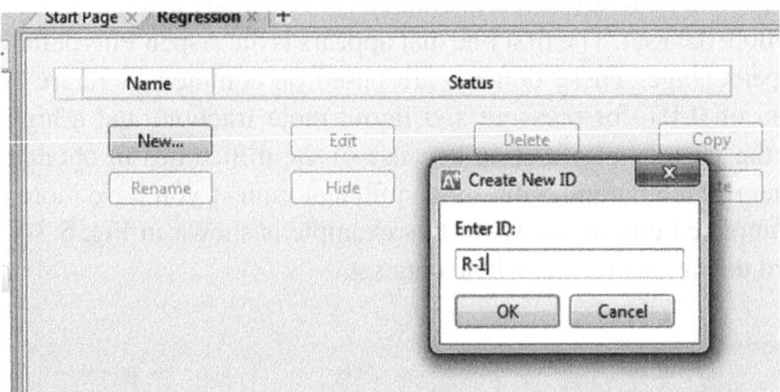

Figure 5.3-10

then click OK. This brings up the window in Fig. 5.3-11,

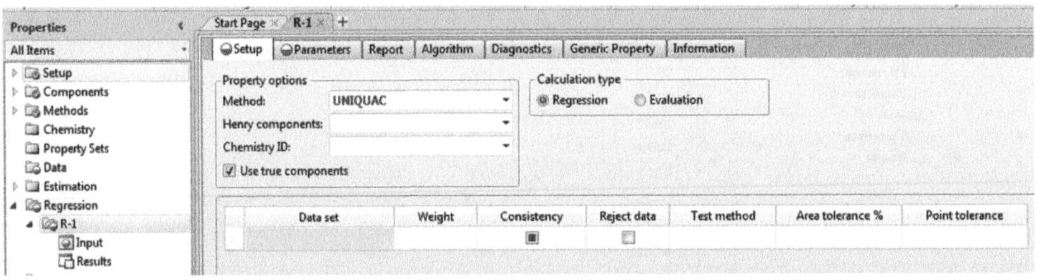

Figure 5.3-11

and then under **Data set** choose **D-1** from the drop-down menu (Fig. 5.3-12).

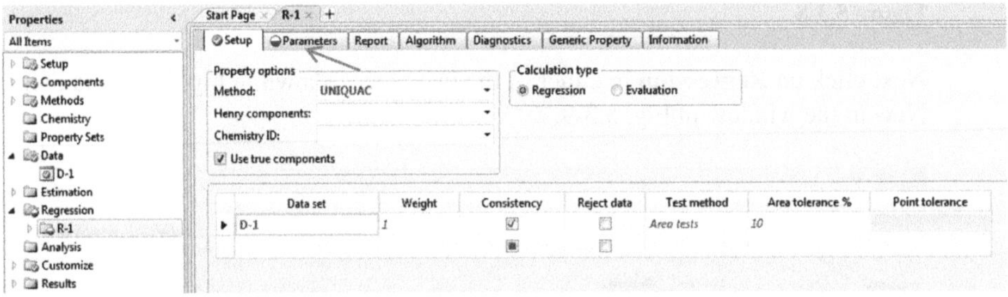

Figure 5.3-12

We will accept the default choices of doing an equal area consistency test (see *Chemical, Biochemical and Engineering Thermodynamics*, 4th ed., S. I. Sandler, John Wiley & Sons, Inc., 2006, p. 537, for an explanation of the test). Then click on the **Parameters** tab. This brings up the table in Fig. 5.3-13 that has been populated as shown.

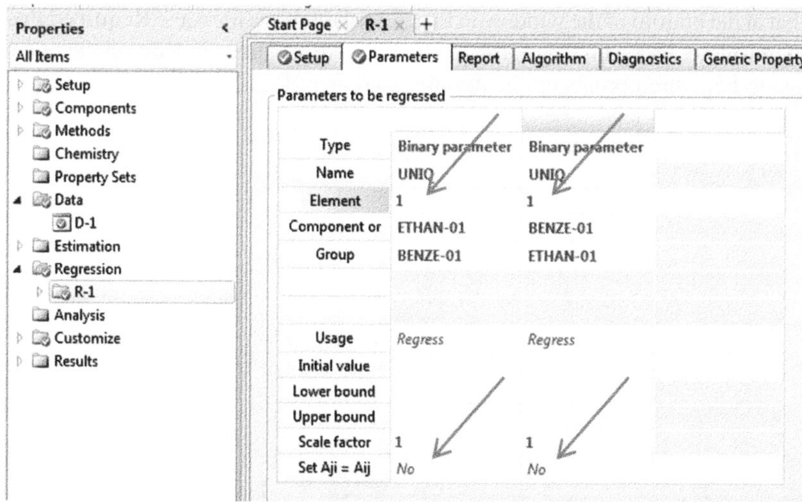

Figure 5.3-13

By completing the table in this way Aspen Plus is being instructed to obtain one (from **Element)** binary parameter in the UNIQUAC model for each binary pair in the mixture and, by entering **No** in the last line of the table (**Set Aij = Aji**), that the two binary parameters are not forced to be equal.

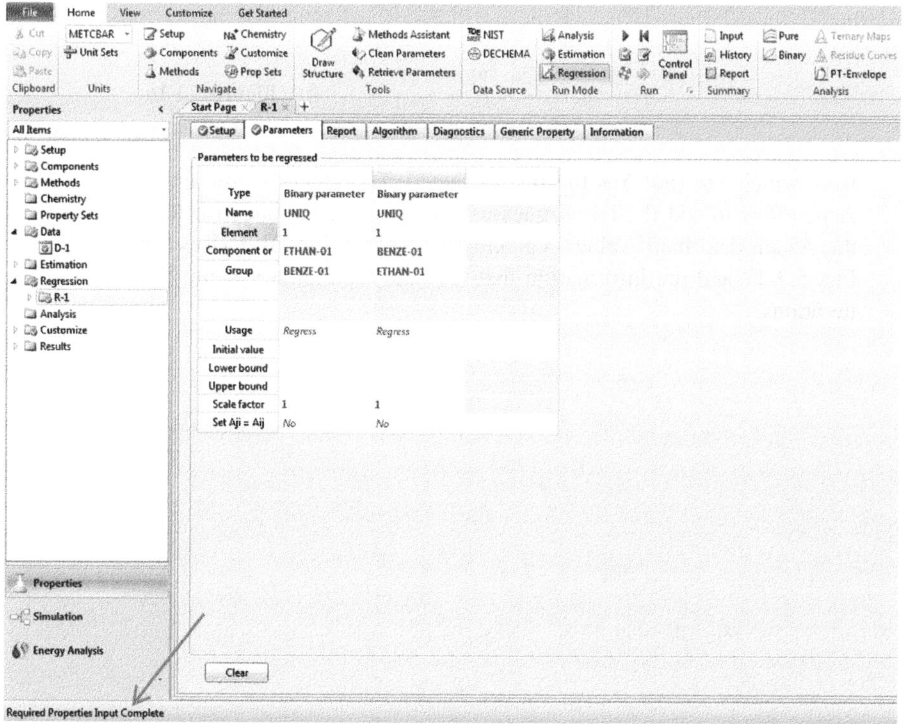

Figure 5.3-14

Note that at the bottom of the window in Fig. 5.3-14 there is a message **Required Properties Input Complete**. Next run the regression by either using the > in **Run** on the main toolbar or pressing **F5**, which brings up the pop-up window of Fig. 5.3-15

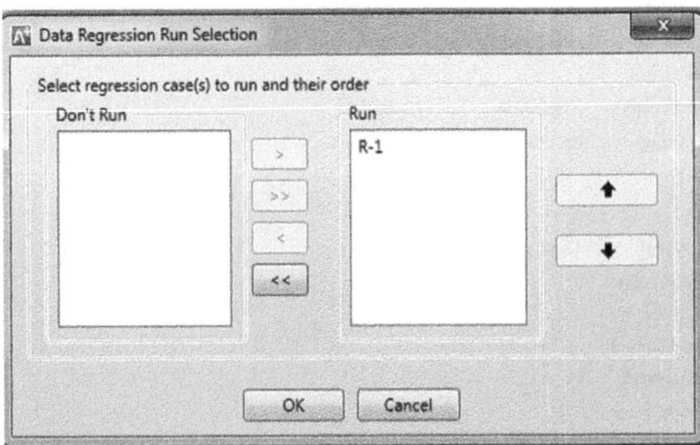

Figure 5.3-15

Click **OK**, which starts the regression. We then obtain the the message in Fig. 5.3-16 for this case,

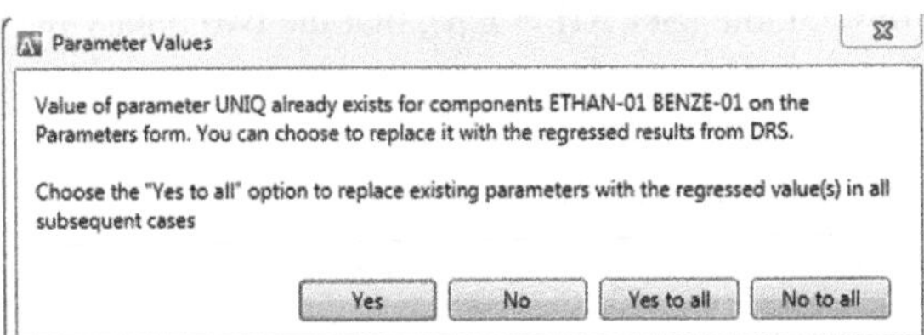

Figure 5.3-16

from which **Yes** (not **Yes to all**) is selected. Then the **Parameters** table is completed by Aspen Plus to add the initial guesses and suggested bounds for the parameters based on the Aspen data bank values. Clicking on **Regression>R-1>Results** gives the results in Fig. 5.3-17 and the information that the data regression system (**DRS**) converged in six iterations.

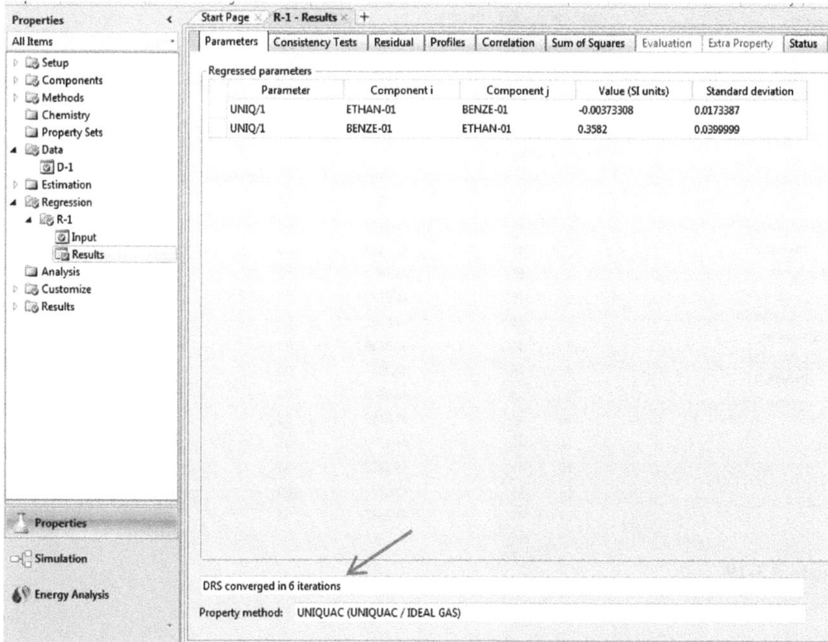

Figure 5.3-17

Note that while the parameters obtained are similar in value to those in the Aspen Plus data bank, they are different, and should provide a better correlation of this particular experimental data set than the default parameters that were obtained from the simultaneous correlation of multiple data sets over a range of temperatures.

Next, clicking on the **Consistency tests** tab produces the window in Fig. 5.3-18 that shows the data passed the thermodynamic area consistency test.

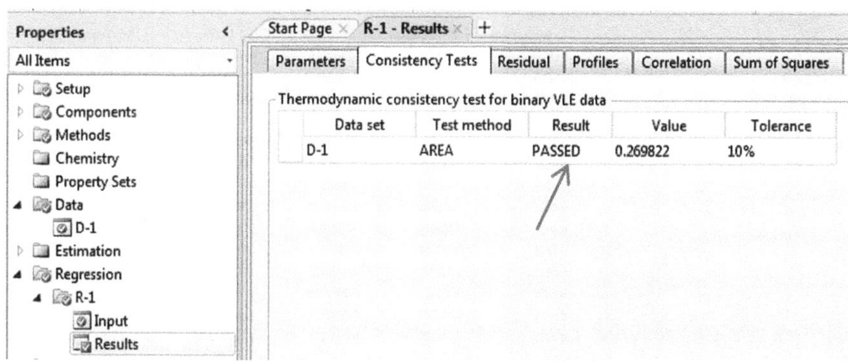

Figure 5.3-18

Clicking on the **Residual** tab we see in the windows of Figs. 5.3-19 to 5.3-24 a collection of windows showing the difference between the measured and correlated temperature, pressure, and vapor and liquid compositions.

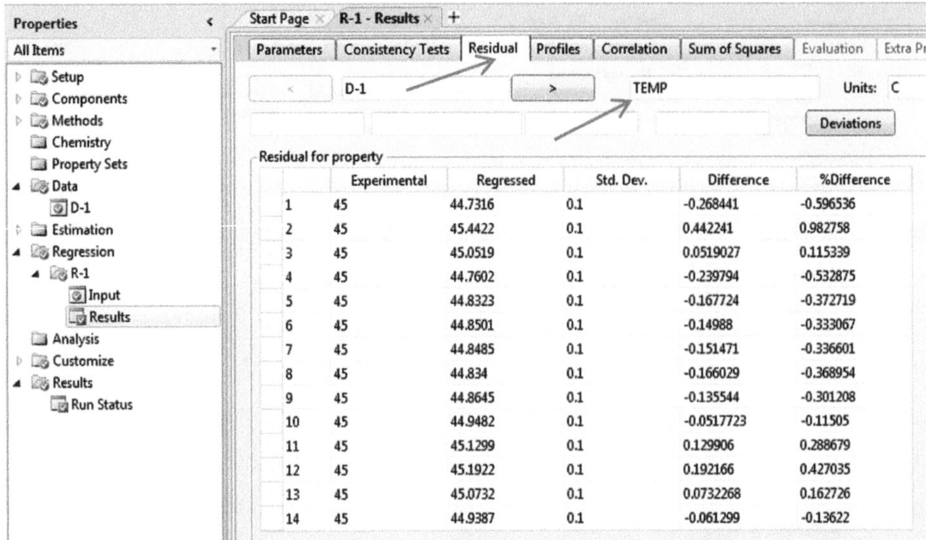

	Experimental	Regressed	Std. Dev.	Difference	%Difference
1	45	44.7316	0.1	-0.268441	-0.596536
2	45	45.4422	0.1	0.442241	0.982758
3	45	45.0519	0.1	0.0519027	0.115339
4	45	44.7602	0.1	-0.239794	-0.532875
5	45	44.8323	0.1	-0.167724	-0.372719
6	45	44.8501	0.1	-0.14988	-0.333067
7	45	44.8485	0.1	-0.151471	-0.336601
8	45	44.834	0.1	-0.166029	-0.368954
9	45	44.8645	0.1	-0.135544	-0.301208
10	45	44.9482	0.1	-0.0517723	-0.11505
11	45	45.1299	0.1	0.129906	0.288679
12	45	45.1922	0.1	0.192166	0.427035
13	45	45.0732	0.1	0.0732268	0.162726
14	45	44.9387	0.1	-0.061299	-0.13622

Figure 5.3-19

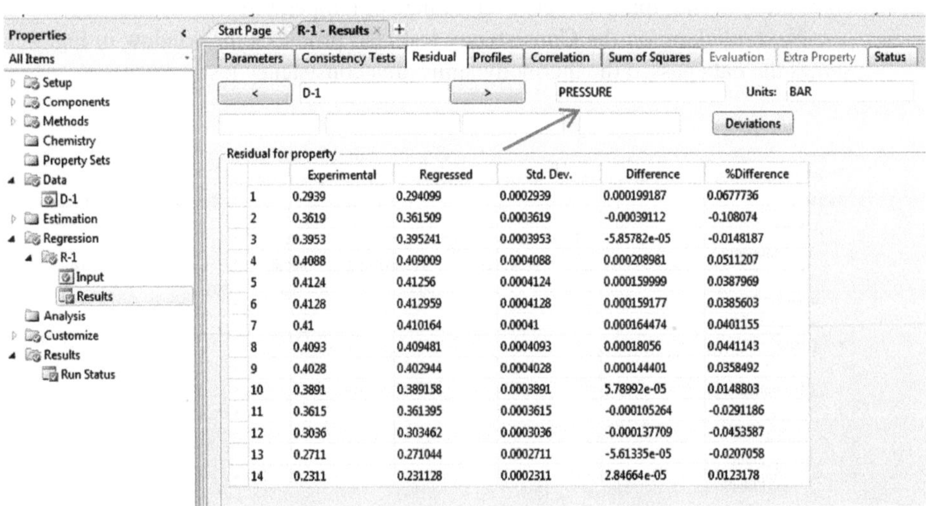

	Experimental	Regressed	Std. Dev.	Difference	%Difference
1	0.2939	0.294099	0.0002939	0.000199187	0.0677736
2	0.3619	0.361509	0.0003619	-0.00039112	-0.108074
3	0.3953	0.395241	0.0003953	-5.85782e-05	-0.0148187
4	0.4088	0.409009	0.0004088	0.000208981	0.0511207
5	0.4124	0.41256	0.0004124	0.000159999	0.0387969
6	0.4128	0.412959	0.0004128	0.000159177	0.0385603
7	0.41	0.410164	0.00041	0.000164474	0.0401155
8	0.4093	0.409481	0.0004093	0.00018056	0.0441143
9	0.4028	0.402944	0.0004028	0.000144401	0.0358492
10	0.3891	0.389158	0.0003891	5.78992e-05	0.0148803
11	0.3615	0.361395	0.0003615	-0.000105264	-0.0291186
12	0.3036	0.303462	0.0003036	-0.000137709	-0.0453587
13	0.2711	0.271044	0.0002711	-5.61335e-05	-0.0207058
14	0.2311	0.231128	0.0002311	2.84664e-05	0.0123178

Figure 5.3-20

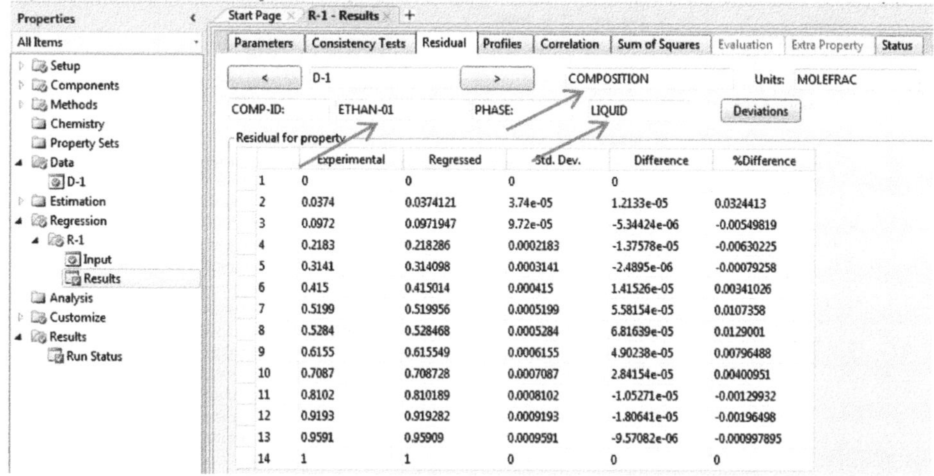

Figure 5.3-21

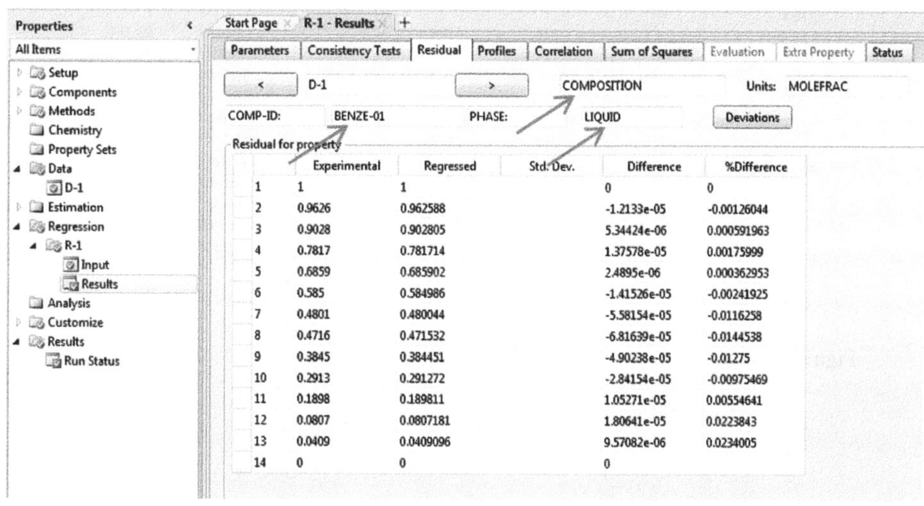

Figure 5.3-22

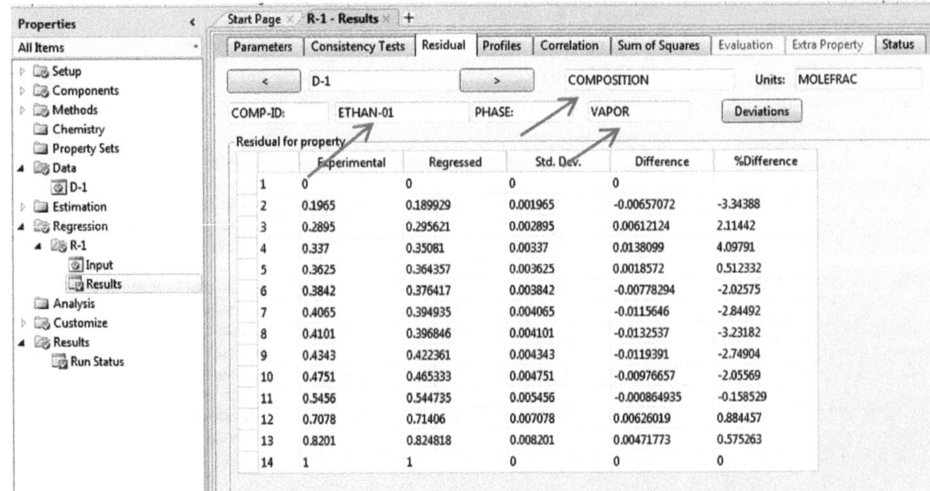

Figure 5.3-23

and finally

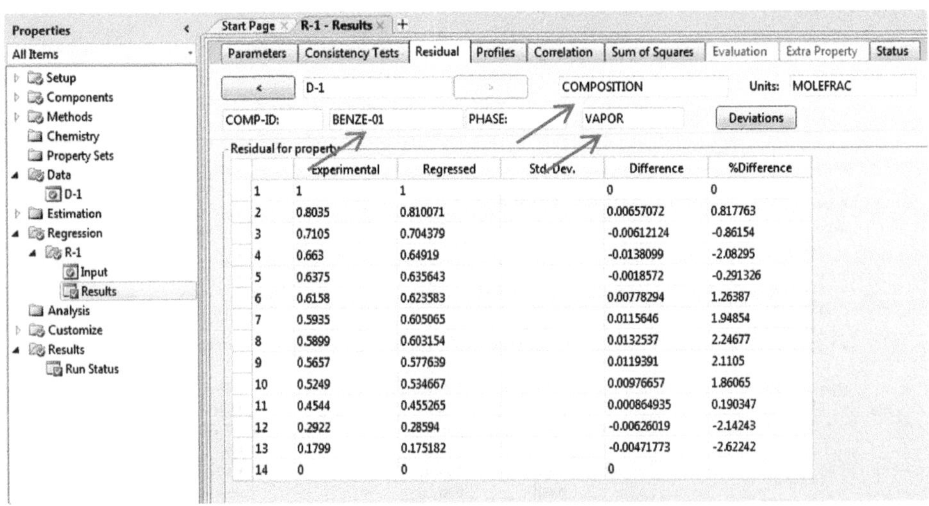

Figure 5.3-24

Clicking on the **Profiles** tab gives the table of results in Fig. 5.3-25

Data set: D-1

Summary of regression results

Exp Val TEMP C	Est Val TEMP C	Exp Val PRES bar	Est Val PRES bar	Exp Val MOLEFRAC X ETHAN-01	Est Val MOLEFRAC X ETHAN-01	Exp Val MOLEFRAC X BENZE-01	Est Val MOLEFRAC X BENZE-01	Exp Val MOLEFRAC Y ETHAN-01	Est Val MOLEFRAC Y ETHAN-01	Exp Val MOLEFRAC Y BENZE-01	Est Val MOLEFRAC Y BENZE-01	Exp Val MOLEFRAC Y-X ETHAN-01	Est Val MOLEFRAC Y-X ETHAN-01	Exp Val MOLEFRAC Y-X BENZE-01	Est Val MOLEFRAC Y-X BENZE-01
45	44.7818	0.2939	0.294099	0	0	1	1	0	0	1	1	0	0	0	0
45	45.4422	0.3819	0.381509	0.0874	0.0874121	0.9126	0.912588	0.1965	0.189929	0.8035	0.810071	0.1091	0.102817	-0.1091	-0.102817
45	45.0519	0.3553	0.356241	0.0972	0.0971946	0.9028	0.902805	0.2895	0.295821	0.7105	0.704179	0.1923	0.198427	-0.1923	-0.198427
45	44.7602	0.4086	0.409009	0.2183	0.218286	0.7817	0.781714	0.337	0.35081	0.663	0.64919	0.1187	0.132524	-0.1187	-0.132524
45	44.8821	0.4124	0.41266	0.3141	0.314098	0.6859	0.685903	0.3625	0.364857	0.6375	0.635643	0.0484	0.0502596	-0.0484	-0.0502596
45	44.8601	0.4128	0.412959	0.415	0.415014	0.585	0.584986	0.3642	0.376417	0.6358	0.623583	-0.0508	-0.038597	0.0508	0.038597
45	44.8485	0.41	0.410165	0.5199	0.519956	0.4801	0.480044	0.4065	0.394935	0.5935	0.605065	-0.1134	-0.12502	0.1134	0.12502
45	44.834	0.4098	0.409481	0.5284	0.528468	0.4716	0.471532	0.4101	0.396846	0.5899	0.603154	-0.1183	-0.131622	0.1183	0.131622
45	44.8645	0.4028	0.402944	0.6155	0.615549	0.3845	0.384451	0.4343	0.422361	0.5657	0.577639	-0.1812	-0.193188	0.1812	0.193188
45	44.9452	0.3891	0.389158	0.7087	0.708728	0.2913	0.291272	0.4751	0.465533	0.5249	0.534467	-0.2336	-0.243195	0.2336	0.243195
45	45.1299	0.3615	0.361335	0.8102	0.81019	0.1898	0.189811	0.5456	0.544735	0.4544	0.455265	-0.2646	-0.265454	0.2646	0.265454
45	45.1922	0.3036	0.303462	0.9193	0.919282	0.0807	0.080718	0.7078	0.71406	0.2922	0.28594	-0.2115	-0.205222	0.2115	0.205222
45	45.0752	0.2711	0.271044	0.9591	0.95909	0.0409	0.0409095	0.8201	0.824818	0.1799	0.175182	-0.139	-0.134272	0.139	0.134272
45	44.9387	0.2311	0.231129	1	1	0	0	1	1	0	0	0	0	0	0

Figure 5.3-25

At each datum point, the best fit was obtained by Aspen Plus by slightly adjusting the temperature, generally within the standard deviation. So even though the specification was that temperature was fixed at 45°C, to obtain the best fit in the correlation of each data point the temperature was allowed to vary slightly.

Now clicking **P-xy** in the **Plot** region of the main toolbar (Fig. 5.3-26) produces the graph in Fig. 5.3-27 of the experimental data (points) and the correlated results (lines)

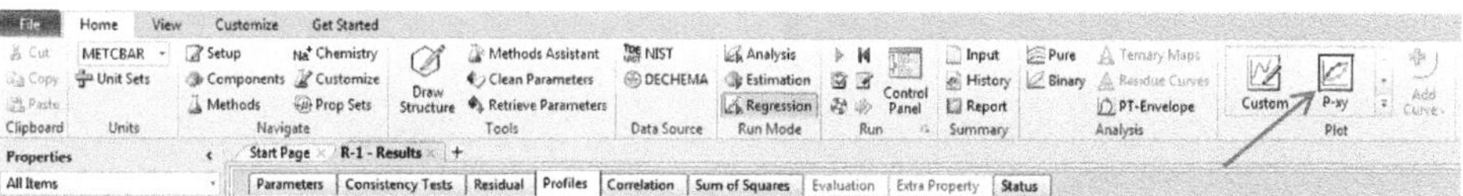

Figure 5.3-26

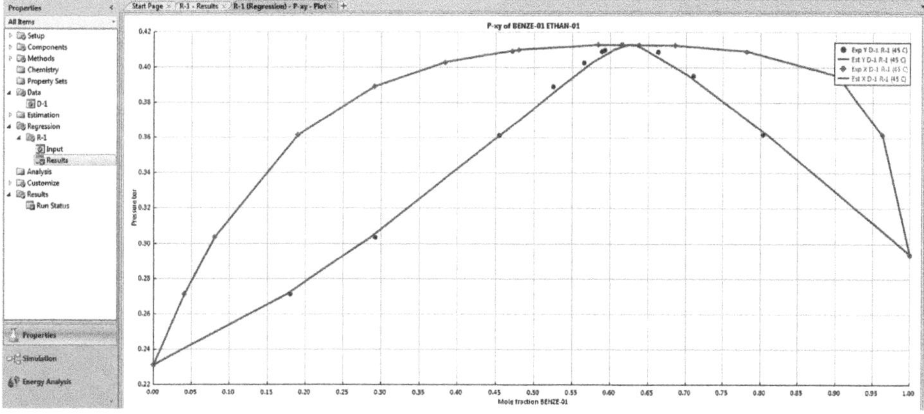

Figure 5.3-27

Having entered the experimental data, it is now relatively simple to correlate the data with other activity coefficient models. For example, the screenshots in Figs. 5.3-28 to 5.3-32 show only the windows that need to be changed to correlate these data with the Wilson equation and the results.

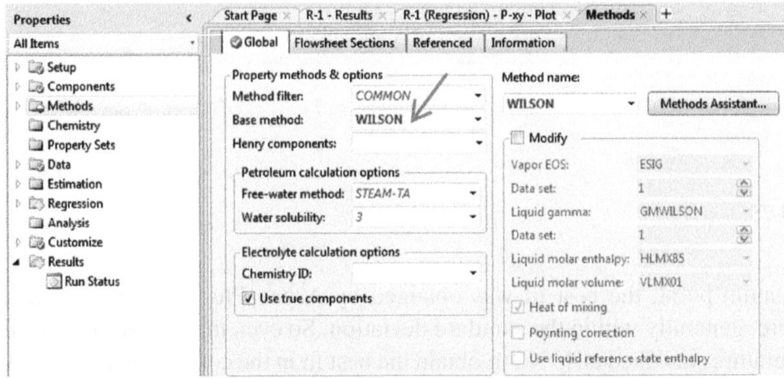

Figure 5.3-28

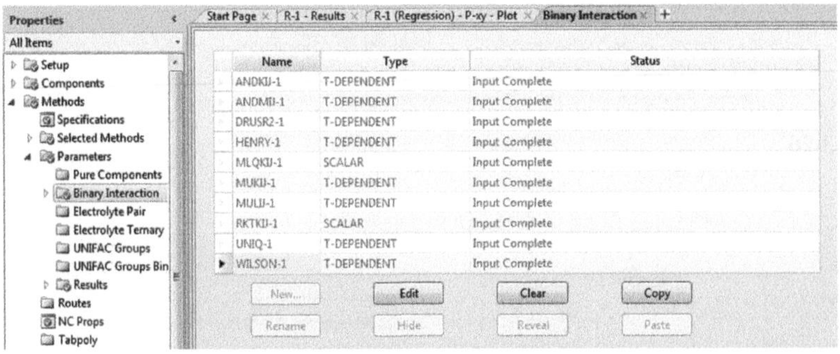

Figure 5.3-29

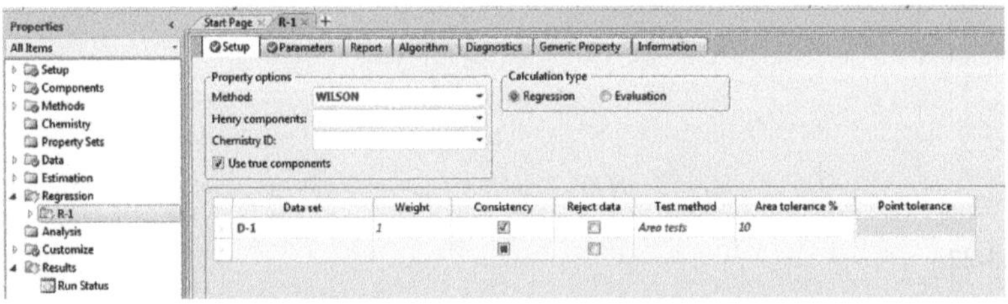

Figure 5.3-30

Properties | All Items
- Setup
- Components
- Methods
- Chemistry
- Property Sets
- Data
- Estimation
- Regression
  - R-1
- Analysis
- Customize
- Results
  - Run Status

Start Page | R-1 | +

Setup | Parameters | Report | Algorithm | Diagnostics | Generic Property | Information

Parameters to be regressed

Type	Binary parameter	Binary parameter	Binary parameter	Binary parameter
Name	WILSON	WILSON	WILSON	WILSON
Element	1	1	2	2
Component or	ETHAN-01	BENZE-01	ETHAN-01	BENZE-01
Group	BENZE-01	ETHAN-01	BENZE-01	ETHAN-01
Usage	Fix	Fix	Regress	Regress
Initial value	1.0936	0.0106	-958.013	-232.531
Lower bound				
Upper bound				
Scale factor			1	1
Set Aji = Aij	No	No	No	No

Figure 5.3-31

Note that the Aspen Plus default is that the first two parameters should be fixed, and the remaining two are to be regressed. We will change that to regress all four parameters (Fig. 5.3-32),

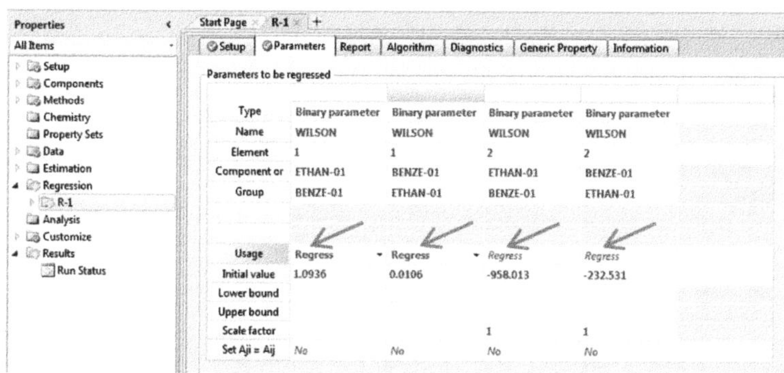

Type	Binary parameter	Binary parameter	Binary parameter	Binary parameter
Name	WILSON	WILSON	WILSON	WILSON
Element	1	1	2	2
Component or	ETHAN-01	BENZE-01	ETHAN-01	BENZE-01
Group	BENZE-01	ETHAN-01	BENZE-01	ETHAN-01
Usage	Regress	Regress	Regress	Regress
Initial value	1.0936	0.0106	-958.013	-232.531
Lower bound				
Upper bound				
Scale factor			1	1
Set Aji = Aij	No	No	No	No

Figure 5.3-32

which produces the results in Figs. 5.3-33 to 5.3-36.

Properties | All Items
- Setup
- Components
- Methods
- Chemistry
- Property Sets
- Data
- Estimation
- Regression
  - R-1
- Analysis
- Customize
- Results
  - Run Status

Start Page | R-1 | +

Setup | Parameters | Report | Algorithm | Diagnostics | Generic Property | Information

Parameters to be regressed

Type	Binary parameter	Binary parameter	Binary parameter	Binary parameter
Name	WILSON	WILSON	WILSON	WILSON
Element	1	1	2	2
Component or	ETHAN-01	BENZE-01	ETHAN-01	BENZE-01
Group	BENZE-01	ETHAN-01	BENZE-01	ETHAN-01
Usage	Regress	Regress	Regress	Regress
Initial value	1.0936	0.0106	-958.013	-232.531
Lower bound	-9.8424	-0.0954	-10538.1	-2557.84
Upper bound	12.0296	0.1166	8622.12	2092.78
Scale factor	1	1	1	1
Set Aji = Aij	No	No	No	No

Figure 5.3-33

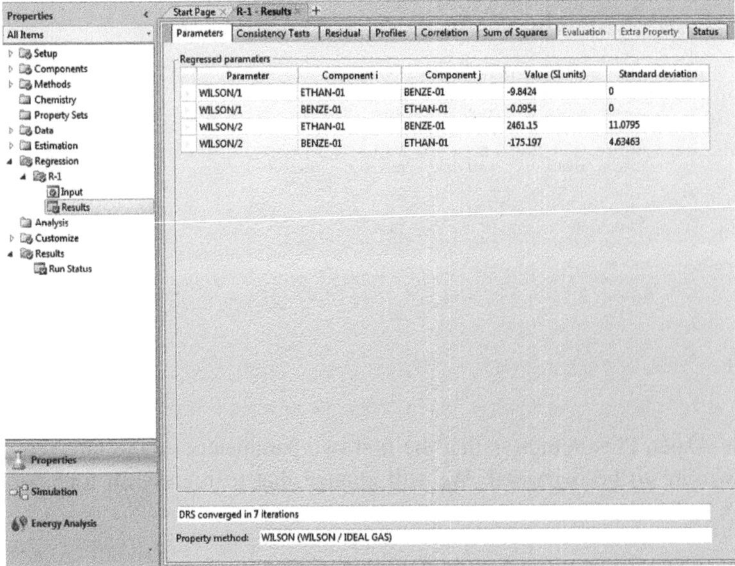

Figure 5.3-34

Data set: D-1

Summary of regression results

Exp Val TEMP	Est Val TEMP	Exp Val PRES	Est Val PRES	Exp Val MOLEFRAC X ETHAN-01	Est Val MOLEFRAC X ETHAN-01	Exp Val MOLEFRAC X BENZE-01	Est Val MOLEFRAC X BENZE-01	Exp Val MOLEFRAC Y ETHAN-01	Est Val MOLEFRAC Y ETHAN-01	Exp Val MOLEFRAC Y BENZE-01	Est Val MOLEFRAC Y BENZE-01	Exp Val MOLEFRAC Y-X ETHAN-01	Est Val MOLEFRAC Y-X ETHAN-01	Exp Val MOLEFRAC Y-X BENZE-01	Est Val MOLEFRAC Y-X BENZE-01
C	C	bar	bar												
45	44.7316	0.2919	0.294099	0	0	1	1	0	0	1	1	0	0	0	0
45	45.0998	0.3629	0.361797	0.0374	0.0373984	0.9626	0.962602	0.1965	0.199274	0.8035	0.800726	0.1591	0.161876	-0.1591	-0.161876
45	45.1259	0.3959	0.395197	0.0972	0.097204	0.9028	0.902796	0.2895	0.287074	0.7105	0.712926	0.1923	0.18987	-0.1923	-0.18987
45	44.9621	0.4069	0.408814	0.2183	0.218298	0.7817	0.781702	0.337	0.338409	0.663	0.661591	0.1187	0.12011	-0.1187	-0.12011
45	45.005	0.4124	0.4124	0.3141	0.314105	0.6859	0.685895	0.3605	0.35913	0.6395	0.64087	0.0464	0.045025	-0.0464	-0.045025
45	45.018	0.4128	0.41279	0.415	0.415012	0.585	0.584988	0.3842	0.379101	0.6158	0.620899	-0.0308	-0.035911	0.0308	0.035911
45	45.0427	0.41	0.409964	0.5199	0.51991	0.4801	0.48009	0.4065	0.403436	0.5935	0.596564	-0.1134	-0.116474	0.1134	0.116474
45	45.0304	0.4093	0.408277	0.5284	0.528418	0.4716	0.471582	0.4101	0.405684	0.5899	0.594316	-0.1183	-0.122734	0.1183	0.122734
45	45.0581	0.4026	0.402749	0.6155	0.615499	0.3845	0.384501	0.4343	0.432998	0.5657	0.567002	-0.1812	-0.182501	0.1812	0.182501
45	45.0746	0.3891	0.389036	0.7087	0.708694	0.2913	0.291306	0.4751	0.474507	0.5249	0.525493	-0.2336	-0.234187	0.2336	0.234187
45	45.0906	0.3615	0.361426	0.8102	0.81019	0.1898	0.18981	0.5456	0.547802	0.4544	0.452198	-0.2646	-0.262388	0.2646	0.262388
45	44.969	0.3036	0.303621	0.9193	0.919102	0.0807	0.0808978	0.7078	0.707129	0.2922	0.292871	-0.2115	-0.211973	0.2115	0.211973
45	44.8745	0.2711	0.271176	0.9591	0.959107	0.0409	0.0408926	0.8201	0.816735	0.1799	0.183265	-0.139	-0.142372	0.139	0.142372
45	44.9387	0.2311	0.231129	1	1	0	0	1	1	0	0	0	0	0	0

Figure 5.3-35

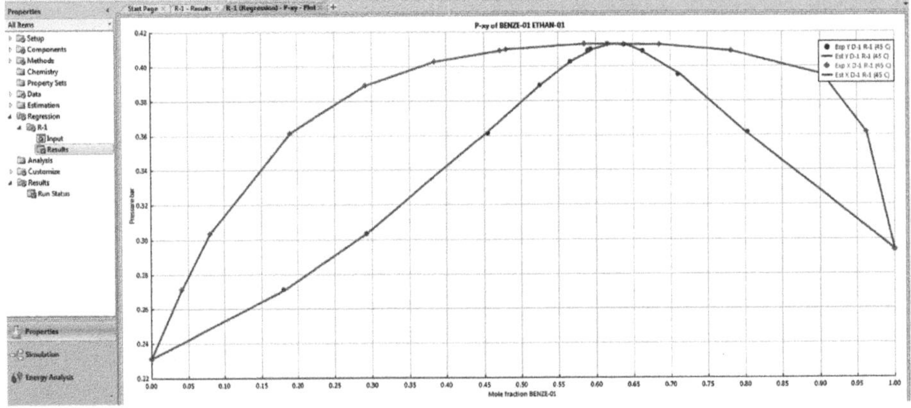

Figure 5.3-36

We see from the table (Fig. 5.3-35) and the graph (Fig. 5.3-36) that the Wilson model also fits the experimental data very well. Indeed, the fit is as good as, and perhaps even better than, with the more complicated UNIQUAC equation.

Repeating the regression, though this time with the NRTL activity coefficient model (Fig. 5.3-37) produces results in Figs. 5.3-38 to 5.3-41.

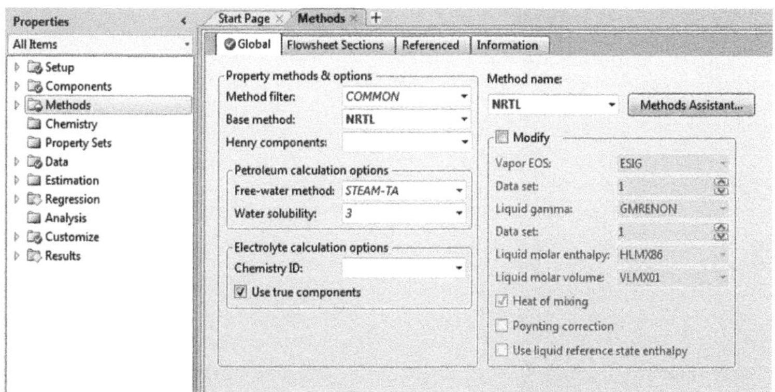

Figure 5.3-37

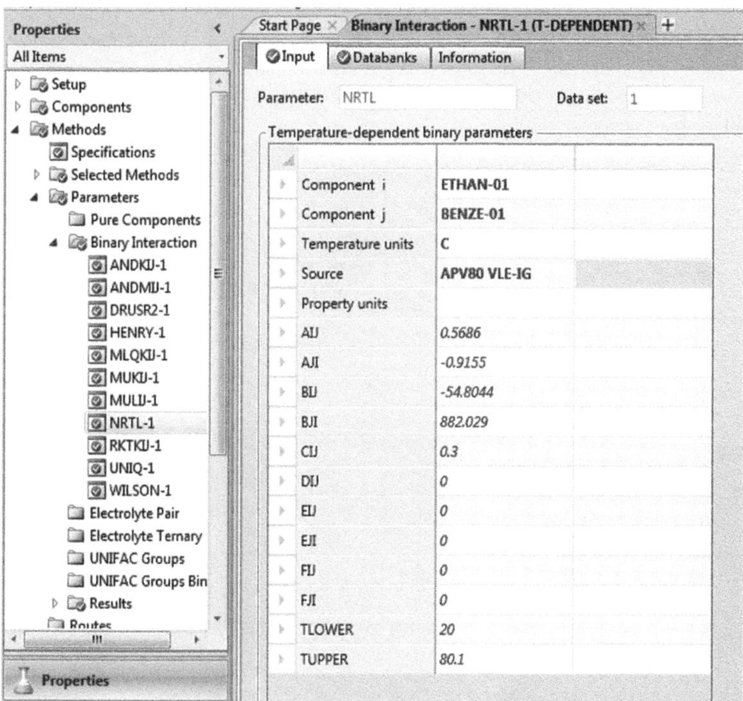

Figure 5.3-38

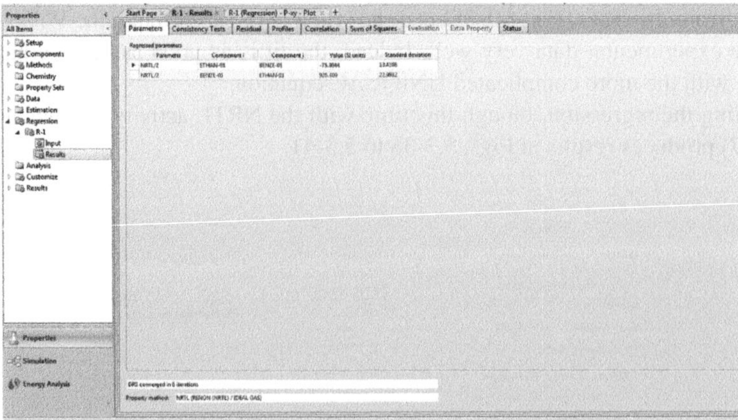

Figure 5.3-39

Figure 5.3-40

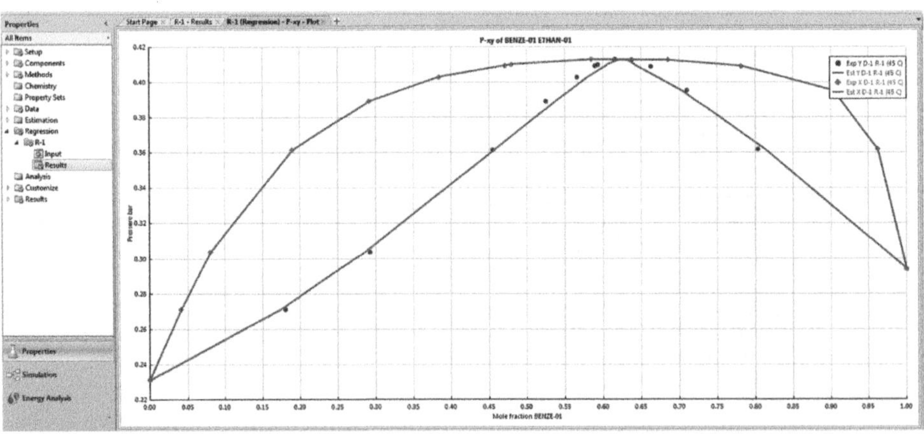

Figure 5.3-41

So that the NRTL model also correlates this data set quite well.

The example so far considered was one in which PTxy data were available. That is, we had experimental data for the temperature, pressure, liquid composition (x), and vapor composition (y). However, there are many cases in the literature in which a static cell or ebulliometer have been used (see *Chemical, Biochemical and Engineering Thermodynamics*, 4th ed., S. I. Sandler, John Wiley & Sons, Inc., 2006, pp. 539–542) for the thermodynamic measurements. In such cases only PTx data, that is, only temperature, pressure, and liquid phase composition have been reported. The regression of such data has to be treated slightly differently than PTxy data. As an example, we will use the data of Hovorka et al., *J. Am. Chem. Soc.*, **58**, 2264 (1935) for the system water (1) and 1,4-dioxane at 353.15 K.

P(mm Hg)	x_1	P(mm Hg)	x_1
382.8	0.000	575.5	0.600
476.0	0.100	569.5	0.700
526.5	0.200	550.0	0.800
556.5	0.300	501.5	0.900
571.0	0.400	355.1	1.000
576.5	0.500		

The procedure is to start Aspen Plus, go to **Components>Specifications**, and add water and 1,4-dioxane in the usual manner. In the **Methods>Specifications** choose some appropriate thermodynamic model; the **NRTL** model will be used here (Fig. 5.3-42).

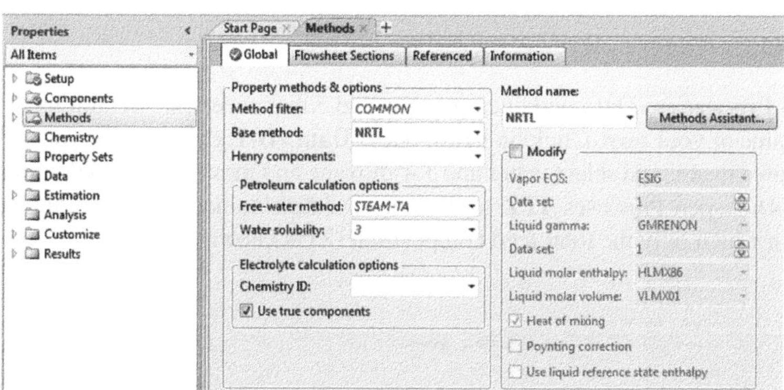

Figure 5.3-42

Next, go to **Parameters>Binary Interaction** (Fig. 5.3-43) and look at NRTL-1 to see the stored parameter values.

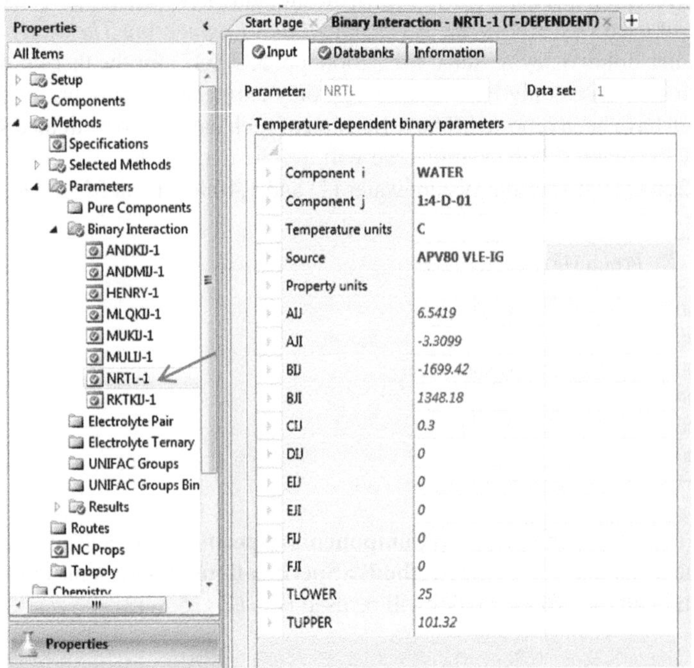

Figure 5.3-43

Then go to **Properties>Data** and click on **New** and either accept the default name **D-1** or choose a name of your own. Click on **Properties>Data>D-1**, choose data type **TPZ** from the drop-down menu, and select water and 1,4-dioxane and move to **Selected components** (Fig. 5.3-44). Aspen Plus uses **TPZ** to designated data obtained using ebulliometry or a static cell in which only the total feed composition (Z) is known.

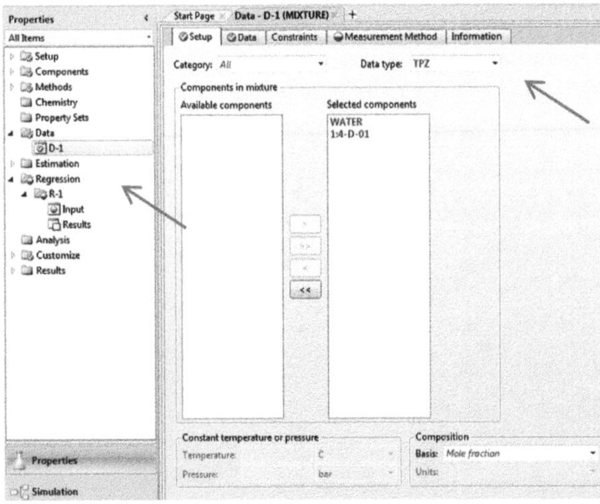

Figure 5.3-44

Next click on the **Data** tab, choose Kelvin as the temperature unit, mm of mercury as the pressure unit, and enter data as shown in Fig. 5.3-45. Note that the Aspen Plus regression program in the **TPZ** option will give an error message if data at the mole fractions of 0 or 1 are included. Therefore, use 0.000001 and 0.999999 for the pure component data points. Pure component limits are included, NOT including the two pure component limits that may cause a problem in the Aspen Plus regression program when using the **TPZ** option.

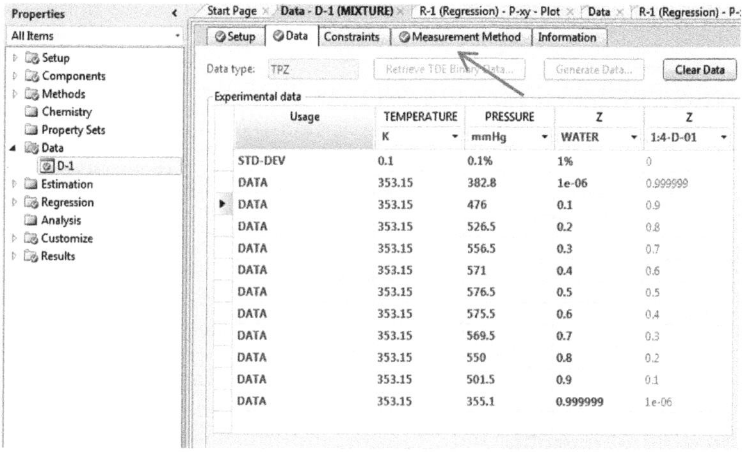

Usage	TEMPERATURE K	PRESSURE mmHg	Z WATER	Z 1:4-D-01
STD-DEV	0.1	0.1%	1%	0
DATA	353.15	382.8	1e-06	0.999999
DATA	353.15	476	0.1	0.9
DATA	353.15	526.5	0.2	0.8
DATA	353.15	556.5	0.3	0.7
DATA	353.15	571	0.4	0.6
DATA	353.15	576.5	0.5	0.5
DATA	353.15	575.5	0.6	0.4
DATA	353.15	569.5	0.7	0.3
DATA	353.15	550	0.8	0.2
DATA	353.15	501.5	0.9	0.1
DATA	353.15	355.1	0.999999	1e-06

Figure 5.3-45

Next click on the **Measurement Method** tab, where you are asked for the **Static-cell constant**, which is not known for this data set. I have used a very small number as seen in Fig. 5.3-46.

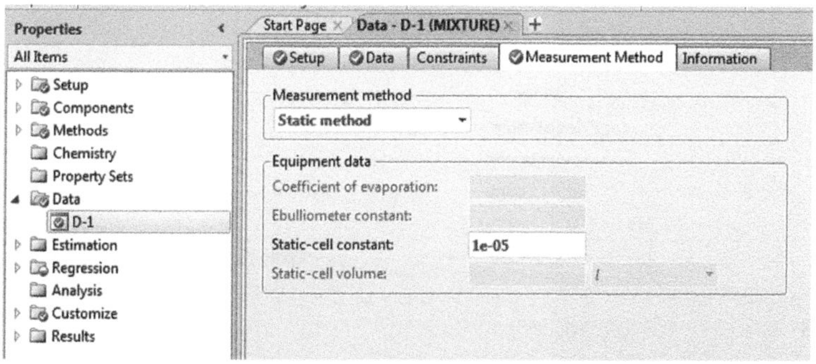

Figure 5.3-46

We are now ready to start the regression by clicking on **Regression**, selecting **New**, and selecting the default name of **R-1** or choosing a name of your own. Click on **R-1 Input**, and choose the **NRTL** model, the **Data set D-1** and do not check the **Consistency> Perform test** box (since a consistency test can only be performed on PTxy data) as shown in Fig. 5.3-47.

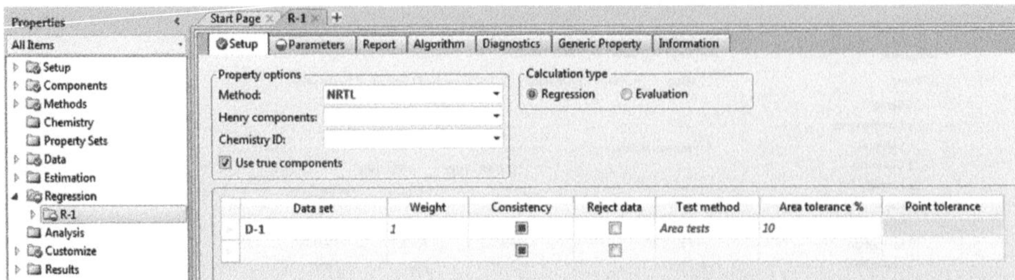

Figure 5.3-47

Next click on the **Parameters** tab in **R-1** and complete as shown in Fig. 5.3-48.

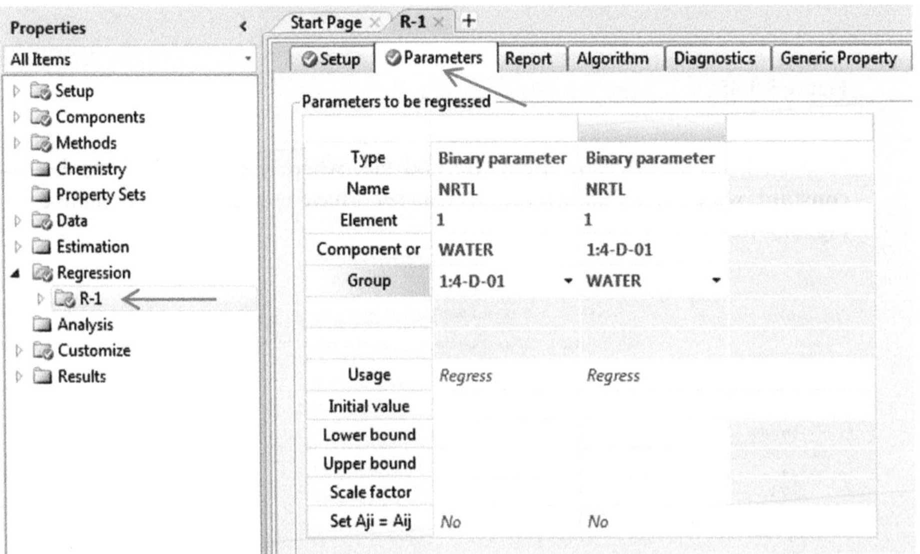

Figure 5.3-48

The **Input** is now complete, run the regression **(F5 or** >), then choose **R-1** and **OK**. The parameter values of a successful run are then found in **Properties>Regression>R-1>Results>Parameters** shown in Fig. 5.3-49. Note that since the standard deviations in the parameter values are much less than the parameter values, these can be taken as meaningful.

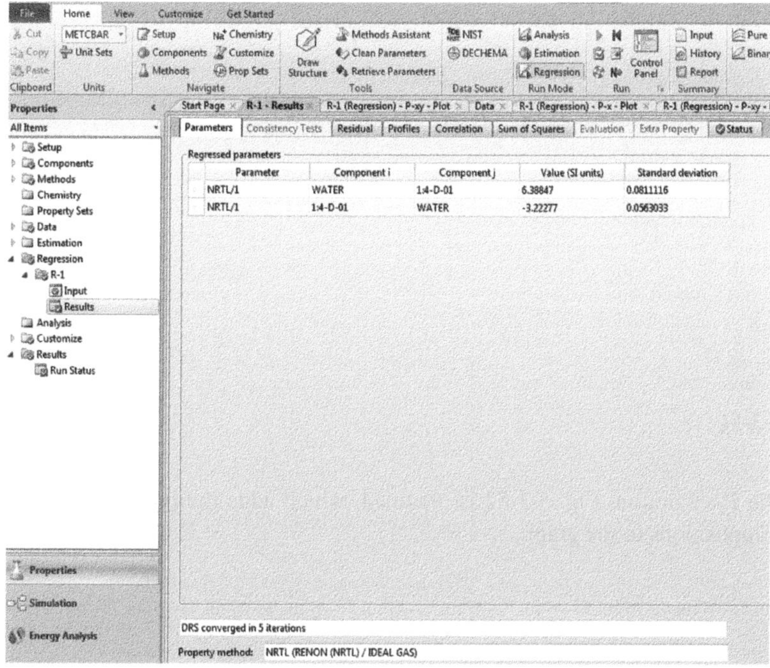

Figure 5.3-49

Then clicking on the **Profile** tabs, we obtain the following results in Fig. 5.3-50.

Data set: D-1

Summary of regression results

Exp Val TEMP	Est Val TEMP	Exp Val PRES	Est Val PRES	Est Val MOLEFRAC X WATER	Est Val MOLEFRAC X 1:4-D-01	Est Val MOLEFRAC Y WATER	Est Val MOLEFRAC Y 1:4-D-01	Exp Val MOLEFRAC Z WATER	Est Val MOLEFRAC Z WATER	Exp Val MOLEFRAC Z 1:4-D-01	Est Val MOLEFRAC Z 1:4-D-01	Est Val MOLEFRAC Y WATER
C	C	bar	bar									
80	80.1693	0.510358	0.510108	9.95145e-07	0.999999	4.50407e-06	0.999996	1e-06	1e-06	0.999999	0.999999	3.50792e-06
80	79.7949	0.634615	0.634985	0.0991347	0.900865	0.276092	0.723908	0.1	0.0991347	0.9	0.900865	0.176957
80	79.745	0.701942	0.70244	0.198089	0.801911	0.394285	0.605715	0.2	0.198089	0.8	0.801911	0.196196
80	79.9665	0.741939	0.742007	0.299765	0.700235	0.457263	0.542737	0.3	0.299765	0.7	0.700235	0.157498
80	80.1624	0.781271	0.780937	0.400664	0.599336	0.492282	0.507718	0.4	0.400664	0.6	0.599336	0.091618
80	80.2898	0.768604	0.768005	0.500155	0.499845	0.512692	0.487308	0.5	0.500155	0.5	0.499845	0.0125365
80	80.2851	0.76727	0.766684	0.599548	0.400452	0.526274	0.473726	0.6	0.599548	0.4	0.400452	-0.0732738
80	80.1945	0.759271	0.738878	0.699541	0.300459	0.540368	0.459632	0.7	0.699464	0.3	0.300536	-0.159173
80	79.904	0.733273	0.733458	0.800408	0.199592	0.567877	0.432123	0.8	0.800408	0.2	0.199592	-0.232532
80	79.7301	0.668612	0.669071	0.901046	0.0989537	0.647735	0.352265	0.9	0.901046	0.1	0.0989537	-0.253511
80	79.9872	0.473428	0.473443	0.999999	9.76585e-07	0.999992	8.31793e-06	0.999999	0.999999	1e-06	1e-06	-7.3413e-06

Figure 5.3-50

Clicking on the **Plot P-x** icon on the **main tool bar**, the graph in Fig. 5.3-51 is obtained.

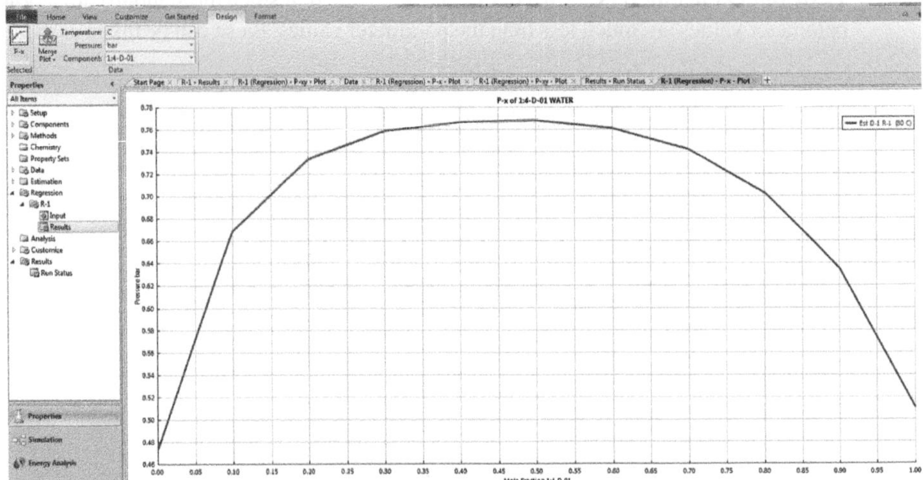

Figure 5.3-51

Using the **P-xy** option, Fig. 5.3-52 is obtained, which adds the predicted values for the vapor compositions to the graph.

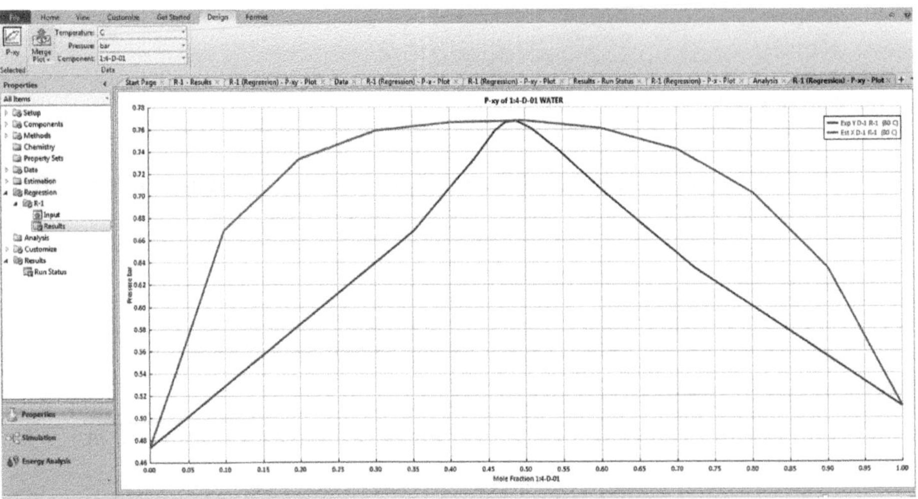

Figure 5.3-52

So we see that the water + 1,4 dioxane system at 353.15 K is predicted to have the equilibrium vapor compositions shown. The results also show that this system has an azeotrope. But you could have determined that only from the P-x data at constant temperature. How?

As the final item of this chapter, it is useful to point out that the **NIST TDE** interfaces directly with the regression module. As an illustration, consider the ethanol–water mixture. Entering those components, then going to **Data>New**, accepting the default name **D-1** and in **Select Type** choosing **MIXTURE** brings up the window in Fig. 5.3-53.

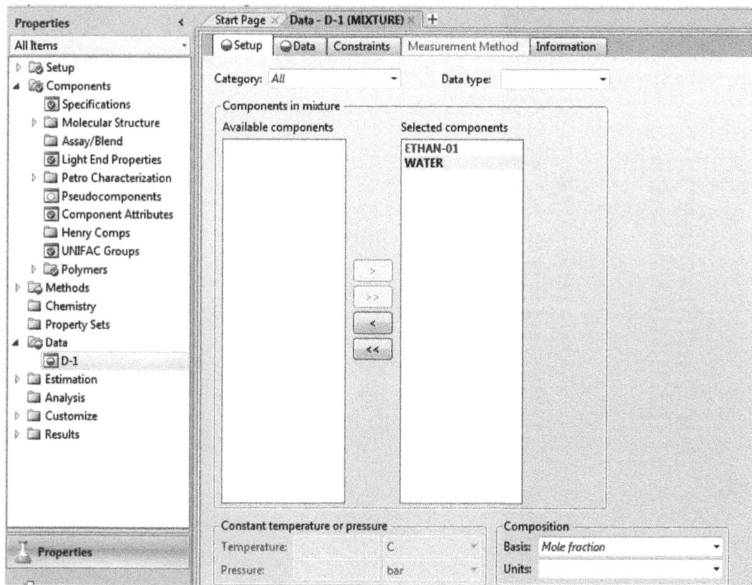

Figure 5.3-53

Then going to the **Data** tab, select **Retrieve TDE Binary Data** as shown in Fig. 5.3-54.

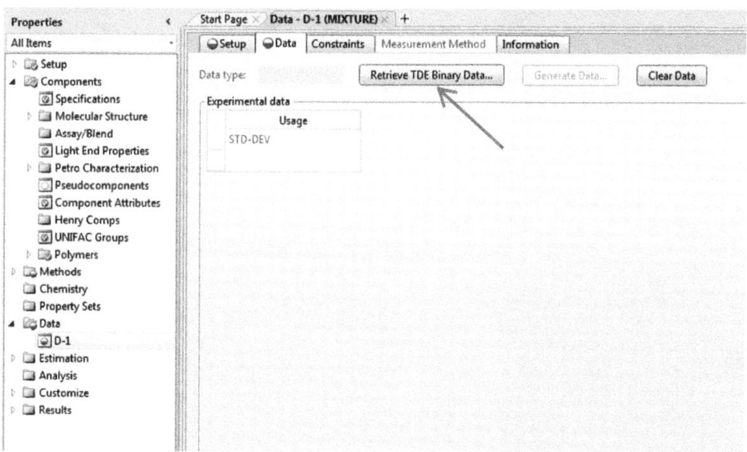

Figure 5.3-54

This brings up the **NIST TDE** with a list of much binary data of all sorts for the ethanol–water system, only some of which are shown in Fig. 5.3-55. Selecting **Binary VLE 076** and then clicking on that item results in the list of data in Fig. 5.3-56. [Note that the data set here designated as **Binary VLE 076** may have a different label in later versions of Aspen Plus as additional data is added to the data bank.]

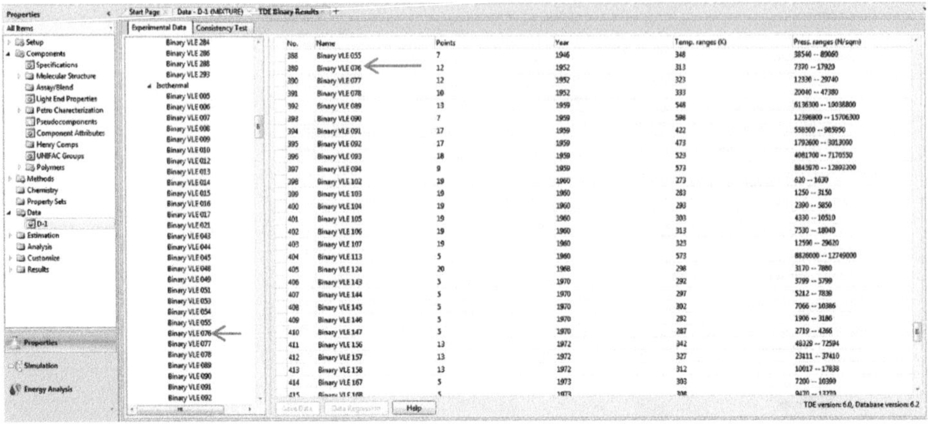

Figure 5.3-55

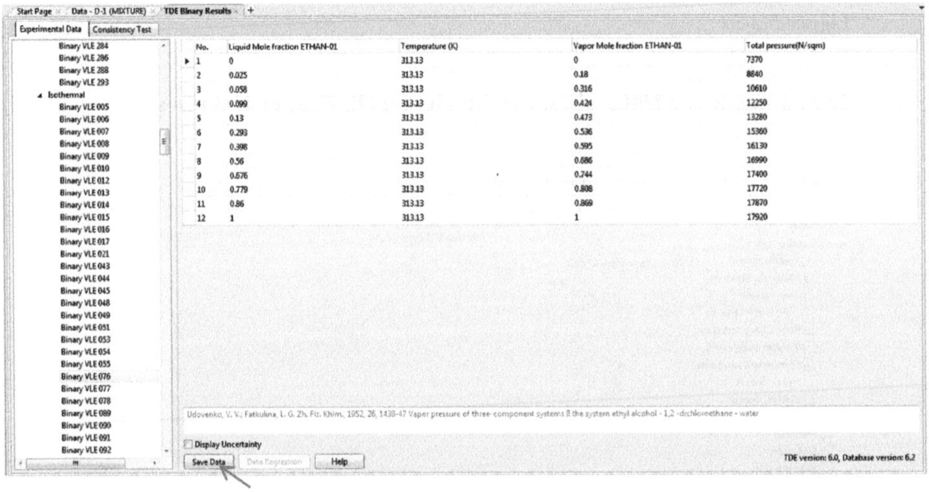

No.	Liquid Mole fraction ETHAN-01	Temperature (K)	Vapor Mole fraction ETHAN-01	Total pressure(N/sqm)
1	0	313.13	0	7370
2	0.025	313.13	0.18	8840
3	0.058	313.13	0.316	10610
4	0.099	313.13	0.424	12250
5	0.13	313.13	0.473	13280
6	0.293	313.13	0.536	15360
7	0.398	313.13	0.595	16130
8	0.56	313.13	0.686	16990
9	0.676	313.13	0.744	17400
10	0.779	313.13	0.808	17720
11	0.86	313.13	0.869	17870
12	1	313.13	1	17920

Figure 5.3-56

Note the item **Save Data** at the bottom of the window in Fig. 5.3-56. Clicking on that produces the result shown in Fig. 5.3-57.

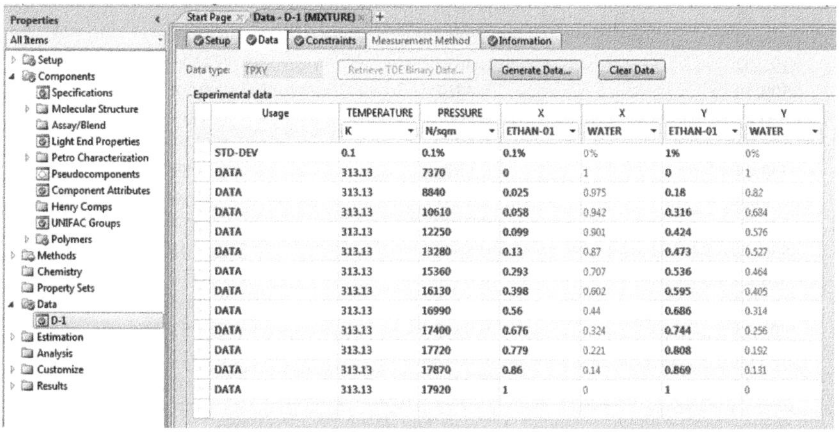

Usage	TEMPERATURE	PRESSURE	X	X	Y	Y
	K	N/sqm	ETHAN-01	WATER	ETHAN-01	WATER
STD-DEV	0.1	0.1%	0.1%	0%	1%	0%
DATA	313.13	7370	0	1	0	1
DATA	313.13	8840	0.025	0.975	0.18	0.82
DATA	313.13	10610	0.058	0.942	0.316	0.684
DATA	313.13	12250	0.099	0.901	0.424	0.576
DATA	313.13	13280	0.13	0.87	0.473	0.527
DATA	313.13	15360	0.293	0.707	0.536	0.464
DATA	313.13	16130	0.398	0.602	0.595	0.405
DATA	313.13	16990	0.56	0.44	0.686	0.314
DATA	313.13	17400	0.676	0.324	0.744	0.256
DATA	313.13	17720	0.779	0.221	0.808	0.192
DATA	313.13	17870	0.86	0.14	0.869	0.131
DATA	313.13	17920	1	0	1	0

Figure 5.3-57

As the data is now loaded as the data set for the regression, one can proceed directly as described earlier to regress model parameters after first going to **Methods** and choosing the model to be used.

PROBLEMS

5.1. Vapor–liquid equilibrium data for the ethanol + benzene system at 45°C were given in a table at the beginning of Section 5.3.

Correlate these data with the following activity coefficient models:

(a) the NRTL model;

(b) the Wilson model; and

(c) the UNIQUAC model.

5.2. Use the UNIFAC model to predict the vapor–liquid equilibria for the system of the previous problem, and compare the predictions with the reported experimental data.

5.3. Fit the vapor–liquid equilibrium data in the text for the water + 1,4 dioxane system with the following activity coefficient models:

(a) the NRTL model;

(b) the Wilson model; and

(c) the UNIQUAC model.

5.4. Use the UNIFAC model to predict the vapor–liquid equilibria for the system of the previous problem, and compare the predictions with the reported experimental data.

5.5. The following vapor–liquid equilibrium data have been reported for the system 1,2-dichloroethane (1) + *n*-heptane (2) at 343.15 K.

Fit these vapor–liquid equilibrium data with the following activity coefficient models:

(a) the NRTL model;

(b) the Wilson model; and

(c) the UNIQUAC model.

P (mm Hg)	x_1	y_1
302.87	0.0000	0.0000
372.62	0.0911	0.2485
429.28	0.1979	0.4174
466.48	0.2867	0.5052
491.02	0.3674	0.5590
509.40	0.4467	0.6078
520.41	0.5044	0.6535
525.71	0.5733	0.6646
533.44	0.6578	0.7089
536.58	0.7644	0.7696
535.78	0.8132	0.7877
530.90	0.8603	0.8201
524.21	0.8930	0.8458
515.15	0.9332	0.8900
505.15	0.9572	0.9253
498.04	0.9812	0.9537
486.41	1.0000	1.0000

Source: R. Eng and S. I. Sandler, *J. Chem. Eng. Data*, **29**, 156 (1984).

5.6. Use the UNIFAC model to predict the vapor–liquid equilibria for the system of the previous problem, and compare the predictions with the reported experimental data.

5.7. Use the NIST TDE to obtain vapor–liquid equilibrium data as a function of temperature for a system of your choice, and correlate data with:

(a) the NRTL equation;

(b) the Wilson model; and

(c) the UNIQUAC equation.

5.8. Use the UNIFAC model to predict the vapor–liquid equilibria for the system of the previous problem, and compare the predictions with the reported experimental data.

5.9. The following P-xy data are available for the carbon tetrachloride + n-heptane system at 50°C. Regress these data with an appropriate thermodynamic model, for example, NRTL, Wilson, or UNIQUAC, and predict the vapor compositions at each of the data points. Also, obtain values for the activity coefficients of each component at each data point.

Mole% of CCl_4	
CCl_4 in the liquid	Pressure (bar)
0.0	0.1873
3.32	0.1956
9.83	0.2131
17.14	0.2320
30.24	0.2649
35.14	0.2765
43.24	0.2943
50.12	0.3097
57.00	0.3263
64.96	0.3425
73.23	0.3616
81.26	0.3765
89.92	0.3939
96.49	0.4055
100.0	0.4113

Source: C. P. Smith and E. W. Engel, *J. Am. Chem. Soc.*, **51**, 2646 (1929).

5.10. Use the UNIFAC model to predict the vapor–liquid equilibria for the system of the previous problem, and compare the predictions with the reported experimental data.

5.11. For a separations process, it is necessary to determine the vapor–liquid equilibrium compositions for a mixture of ethyl bromide and n-heptane at 30°C. Calculate the vapor composition in equilibrium with a liquid containing 47.23 mol% ethyl bromide using

(a) the Wilson model;

(b) the NRTL model; and

(c) the UNIFAC model.

5.12. Benzene and ethanol form azeotropic mixtures. Consequently, benzene is sometimes added to solvent grades of ethanol to prevent industrious chemical engineering students from purifying solvent-grade ethanol by distillation for use at an after-finals party. Prepare an x-y and P-x diagram for the benzene–ethanol system at 45°C assuming

(a) the mixture is described by the NRTL model, and

(b) the mixture is described by the UNIFAC model.

Compare the results obtained with the experimental data in the following table.

x_{EA}	y_{EA}	P(bar)
0	0	0.2939
0.0374	0.1965	0.3613
0.0972	0.2895	0.3953
0.2183	0.3370	0.4088
0.3141	0.3625	0.4124
0.4150	0.3842	0.4128
0.5199	0.4065	0.4100
0.5284	0.4101	0.4093
0.6155	0.4343	0.4028
0.7087	0.4751	0.3891
0.8102	0.5456	0.3615
0.9193	0.7078	0.3036
0.9591	0.8201	0.2711
1.00	1.00	0.2321

Source: Data from I. Brown and F. Smith, *Aust. J. Chem.*, **7**, 264 (1954).

5.13. Use the UNIFAC model to predict the vapor–liquid equilibria for the acetone + water system at 25°C, and compare the results with the experimental data that you can find in the NIST TDE (Chapter 4).

5.14. The following vapor–liquid equilibrium data have been reported for the system water (1) + 1,4-dioxane (2) at 323.15 K.

(a) Compare the experimental data with the UNIQUAC predictions for this system using the Aspen parameters.

(b) Compare the experimental data with the NRTL predictions for this system using the Aspen parameters.

(c) Compare the experimental data with the Wilson predictions for this system using the Aspen parameters.

(d) Compare the experimental data with UNIFAC predictions for this system.

P (mm Hg)	x_1	y_1
120.49	0.0000	0.0000
140.85	0.0560	0.1920
151.16	0.0970	0.2680
159.17	0.1700	0.3450
164.57	0.2160	0.3830
165.65	0.2980	0.4030
167.89	0.3660	0.4250
167.74	0.4400	0.4430
167.79	0.4460	0.4460
167.95	0.4840	0.4510
166.84	0.5390	0.4550
165.48	0.6290	0.4660
160.82	0.7490	0.4950
155.14	0.8110	0.5430
142.64	0.8900	0.6040
114.76	0.9670	0.7950
92.51	1.0000	1.0000

Source: G. Kortum and V. Valent, *Ber. Bunsenges Phys. Chem.*, **81**, 752 (1977).

5.15. Correlate the binary VLE data for the ethanol + water mixture you found in Problem 4.5 using the NIST TDE with the UNIQUAC and Wilson models. Which of these models correlates the data better?

5.16. Correlate the binary VLE data for the furfural + water mixture you found in Problem 4.6 using the NIST TDE with the UNIQUAC and Wilson models. Which of these models correlates the data better?

5.17. Correlate the binary VLE data for the *n*-heptane + toluene mixture you found in Problem 4.7 using the NIST TDE with the UNIQUAC and Wilson models. Which of these models correlates the data better?

5.18. Correlate the binary isobaric VLE data for the carbon tetrachloride + *n*-heptane mixture you found in Problem 4.8 using the NIST TDE with the UNIQUAC and Wilson models. Which of these models correlates the data better?

Chapter 6

Vapor–Liquid Equilibrium Calculations Using an Equation of State

In this chapter we consider vapor–liquid equilibrium (VLE) calculations with an equation of state using Aspen Plus®. The organization is very similar to the previous chapter in that first we consider equation of state calculations using the property analysis method, followed by the simulation method, and finally the regression of experimental binary VLE data to obtain values for the binary parameter in an equation of state. Also, there is a comparison of the results with those for the benzene–decane system considered in the last chapter using an activity coefficient model so that the results of using an activity coefficient and equation of state for a simple system can be compared. [Jumping ahead, it is found that the results of both sets of calculations are in reasonable agreement for this simple system; however, that would not be the case for mixtures that contained polar components or a mixture that is at high pressure.]

Note that here it will be assumed that the reader is familiar with equations of state, such as the Peng–Robinson, the Soave–Redlich–Kwong and others, and also with the mixing rules when used with mixtures. If this is not the case, the reader should refer to a standard chemical engineering thermodynamics textbook.

As before, the **Property Analysis** method to calculate VLE has the advantage that it can easily be used to generate tables and graphs of Pxy and/or Txy VLE and Gibbs energy of mixing as a function of composition. The disadvantages are that it can only be used for binary mixtures, and that it can only be used for calculations at a fixed temperature or pressure, but not, for example, for a Joule–Thomson expansion in which only the final pressure and enthalpy are known. An alternative is to do phase equilibrium calculations in the context of a process simulation using a separator such as **Flash2** (for vapor–liquid equilibria).

As an example of the two main types of vapor–liquid equilibrium calculations, property analysis, and simulation, the VLE of the benzene + n-decane system will be considered here.

Using Aspen Plus® in Thermodynamics Instruction: A Step-by-Step Guide, First Edition. Stanley I. Sandler.

6.1 THE PROPERTY ANALYSIS METHOD

Set up Aspen as usual by going through **Setup**, **Components**, and **Methods** as shown in Figs. 6.1-1 to 6.1-3.

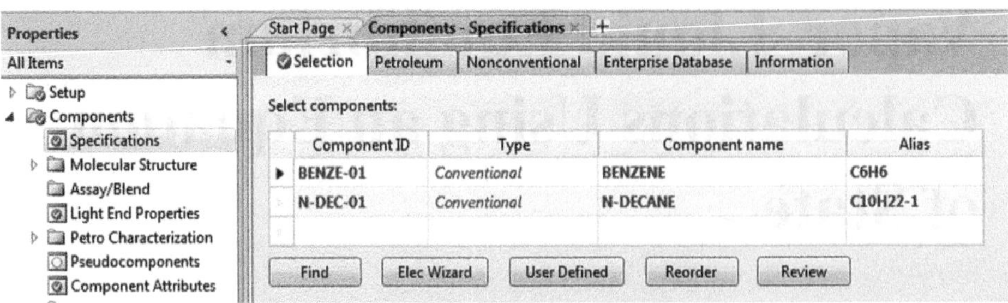

Figure 6.1-1

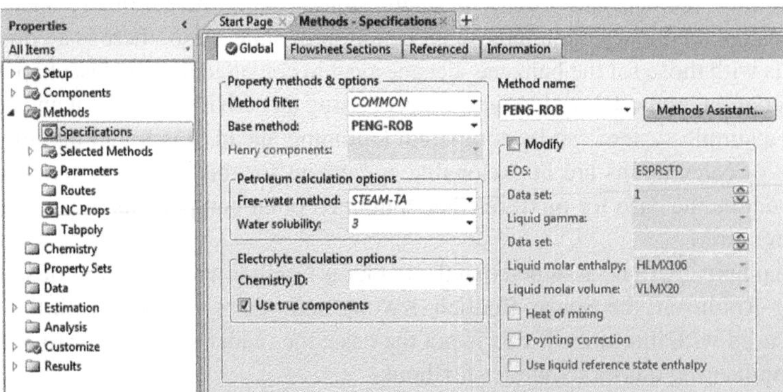

Figure 6.1-2

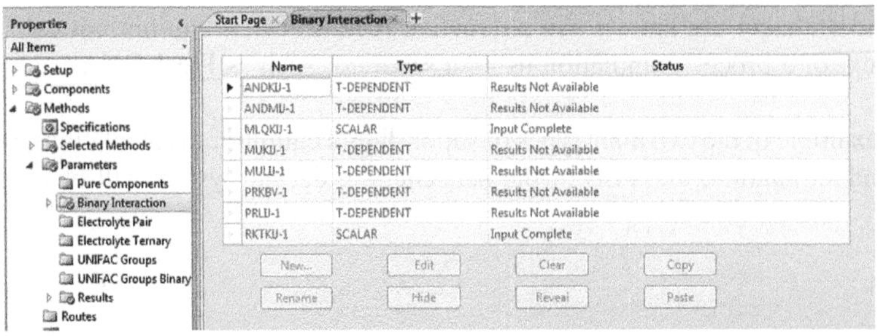

Properties | All Items | Setup | Components | Methods | Specifications | Selected Methods | Parameters | Pure Components | Binary Interaction | Electrolyte Pair | Electrolyte Ternary | UNIFAC Groups | UNIFAC Groups Binary | Results | Routes

Start Page | Binary Interaction

Name	Type	Status
ANDKIJ-1	T-DEPENDENT	Results Not Available
ANDMIJ-1	T-DEPENDENT	Results Not Available
MLQKIJ-1	SCALAR	Input Complete
MUKIJ-1	T-DEPENDENT	Results Not Available
MULIJ-1	T-DEPENDENT	Results Not Available
PRKBV-1	T-DEPENDENT	Results Not Available
PRLIJ-1	T-DEPENDENT	Results Not Available
RKTKIJ-1	SCALAR	Input Complete

New... | Edit | Clear | Copy | Rename | Hide | Reveal | Paste

Figure 6.1-3

Note, however, that in the Aspen Plus database there is no Peng–Robinson binary parameter available for the benzene + decane binary system. Use **Binary>Pxy** as shown in Fig. 6.1-4

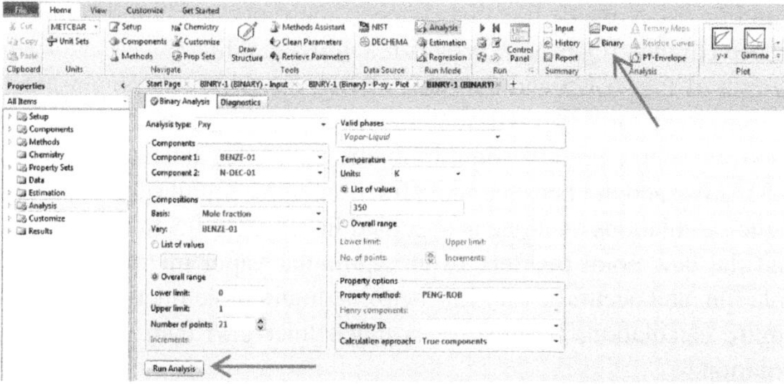

Figure 6.1-4

and then **Run Analysis** to give the graph of Fig. 6.1-5,

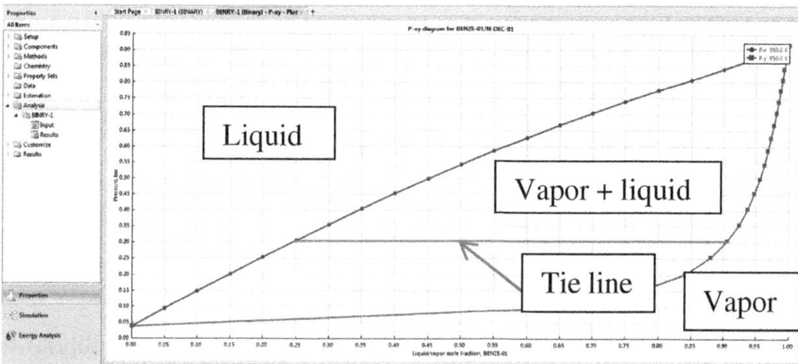

Figure 6.1-5

(in which the phase labels and tie line have been added manually), and the table of Fig. 6.1-6.

Binary analysis results

TEMP	MOLEFRAC BENZE-01	TOTAL PRES	TOTAL KVL BENZE-01	TOTAL KVL N-DEC-01	LIQUID GAMMA BENZE-01	LIQUID GAMMA N-DEC-01	VAPOR MOLEFRAC BENZE-01	VAPOR MOLEFRAC N-DEC-01	LIQUID MOLEFRAC BENZE-01	LIQUID MOLEFRAC N-DEC-01
K		bar								
350	0	0.0370551	31.1097	1	1.2961	1	0	1	0	1
350	0.05	0.0926562	12.3595	0.402133	1.28457	1.00023	0.617974	0.382026	0.05	0.95
350	0.1	0.147383	7.71198	0.254225	1.27254	1.001	0.771198	0.228802	0.1	0.9
350	0.15	0.201156	5.60473	0.187402	1.25999	1.00243	0.840709	0.159291	0.15	0.85
350	0.2	0.253892	4.40206	0.149486	1.2469	1.00466	0.880411	0.119589	0.2	0.8
350	0.25	0.305501	3.62444	0.125188	1.23323	1.00789	0.906109	0.0938912	0.25	0.75
350	0.3	0.355889	3.08038	0.10641	1.21898	1.01236	0.924113	0.075887	0.3	0.7
350	0.35	0.404958	2.6784	0.0962462	1.20412	1.01838	0.93744	0.06256	0.35	0.65
350	0.4	0.452608	2.36928	0.0871443	1.18863	1.02634	0.947713	0.0522865	0.4	0.6
350	0.45	0.498738	2.12419	0.0802065	1.17253	1.03678	0.955886	0.0441136	0.45	0.55
350	0.5	0.543254	1.92511	0.0748876	1.15581	1.05038	0.962556	0.0374438	0.5	0.5
350	0.55	0.586068	1.76021	0.070849	1.1385	1.06809	0.968118	0.031882	0.55	0.45
350	0.6	0.627115	1.62141	0.0678856	1.12067	1.09121	0.972846	0.0271542	0.6	0.4
350	0.65	0.666365	1.50298	0.0658884	1.10242	1.12139	0.976939	0.0230609	0.65	0.35
350	0.7	0.703844	1.40079	0.0648294	1.08393	1.16186	0.980551	0.0194488	0.7	0.3
350	0.75	0.739671	1.31175	0.0647627	1.06545	1.21597	0.983809	0.0161906	0.75	0.25
350	0.8	0.774118	1.23354	0.0658451	1.0474	1.29003	0.986831	0.013169	0.8	0.2
350	0.85	0.807696	1.1644	0.0683838	1.03045	1.39384	0.989742	0.0102575	0.85	0.15
350	0.9	0.84131	1.10301	0.0729376	1.01563	1.54406	0.992706	0.00729376	0.9	0.1
350	0.95	0.87651	1.04839	0.0805312	1.00457	1.77081	0.995973	0.00402656	0.95	0.05
350	1	0.91593	1	0.0931424	1	2.13309	1	0	1	0

Figure 6.1-6

In the diagram on Fig. 6.1-5 one tie line that connects the vapor and liquid compositions in equilibrium has been drawn. [Question: Why is it a horizontal line?]

From the graph of Fig. 6.1-5 and the table of Fig. 6.1-6 we can obtain much of the information we may want for this mixture at the specified temperature. For example, the bubble point pressure of an equimolar liquid mixture at 350 K (indicated by the arrow in Fig. 6.1-6) is 0.5432 bar and the benzene mole fraction in the equilibrium vapor is 0.9626. Note that since the calculations are done at specified liquid mole fractions, not vapor mole fractions, to find the dew point pressure of an equimolar vapor mixture from this table requires interpolation, and occurs at very low benzene liquid mole fractions of below 0.05. This would require calculations be done on a much finer grid than was done here for accurate interpolation.

With the Peng–Robinson equation of state prediction above, the bubble point pressure of an equimolar liquid mixture at 350 K is 0.5432 bar (compared to 0.5206 for the NRTL model of the previous chapter) and the benzene mole fraction in the equilibrium vapor is 0.9626 (compared with 0.9660 with the NRTL model). Thus the results are close, as would be expected for this hydrocarbon–hydrocarbon mixture.

We see that for this case the predictions of using the Peng–Robinson equation of state and the NRTL model are quite similar, but not in exact agreement. Which is correct? That can only be resolved by comparison with experimental data. Unfortunately, there is only bubble point data and not Pxy data in the NIST TDE data bank for this system, and none of those data are at 350 K, so we cannot check the results.

6.2 THE SIMULATION METHOD

A direct method to find the dew point pressure (and other phase equilibria) for mixtures that does not require interpolation is to use the **Flash2** separator. In **Setup>Report Options>Stream** click on **Fraction basis>Mole**, and set up the rest of the simulation as shown in Figs. 6.2-1 and 6.2-2.

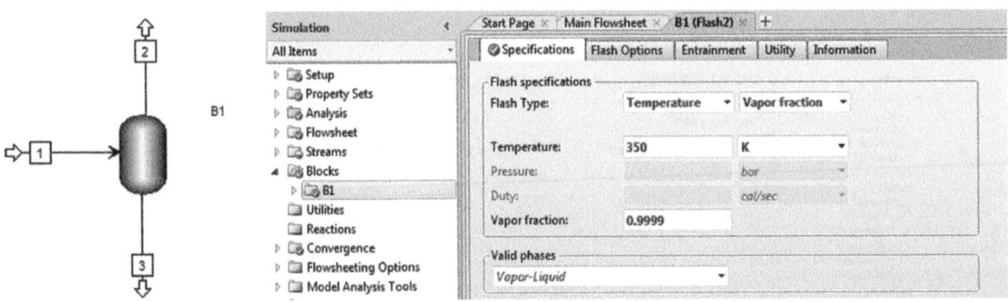

Figure 6.2-1

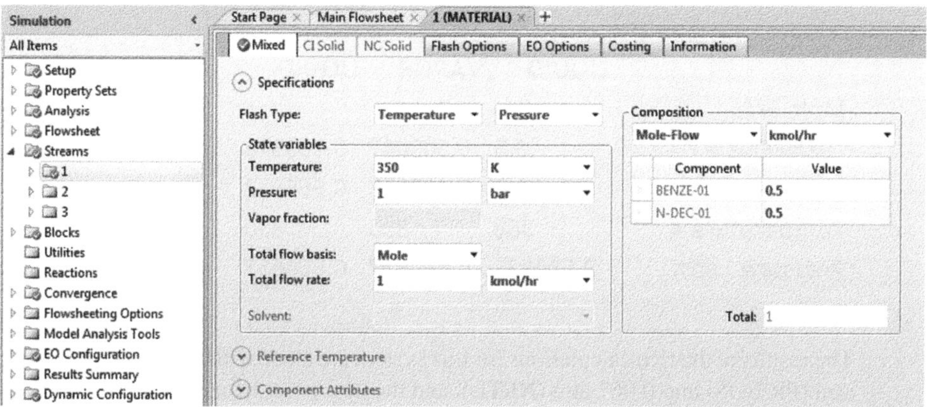

Figure 6.2-2

Then running the simulation results in the table in Fig. 6.2-3 (which here shows only part of the table).

Display: All streams Format: FULL Stream Table Copy All

	1	2	3
Substream: MIXED			
Mole Flow kmol/hr			
BENZE-01	0.5	3.13761e-06	0.499997
N-DEC-01	0.5	9.68624e-05	0.499903
Mole Frac			
BENZE-01	0.5	0.031376 ←	0.500047
N-DEC-01	0.5	0.968624	0.499953
Total Flow kmol/hr	1	0.0001	0.9999
Total Flow kg/hr	110.199	0.0140271	110.185
Total Flow l/min	2.59733	0.000339847	6702.36
Temperature C	76.85	76.85	76.85
Pressure bar	1	0.072044 ←	0.072044
Vapor Frac	0	0	1
Liquid Frac	1	1	0
Solid Frac	0	0	0

Figure 6.2-3

So with the Peng–Robinson equation, for an equimolar mixture at 350 K, the dew point pressure is predicted to be 0.072 bar with an equilibrium liquid benzene mole fraction of 0.0314. For comparison, using the NRTL model (Fig. 6.2-4) gives

	1	2	3
	LIQUID	VAPOR	LIQUID
Mole Frac			
BENZENE	0.5	0.500047	0.033378
DECANE	0.5	0.499953	0.966622
Temperature K	350	350	350
Pressure atm	0.986923	0.066657	0.066657

Figure 6.2-4

The results of the two calculations for this system are close with dew point pressure of 0.071 atm (PR EOS) and 0.067 atm (NRTL), and the dew point benzene liquid mole fractions of 0.0314 (PR EOS) and 0.334 (NRTL). As the two thermodynamic models are not identical, the results should not be expected to be so.

An adiabatic flash calculation cannot be done using **Properties>Binary** method, but is easily done using the **Flash2** block. As an example, consider starting with an equimolar mixture of benzene and *n*-decane at 400 K and 10 bar and flashing it to 0.5 bar. The input for **Streams>1** is (Fig. 6.2-5)

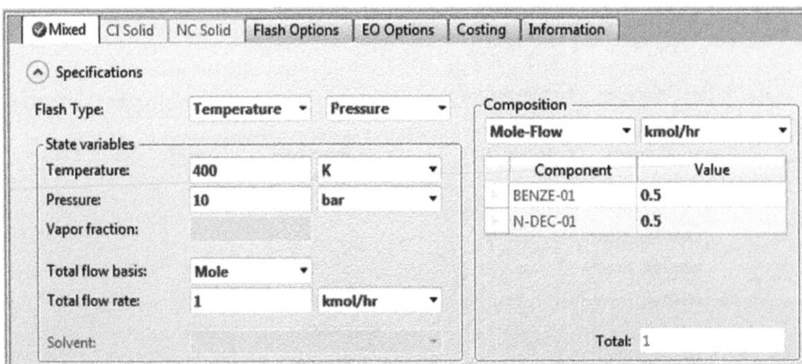

Figure 6.2-5

and the specification for **Block 1** is (Fig. 6.2-6),

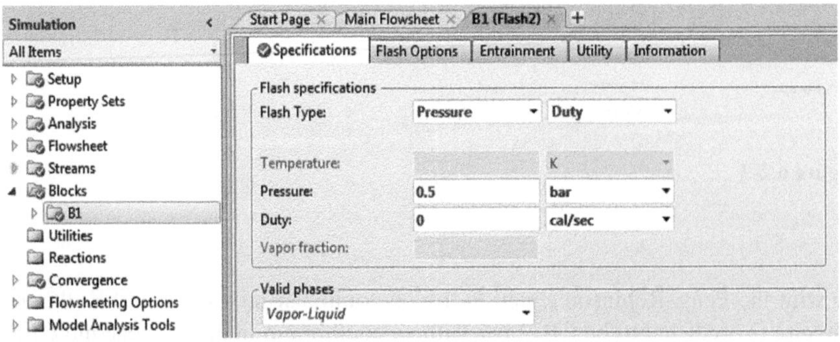

Figure 6.2-6

giving the results in Fig. 6.2-7.

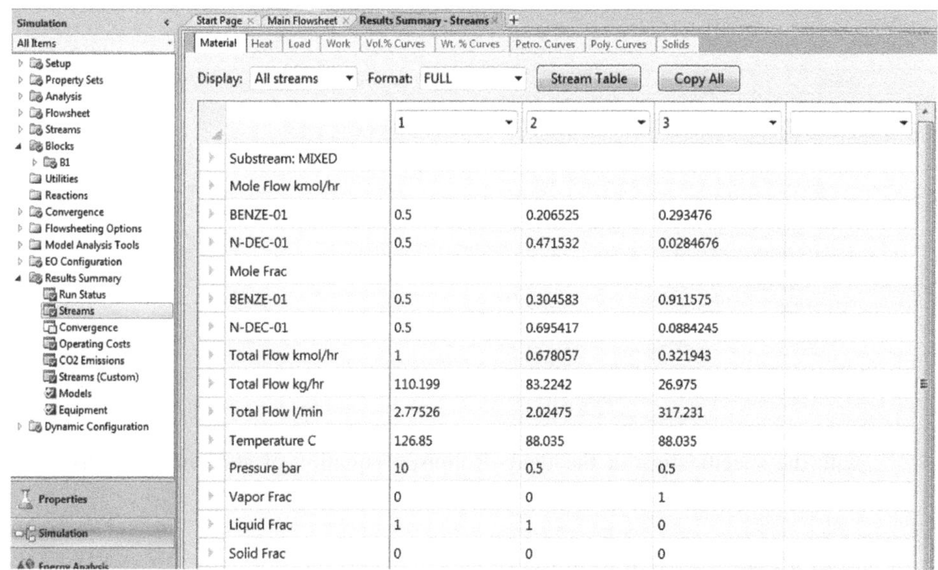

	1	2	3
Substream: MIXED			
Mole Flow kmol/hr			
BENZE-01	0.5	0.206525	0.293476
N-DEC-01	0.5	0.471532	0.0284676
Mole Frac			
BENZE-01	0.5	0.304583	0.911575
N-DEC-01	0.5	0.695417	0.0884245
Total Flow kmol/hr	1	0.678057	0.321943
Total Flow kg/hr	110.199	83.2242	26.975
Total Flow l/min	2.77526	2.02475	317.231
Temperature C	126.85	88.035	88.035
Pressure bar	10	0.5	0.5
Vapor Frac	0	0	1
Liquid Frac	1	1	0
Solid Frac	0	0	0

Figure 6.2-7

The abbreviated results for the adiabatic flash calulation (after being cut and pasted into Excel) for both the PR EOS and NRTL models are shown side-by-side in Fig. 6.2-8.

		PR EOS			NRTL	
		2	3		2	3
	FEED	VAPOR	LIQUID		VAPOR	LIQUID
Mole Flow kmol/hr						
BENZENE	0.5000	0.2935	0.2065		0.2914	0.2086
DECANE	0.5000	0.0285	0.4715		0.0268	0.4732
Mole Frac						
BENZENE	0.5000	0.9115	0.3045		0.9157	0.3059
DECANE	0.5000	0.0885	0.6955		0.0843	0.6941
Temperature K	400.0000	361.1887	361.1887		362.3468	362.3468
Pressure atm	9.8692	0.4935	0.4935		0.4935	0.4935
Vapor Frac	0	1	0		1	0
Liquid Frac	1	0	1		0	1

Figure 6.2-8

We see that the results of the two models are again close. For this adiabatic flash the exit streams are at 361 K (PR) or 362 K (NRTL) compared to the input feed stream temperature of 400 K, and compared to the feed stream that had a mole fraction of benzene of 0.5, the mole fractions of benzene are 0.912 in the vapor, 0.305 in the liquid (PR EOS) and 0.915 in the vapor, 0.306 in the liquid (NRTL).

While so far we have only examined binary mixtures, an important feature of using the **FLASH2** module is that multicomponent mixtures are easily considered. As an example, consider flash calculations involving the six-component hydrocarbon mixture of Fig. 6.2-9.

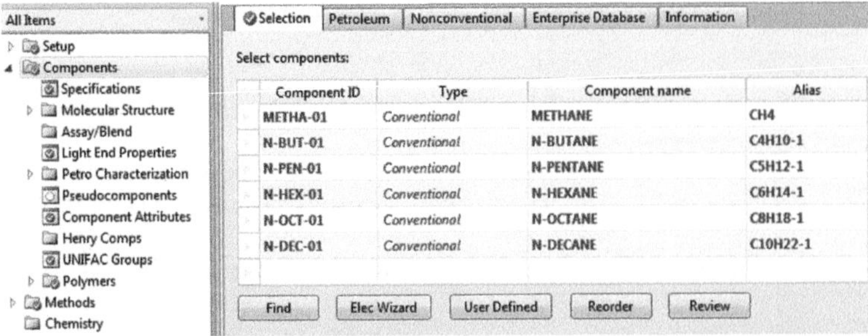

Figure 6.2-9

with the specification of the Peng–Robinson equation of state in **Methods** as shown in Fig. 6.2-10.

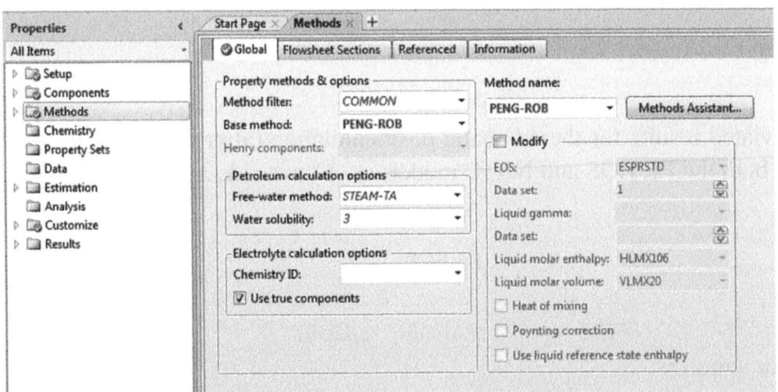

Figure 6.2-10

Checking on the availability of the binary interaction parameters (Fig. 6.2-11), we see that many, but not all are available, and that they are very small in value. Therefore, using a zero value for the others is likely to be satisfactory.

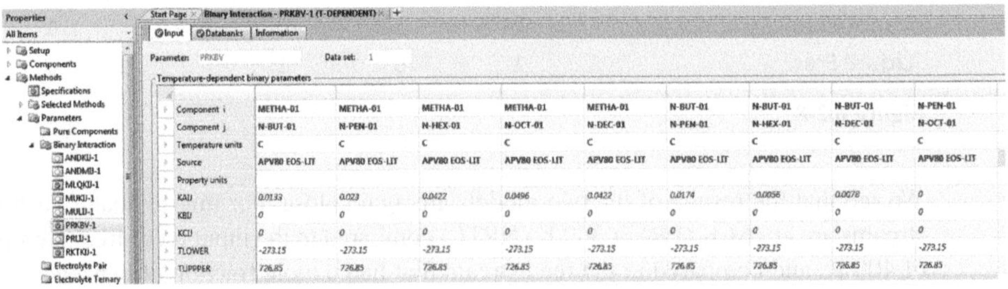

Start Page | Binary Interaction - PRKBV-1 (T-DEPENDENT)

Input | Databanks | Information

Parameter: PRKBV Data set: 1

Temperature-dependent binary parameters

Component i	METHA-01	METHA-01	METHA-01	METHA-01	METHA-01	N-BUT-01	N-BUT-01	N-BUT-01	N-PEN-01
Component j	N-BUT-01	N-PEN-01	N-HEX-01	N-OCT-01	N-DEC-01	N-PEN-01	N-HEX-01	N-DEC-01	N-OCT-01
Temperature units	C	C	C	C	C	C	C	C	C
Source	APV80 EOS-LIT	APV80 EOS-LIT	APV80 EOS-LIT	APV80 EOS-LIT	APV80 EOS-LIT	APV80 EOS-LIT	APV80 EOS-LIT	APV80 EOS-LIT	APV80 EOS-LIT
Property units									
KAIJ	0.0133	0.023	0.0422	0.0496	0.0422	0.0174	-0.0056	0.0078	0
KBIJ	0	0	0	0	0	0	0	0	0
KCIJ	0	0	0	0	0	0	0	0	0
TLOWER	-273.15	-273.15	-273.15	-273.15	-273.15	-273.15	-273.15	-273.15	-273.15
TUPPER	726.85	726.85	726.85	726.85	726.85	726.85	726.85	726.85	726.85

Figure 6.2-11

First, doing an isothermal flash at 100°C and 5 bar (Fig. 6.2-12) produces the abbreviated results in Fig. 6.2-13 (the rest of the results can be seen by scrolling down in the simulation).

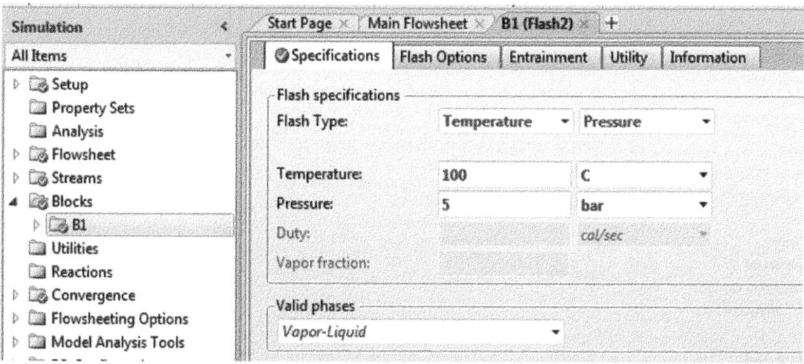

Figure 6.2-12

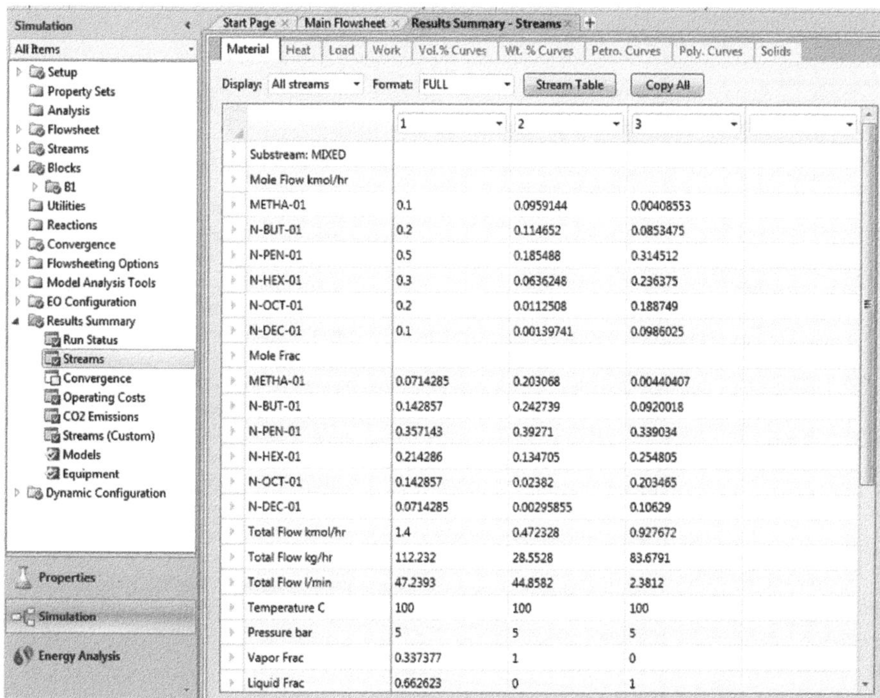

	1	2	3
Substream: MIXED			
Mole Flow kmol/hr			
METHA-01	0.1	0.0959144	0.00408553
N-BUT-01	0.2	0.114652	0.0853475
N-PEN-01	0.5	0.185488	0.314512
N-HEX-01	0.3	0.0636248	0.236375
N-OCT-01	0.2	0.0112508	0.188749
N-DEC-01	0.1	0.00139741	0.0986025
Mole Frac			
METHA-01	0.0714285	0.203068	0.00440407
N-BUT-01	0.142857	0.242739	0.0920018
N-PEN-01	0.357143	0.39271	0.339034
N-HEX-01	0.214286	0.134705	0.254805
N-OCT-01	0.142857	0.02382	0.203465
N-DEC-01	0.0714285	0.00295855	0.10629
Total Flow kmol/hr	1.4	0.472328	0.927672
Total Flow kg/hr	112.232	28.5528	83.6791
Total Flow l/min	47.2393	44.8582	2.3812
Temperature C	100	100	100
Pressure bar	5	5	5
Vapor Frac	0.337377	1	0
Liquid Frac	0.662623	0	1

Figure 6.2-13

As would be expected, the vapor phase is enriched in the light hydrocarbons methane and butane, and the liquid phase is enriched in the heavier hydrocarbons, especially octane and decane.

Instead, doing an adiabatic flash (Joule–Thomson expansion) to 1 bar (Fig. 6.2-14)

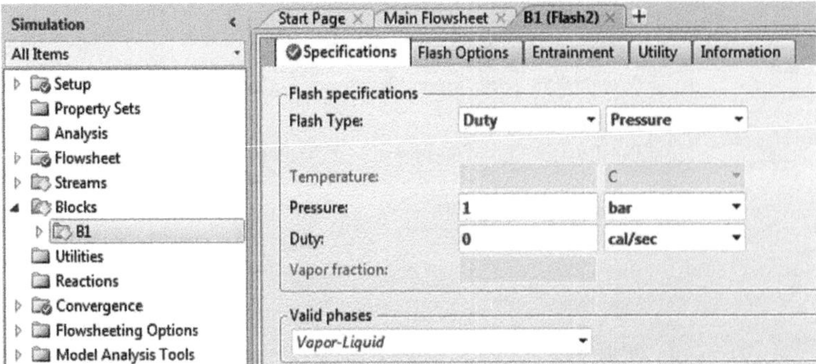

Figure 6.2-14

gives the results in Fig. 6.2-15.

	1	2	3
Substream: MIXED			
Mole Flow kmol/hr			
METHA-01	0.1	0.0997295	0.000270415
N-BUT-01	0.2	0.181061	0.0189386
N-PEN-01	0.5	0.385633	0.114367
N-HEX-01	0.3	0.167216	0.132784
N-OCT-01	0.2	0.0311181	0.168882
N-DEC-01	0.1	0.00288054	0.0971194
Mole Frac			
METHA-01	0.0714285	0.114944	0.000507955
N-BUT-01	0.142857	0.208683	0.0355747
N-PEN-01	0.357143	0.444463	0.214829
N-HEX-01	0.214286	0.192726	0.249424
N-OCT-01	0.142857	0.0358653	0.317232
N-DEC-01	0.0714285	0.00331998	0.182432
Total Flow kmol/hr	1.4	0.867639	0.532361
Total Flow kg/hr	112.232	58.3221	53.9098
Total Flow l/min	47.2393	390.012	1.39032
Temperature C	100	60.1877	60.1877
Pressure bar	5	1	1
Vapor Frac	0.337377	1	0
Liquid Frac	0.662623	0	1

Figure 6.2-15

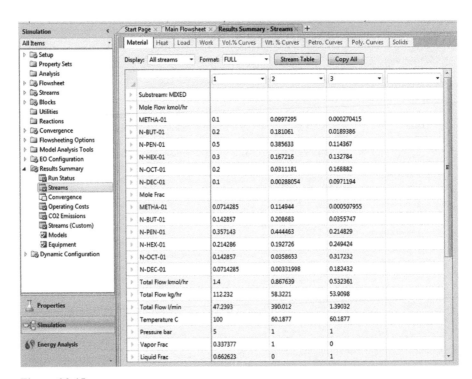

So we see that in the Joule–Thomson expansion from 5 bar and 100°C to 1 bar the exit temperature is reduced to 60.2°C, and again the light hydrocarbons are concentrated in the vapor phase and the heavier hydrocarbons in the liquid phase. The main point, however, is that multicomponent mixtures can be treated in the **FLASH2** separator model just as easily as a single or binary component system.

So in summary, two methods of predicting vapor–liquid equilibrium have been considered. The first is the **Tools>Analysis>Property>Binary>Pxy** (or **Txy**) method that allows quick calculations of either Pxy or Txy diagrams for binary mixtures on a grid of liquid compositions. The second method uses the simulation capability of Aspen Plus with a **Flash2** separator that provides greater flexibility in that any specified composition can be considered, that binary and multicomponent mixtures can be analyzed, and that a variety of flash calculations can be done, including bubble point and dew point pressures or temperatures, as well as adiabatic flash calculations.

6.3 REGRESSION OF BINARY VLE DATA WITH AN EQUATION OF STATE

The last subject to be considered in this chapter is using Aspen Plus to regress binary VLE data to obtain the binary interaction parameter in an equation of state model. In this example, we will consider an ethane–propylene mixture, so the starting point is to go to **Components** and add ethane and propylene (Fig. 6.3-1).

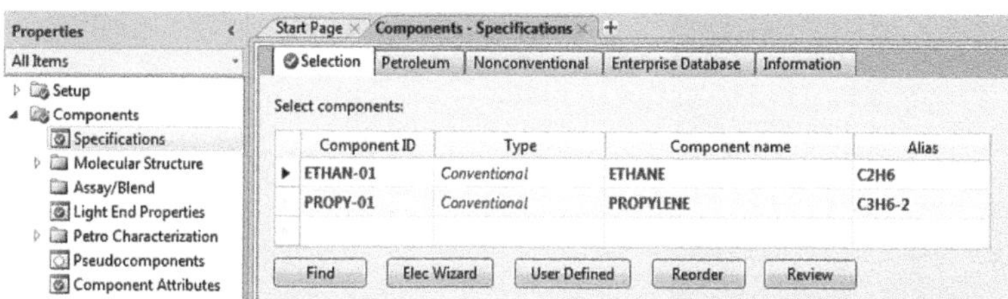

Figure 6.3-1

Next go to the **Methods>Specifications** window of Fig. 6.3-2. Since the data for this system are at high pressure (unless the temperature is very low), an equation of state model will be used.

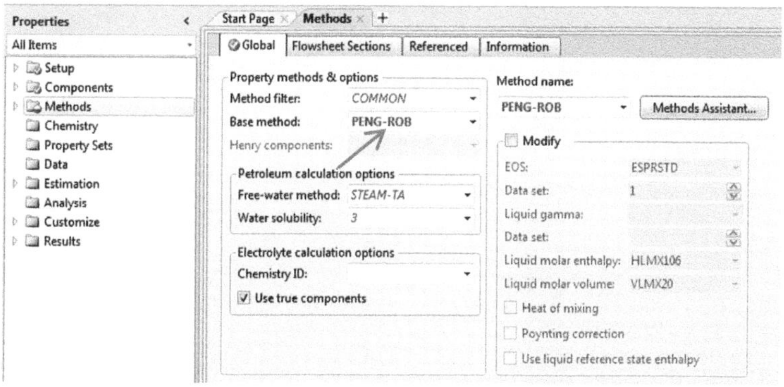

Figure 6.3-2

The Peng–Robinson equation of state has been chosen by setting **Base method>PENG-ROB.** Then going to **Parameters>Binary Interaction>PRKBV-1** brings up the window of Fig. 6.3-3 showing that the Aspen default value for the binary interaction parameter is very small, 0.0089.

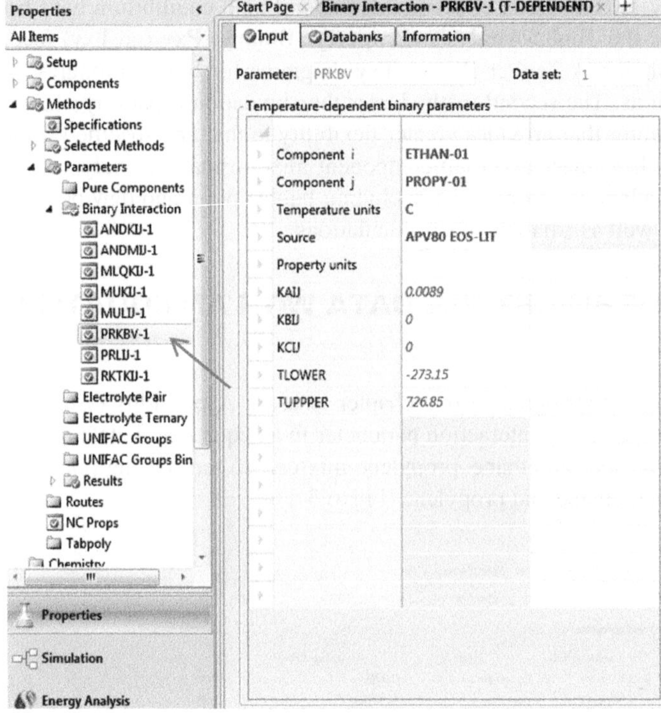

Figure 6.3-3

The next item is **Data**. This is where you enter a data set that will be used in the regression. If you have your own data, you can enter it. Here the **NIST TDE** described in Chapter 4 will be used to find a set of data to analyze. We start with identifying a data set using the default name ID of **D-1** and select the components (Fig. 6.3-4).

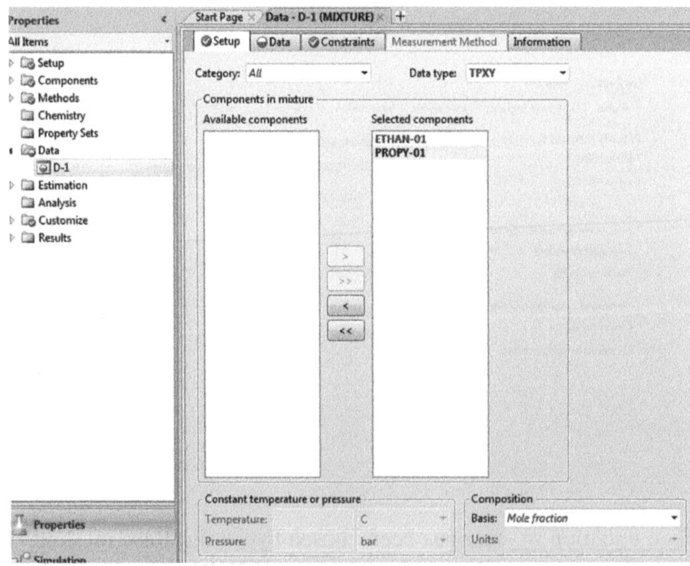

Figure 6.3-4

Next click on the **Data** tab and then **Retrieve TDE Binary Data...** (Fig. 6.3-5),

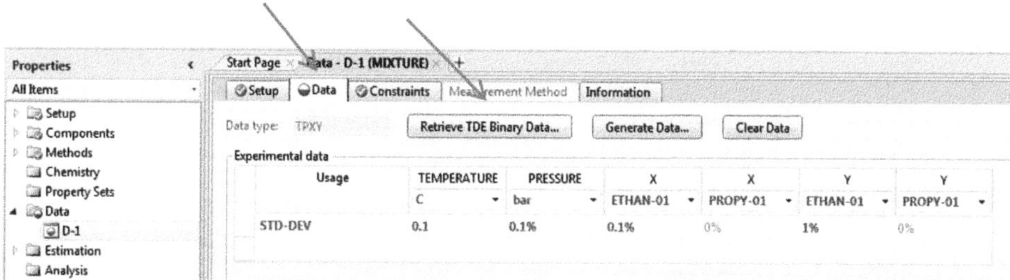

Figure 6.3-5

which brings up the list of data sets in Fig. 6.3-6 available for this mixture, only some of which are visible in the window here.

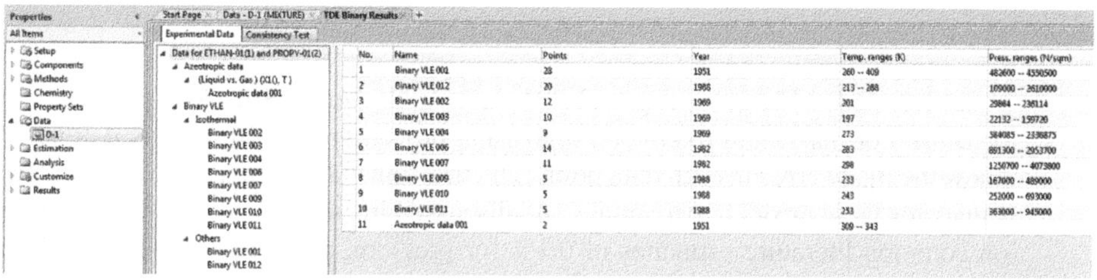

Figure 6.3-6

Clicking, in the left window, on **Data>Binary VLE>Isothermal>Binary VLE 004** gives the data set shown in Fig. 6.3-7.

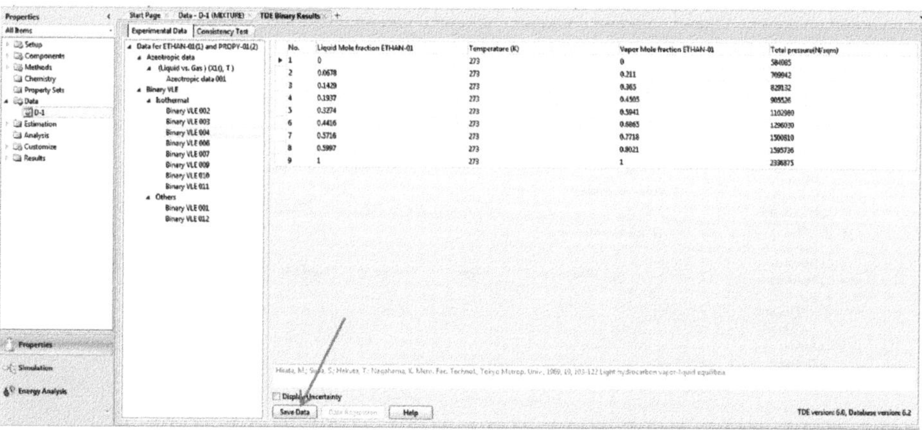

Figure 6.3-7

These are the data that will be used in the regression. Click on **Save Data** and the data set **D-1** is filled in using this data set as shown in Fig. 6.3-8.

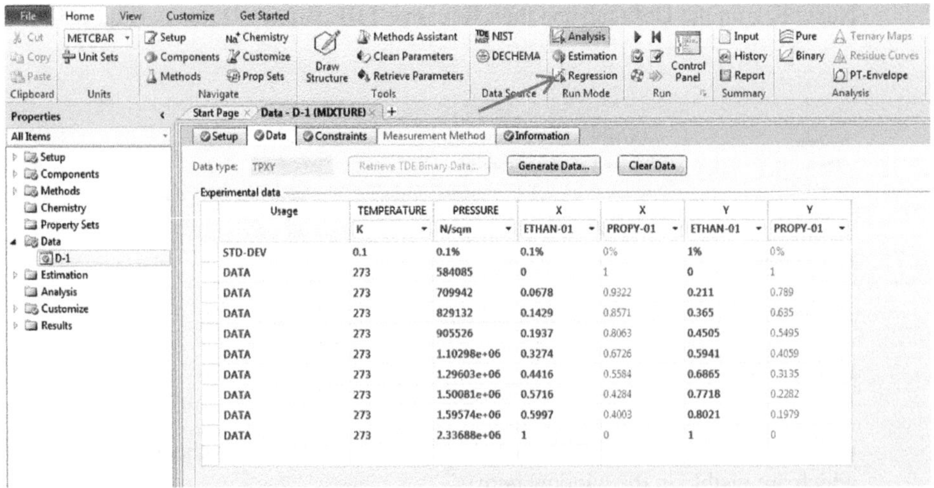

Usage	TEMPERATURE K	PRESSURE N/sqm	X ETHAN-01	X PROPY-01	Y ETHAN-01	Y PROPY-01
STD-DEV	0.1	0.1%	0.1%	0%	1%	0%
DATA	273	584085	0	1	0	1
DATA	273	709942	0.0678	0.9322	0.211	0.789
DATA	273	829132	0.1429	0.8571	0.365	0.635
DATA	273	905526	0.1937	0.8063	0.4505	0.5495
DATA	273	1.10298e+06	0.3274	0.6726	0.5941	0.4059
DATA	273	1.29603e+06	0.4416	0.5584	0.6865	0.3135
DATA	273	1.50081e+06	0.5716	0.4284	0.7718	0.2282
DATA	273	1.59574e+06	0.5997	0.4003	0.8021	0.1979
DATA	273	2.33688e+06	1	0	1	0

Figure 6.3-8

Note that under **Usage** in Fig. 6.3-8 there is a choice of **Std-Dev** (standard deviation) or **Data**. For maximum likelihood regression procedures in Aspen Plus it is necessary to know the likely error in each data point. The first line that appears is the Aspen Plus default values for these errors in percentage. As discussed in Section 5.2, these defaults are based on common literature estimates of 0.1% for pressure, pressure and liquid mole fraction, and a higher error estimate of 1% for the vapor mole fraction because of the difficulties in obtaining good vapor samples. Generally the default values are sufficient, unless you have other information. The remaining entries are the data retrieved from the **NIST-TDE**.

Next go to **Regression** under the **Setup tab** and click on **New,** accept the default ID of **R-1** (or enter your own ID) and click on **OK** (Fig. 6.3-9). Remaining on the **Setup** tab, enter the data set **D-1** from the drop-down menu under **Data set**, and click on **Regression** for calculation type.

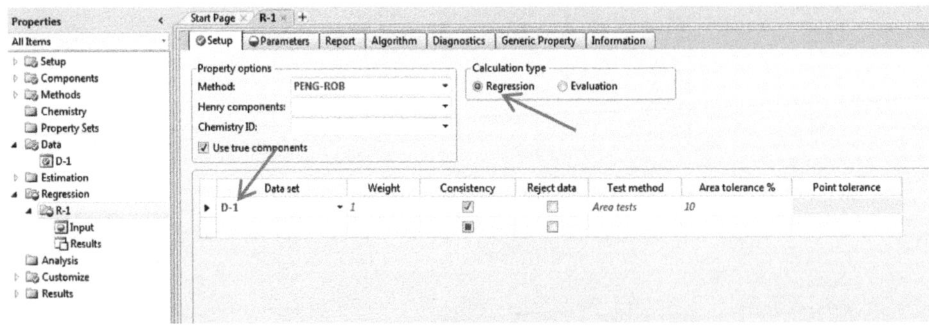

Figure 6.3-9

Then go to **Parameters** tab and populate it as shown, paying careful attention to choose **Yes** on the **Set Aij = Aji** line (Fig. 6.3-10).

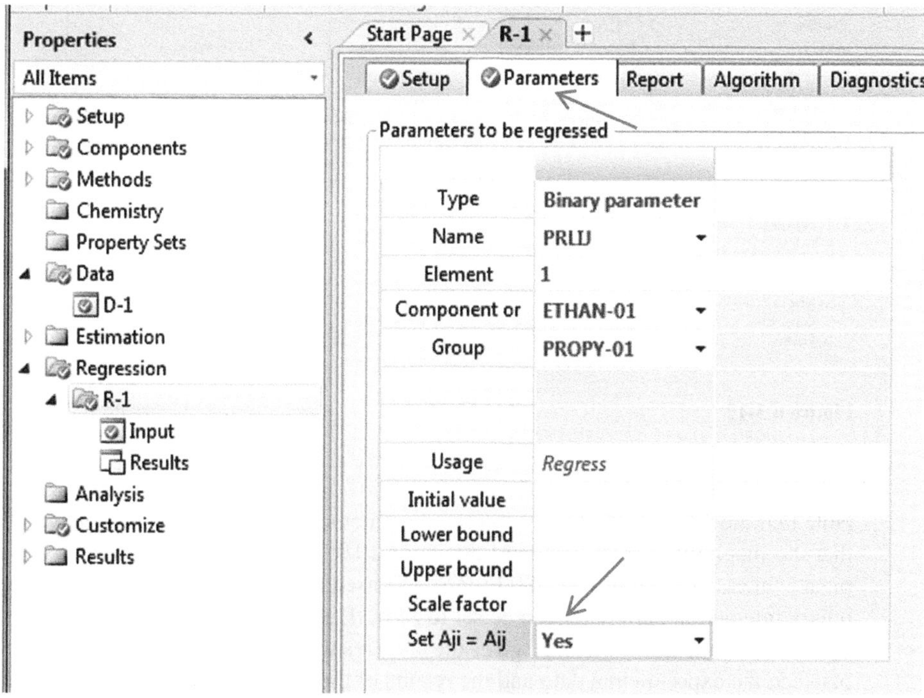

Figure 6.3-10

Finally, on the main toolbar click > to see the window in Fig. 6.3-11 and then click on **OK**.

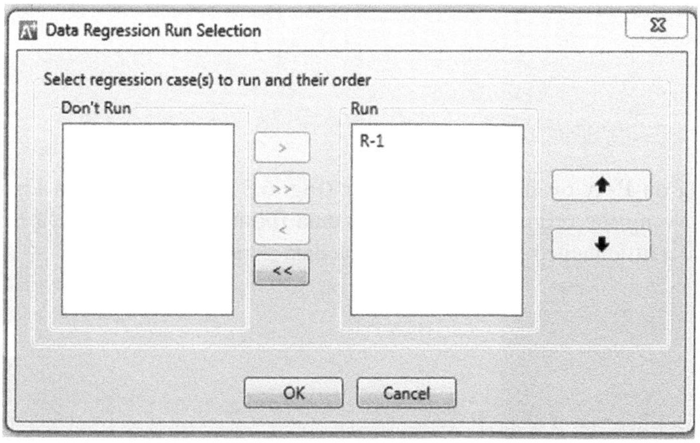

Figure 6.3-11

Then click on **Regression>R-1>Results** to see the results for the parameters in Fig. 6.3-12.

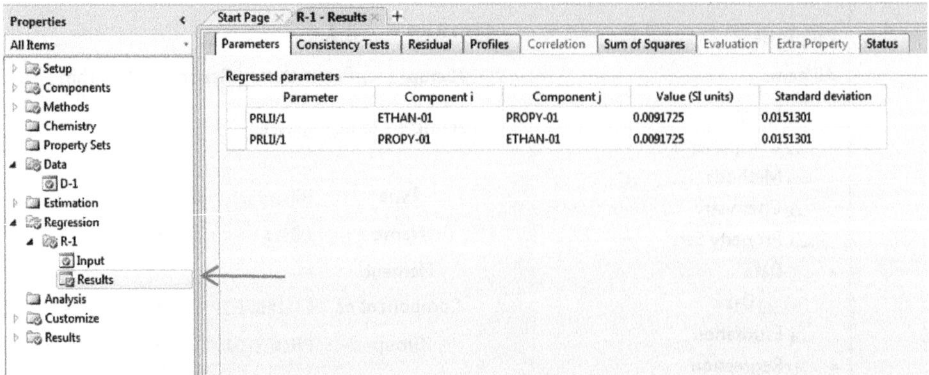

Figure 6.3-12

Note that the value of the binary interaction parameter for this mixture is quite small, and that the magnitude of the standard deviation (0.0151301) is larger than the value of the binary interaction parameter (0.0091725). Consequently, for this system the value of the binary interaction parameter can be set to zero, if desired.

Clicking on the **Profiles** tab gives the table results in Fig. 6.3-13 showing the difference between the experimental data and the results of the correlation at each data point.

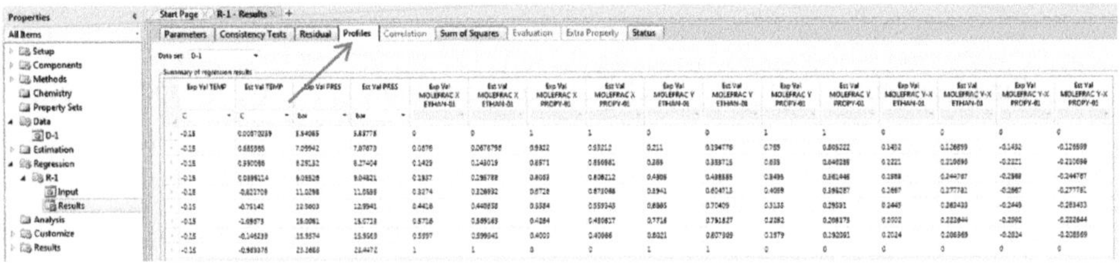

Figure 6.3-13

Next, by clicking on **P-xy** on the **Main Toolbar** (Fig. 6.3-14), the graph in Fig. 6.3-15 is obtained containing the original experimental data (points) and the correlated results (lines), which are a little jagged because of the few data points.

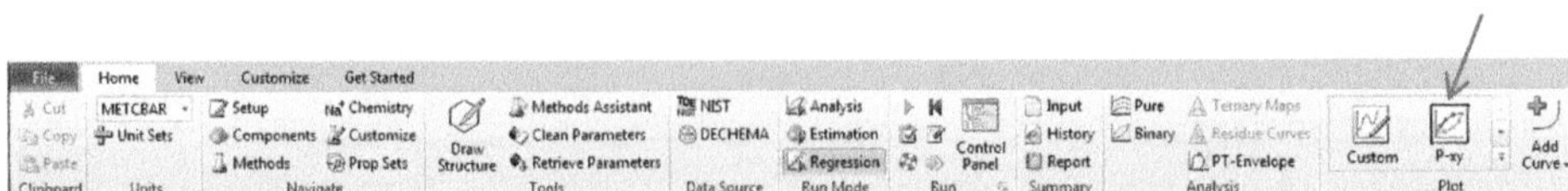

Figure 6.3-14

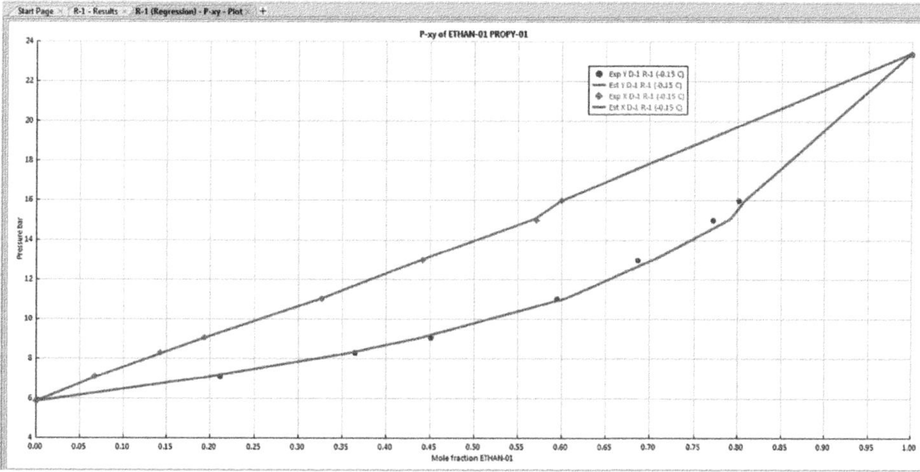

Figure 6.3-15

Following the same procedure for the carbon dioxide–hexane system, which is somewhat more interesting since the value of the binary interaction parameter is larger, leads to the sequence of windows in Figs. 6.3-16 to 6.3-26 (largely without intermediate instructions as the method is identical to what has been done previously in this chapter).

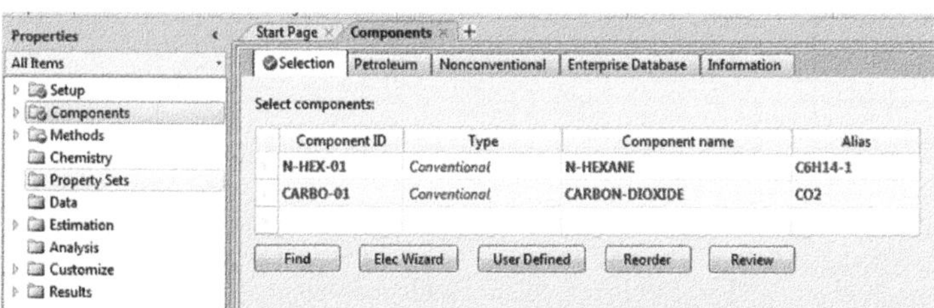

Figure 6.3-16

Figure 6.3-17

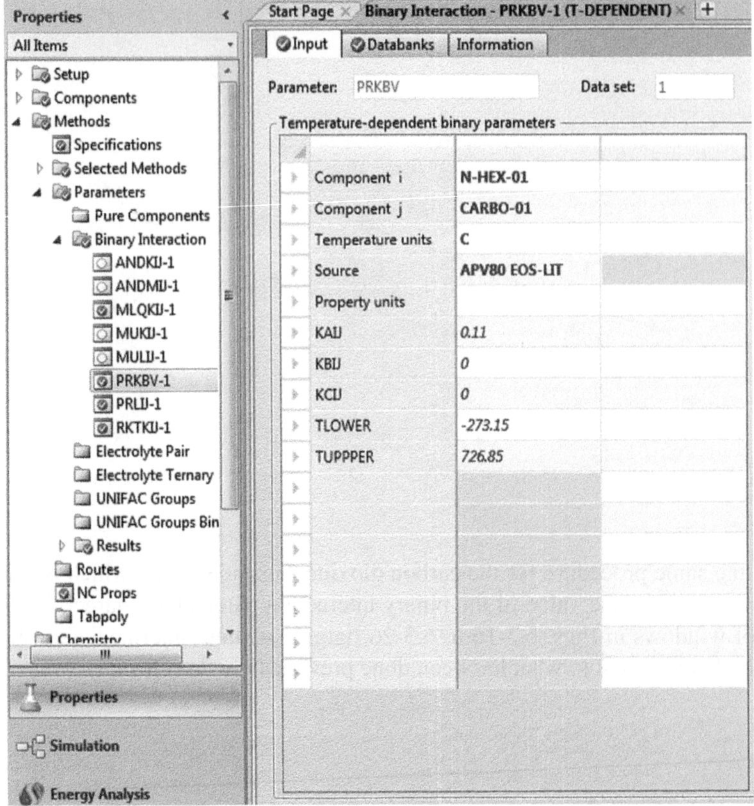

Figure 6.3-18

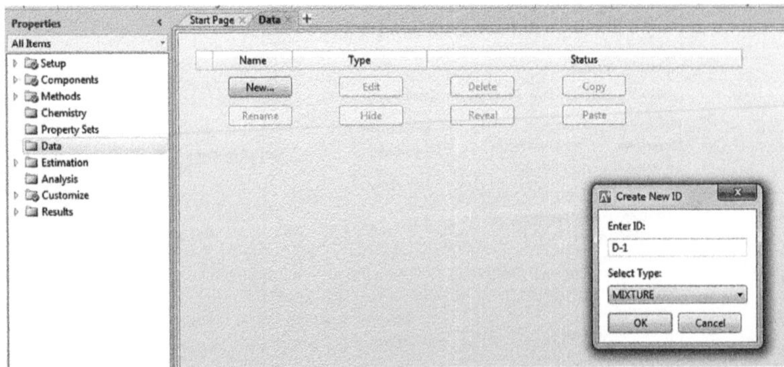

Figure 6.3-19

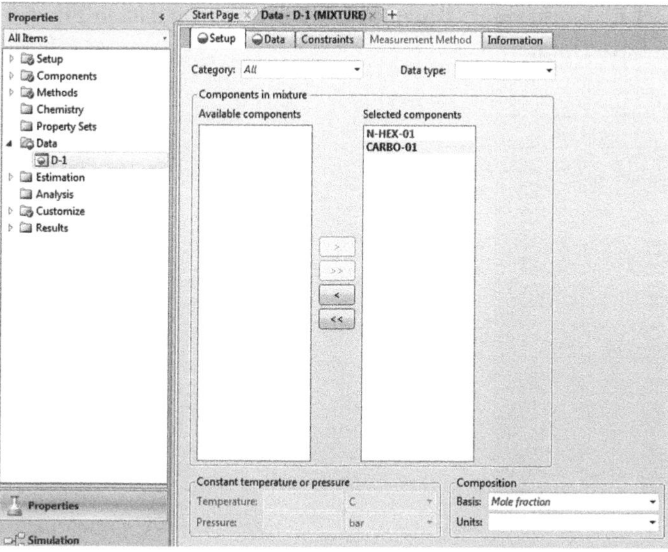

Figure 6.3-20

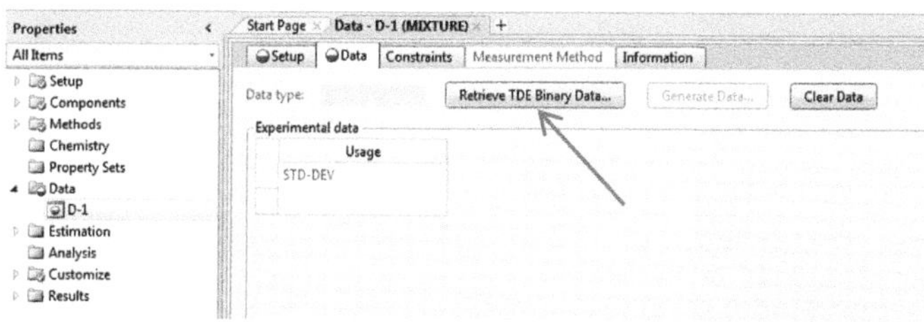

Figure 6.3-21

Next, clicking on **Retrieve TDE Binary Data...** as described in Chapter 4 provides the data sets in Fig. 6.3-22.

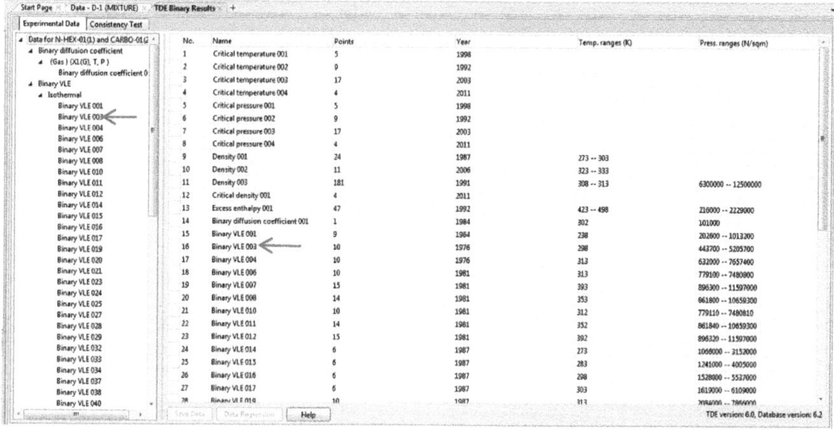

Figure 6.3-22

Choosing **Binary VLE>Isothermal>Binary VLE 003** gives the data in Fig. 6.3-23 and its source. Click on **Save Data.**

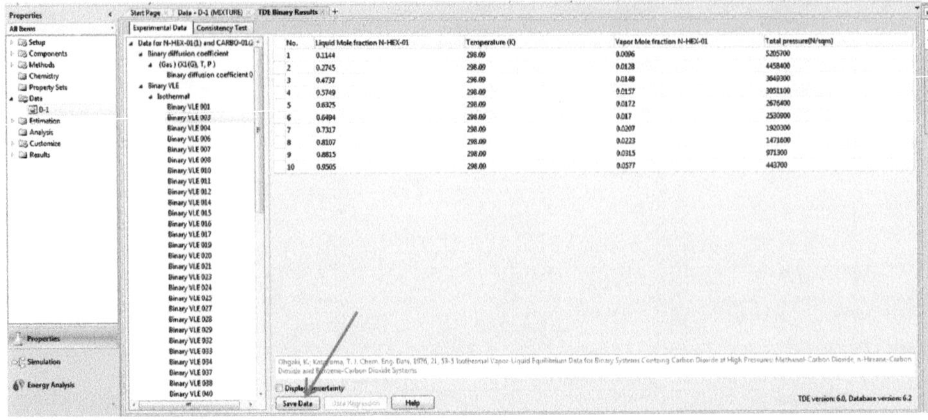

Figure 6.3-23

Clicking on **Plot P-xy** (Fig. 6.3-24) produces the graph of the experimental data in Fig. 6.3-25.

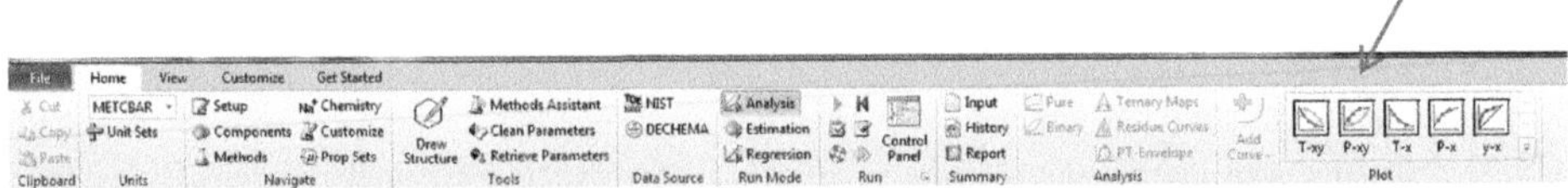

Figure 6.3-24

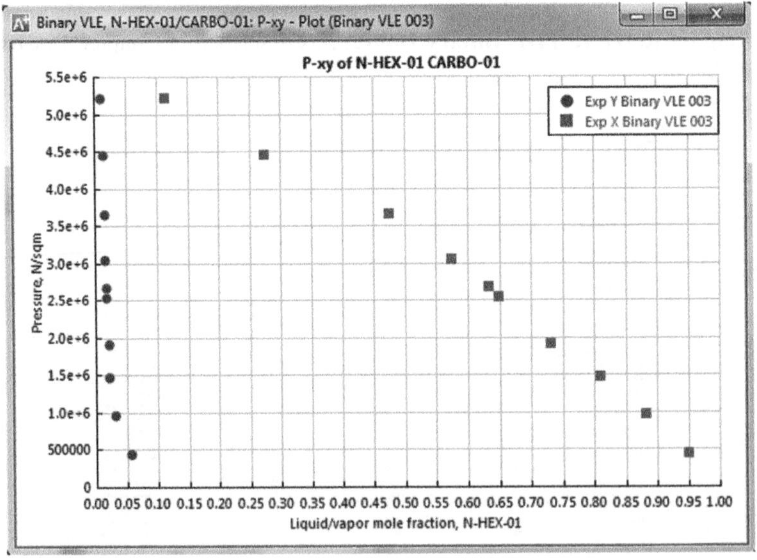

Figure 6.3-25

Clicking on **Save Data** uses these data as data set **D-1**, as seen in Fig. 6.3-26.

Usage	TEMPERATURE K	PRESSURE N/sqm	X N-HEX-01	X CARBO-01	Y N-HEX-01	Y CARBO-01
STD-DEV	0.1	0.1%	0.1%	0%	1%	0%
DATA	298.09	5.2057e+06	0.1144	0.8856	0.0096	0.9904
DATA	298.09	4.4584e+06	0.2745	0.7255	0.0128	0.9872
DATA	298.09	3.6493e+06	0.4737	0.5263	0.0148	0.9852
DATA	298.09	3.0511e+06	0.5749	0.4251	0.0157	0.9843
DATA	298.09	2.6764e+06	0.6325	0.3675	0.0172	0.9828
DATA	298.09	2.5309e+06	0.6494	0.3506	0.017	0.983
DATA	298.09	1.9203e+06	0.7317	0.2683	0.0207	0.9793
DATA	298.09	1.4716e+06	0.8107	0.1893	0.0223	0.9777
DATA	298.09	971300	0.8815	0.1185	0.0315	0.9685
DATA	298.09	443700	0.9505	0.0495	0.0577	0.9423

Figure 6.3-26

Next, clicking on **Regression** on the main toolbar as shown in Fig. 6.3-27 (or the folder list in Fig. 6.3-27) and then **New** brings up the window in Fig. 6.3-28.

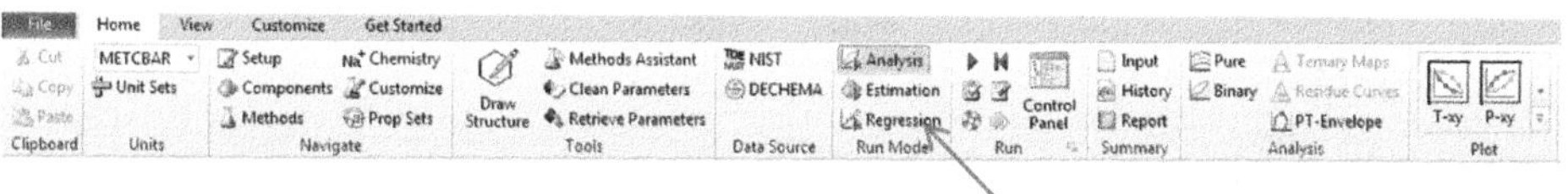

Figure 6.3-27

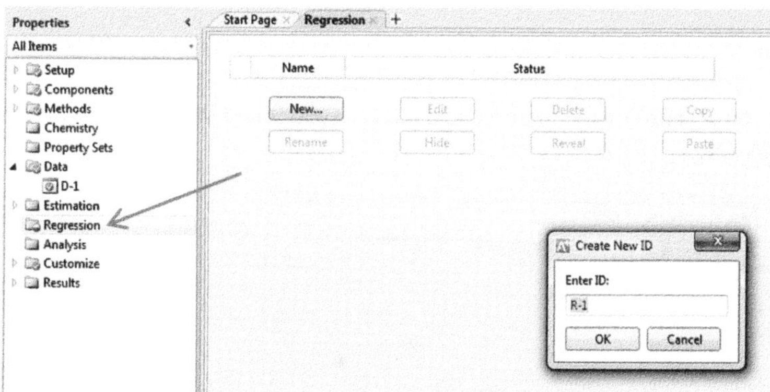

Figure 6.3-28

Next, enter **D-1** as the **Data set** from the drop-down list in the window in Fig. 6.3-29 and uncheck consistency test as this is not done on high pressure vapor–liquid equilibrium data. Consistency tests generally assume an ideal vapor phase, which is not the case for high pressure data.

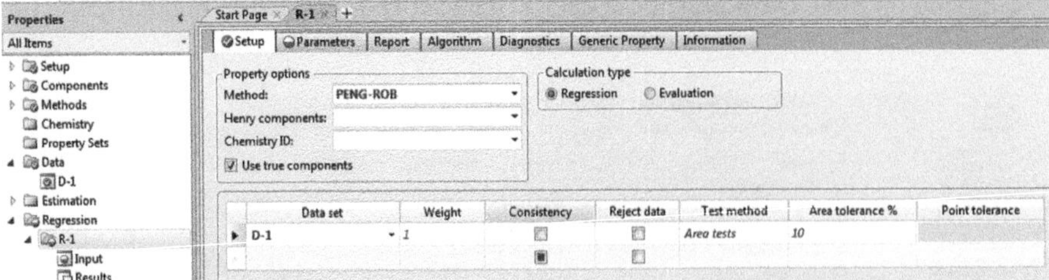

Figure 6.3-29

Then go to the **Parameters** tab and complete it as shown in Fig. 6.3-30 for a single binary parameter.

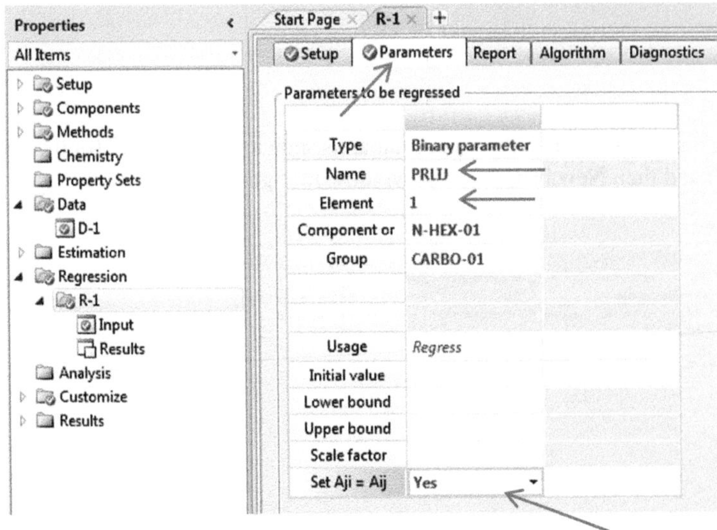

Figure 6.3-30

Finally, click on > **(run)** on the main toolbar (or **N>**) and select **R-1** as shown in Fig. 6.3-31 and then click **OK**.

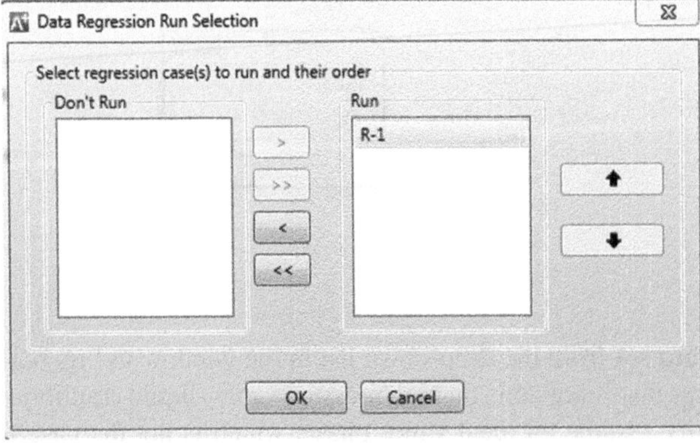

Figure 6.3-31

After the regression is complete go to **Properties>Regression>R-1>Results>Parameters**, which gives the results shown in Fig. 6.3-32.

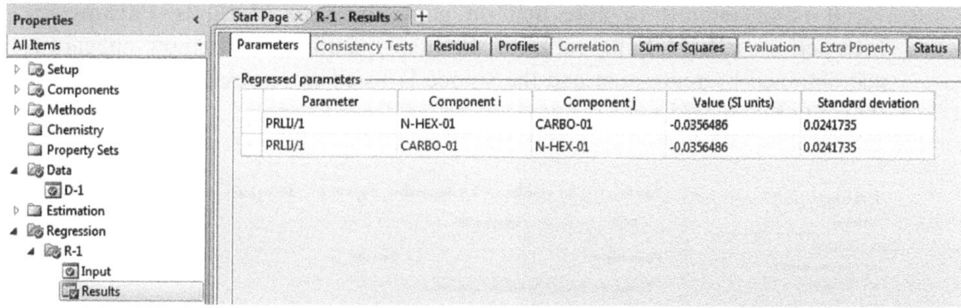

Parameter	Component i	Component j	Value (SI units)	Standard deviation
PRLIJ/1	N-HEX-01	CARBO-01	-0.0356486	0.0241735
PRLIJ/1	CARBO-01	N-HEX-01	-0.0356486	0.0241735

Figure 6.3-32

The value of the binary parameter found here is −0.03565, somewhat larger than the standard deviation 0.0242, so the value found is meaningful. The point-by-point results obtained by clicking on the **Profiles** tab are as shown in Fig. 6.3-33.

Data set: D-1

Summary of regression results

Exp Val TEMP (C)	Est Val TEMP (C)	Exp Val PRES (bar)	Est Val PRES (bar)	Exp Val MOLEFRAC X N-HEX-01	Est Val MOLEFRAC X N-HEX-01	Exp Val MOLEFRAC X CARBO-01	Est Val MOLEFRAC X CARBO-01	Exp Val MOLEFRAC Y N-HEX-01	Est Val MOLEFRAC Y N-HEX-01	Exp Val MOLEFRAC Y CARBO-01	Est Val MOLEFRAC Y CARBO-01	Exp Val MOLEFRAC Y-X N-HEX-01	Est Val MOLEFRAC Y-X N-HEX-01	Exp Val MOLEFRAC Y-X CARBO-01	Est Val MOLEFRAC Y-X CARBO-01
24.94	22.8903	52.057	52.7611	0.1144	0.114554	0.8856	0.885446	0.0096	0.00859241	0.9904	0.991408	-0.1048	-0.105962	0.1048	0.105962
24.94	21.87	44.884	45.6859	0.2748	0.276368	0.7252	0.723632	0.0128	0.00874839	0.9872	0.991252	-0.262	-0.26762	0.262	0.26762
24.94	23.8897	36.493	37.0443	0.4737	0.479917	0.5263	0.520083	0.0148	0.0102413	0.9852	0.989759	-0.4589	-0.469676	0.4589	0.469676
24.94	24.9209	30.511	30.7387	0.5749	0.58147	0.4251	0.41853	0.0157	0.0114026	0.9843	0.988597	-0.5592	-0.570068	0.5592	0.570068
24.94	25.0822	26.784	26.9544	0.6325	0.635668	0.3675	0.364332	0.0172	0.0121805	0.9828	0.98782	-0.6153	-0.623487	0.6153	0.623487
24.94	24.6449	25.309	25.5459	0.6494	0.65315	0.3506	0.34685	0.017	0.0123129	0.983	0.987687	-0.6324	-0.640837	0.6324	0.640837
24.94	24.1559	19.205	19.4733	0.7317	0.73597	0.2683	0.26403	0.0207	0.0141796	0.9793	0.98582	-0.711	-0.72179	0.711	0.72179
24.94	26.8299	14.716	14.621	0.8107	0.809732	0.1893	0.190268	0.0223	0.018988	0.9777	0.981012	-0.7884	-0.790744	0.7884	0.790744
24.94	27.8222	9.719	9.55306	0.8815	0.879716	0.1185	0.120284	0.0315	0.027984	0.9685	0.972016	-0.85	-0.851732	0.85	0.851732
24.94	28.9481	4.407	4.36587	0.9505	0.949076	0.0495	0.0509239	0.0577	0.0602549	0.9423	0.939745	-0.8928	-0.888821	0.8928	0.888821

Figure 6.3-33

Now plotting the results using the **Plot P-xy** on the main toolbar produces the graph in Fig. 6.3-34.

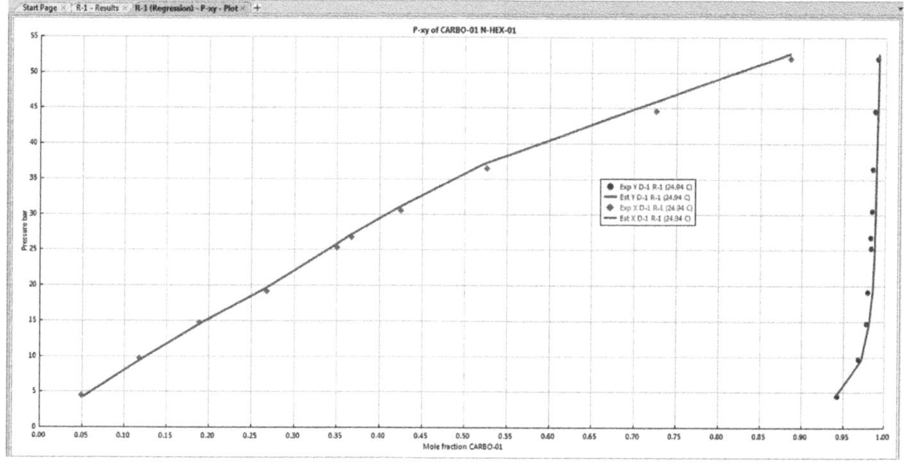

Figure 6.3-34

This figure, and the results on which it is based, show that the Peng–Robinson equation of state provides a good fit of this set of experimental data with a binary interaction parameter value of $k_{ij} = -0.03565$, even though the default value in the Aspen database is $k_{ij} = 0.11$.

It is of interest to note that on going back to **Methods>Parameters>Binary Interaction>PRLIJ-1** the value of the binary parameter that now appears is the one just obtained by regression and the source is listed as regression R-1 (that is **R-R-1**) as shown in Fig. 6.3-35.

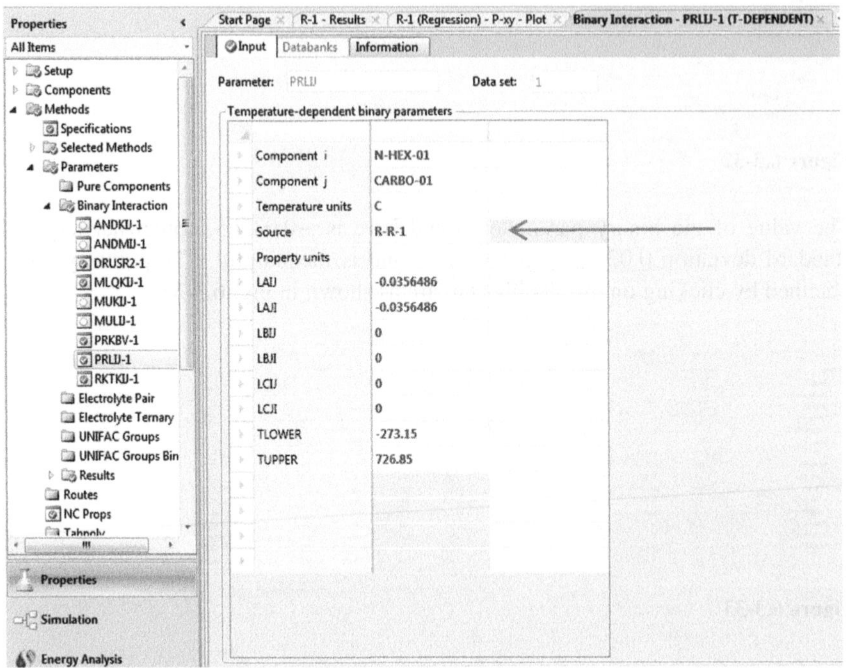

Figure 6.3-35

PROBLEMS

6.1. The following vapor–liquid equilibrium data are available for ethane–ethylene mixtures at −0.01°C.

Mol% ethane

Pressure (bar)	Liquid	Vapor
7.8	6.2	39.73
22.8	19.7	37.07
30.3	25.5	35.60
59.0	53.1	32.13
89.0	85.4	25.45

Find the value of the binary interaction parameter in the Peng–Robinson equation of state with the van der Waals one-fluid mixing rules that best fits these data.

6.2. Repeat Problem 6.1 using the Soave–Redlich–Kwong equation of state.

6.3. The following vapor–liquid equilibrium data are available for the system carbon dioxide (1) + isobutane (2) at 273.15 K.

P (bar)	x_1	y_1	P (bar)	x_1	y_1
1.57	0.000	0.000	14.793	0.317	0.877
2.736	0.022	0.422	16.718	0.378	0.899
3.546	0.037	0.541	17.63	0.403	0.913
4.256	0.038	0.623	18.34	0.416	0.914
5.167	0.053	0.690	21.379	0.513	0.922
6.282	0.073	0.744	23.811	0.598	0.925
7.498	0.098	0.783	26.445	0.697	0.947
8.511	0.149	0.800	29.485	0.817	0.963
9.93	0.166	0.830	32.423	0.917	0.981
11.855	0.224	0.863	34.855	1.000	1.000

Find the value of the binary interaction parameter in the Peng–Robinson equation of state with the van der Waals one-fluid mixing rules that best fits these data.

6.4. Repeat Problem 6.3 using the Soave–Redlich–Kwong equation of state.

6.5. Find the value of the binary interaction parameter in the Peng–Robinson equation of state with the van der Waals one-fluid mixing rules that best fits the data for the carbon dioxide + propane mixture at 270 K that you found in Problem 4.9.

6.6. Find the value of the binary interaction parameter in the Peng–Robinson equation of state with the van der Waals one-fluid mixing rules that best fits the data for the carbon dioxide + n-butane mixture at 310.9 K that you found in Problem 4.10.

6.7. Use the NIST TDE to find vapor–liquid equilibrium data for a binary system of your choice containing carbon dioxide as one of the components. Find the value of the binary interaction parameter in the Peng–Robinson equation of state with the van der Waals one-fluid mixing rules that best fits these data.

6.8. Repeat Problem 6.5 using the Soave–Redlich–Kwong equation of state.

Chapter 7

Regression of Liquid–Liquid Equilibrium (LLE) Data and Vapor–Liquid–Liquid Equilibrium (VLLE) and Predictions

In this chapter we consider liquid–liquid equilibrium (LLE) and vapor–liquid–liquid equilibrium (VLLE). Since LLE and VLLE only occur in very nonideal mixtures, generally involving very different chemical species, activity coefficient models are used in this chapter. In Section 7.1 the regression of LLE data to obtain parameters in activity coefficient models is examined. The prediction of liquid–liquid and vapor–liquid–liquid equilibria is considered in Section 7.2. Finally, in Section 7.3 the case of high pressure VLLE is examined in which an activity coefficient model is used to describe the two liquid phases, but because of the pressures involved, an equation of state is used for the vapor phase.

7.1 LIQUID–LIQUID DATA REGRESSION

In this section we consider the regression of binary liquid–liquid equilibrium (LLE) data to obtain values of the parameters in activity models. This is being considered separately from the regression of vapor–liquid equilibrium data since it can be a little more difficult to implement correctly, and the results from the various models can be much more different than is the case for vapor–liquid equilibria. This is because in VLE the equality of the fugacities of each species in each phase leads to $f_i^{\mathrm{L}} = x_i\gamma_i P^{\mathrm{vap}} = f_i^{\mathrm{V}} = y_i P$ or $\frac{y_i}{x_i} = \frac{\gamma_i P^{\mathrm{vap}}}{P}$, so that the ratio of the mole fractions in the two phases is a result of the combined effects of the activity coefficient in the liquid phase and the ratio of the vapor pressure to the total pressure. However, in LLE the equality of fugacities of each species in each phase leads to $f_i^{\mathrm{I}} = x_i^{\mathrm{I}}\gamma_i^{\mathrm{I}} P^{\mathrm{vap}} = f_i^{\mathrm{II}} = x_i^{\mathrm{II}}\gamma_i^{\mathrm{II}} P^{\mathrm{vap}}$ or $\frac{x_i^{\mathrm{II}}}{x_i^{\mathrm{I}}} = \frac{\gamma_i^{\mathrm{I}}}{\gamma_i^{\mathrm{II}}}$, so that the ratio of the species mole fractions in the two phases in LLE depends completely on the ratio of activity coefficients. As a result, the calculation of liquid–liquid equilibrium is very sensitive to the thermodynamic models used, the values of the activity coefficients, and their accuracy over the whole concentration

Using Aspen Plus® in Thermodynamics Instruction: A Step-by-Step Guide, First Edition. Stanley I. Sandler.

range. Small differences in activity coefficient values may have only a small effect on the predicted VLE, but result in a large change in the calculated LLE. Also, the use of different activity coefficient models may result in VLE correlations of similar accuracy, but quite different LLE predictions.

In fact, what we see in this chapter is that there can be some qualitative and quantitative inconsistencies when correlating liquid–liquid equilibrium data and then using the parameters so obtained to predict vapor–liquid–liquid equilibria. This is why in the predictive UNIFAC model, there is one set of parameters only for use in vapor–liquid equilibrium predictions and a separate UNIFAC-LLE parameter set only for LLE predictions.

To correlate liquid–liquid equilibrium data using Aspen Plus®, the starting point is, as usual, to begin at **Setup**. Then add a title (if you wish), and choose the units (I have chosen metric METCBAR). We will start with the UNIQUAC model and then move on the NRTL model; the Wilson model will not be used since it does not predict LLE.

Next, use the **Components Specifications>Selection** described in Chapter 1 to enter the components, here methanol and *n*-hexane; this mixture is known to exhibit LLE and will be used as an example (see Fig. 7.1-1)

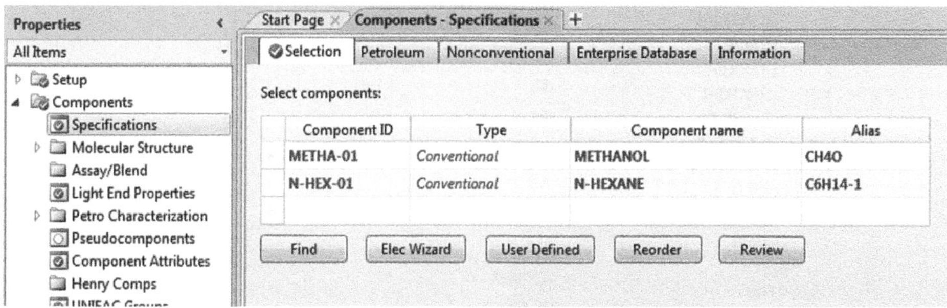

Figure 7.1-1

and through **Properties>Methods** the **UNIQUAC** model is chosen in Fig. 7.1-2.

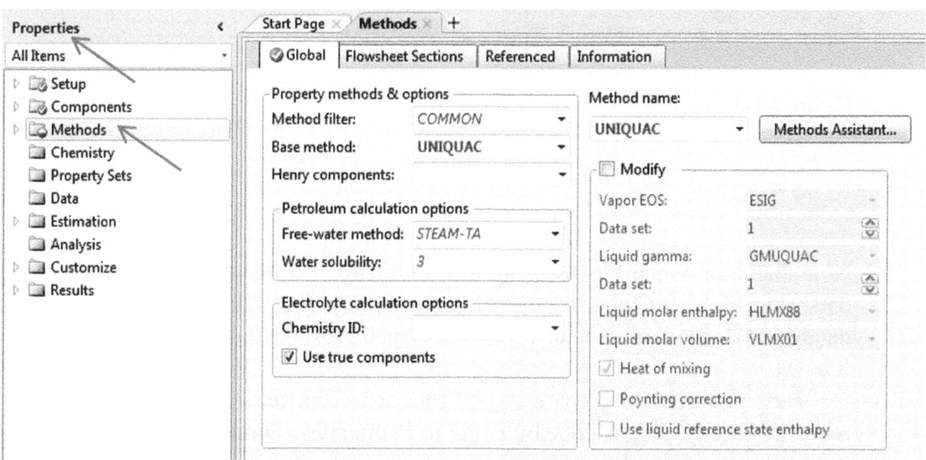

Figure 7.1-2

Choosing **Methods>Parameters>Binary Interaction** brings up the parameter list in Fig. 7.1-3, and clicking on **UNIQ-1** displays the values in the Aspen data bank. These will be used as the initial guess.

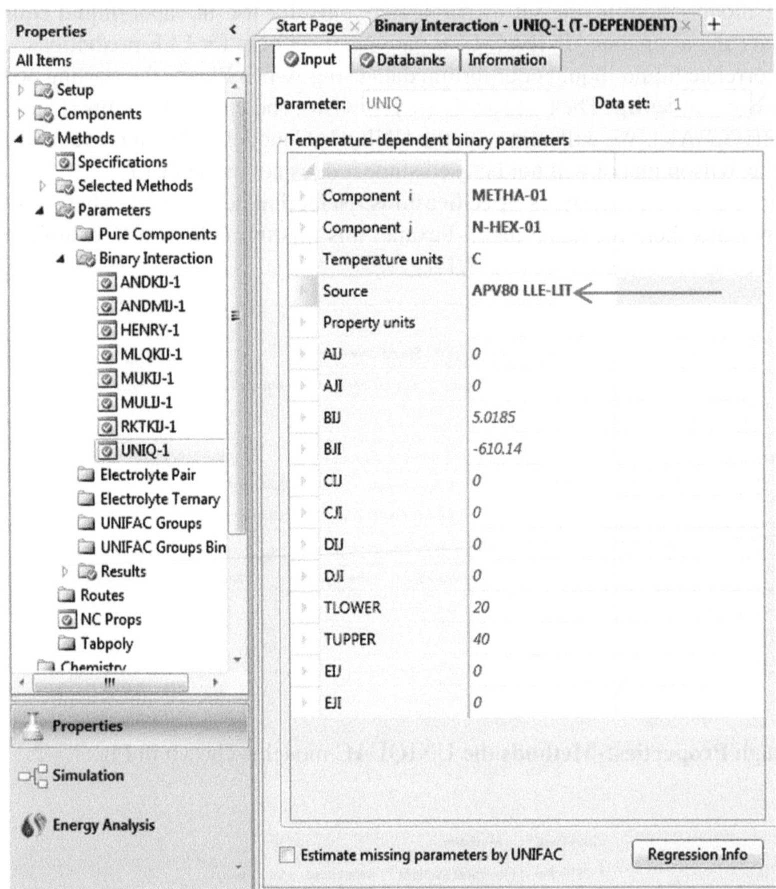

Figure 7.1-3

Note that here Aspen Plus is using the UNIQUAC parameters from its APV80 LLE-LIT database, while in Chapter 5 the VLE parameters were taken from the APV80 VLE-IG database. [Depending on your version of Aspen Plus, the database may be APV 8X where $X > 0$.]

Next you need to prepare Aspen Plus to accept the experimental data that will be retrieved shortly. This is done by going to **Properties>Data** and setting up a new data set that will contain the data to be regressed. The default name of the data set **D-1** has been used in Fig. 7.1-4.

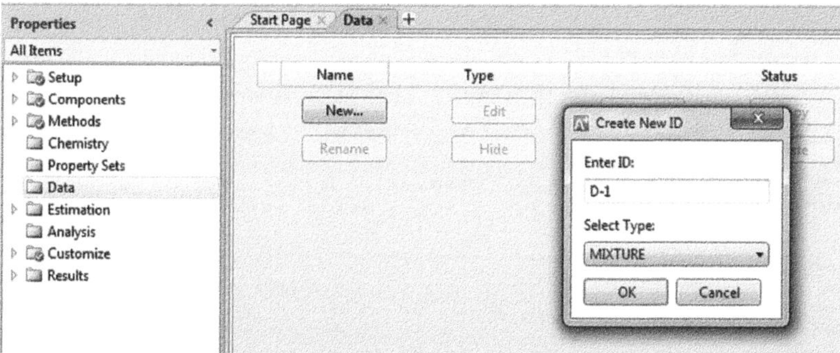

Figure 7.1-4

Add both available components identified in the **Data>D-1>Setup** step as shown in Fig. 7.1-5, choose the **Category** as **Phase equilibrium** and the **Data type** as **TXX.**

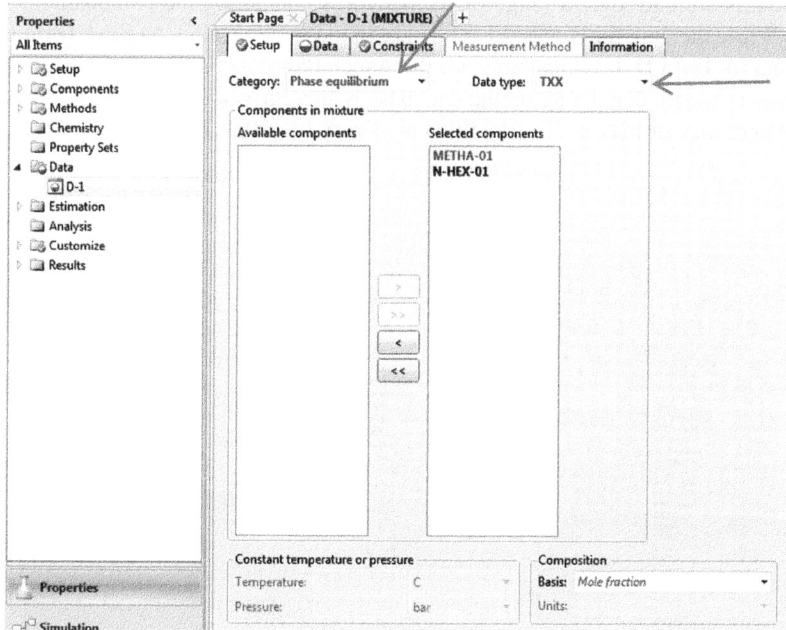

Figure 7.1-5

To proceed further, we need experimental data to regress—the NIST Thermodynamic Data Engine (TDE) accessed from **Retrieve TDE Binary Data**. This is done as follows. Click on the **Data** tab and then **Retrieve TDE Binary Data . . .** as shown in Fig. 7.1-6.

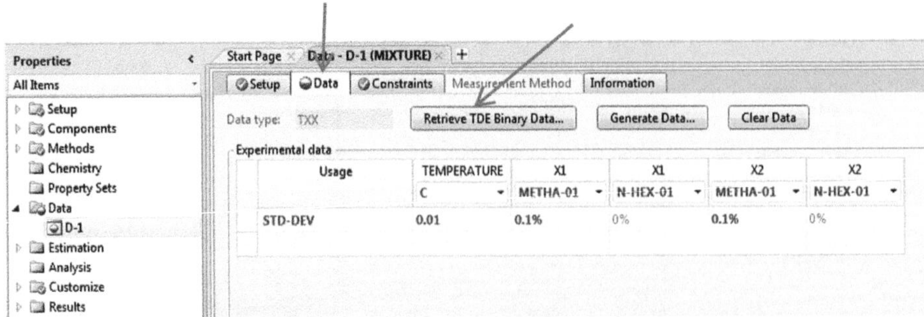

Figure 7.1-6

This brings up, in Fig. 7.1-7, a number of liquid–liquid equilibrium data sets that are available. A reasonably small set of eight points, **Binary LLE 027**, contains the compositions for both liquid phases and has been chosen for this example (Fig. 7.1-8). Note that some of the data sets have the compositions of the two phases but at only a single temperature, and others have numerous data points but report the composition of only one of the liquid phases. Such data sets are not as useful for the determination of binary parameters by regression as are data over a range of temperatures containing the compositions of both phases. While the data for Binary LLE 027 are in the window click on **Save Data** that fills in the data for **Data D-1.** [Note that since the Aspen Plus data bank is periodically updated and data sets added, depending on the data base with your version of Aspen Plus, the data set used here may or may not be identified as Binary LLE 027.]

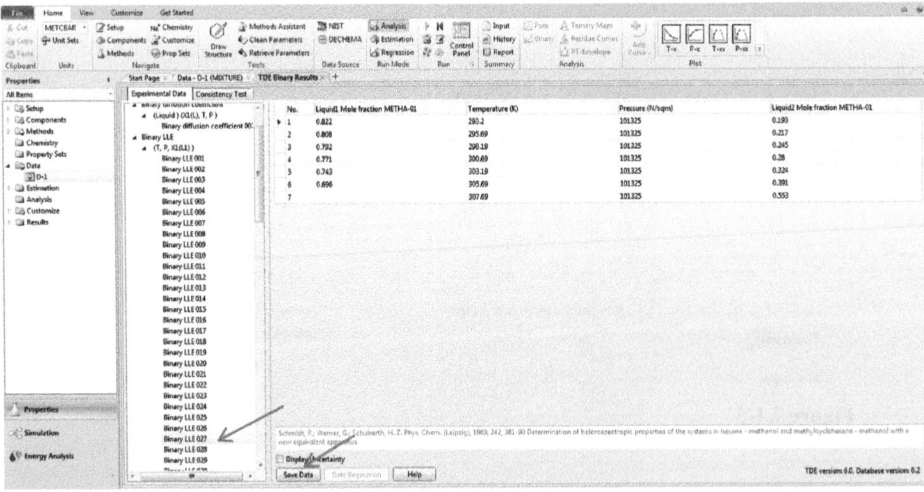

Figure 7.1-7

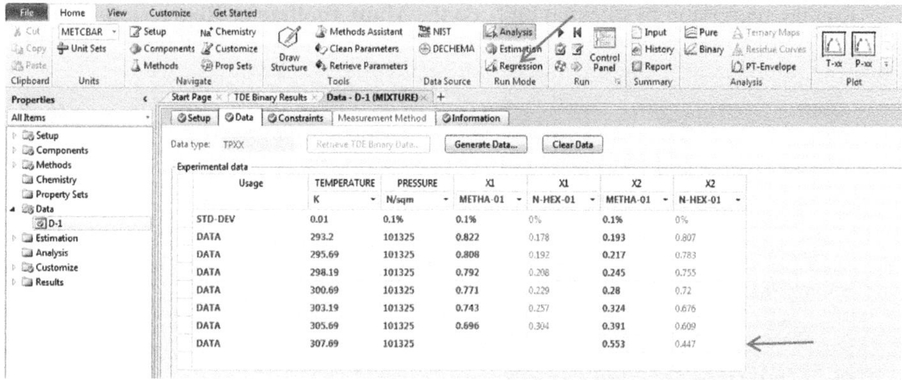

Figure 7.1-8

In order to simplify the regression, the last (highest temperature) data point that is in the one-phase region will be eliminated by clicking on it and then on **Delete Row** (Fig. 7.1-9).

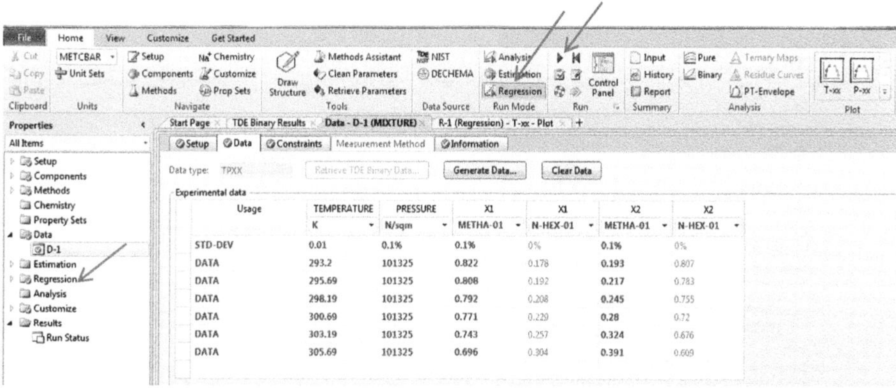

Figure 7.1-9

Next click on **Regression** on the toolbar shown in Fig. 7.1-9, and then **New** and create an ID for the regression calculation. Here the default name **R-1** has been used in Fig. 7.1-10, and then click on **OK**.

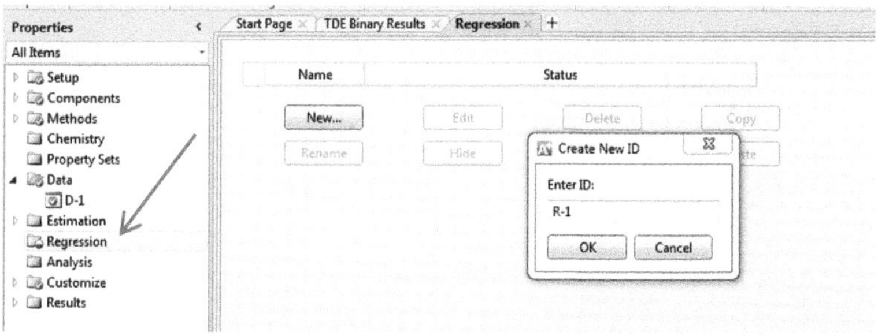

Figure 7.1-10

Run the regression by clicking on > on the **Main Toolbar** or **Regression** on the left.

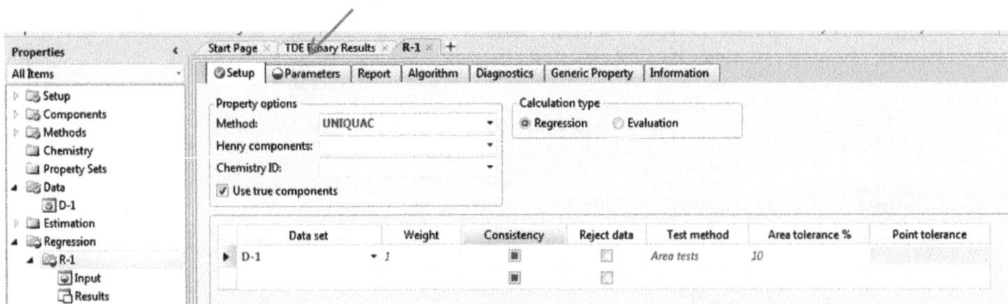

Figure 7.1-11

On the screen in Fig. 7.1-11, click on the **Parameters tab** to see the parameters to be regressed and, depending on the system, possibly the Aspen Plus initial guesses (from the Aspen Plus data bank) for the parameters as shown below. Note that here, since a range of temperatures are involved, two interaction parameters (referred to as elements) for each binary pair will be used (Fig. 7.1-12).

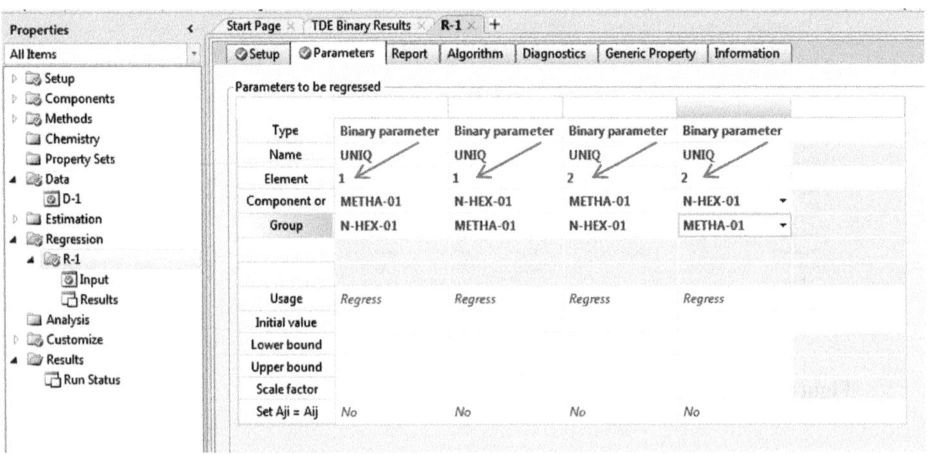

Figure 7.1-12

Then click on > on the **Main Toolbar** or Shift F5 (Fig. 7.1-13).

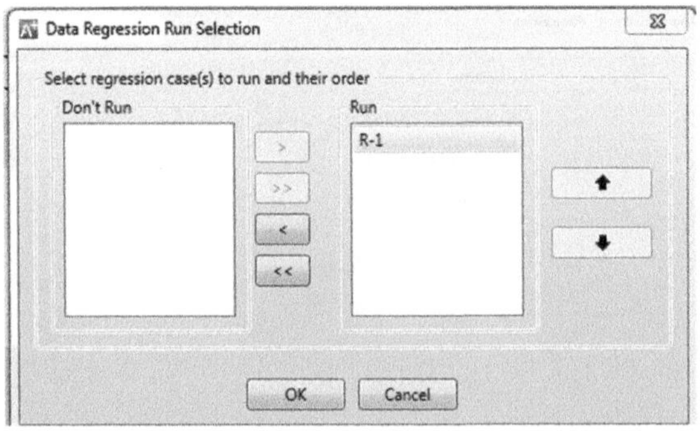

Figure 7.1-13

Click **OK** to run the regression. The pop-up window of Fig. 7.1-14 will appear and click **Yes**.

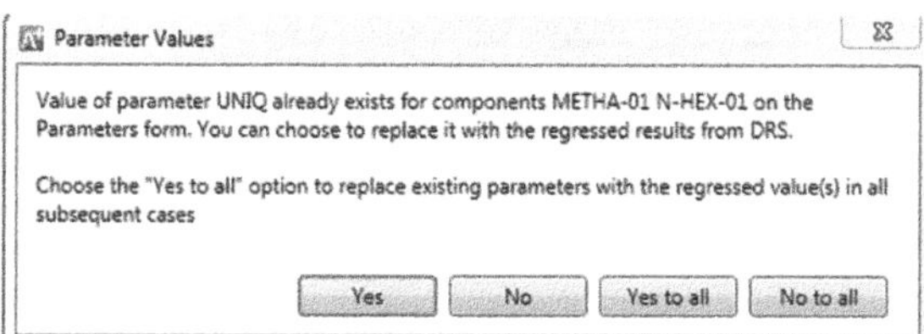

Figure 7.1-14

Clicking on **Regression>R-1>Results** produces the window of Fig. 7.1-15.

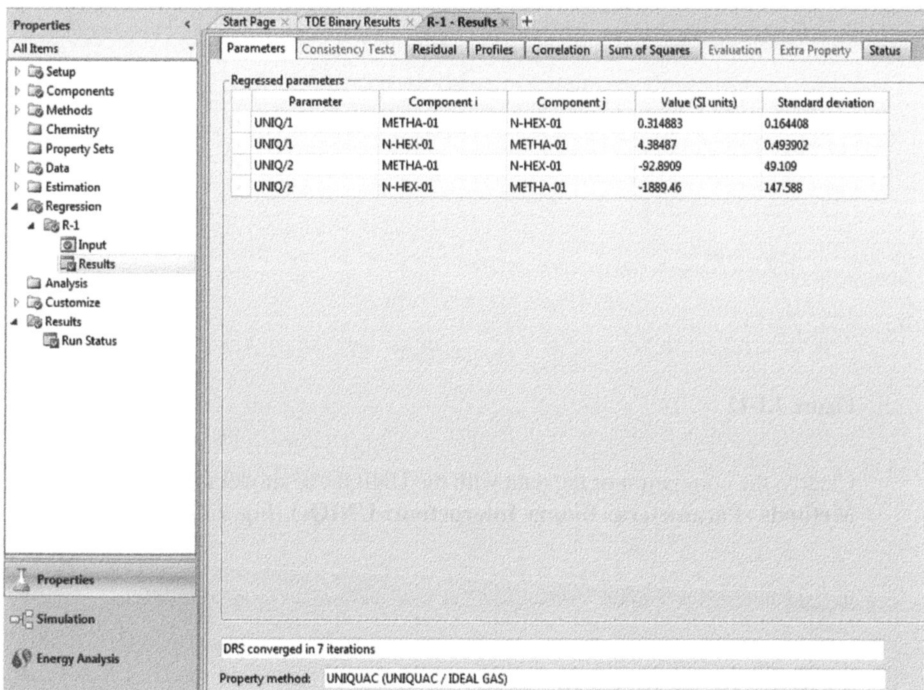

Parameter	Component i	Component j	Value (SI units)	Standard deviation
UNIQ/1	METHA-01	N-HEX-01	0.314883	0.164408
UNIQ/1	N-HEX-01	METHA-01	4.38487	0.493902
UNIQ/2	METHA-01	N-HEX-01	-92.8909	49.109
UNIQ/2	N-HEX-01	METHA-01	-1889.46	147.588

Figure 7.1-15

We see in Fig. 7.1-15 that the regression converged in 7 iterations (without any errors or warnings) and that the parameters obtained are all larger than their standard deviations and therefore are meaningful. Clicking on the **Profiles** tab in Fig. 7.1-16 shows that the results are quite good.

Data set: D-1

Summary of regression results

Exp Val TEMP	Est Val TEMP	Exp Val PRES	Est Val PRES	Exp Val MOLEFRAC X1 METHA-01	Est Val MOLEFRAC X1 METHA-01	Exp Val MOLEFRAC X1 N-HEX-01	Est Val MOLEFRAC X1 N-HEX-01	Exp Val MOLEFRAC X2 METHA-01	Est Val MOLEFRAC X2 METHA-01	Exp Val MOLEFRAC X2 N-HEX-01	Est Val MOLEFRAC X2 N-HEX-01
C	C	bar	bar								
20.05	20.116	1.01325	1.01325	0.822	0.82315	0.178	0.17685	0.193	0.191272	0.807	0.808728
22.54	22.5535	1.01325	1.01325	0.808	0.808458	0.192	0.191542	0.217	0.216776	0.783	0.783224
25.04	24.9909	1.01325	1.01325	0.792	0.791007	0.208	0.208993	0.245	0.246429	0.755	0.753571
27.54	27.4675	1.01325	1.01325	0.771	0.76933	0.229	0.23067	0.28	0.282082	0.72	0.717918
30.04	29.9691	1.01325	1.01325	0.743	0.741291	0.257	0.25871	0.324	0.326022	0.676	0.673979
32.54	32.6531	1.01325	1.01325	0.696	0.698646	0.304	0.301354	0.391	0.387931	0.609	0.612069

Figure 7.1-16

Next, clicking on **Plot>T-xx** (use the drop-down menu to get to **T-xx** in the **Plot** region of the **Main Toolbar**) in Fig. 7.1-16 produces Fig. 7.1-17 showing the experimental data and the correlated results.

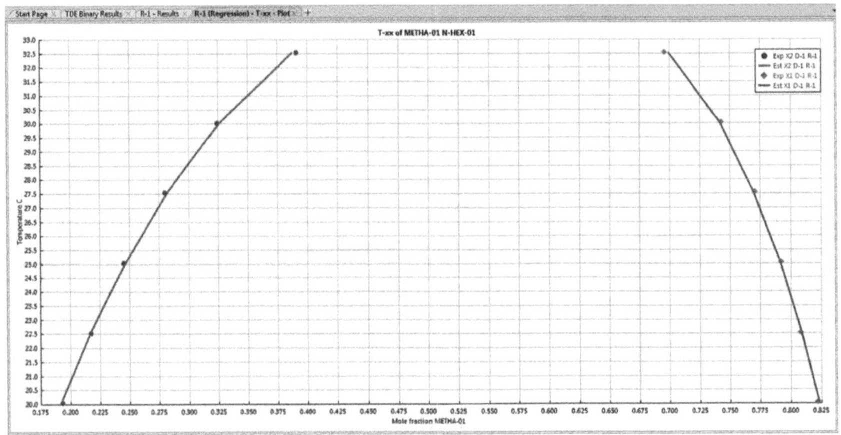

Figure 7.1-17

Clearly, the correlation of the data with the UNIQUAC model is very good. Going back to **Methods>Parameters>Binary Interaction>UNIQ-1** (Fig. 7.1-18) brings up Fig. 7.1-18.

Name	Type	Status
ANDKIJ-1	T-DEPENDENT	Results Not Available
ANDMIJ-1	T-DEPENDENT	Results Not Available
DRUSR2-1	T-DEPENDENT	Results Not Available
HENRY-1	T-DEPENDENT	Results Not Available
MLQKIJ-1	SCALAR	Input Complete
MUKIJ-1	T-DEPENDENT	Results Not Available
MULIJ-1	T-DEPENDENT	Results Not Available
RKTKIJ-1	SCALAR	Input Complete
UNIQ-1	T-DEPENDENT	Input Complete

Figure 7.1-18

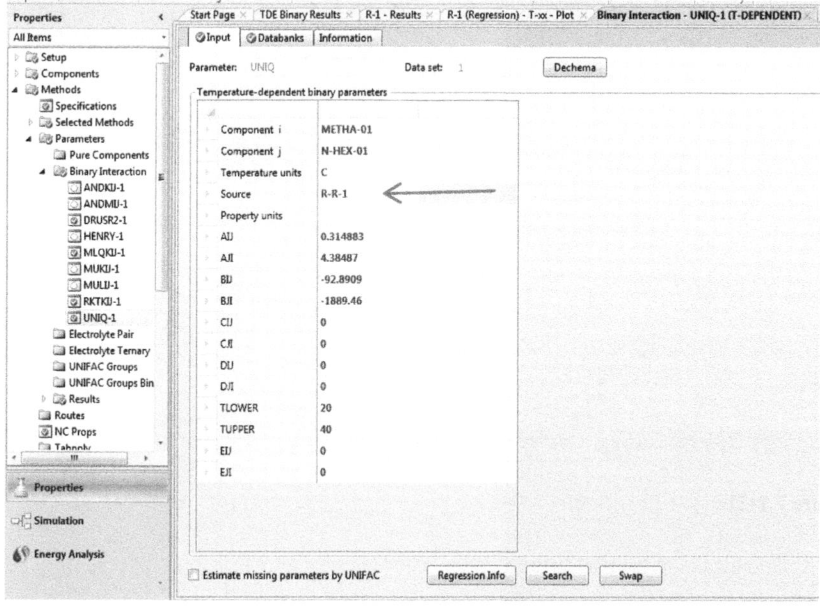

Figure 7.1-19

The table in Fig. 7.1-19 shows that the parameters just obtained will now be used for all further calculations, and the source is no longer the Aspen APV80 LLE-LIT database, but the values from the just completed regression, and denoted as **R-R-1**. Next choose **Binary** under **Analysis** on the main toolbar, **Vapor–Liquid–Liquid** under **Valid phases**, and then the other choices as shown in Fig. 7.1-20, and then **Run Analysis** produces Fig. 7.1-21

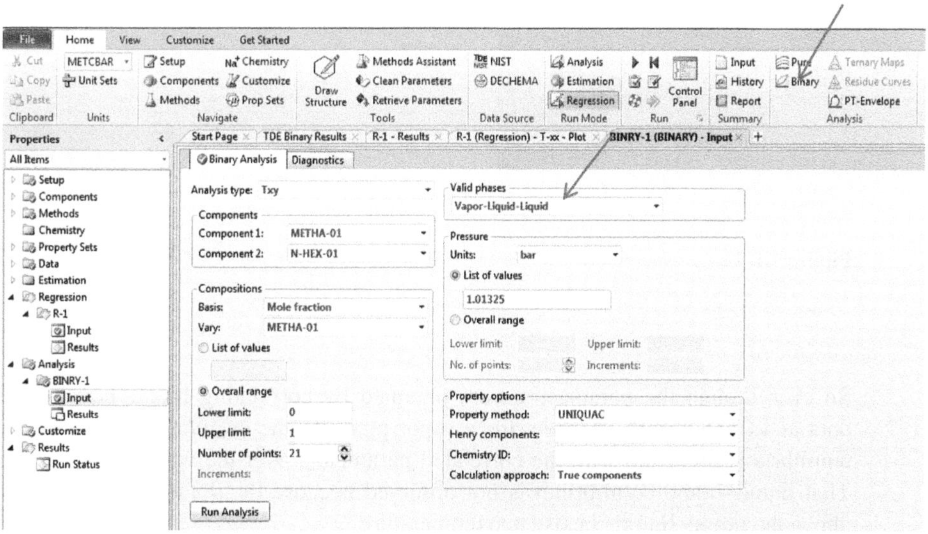

Figure 7.1-20

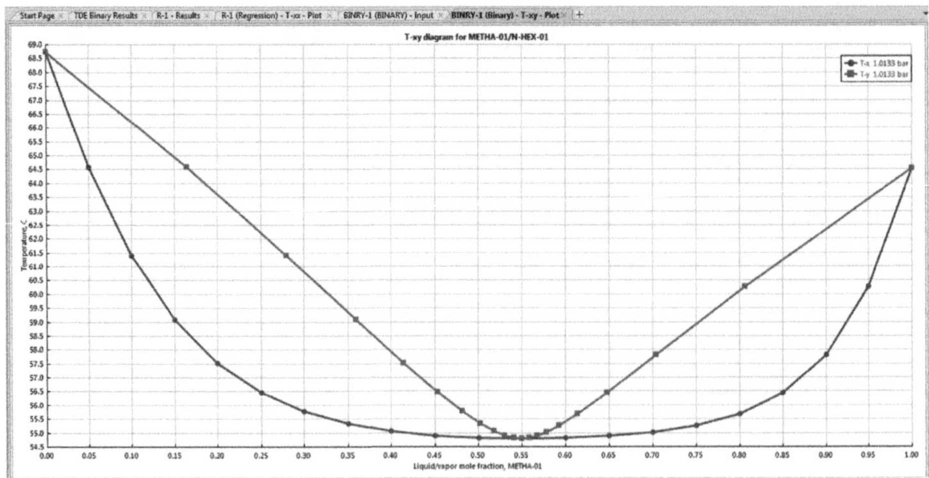

Figure 7.1-21

and the table of results in Fig. 7.1-22.

Binary analysis results

PRES	MOLEFRAC METHA-01	TOTAL TEMP	TOTAL KVL METHA-01	TOTAL KVL N-HEX-01	LIQUID1 GAMMA METHA-01	LIQUID1 GAMMA N-HEX-01	LIQUID2 GAMMA METHA-01	LIQUID2 GAMMA N-HEX-01	TOTAL BETA	VAPOR MOLEFRAC METHA-01	VAPOR MOLEFRAC N-HEX-01	LIQUID1 MOLEFRAC METHA-01	LIQUID1 MOLEFRAC N-HEX-01	LIQUID2 MOLEFRAC METHA-01	LIQUID2 MOLEFRAC N-HEX-01
bar		C													
1.01325	0	68.7345	3.86842	1	3.28678	1			1	0	1	0	1		
1.01325	0.05	64.5767	3.29728	0.879080	3.29184	1.0026			1	0.16486	0.83514	0.05	0.95		
1.01325	0.1	61.3862	2.80261	0.799715	3.17639	1.01155			1	0.280253	0.719747	0.1	0.9		
1.01325	0.15	59.0996	2.3967	0.753546	2.96011	1.0281			1	0.359466	0.640534	0.15	0.85		
1.01325	0.2	57.525	2.07147	0.732138	2.74772	1.06309			1	0.414286	0.585714	0.2	0.8		
1.01325	0.25	56.4704	1.81197	0.729366	2.51086	1.08727			1	0.452985	0.547015	0.25	0.75		
1.01325	0.3	55.7787	1.60339	0.74135	2.28701	1.11164			1	0.481025	0.518975	0.3	0.7		
1.01325	0.35	55.3327	1.43398	0.766309	2.08385	1.16757			1	0.501896	0.498104	0.35	0.65		
1.01325	0.4	55.0604	1.29468	0.80365	1.90291	1.25702			1	0.517868	0.482132	0.4	0.6		
1.01325	0.45	54.9012	1.17888	0.851687	1.74418	1.34289			1	0.53048	0.46952	0.45	0.55		
1.01325	0.5	54.823	1.08166	0.918283	1.60877	1.44828			1	0.540846	0.459154	0.5	0.5		
1.01325	0.55	54.8009	0.999783	1.00027	1.4656	1.57878			1	0.54988	0.45012	0.55	0.45		
1.01325	0.6	54.8222	0.930713	1.10392	1.38173	1.7411			1	0.558429	0.441571	0.6	0.4		
1.01325	0.65	54.8896	0.871009	1.23586	1.2934	1.94468			1	0.567454	0.432546	0.65	0.35		
1.01325	0.7	55.0183	0.826015	1.40897	1.21618	2.20261			1	0.57821	0.42179	0.7	0.3		
1.01325	0.75	55.2534	0.790152	1.6296	1.15187	2.53239			1	0.592608	0.407392	0.75	0.25		
1.01325	0.8	55.8739	0.767217	1.9812	1.09904	2.96814			1	0.613766	0.386234	0.8	0.2		
1.01325	0.85	56.4335	0.76168	2.35134	1.05704	3.50942			1	0.647372	0.352628	0.85	0.15		
1.01325	0.9	57.8011	0.782402	2.95679	1.02608	4.21648			1	0.704136	0.295864	0.9	0.1		
1.01325	0.95	60.289	0.848728	3.87417	1.00872	5.08896			1	0.806291	0.193709	0.95	0.05		
1.01325	1	64.5348	1	5.29995	1	6.04152			1	1	0	1	0		

Figure 7.1-22

So even though the parameters were obtained by correlating liquid–liquid equilibrium data at 1.0135 bar, if we allow for a vapor phase in the calculations, only vapor–liquid equilibria is predicted with the correlated parameters over the whole composition range. That liquid–liquid equilibrium is not observed because the default temperature range is above the liquid–liquid coexistence temperature.

For a comparison, we repeat the regression, but now with the NRTL model by going back to **Methods>Parameters>Binary Interaction>NRTL-1** and then proceeding as shown in Figs. 7.1-23 to 7.1-28.

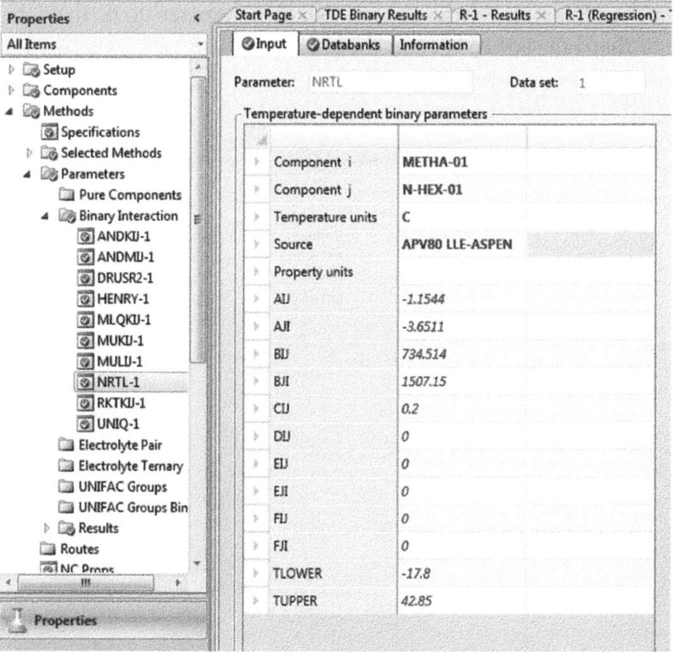

Figure 7.1-23

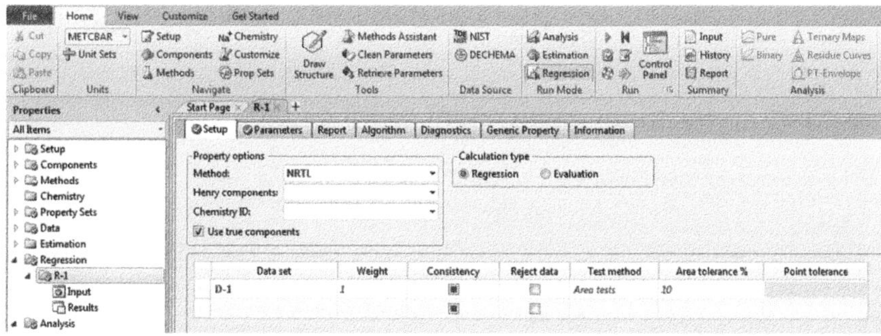

Figure 7.1-24

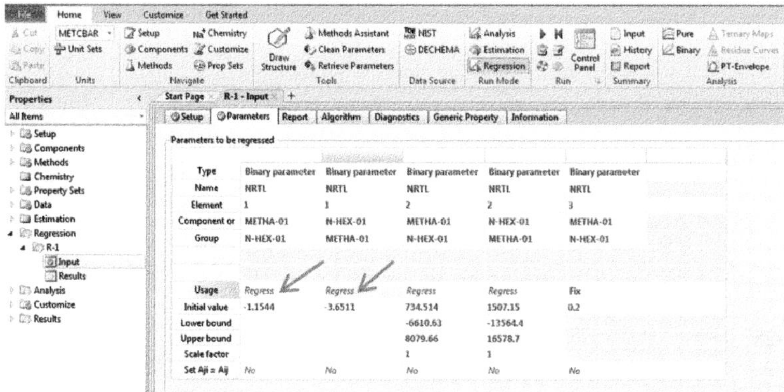

Figure 7.1-25

Note that the Aspen Plus default is for the first two NRTL parameters to be fixed, but I have changed the usage to ***Regress***. The results for the parameters are

Regressed parameters

Parameter	Component i	Component j	Value (SI units)	Standard deviation
NRTL/1	METHA-01	N-HEX-01	3.06781	0.894046
NRTL/1	N-HEX-01	METHA-01	-8.9614	0.881253
NRTL/2	METHA-01	N-HEX-01	-475.87	266.958
NRTL/2	N-HEX-01	METHA-01	3007.49	263.276

Figure 7.1-26

Data set: D-1

Summary of regression results

Exp Val TEMP	Est Val TEMP	Exp Val PRES	Est Val PRES	Exp Val MOLEFRAC X1 METHA-01	Est Val MOLEFRAC X1 METHA-01	Exp Val MOLEFRAC X1 N-HEX-01	Est Val MOLEFRAC X1 N-HEX-01	Exp Val MOLEFRAC X2 METHA-01	Est Val MOLEFRAC X2 METHA-01	Exp Val MOLEFRAC X2 N-HEX-01	Est Val MOLEFRAC X2 N-HEX-01
C	C	bar	bar								
20.05	20.1119	1.01325	1.01325	0.822	0.822991	0.178	0.177009	0.193	0.1913	0.807	0.8087
22.54	22.5541	1.01325	1.01325	0.808	0.80848	0.192	0.19152	0.217	0.216765	0.783	0.783236
25.04	24.994	1.01325	1.01325	0.792	0.79113	0.208	0.20887	0.245	0.246394	0.755	0.753606
27.54	27.4708	1.01325	1.01325	0.771	0.769468	0.229	0.230532	0.28	0.282039	0.72	0.717961
30.04	29.9703	1.01325	1.01325	0.743	0.741336	0.257	0.258664	0.324	0.32599	0.676	0.67401
32.54	32.6491	1.01325	1.01325	0.696	0.698443	0.304	0.301557	0.391	0.387945	0.609	0.612055

Figure 7.1-27

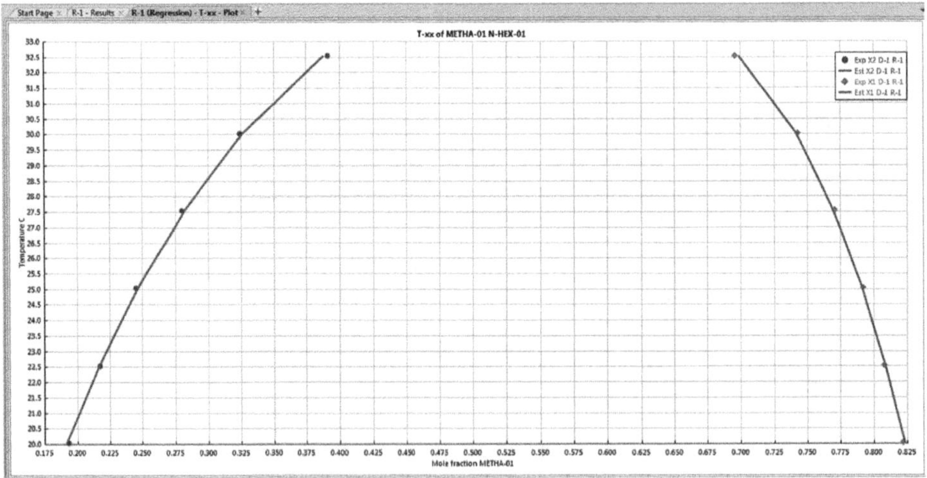

Figure 7.1-28

So the results of using the NRTL model are good, and comparable to those obtained with the UNIQUAC model. Repeating the steps, but now with the NRTL model produces the results in Figs. 7.1-29 and 7.1-30.

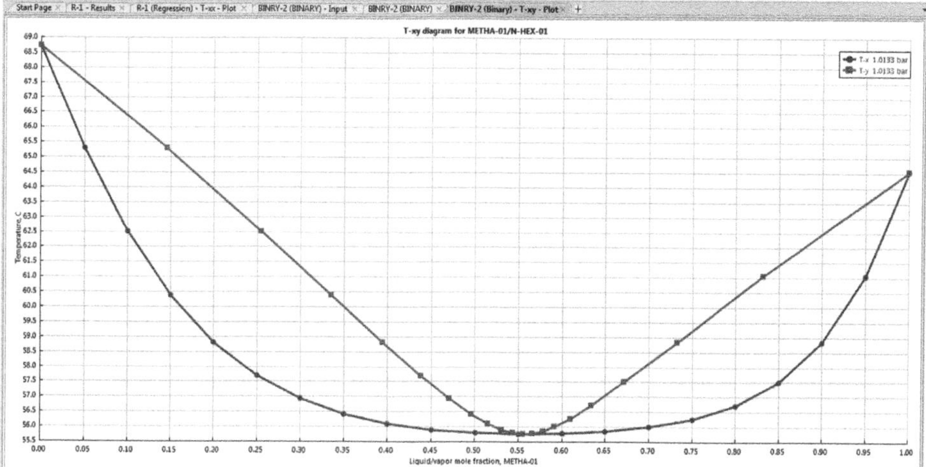

Figure 7.1-29

Binary analysis results

PRES	MOLEFRAC METHA-01	TOTAL TEMP	TOTAL KVL METHA-01	TOTAL KVL N-HEX-01	LIQUID1 GAMMA METHA-01	LIQUID1 GAMMA N-HEX-01	LIQUID2 GAMMA METHA-01	LIQUID2 GAMMA N-HEX-01	TOTAL BETA	VAPOR MOLEFRAC METHA-01	VAPOR MOLEFRAC N-HEX-01	LIQUID1 MOLEFRAC METHA-01	LIQUID1 MOLEFRAC N-HEX-01	LIQUID2 MOLEFRAC METHA-01
bar		C												
1.01325	0	68.7345	3.31004	1	2.81236	1			1	0	1	0	1	
1.01325	0.05	65.5077	2.91196	0.893368	2.82485	1.00202			1	0.143597	0.854403	0.05	0.95	
1.01325	0.1	62.5172	2.54976	0.827771	2.76193	1.00905			1	0.25501	0.74499	0.1	0.9	
1.01325	0.15	60.3789	2.23599	0.78188	2.63937	1.02238			1	0.335399	0.664601	0.15	0.85	
1.01325	0.2	58.8093	1.97132	0.757103	2.48088	1.04302			1	0.394273	0.605727	0.2	0.8	
1.01325	0.25	57.7005	1.7515	0.74965	2.30654	1.07192			1	0.437822	0.562178	0.25	0.75	
1.01325	0.3	56.9289	1.56822	0.756431	2.13211	1.1102			1	0.470481	0.529519	0.3	0.7	
1.01325	0.35	56.4157	1.41546	0.77618	1.96586	1.15921			1	0.495451	0.504849	0.35	0.65	
1.01325	0.4	56.0896	1.28742	0.808375	1.81231	1.22078			1	0.514973	0.485028	0.4	0.6	
1.01325	0.45	55.8923	1.17927	0.853321	1.67396	1.29736			1	0.530672	0.469328	0.45	0.55	
1.01325	0.5	55.7881	1.08756	0.912433	1.55051	1.39219			1	0.543782	0.456218	0.5	0.5	
1.01325	0.55	55.7518	1.00969	0.988131	1.44169	1.50956			1	0.555337	0.444663	0.55	0.45	
1.01325	0.6	55.7716	0.943905	1.08417	1.34664	1.65316			1	0.566337	0.433663	0.6	0.4	
1.01325	0.65	55.8468	0.889077	1.20605	1.26443	1.83649			1	0.57789	0.42211	0.65	0.35	
1.01325	0.7	55.9952	0.844924	1.36186	1.19422	2.06328			1	0.591444	0.408557	0.7	0.3	
1.01325	0.75	56.2605	0.812158	1.5637	1.13529	2.34774			1	0.609092	0.390908	0.75	0.25	
1.01325	0.8	56.7136	0.792492	1.82939	1.08712	2.70464			1	0.634059	0.365941	0.8	0.2	
1.01325	0.85	57.5049	0.790494	2.18719	1.04943	3.14814			1	0.671922	0.328079	0.85	0.15	
1.01325	0.9	58.8384	0.813354	2.6794	1.02219	3.68766			1	0.73204	0.26796	0.9	0.1	
1.01325	0.95	61.047	0.875214	3.37093	1.0056	4.31174			1	0.831453	0.168547	0.95	0.05	
1.01325	1	64.5348	1	4.34912	1	4.98686			1	1	0	1	0	

Figure 7.1-30

There is an important observation from this graph and table. It is that the NRTL model, like the UNIQUAC model discussed before it, with parameters obtained by correlating liquid–liquid equilibrium data when the allowed phases were restricted to only two liquids leads to predictions of only vapor–liquid equilibrium, not LLE when a vapor phase is also allowed. Again, this is because of the default temperature range considered in the calculations being above the liquid–liquid coexistence temperature.

These results are a poignant reminder of why one should not blindly accept the results from Aspen Plus just because they come from a computer. Rather, one needs to do their own (at least mental) analysis of whether the results so obtained make sense.

7.2 THE PREDICTION OF LIQUID–LIQUID AND VAPOR–LIQUID–LIQUID EQUILIBRIUM

Liquid–liquid equilibrium (LLE) and vapor–liquid–liquid equilibrium (VLLE) calculations can easily be done in Aspen Plus when used in the **Simulation/Flowsheet** mode or to obtain a ternary diagram using **Properties>Analysis>Ternary>Maps** option from the main toolbar. Both of these will be illustrated here.

Before going to the use of Aspen Plus for VLLE calculations, there is an item of thermodynamics to be considered, the Gibbs phase rule

$$F = C - P - R + 2$$

where F is the number of degrees of freedom (e.g., from among temperature, pressure, and compositions) that are free to be specified, C is the number of components, P is the number of phases, and R is the number of independent reactions. For a nonreacting ($R = 0$) three-phase ($P = 3$) system, $F = C - 3 - 0 + 2 = C - 1$. So in a binary system, where $C = 2$, there is only one degree of freedom, $F = 1$. That is, at each fixed composition (the one degree of freedom) there is only one temperature–pressure point at which three phases coexist. It would take a considerable number of trials to hit upon this one state in the two-dimensional (2D) T–P space. Furthermore, usually one is not interested in only this one point. However, if there are three components, $F = 2$, so there are two degrees of freedom, identifying the VLLE state at each composition is a much easier 1D search. For example, finding for the equilibrium pressure at fixed temperature and composition. Also, the ternary phase diagram is much more useful in developing separations processes (and especially extraction processes). Therefore, VLLE calculations are not usually performed unless three (or more) components are present. We consider one such example here.

We start by using the **Properties>Analysis>Ternary Maps** option here, but first go through the specifications in the **Setup**, **Components**, **Methods,** and **Parameters** steps as described a number of times previously. For the example here the water + *n*-octanol + *n*-hexane system is considered as it is known to exhibit LLE and VLLE and the UNIQUAC model will be used to describe this system. Using **Tools>Analysis>Property>Ternary** option from the main toolbar then brings up the menu in Fig. 7.2-1.

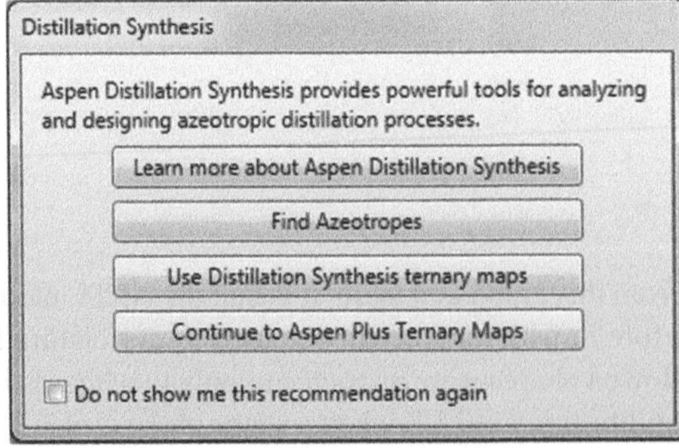

Figure 7.2-1

Continue by clicking on **Continue to Aspen Plus Ternary Maps** and the window in Fig. 7.2-2 appears. [You will learn about the other options in your later courses on mass transfer and separations.]

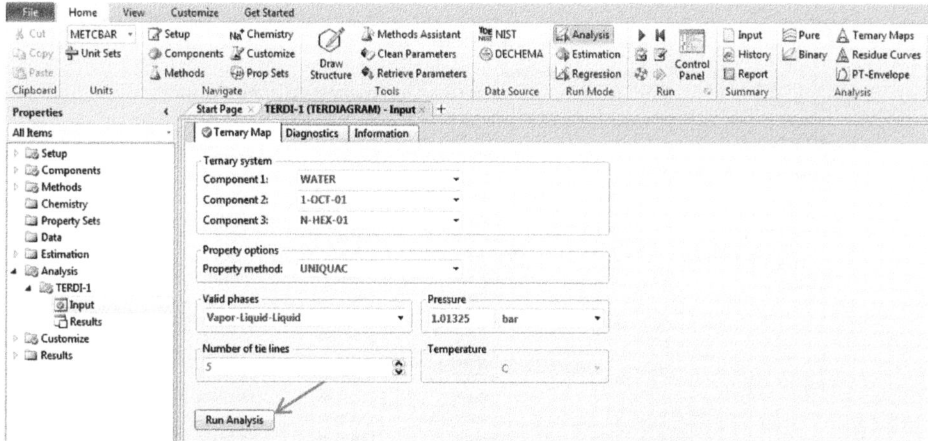

Figure 7.2-2

Clicking on **Run Analysis** produces the ternary diagram at 1 atm total pressure in Fig. 7.2-3 computed with the UNIQUAC model and the default parameters for this system. [Aspen Plus does not allow specifying temperature that can be more useful than specifying pressure, especially since LLE is only very weakly dependent on pressure.]

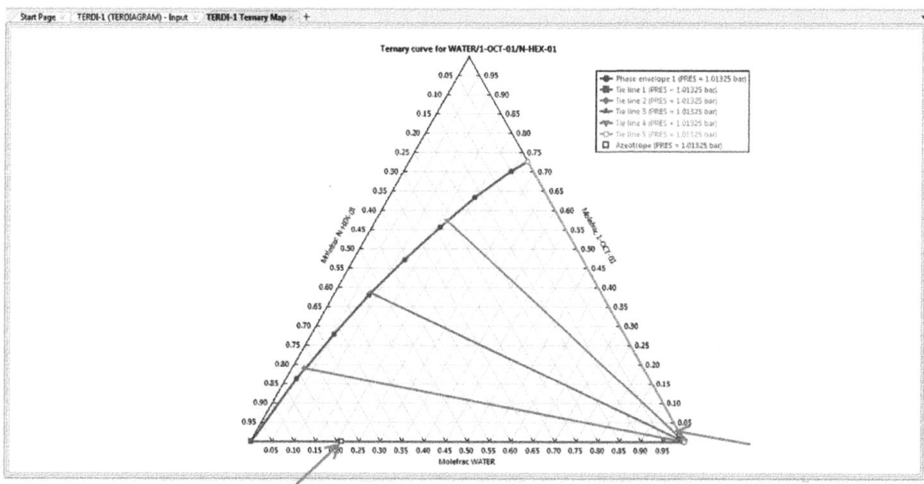

Figure 7.2-3

Clicking on **Analysis>TERDI-1>Results** and then clicking on the **Phase Envelope** tab gives the table in Fig. 7.2-4.

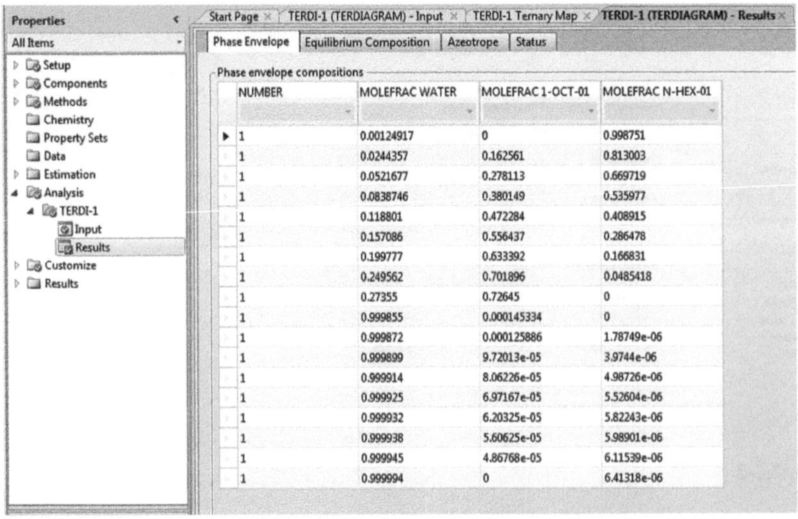

Phase envelope compositions

NUMBER	MOLEFRAC WATER	MOLEFRAC 1-OCT-01	MOLEFRAC N-HEX-01
1	0.00124917	0	0.998751
1	0.0244357	0.162561	0.813003
1	0.0521677	0.278113	0.669719
1	0.0838746	0.380149	0.535977
1	0.118801	0.472284	0.408915
1	0.157086	0.556437	0.286478
1	0.199777	0.633392	0.166831
1	0.249562	0.701896	0.0485418
1	0.27355	0.72645	0
1	0.999855	0.000145334	0
1	0.999872	0.000125886	1.78749e-06
1	0.999899	9.72013e-05	3.9744e-06
1	0.999914	8.06226e-05	4.98726e-06
1	0.999925	6.97167e-05	5.52604e-06
1	0.999932	6.20325e-05	5.82243e-06
1	0.999938	5.60625e-05	5.98901e-06
1	0.999945	4.86768e-05	6.11539e-06
1	0.999994	0	6.41318e-06

Figure 7.2-4

Then clicking on the **Equilibrium Composition** tab gives, in Fig. 7.2-5, the compositions of the two liquids and the vapor that coexist at each tie line shown in Fig. 7.2-3.

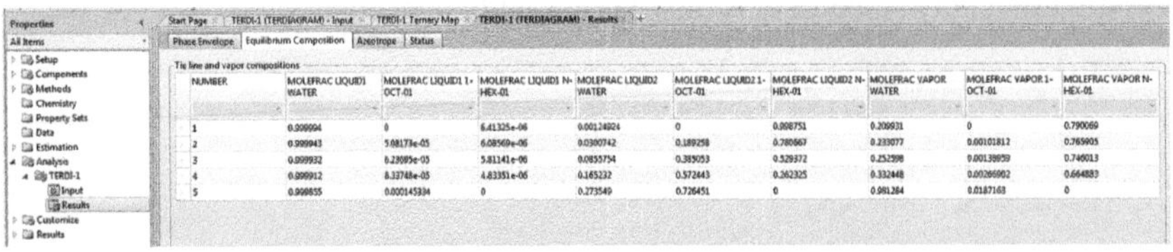

Tie line and vapor compositions

NUMBER	MOLEFRAC LIQUID1 WATER	MOLEFRAC LIQUID1 1-OCT-01	MOLEFRAC LIQUID1 N-HEX-01	MOLEFRAC LIQUID2 WATER	MOLEFRAC LIQUID2 1-OCT-01	MOLEFRAC LIQUID2 N-HEX-01	MOLEFRAC VAPOR WATER	MOLEFRAC VAPOR 1-OCT-01	MOLEFRAC VAPOR N-HEX-01
1	0.999994	0	6.41325e-06	0.00124924	0	0.998751	0.209931	0	0.790069
2	0.999943	5.08173e-05	6.08569e-06	0.0300742	0.189258	0.780667	0.233077	0.00101812	0.765905
3	0.999932	6.23085e-05	5.81141e-06	0.0855754	0.385053	0.529372	0.252598	0.00138959	0.746013
4	0.999912	8.33748e-05	4.83351e-06	0.165232	0.572443	0.262325	0.332448	0.00266902	0.664883
5	0.999855	0.000145334	0	0.273549	0.726451	0	0.981284	0.0187163	0

Figure 7.2-5

Further, clicking on the **Azeotrope** tab gives, in Fig. 7.2-6, the compositions of the vapor–liquid azeotropes that are formed in this system. For this mixture there are two binary mixture azeotropes (at the compositions indicated by the arrows in the ternary diagram of Fig. 7.2-3), and there are no ternary azeotropes. Indeed, ternary, and multicomponent azeotropes are not very common. Can you posit why? Also, the temperatures of these binary azeotropes are reported in Fig. 7.2-6; unfortunately Aspen Plus does not give temperatures elsewhere in this analysis, for example, for the tie lines.

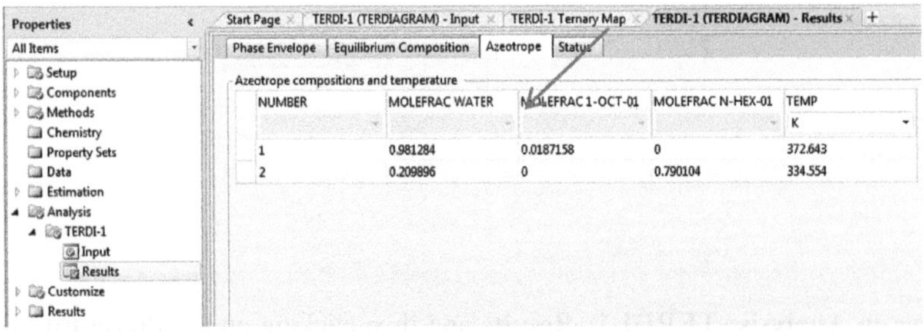

Azeotrope compositions and temperature

NUMBER	MOLEFRAC WATER	MOLEFRAC 1-OCT-01	MOLEFRAC N-HEX-01	TEMP
				K
1	0.981284	0.0187158	0	372.643
2	0.209896	0	0.790104	334.554

Figure 7.2-6

Before leaving the **Analysis>Ternary Maps** method of ternary analysis, it is useful to determine the change in the predictions resulting from using other thermodynamic models with the Aspen Plus default database parameters. Here, as an example, we recalculate the phase behavior using the **NRTL** model. To do this, we first click on **Ternary Maps** in Fig. 7.2-7 to set up a new analysis **TERDI-2**; go to **Methods>Specifications** and choose **NRTL**, then go to **Parameters>Binary Interaction** and check the values for **NRTL-1**, and then return to **TERDI-1>Input** that is set as shown in Fig. 7.2-7.

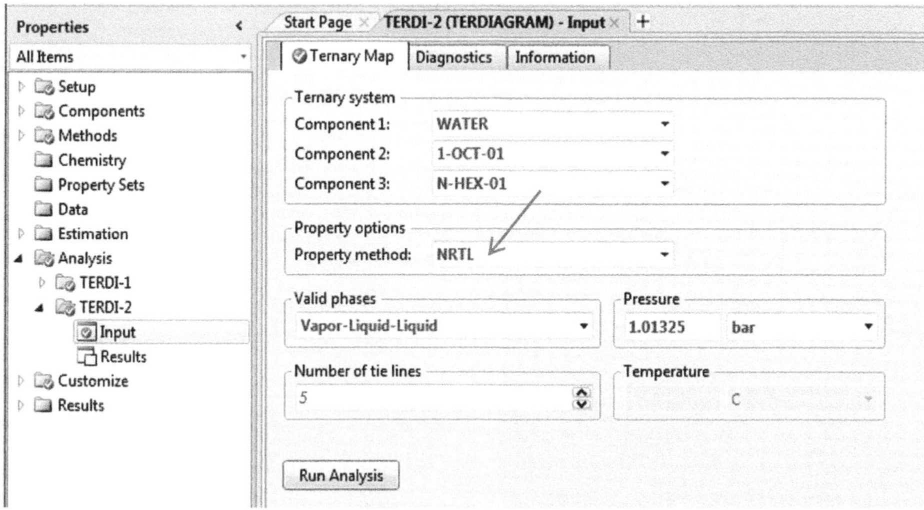

Figure 7.2-7

Then **Run Analysis** following the same procedures as in Figs. 7.2-3 to 7.2-6 produces Figs. 7.2-8 to 7.2-11.

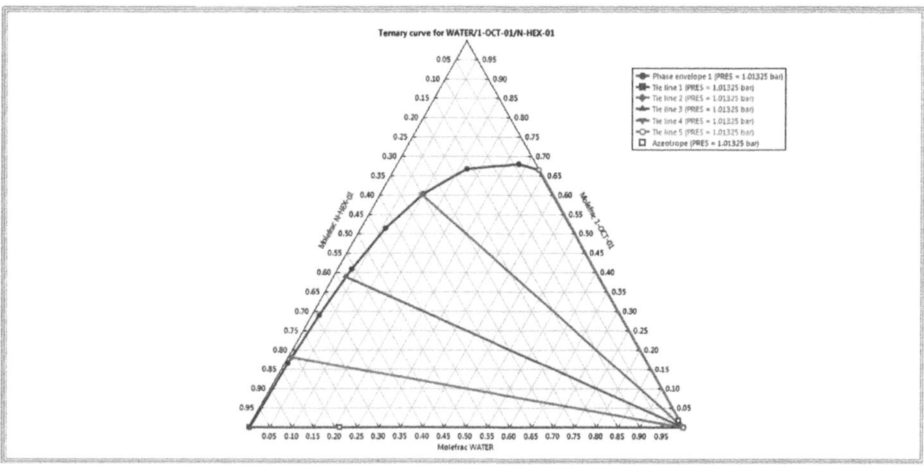

Figure 7.2-8

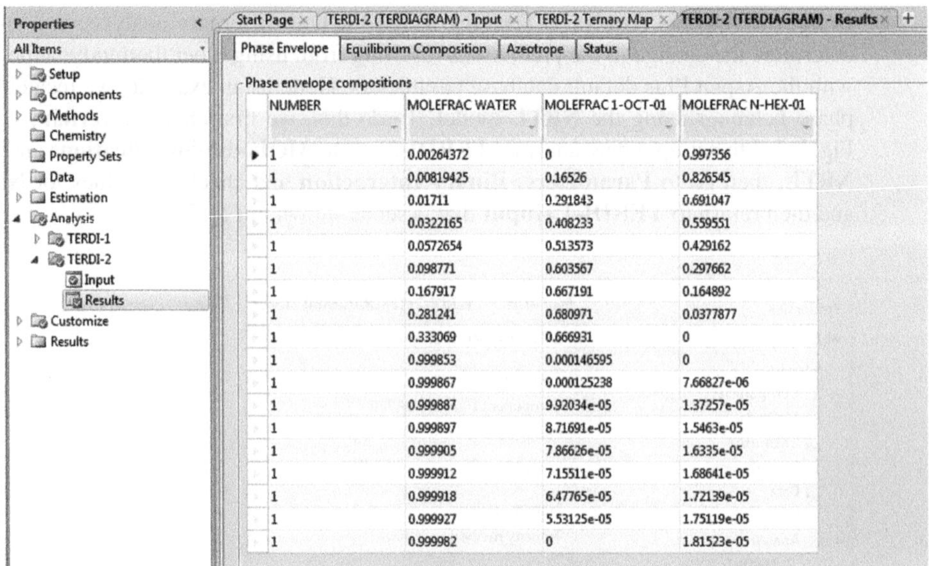

Phase envelope compositions

NUMBER	MOLEFRAC WATER	MOLEFRAC 1-OCT-01	MOLEFRAC N-HEX-01
1	0.00264372	0	0.997356
1	0.00819425	0.16526	0.826545
1	0.01711	0.291843	0.691047
1	0.0322165	0.408233	0.559551
1	0.0572654	0.513573	0.429162
1	0.098771	0.603567	0.297662
1	0.167917	0.667191	0.164892
1	0.281241	0.680971	0.0377877
1	0.333069	0.666931	0
1	0.999853	0.000146595	0
1	0.999867	0.000125238	7.66827e-06
1	0.999887	9.92034e-05	1.37257e-05
1	0.999897	8.71691e-05	1.5463e-05
1	0.999905	7.86626e-05	1.6335e-05
1	0.999912	7.15511e-05	1.68641e-05
1	0.999918	6.47765e-05	1.72139e-05
1	0.999927	5.53125e-05	1.75119e-05
1	0.999982	0	1.81523e-05

Figure 7.2-9

Tie line and vapor compositions

NUMBER	MOLEFRAC LIQUID1 WATER	MOLEFRAC LIQUID1 1-OCT-01	MOLEFRAC LIQUID1 N-HEX-01	MOLEFRAC LIQUID2 WATER	MOLEFRAC LIQUID2 1-OCT-01	MOLEFRAC LIQUID2 N-HEX-01	MOLEFRAC VAPOR WATER	MOLEFRAC VAPOR 1-OCT-01	MOLEFRAC VAPOR N-HEX-01
1	0.999982	0	1.81518e-05	0.00264385	0	0.997356	0.210219	0	0.789781
2	0.999926	5.68536e-05	1.74739e-05	0.00899822	0.180611	0.810391	0.233065	0.00114489	0.765791
3	0.999913	7.0399e-05	1.6933e-05	0.0290681	0.389082	0.58185	0.251022	0.00156015	0.747418
4	0.999898	8.69857e-05	1.54836e-05	0.0978027	0.602089	0.300108	0.301422	0.00244394	0.696134
5	0.999853	0.000146596	0	0.333094	0.666907	0	0.981715	0.0182849	0

Figure 7.2-10

and

Azeotrope compositions and temperature

NUMBER	MOLEFRAC WATER	MOLEFRAC 1-OCT-01	MOLEFRAC N-HEX-01	TEMP K
1	0.981716	0.0182838	0	372.655
2	0.21022	0	0.78978	334.584

Figure 7.2-11

We see that the predictions of the UNIQUAC and NRTL models, based on the binary parameters for each model in the Aspen Plus data bank are qualitatively the same, though there are quantitative differences between the two, especially in the shape of the liquid–liquid phase boundary (dark blue line).

While phase envelope calculations such as above provide some guidance for the design of vapor–liquid–liquid equilibrium processes by determining the likely phases present for the given feed composition, the actual design of such separation processes is done using process blocks in the simulation mode. Consequently, in simulations using Aspen Plus for the design of separation processes that involves or may involve vapor–liquid–liquid equilibrium, the **Flash3** module from the **Separators** tab on the **Model Palette** of the **Flowsheet** window (shown in Fig. 7.2-12) must be used. For liquid–liquid equilibrium calculations with no vapor phase, either the **Flash2** or the **Decanter** modules may be used, or the **Flash3** module can be used though in such cases it will return a zero vapor phase.

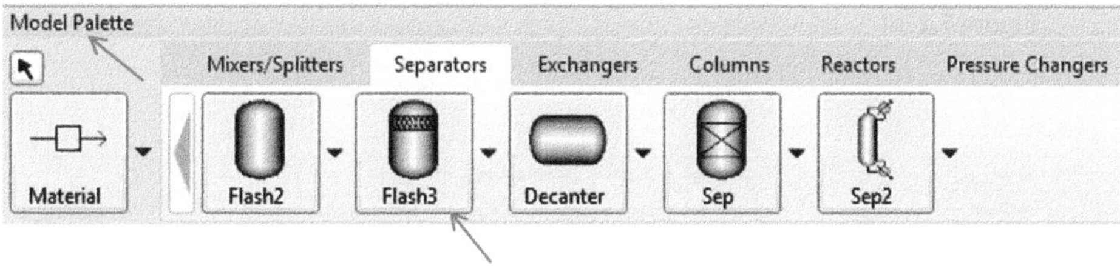

Figure 7.2-12

To illustrate the use of all of these modules, the flow sheet in Fig. 7.2-13 was prepared where, for simplicity, a new process element, the splitter, was introduced so that the three process units (here labeled as **2DECANT**, **3FLASH3**, and **4FLASH3**) would receive the same feed. [Note that I generally number the process units in the order in which they appear. This is so that they appear in that order in the output. Otherwise the units appear in alphabetical order that is less convenient.]

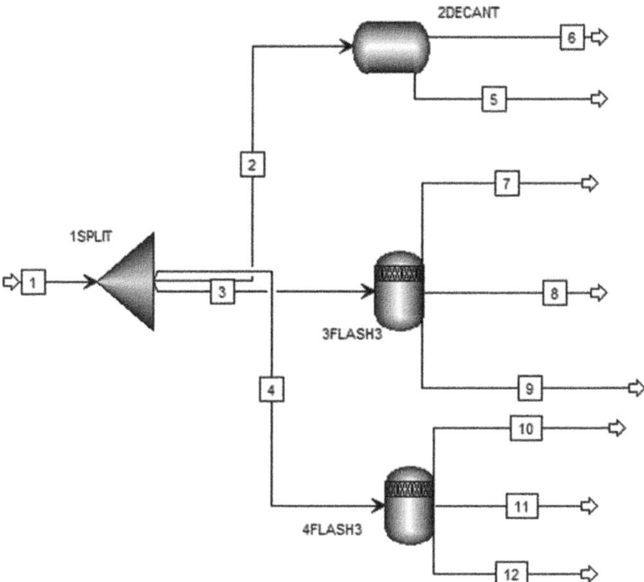

Figure 7.2-13

The inputs in Figs. 7.2-14 to 7.2-20 were used for the water + n-octane + n-hexane system.

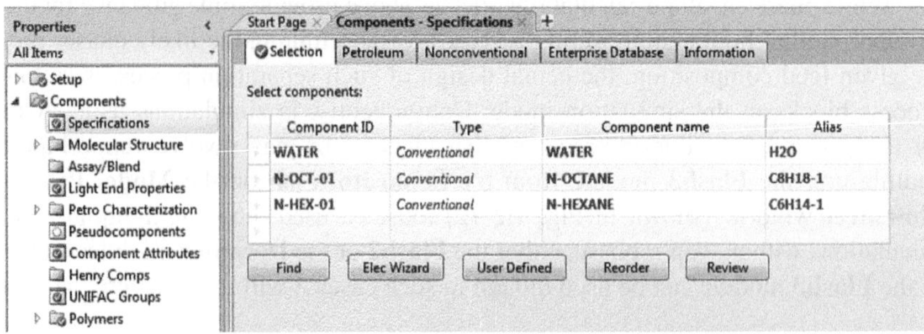

Figure 7.2-14

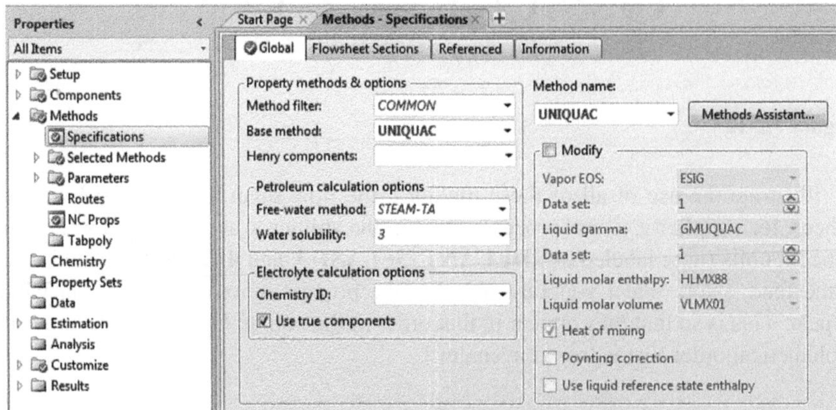

Figure 7.2-15

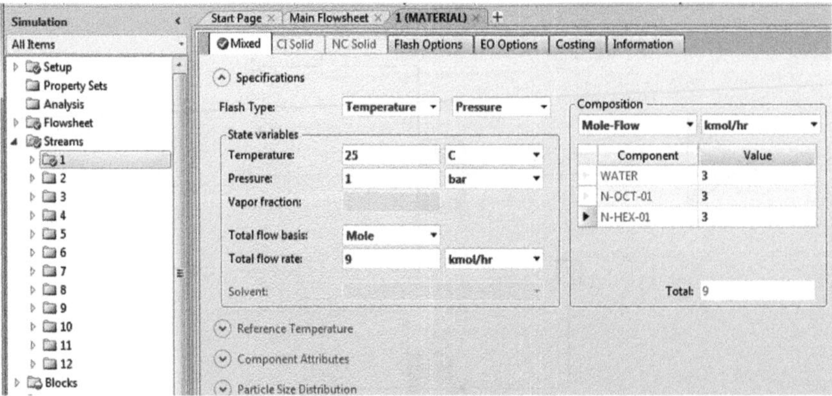

Figure 7.2-16 Stream 1

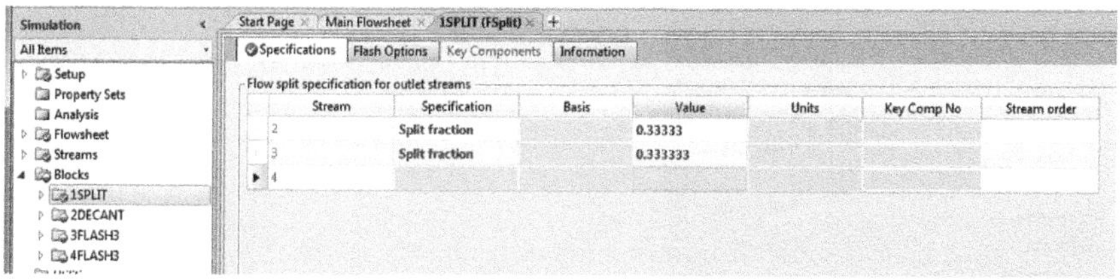

Flow split specification for outlet streams

Stream	Specification	Basis	Value	Units	Key Comp No	Stream order
2	Split fraction		0.33333			
3	Split fraction		0.333333			
4						

Figure 7.2-17 1SPLIT

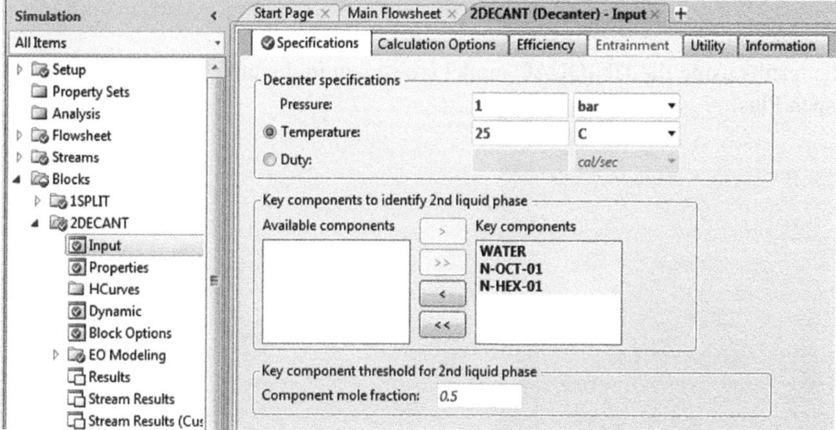

Figure 7.2-18 2DECANT

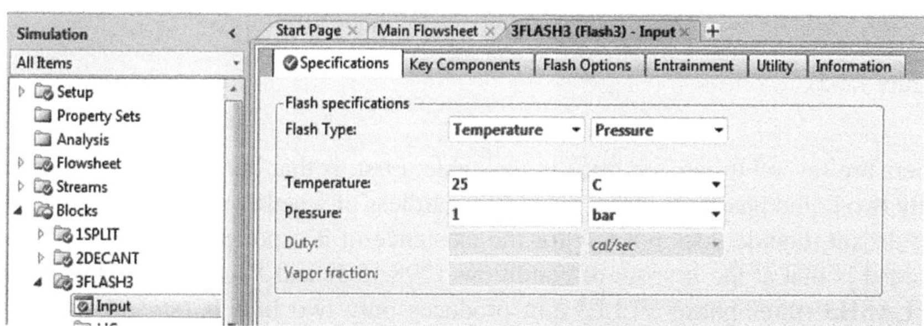

Figure 7.2-19 3FLASH3

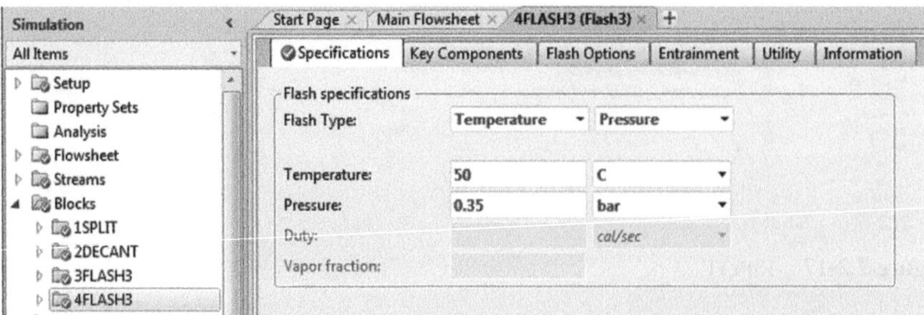

Figure 7.2-20 4FLASH3 (same as 3Flash3 except at a higher temperature and lower pressure to ensure a three-phase split)

The results using the UNIQUAC model are shown in the table in Fig. 7.2-21 produced by Aspen Plus.

Display: All streams Format: FULL Stream Table Copy All

	1	2	3	4	5	6	7	8	9	10	11	12
Substream: MIXED												
Mole Flow kmol/hr												
WATER	3	0.99999	0.999999	1.00001	0.99857	0.00141996	0	0.00141999	0.998579	0.326892	0.00171192	0.671407
N-OCT-01	3	0.99999	0.999999	1.00001	5.42622e-08	0.99999	0	0.999999	5.42826e-08	0.110648	0.889368	9.80537e-08
N-HEX-01	3	0.99999	0.999999	1.00001	1.38768e-06	0.999989	0	0.999998	1.3677e-06	0.488744	0.511268	1.15019e-06
Mole Frac												
WATER	0.333333	0.333333	0.333333	0.333333	0.999999	0.000709484	0	0.000709493	0.999999	0.352908	0.00122076	0.999998
N-OCT-01	0.333333	0.333333	0.333333	0.333333	5.43599e-08	0.499646	0	0.499646	5.43598e-08	0.119449	0.6342	1.46101e-07
N-HEX-01	0.333333	0.333333	0.333333	0.333333	1.36964e-06	0.499645	0	0.499645	1.36964e-06	0.527642	0.364579	1.7131e-06
Total Flow kmol/hr	9	2.99997	3	3.00003	0.998571	2.0014	0	2.00142	0.99858	0.926279	1.40235	0.671409
Total Flow kg/hr	655.27	218.421	218.423	218.426	17.9896	200.432	0	200.433	17.9898	60.6465	145.684	12.0957
Total Flow l/min	15.6262	5.20867	5.20872	5.20878	0.301651	4.90702	0	4.90706	0.301694	1185.1	3.64992	0.207937
Temperature C	25	25	25	25	25	25		25	25	50	50	50
Pressure bar	1	1	1	1	1	1	1	1	1	0.35	0.35	0.35
Vapor Frac	0	0	0	0	0	0		0	0	1	0	0
Liquid Frac	1	1	1	1	1	1		1	1	0	1	1
Solid Frac	0	0	0	0	0	0		0	0	0	0	0
Enthalpy cal/mol	-58332.1	-58332.1	-58332.1	-58332.1	-68232.1	-53392.6		-53392.6	-68232.1	-46705.3	-53671.1	-67810.6
Enthalpy cal/gm	-801.179	-801.179	-801.179	-801.179	-3787.44	-533.149		-533.149	-3787.44	-713.35	-516.637	-3764.03
Enthalpy cal/sec	-145830	-48609.6	-48610	-48610.6	-18926.3	-29683.3		-29683.6	-18926.5	-12017.5	-20907.1	-12646.8
Entropy cal/mol-K	-131.923	-131.923	-131.923	-131.923	-38.8502	-178.361		-178.361	-38.8502	-89.6038	-180.585	-37.5027
Entropy cal/gm-K	-1.81194	-1.81194	-1.81194	-1.81194	-2.1565	-1.78101		-1.78101	-2.1565	-1.36856	-1.7383	-2.0817
Density mol/cc	0.00959928	0.00959928	0.00959928	0.00959928	0.0551725	0.00679774		0.00679774	0.0551725	1.30268e-05	0.00640355	0.053815
Density gm/cc	0.698903	0.698903	0.698903	0.698903	0.993954	0.680765		0.680765	0.993954	0.000852905	0.665227	0.9695
Average MW	72.8078	72.8078	72.8078	72.8078	18.0154	100.146		100.146	18.0154	65.4733	103.886	18.0154
Liq Vol 60F l/min	15.5112	5.17038	5.17043	5.17049	0.300406	4.86998	0	4.87002	0.300409	1.45722	3.51129	0.201985

Figure 7.2-21

There are several things to notice in this table. First, is that the **2DECANT** unit produces only two liquid phases (streams 5 and 6) regardless of whether a vapor phase should exist; the decant module does not test for the existence of a vapor phase in the computations. Second is that at the first set of conditions (298.15 K and P = 1 bar = 0.9869 atm), the **3FLASH3** (three-phase VLLE) unit produces only two liquids (streams 8 and 9), and reports that the vapor phase (stream 7) is not present. Also, to within round off errors (or difference in convergence criteria in the **DECANT** and **FLASH3** blocks), the two results are in complete agreement. Finally, the **4FLASH3** VLLE calculation at a higher temperature and lower pressure reports that a vapor and two liquid phases are present with the UNIQUAC model.

For comparison, using the NRTL model, we obtain the table of results in Fig. 7.2-22. Note that there are quantitative differences in these results and those using the UNIQUAC model. The reasons for this were explained at the beginning of this chapter. These results point out once again that one has to be careful in not accepting computer predictions without some form of verification with experimental data, or data for similar mixtures. When dealing with predictions from thermodynamic models, the user should always beware.

Display: All streams Format: FULL Stream Table Copy All

	1	2	3	4	5	6	7	8	9	10	11	12
Substream: MIXED												
Mole Flow kmol/hr												
WATER	3	0.99995	0.999999	1.00001	0.997557	0.00243177	0	0.00243367	0.997565	0.326188	0.00381344	0.67001
N-OCT-01	3	0.99999	0.999999	1.00001	1.02995e-06	0.999989	0	0.999998	1.02996e-06	0.110041	0.888968	1.84885e-08
N-HEX-01	3	0.99995	0.999999	1.00001	2.9237e-06	0.999987	0	0.999986	2.91372e-06	0.488062	0.511947	2.99304e-06
Mole Frac												
WATER	0.333333	0.333335	0.333333	0.333333	0.999996	0.00121542	0	0.00121536	0.999996	0.352906	0.00271279	0.999993
N-OCT-01	0.333333	0.333333	0.333333	0.333333	1.03247e-06	0.499393	0	0.499395	1.03247e-06	0.119064	0.633101	2.75942e-08
N-HEX-01	0.333333	0.333333	0.333333	0.333333	2.92062e-06	0.499392	0	0.499392	2.92082e-06	0.528039	0.364158	4.46713e-06
Total Flow kmol/hr	9	2.99997	3	3.00003	0.997561	2.00241	0	2.00243	0.997569	0.92429	1.40573	0.670014
Total Flow kg/hr	655.27	218.421	218.423	218.426	17.9716	200.45	0	200.451	17.9718	60.5062	145.849	12.0709
Total Flow l/min	15.6255	5.20848	5.2086	5.20856	0.301161	4.9071	0	4.90715	0.301384	1182.56	3.05337	0.207811
Temperature C	25	25	25	25	25	25		25	25	50	50	50
Pressure bar	1	1	1	1	1	1	1	1	1	0.35	0.35	0.35
Vapor Frac	0	0	0	0	0	0		0	0	1	0	0
Liquid Frac	1	1	1	1	1	1		1	1	0	1	1
Solid Frac	0	0	0	0	0	0		0	0	0	0	0
Enthalpy cal/mol	-58330.8	-58330.8	-58330.8	-58330.8	-68232.1	-53398.2		-53398.2	-68232.1	-46701.5	-53673.1	-67810.4
Enthalpy cal/gm	-801.162	-801.162	-801.162	-801.162	-3787.39	-533.428		-533.426	-3787.39	-713.41	-517.315	-3763.93
Enthalpy cal/sec	-145830	-48608.5	-48609	-48609.5	-18907.1	-29701.4		-29701.7	-18907.3	-11990.5	-20958.3	-12620.5
Entropy cal/mol-K	-131.918	-131.918	-131.918	-131.918	-38.9506	-178.283		-178.283	-38.8506	-88.5872	-180.306	-37.5033
Entropy cal/gm-K	-1.81187	-1.81187	-1.81187	-1.81187	-2.15849	-1.78097		-1.78097	-2.15849	-1.36853	-1.73784	-2.08168
Density mol/cc	0.00958968	0.00958968	0.00958968	0.00958968	0.0551716	0.00680106		0.00680106	0.0551716	1.30266e-05	0.00641294	0.053613
Density gm/cc	0.698332	0.698332	0.698332	0.698332	0.993948	0.680814		0.680814	0.993948	0.000652702	0.665362	0.969487
Average MW	72.8078	72.8078	72.8078	72.8078	18.0156	100.104		100.104	18.0156	65.4623	103.753	18.0159
Liq Vol 60F l/min	15.5113	5.17036	5.17043	5.17049	0.300107	4.87027	0	4.87032	0.30011	1.4539	3.51902	0.201573

Figure 7.2-22

7.3 HIGH PRESSURE VAPOR–LIQUID–LIQUID EQUILIBRIUM

The final type of VLLE (or VLE) that will be considered is that in which the pressure is sufficiently high that the vapor phase cannot be considered to be an ideal gas. In this case an equation of state must be used for the vapor, but the liquid phase is, or phases are, sufficiently nonideal that an activity coefficient model is needed. One example of this that will be used here is of an enhanced oil recovery where high pressure carbon dioxide is pumped into an oil reservoir and some of that dissolves into the hydrocarbon phase increasing its volume and reducing its viscosity, resulting in increased hydrocarbon recovery. However, water is also typically present in oil reservoirs, so the likely reservoir phase behavior is liquid (hydrocarbon-rich)-liquid (water-rich)-vapor (carbon dioxide-rich) equilibium. Since equations of state with the usual mixing rules do not describe liquid–liquid equilibrium well, what is typically done in situations such as this is to use an activity coefficient model for the liquid phase or phases and an equation of state for the vapor phase.

As an example, consider an equimolar mixture of each of *n*-decane (to represent the hydrocarbon reservoir fluid), water, and carbon dioxide at the approximate reservoir conditions of 75 bar and 415 K. The **Flash3** separator will be used with the UNIQUAC activity coefficient model first assuming ideal gas behavior. The calculation will then be repeated using the Peng–Robinson equation of state for the vapor and the differences observed. Start as shown in Figs. 7.3-1 and 7.3-2. Start from **Properties>Components>Specifications** and enter the components as shown in Fig. 7.3-1.

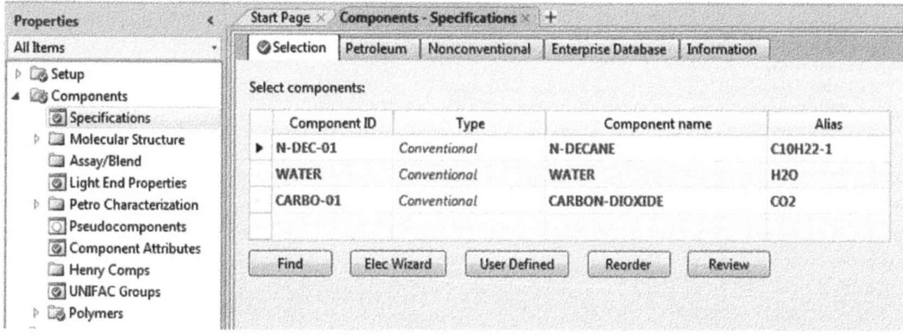

Figure 7.3-1

Next go to **Properties** and choose the **UNIQUAC** model.

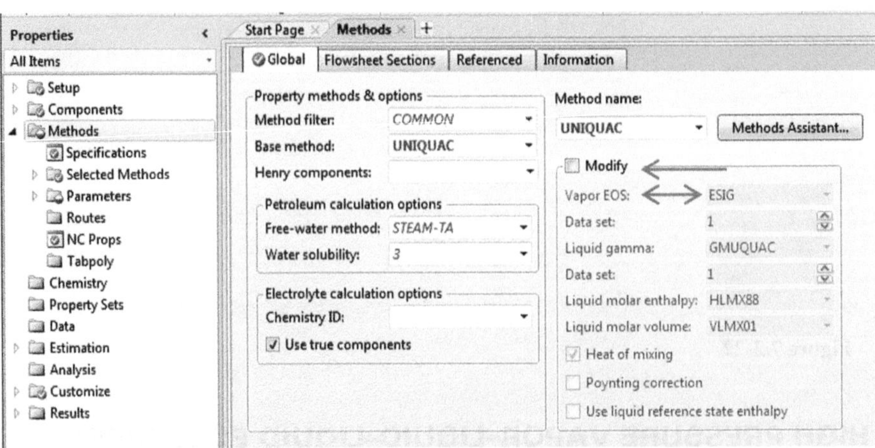

Figure 7.3-2

The usual next step is to choose the UNIQUAC binary parameters. However, before doing so direct your attention to the box **Modify** indicated by the arrow in Fig. 7.3-2 and below it the **Vapor EOS,** followed by the box containing **ESIG**, which is Aspen Plus notation for use of the ideal gas equation of state. At present we will accept the use of the ideal gas equation of state, though we will revisit this assumption shortly.

Now going to **Properties>Parameters>Binary Interaction** and clicking on **UNIQ-1** brings up the parameter information in Fig. 7.3-3.

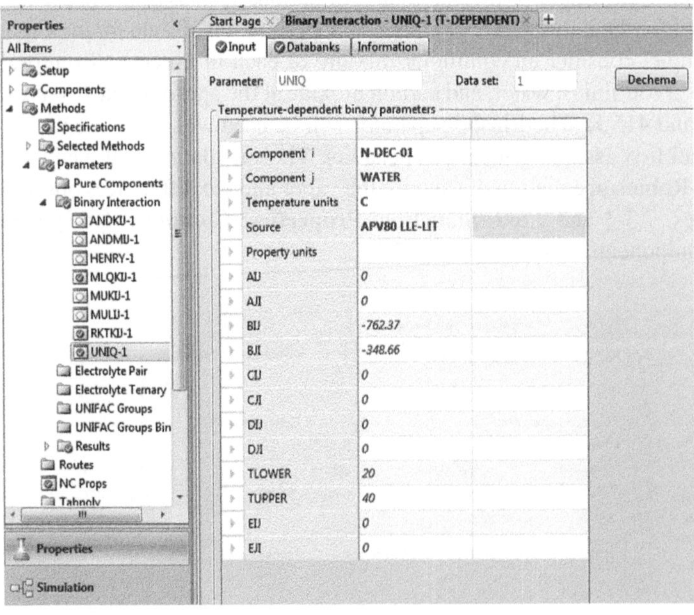

Figure 7.3-3

This indicates that there are water–decane parameters available, but also that we will be operating above the temperature range at which these parameters were obtained (415 K is slightly above TUPPER of 40°C of this parameter set). Nonetheless, for the purpose of this illustration we will use these parameters.

The next step is to go **Simulation** and insert the **Flash3** separator in the **Main Flowsheet** (Fig. 7.3-4).

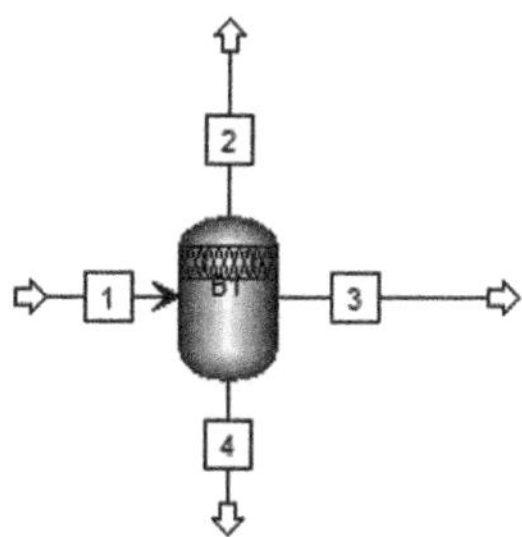

Figure 7.3-4

Then go to **Streams>1>Input** and enter the information in Fig. 7.3-5, which for this example is an equimolar mixture of the three components at the conditions specified.

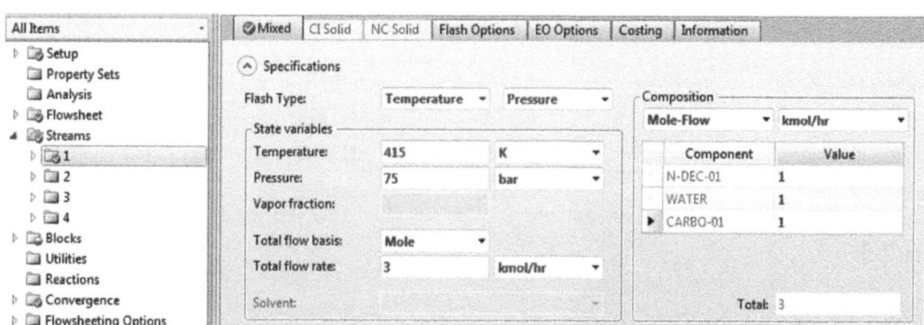

Figure 7.3-5

Next go to **Blocks>B1>Input** and enter the **Flash3** operating conditions (shown in Fig. 7.3-6).

Simulation
All Items
Setup
Property Sets
Analysis
Flowsheet
Streams
Blocks
B1
Utilities
Reactions
Convergence
Start Page
Main Flowsheet
B1 (Flash3)
Specifications
Key Components
Flash Options
Entrainment
Utility
Information
Flash specifications
Flash Type:
Temperature
Pressure
Temperature:
415
K
Pressure:
75
bar
Duty:
cal/sec
Vapor fraction:

Figure 7.3-6

Next, run the simulation by clicking on > on the main toolbar. The results with the ideal gas assumption are shown in the table in Fig. 7.3-7.

	1	2	3	4
Substream: MIXED				
Mole Flow kmol/hr				
N-DEC-01	1	0.00153771	0.998434	2.78497e-05
WATER	1	0.0206604	0.104753	0.874587
CARBO-01	1	0.427437	0.475811	0.0967527
Mole Frac				
N-DEC-01	0.333333	0.00341991	0.632322	2.86706e-05
WATER	0.333333	0.0459493	0.0663412	0.900367
CARBO-01	0.333333	0.950631	0.301337	0.0996046
Total Flow kmol/hr	3	0.449635	1.579	0.971367
Total Flow kg/hr	204.31	19.4024	164.889	20.018
Total Flow l/min	6.143	3.44765	4.33707	0.37821
Temperature C	141.85	141.85	141.85	141.85
Pressure bar	75	75	75	75
Vapor Frac	0.0482635	1	0	0
Liquid Frac	0.951736	0	1	1
Solid Frac	0	0	0	0
Enthalpy cal/mol	-73350.6	-91078.2	-71567.7	-68719.4
Enthalpy cal/gm	-1077.05	-2110.66	-685.339	-3334.59
Enthalpy cal/sec	-61125.5	-11375.5	-31390.3	-18542.2
Entropy cal/mol-K	-87.5606	-5.55852	-146.769	-29.8244

Figure 7.3-7

These are the results assuming that the vapor phase is an ideal gas. Next, because of the pressure involved, the description of the vapor phase will be changed to the Peng–Robinson equation of state. To do this, one goes back to **Properties>Methods>Specifications** and clicks the **Modify property models**, which brings up the **Modify Property Method** window (Fig. 7.3-8) into which the name ***PRUNIQ*** (or you can choose another name) is entered.

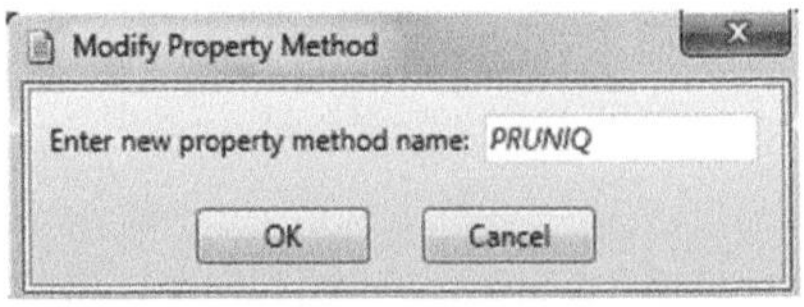

Modify Property Method

Enter new property method name: *PRUNIQ*

OK Cancel

Figure 7.3-8

Clicking on **OK** leads to Fig. 7.3-9.

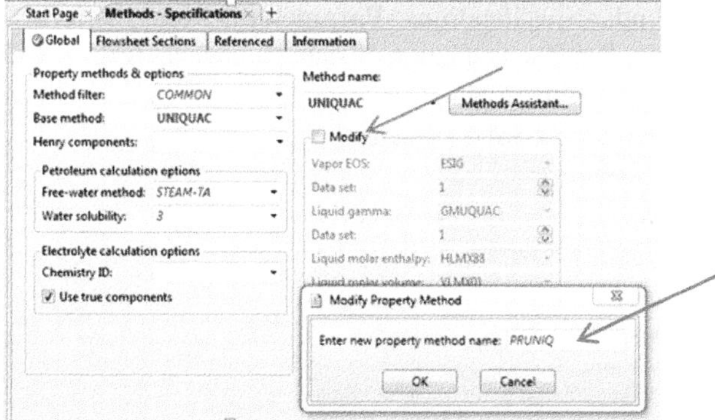

Figure 7.3-9

This brings up the window in Fig. 7.3-10 with the ideal gas equation of state (ESIG) still chosen.

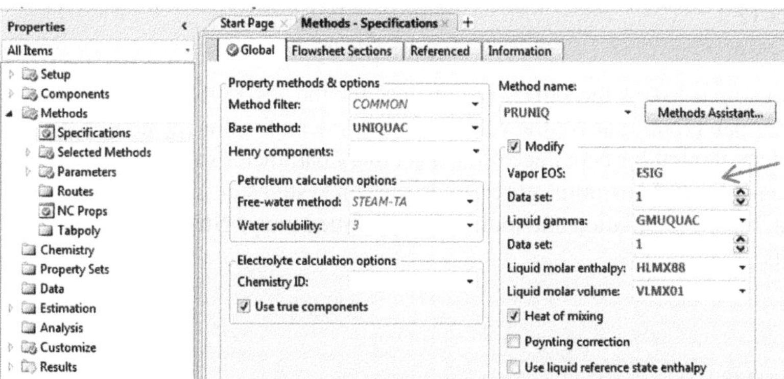

Figure 7.3-10

Click on the drop-down menu at **ESIG** and scroll down to **ESPR**, the Peng–Robinson equation of state (or another equation of state of your choice), (Fig. 7.3-11).

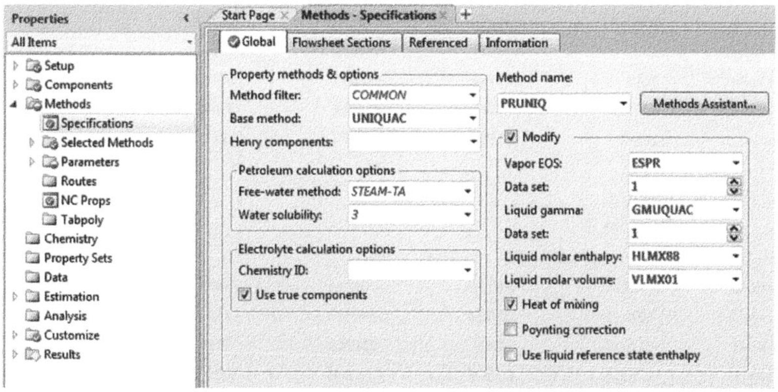

Figure 7.3-11

Next go to **Properties>Methods>Specifications>Parameters>Binary Interaction** and choose **PRKBV-1** for the Peng–Robinson equation of state parameters (Fig. 7.3-12).

Parameter: PRKBV Data set: 1

Temperature-dependent binary parameters

Component i	N-DEC-01	WATER
Component j	CARBO-01	CARBO-01
Temperature units	C	C
Source	APV80 EOS-LIT	APV80 EOS-LIT
Property units		
KAIJ	0.1141	0.12
KBIJ	0	0
KCIJ	0	0
TLOWER	-273.15	-273.15
TUPPPER	726.85	726.85

Figure 7.3-12

This brings up a list of the binary parameters for the PR EOS for the decane–CO_2 and water–CO_2 pairs (but none for the water–decane system, for this illustration we will ignore this problem, though for better accuracy this binary parameter should be specified, perhaps by looking at other, but similar water–hydrocarbon systems or in other databases.)

Next, run the simulation and look at **Results Summary>Streams** of Fig. 7.3-13 to see the results.

	1	2	3	4
Substream: MIXED				
Mole Flow kmol/hr				
N-DEC-01	1	0.00778808	0.992195	2.10833e-05
WATER	1	0.0396583	0.0932465	0.867098
CARBO-01	1	0.530694	0.390609	0.0786899
Mole Frac				
N-DEC-01	0.333333	0.0134709	0.672196	2.22913e-05
WATER	0.333333	0.0685964	0.063173	0.916779
CARBO-01	0.333333	0.917933	0.264631	0.0831985
Total Flow kmol/hr	3	0.57814	1.47605	0.945809
Total Flow kg/hr	204.31	25.1783	160.045	19.0872
Total Flow l/min	7.10835	3.75683	4.21339	0.36063
Temperature C	141.85	141.85	141.85	141.85
Pressure bar	75	75	75	75
Vapor Frac	0.121805	1	0	0
Liquid Frac	0.878195	0	1	1
Solid Frac	0	0	0	0
Enthalpy cal/mol	-73272.6	-90383.8	-70426.6	-68279.3
Enthalpy cal/gm	-1075.91	-2075.38	-649.527	-3383.39
Enthalpy cal/sec	-61060.5	-14515.1	-28875.9	-17938.7
Entropy cal/mol-K	-87.3455	-8.67548	-155.548	-30.2852

Figure 7.3-13

Recapitulating, the results of using the ideal gas assumption are (mole fraction in order of feed, vapor, liquid 1 and liquid 2)

Mole Frac				
N-DEC-01	0.333333	0.00341991	0.632322	2.86706e-05
WATER	0.333333	0.0459493	0.0663412	0.900367
CARBO-01	0.333333	0.950631	0.301337	0.0996046

While using the Peng–Robinson model for the vapor phase gives

Mole Frac				
N-DEC-01	0.333333	0.0134709	0.672196	2.22913e-05
WATER	0.333333	0.0685964	0.063173	0.916779
CARBO-01	0.333333	0.917933	0.264631	0.0831985

We see that, as to be expected, there are differences in the results of the two models. For example, the initial feed is only 4.8 mol% vapor with the ideal gas model, but 12.2 mol% with the PR model. Also, all the mole fractions in each of the three phases are different between the two models. The UNIQ+PR results should be the more accurate.

What has been done in this last case is to use two completely different models, an activity coefficient model for the liquid phases and an equation of state for the vapor phase. Since the vapor-phase and liquid-phase models are completely different, there is no set of conditions at which they will be the same or produce identical results. However, at the critical point of a real mixture, the vapor and liquid phases of the mixture are identical. Therefore, neither the UNIQ + IG nor the more accurate UNIQ + PR models will predict the occurrence of a critical point for this (or any other) system. Indeed, no combination of an activity coefficient model and an equation of state model with the van der Waals one-fluid mixing rules will lead to the prediction of a mixture critical point.

There is a way of combining an activity coefficient model and a cubic equation of state in a unified way that will lead to the prediction of a mixture critical point. The method is discussed in Section 9.9 of the textbook, *Chemical, Biochemical and Engineering Thermodynamics*, 4th ed., S. I. Sandler, (John Wiley & Sons, Inc., 2006). This method has been incorporated into the Aspen Plus process simulator, but will not be discussed here.

PROBLEMS

7.1. The following liquid–liquid equilibrium data have been reported (J. M. Sorenson and W. Arlt, *Liquid–Liquid Equilibrium Data Collection: 1. Binary Systems*, DECHEMA Chemistry Data

Series, Vol. V, 1979, Frankfurt, p. 33) for the system nitromethane (1) + cyclohexane (2) as a function of temperature.

T(°C)	Mol% 1 in L1	Mol% 2 in L2
15	2.76	2.90
20	3.20	3.33
25	3.72	3.81
30	4.33	4.35
40	5.81	5.52
50	7.77	7.38
60	10.6	9.52

Fit these data with the NRTL model with temperature-dependent parameters.

7.2. Repeat the calculation of the previous problem with the UNIQUAC equation.

7.3. The following smoothed liquid–liquid equilibrium data have been reported (J. M. Sorenson and W. Arlt, *Liquid–Liquid Equilibrium Data Collection: 1. Binary Systems*, DECHEMA Chemistry Data Series, Vol. V, 1979, Frankfurt, p. 297.) for the system ethyl ester propanoic acid (1) + water (2) as a function of temperature.

T(°C)	Mol% 1 in L1	Mol% 2 in L2
20	0.347	8.17
25	0.351	8.67
30	0.358	9.30
40	0.379	10.9
50	0.409	12.7
60	0.452	14.8
70	0.519	17.0
80	0.659	19.2

Fit these data with the NRTL model with temperature-dependent parameters.

7.4. Repeat the calculation of Problem 7.3 with the UNIQUAC equation.

7.5. Use the NIST TDE to obtain liquid–liquid equilibrium data as a function of temperature for a system of your choice, and correlate data with:

(a) the NRTL equation; and

(b) the UNIQUAC equation.

Chapter 8

The Property Methods Assistant and Property Estimation

One of the important decisions to be made in any property analysis or process simulation is the choice of thermodynamic model to be used. To repeat the old computer acronym, GIGO, for garbage in, garbage out. The relevance here is that the initial choice of an inappropriate thermodynamic model will lead to incorrect results. For example, in the simple liquefaction of propane in Chapter 2, had one chosen ideal gas as the thermodynamic model, the calculation would have been completed with incorrect answers in every piece of process equipment, and no liquid would have been produced. This is nonsense, but it is the result that would have been obtained. Therefore here, as with every other computer calculation, the engineer needs to look carefully at the results and ask whether they make sense. The user should not accept the result without critical analysis just because it came from the computer.

So the question arises as to how to choose the appropriate thermodynamic model. This is briefly discussed in some thermodynamics textbooks (see, for example, Section 9.11 in *Chemical, Biochemical and Engineering Thermodynamics* 4th ed., S. I. Sandler, John Wiley & Sons, Inc., 2006). Two tables from that book, one for choosing activity coefficient models and another for equations of state and their mixing rules are reproduced as Tables 8.1-1 and 8.1-2. These tables provide some guidance in making the appropriate mixture model choices. There are some general guidelines, such as an equation of state is the appropriate model for high pressure systems, but there are numerous ones to choose from, and for mixtures one also has to choose an appropriate mixing rule and binary parameter(s). At ambient, low, and slightly elevated pressures, activity coefficient models are generally used, but again there are a number of models that can be used, and once a model has been chosen parameter values are also needed.

8.1 THE PROPERTY METHODS ASSISTANT

Aspen Plus® provides the interactive **Property Methods Assistant** tool to assist the user in choosing the appropriate class of thermodynamic or transport property models for the simulation or property analysis. This tool is generally useful in helping the user choose a class of models (e.g., equations of state or activity coefficient models) for a given type

Using Aspen Plus® in Thermodynamics Instruction: A Step-by-Step Guide, First Edition. Stanley I. Sandler.

of simulation based on the components and conditions, and then in providing some information on specific models within that class. However, it is still left to the user to decide which model within that class to use. Once an activity coefficient or equation of state model has been chosen, an important advantage of Aspen Plus is that it provides it default values for the parameters in the model. Alternatively, if the user has their own experimental data, the modules discussed in Chapters 5–7 (and in Section 8.3, if only infinite dilution activity coefficient data are available) can be used to obtain values of the model parameters.

TABLE 8.1-1 Recommended activity coefficient models

Nonpolar + nonpolar compounds: All of the models (Margules, Van Laar, Wilson, UNIQUAC and NRTL) will give good correlations of data for these mixturess.

Nonpolar + weakly polar compounds: All of the models (Margules, Van Laar, Wilson, UNIQUAC and NRTL) can be used, though the UNIQUAC model is better for mixtures that are more nonideal.

Nonpolar + strongly polar compounds: While all of the models (Margules, Van Laar, Wilson, UNIQUAC and NRTL) can be used for mixtures that are not too nonideal, the UNIQUAC model appears to give the best correlation for slightly, nonideal systems, while the Wilson model may be better for mixtures that are more nonideal (but not so nonideal as to result in liquid–liquid immiscibility. See Section 11.2).

Weakly polar + weakly polar compounds: All of the models (Margules, Van Laar, Wilson, UNIQUAC and NRTL) can be used, though it appears that the UNIQUAC model is better for mixtures that are more nonideal.

Weakly polar + strongly polar compounds: All of the models (Margules, Van Laar, Wilson, UNIQUAC and NRTL) can be used, though it appears that the UNIQUAC model is better for mixtures that are more nonideal.

Strongly polar + strongly polar compounds: The UNIQUAC model appears to best correlate data, though all models give reasonable results.

Water + nonpolar compounds: These mixtures generally have limited mutual solubility, and are discussed in Section 11.2.

Water + weakly polar compounds: These mixtures generally have limited mutual solubility, and are discussed in Section 11.2.

Water + strongly polar compounds: The UNIQUAC model appears to best correlate data for aqueous mixtures.

Solutions containing carboxylic acids: The Wilson model appears to best correlate data for mixtures containing carboxylic acids if the components are mutually soluble. (The Wilson model does not predict liquid–liquid phase splitting as discussed in Section 11.2). Otherwise, the UNIQUAC, Van Laar or NRTL models should be used.

Solutions containing polymers: Use the Flory–Huggins model.

Solutions containing ionizable salts, strong acids or bases: Use the extended Debye–Hückel model.

Multicomponent mixtures: Choose a model that based on the suggestions above best describes the dominant components in the mixture.

[Note that the chapter and section references in this table and in Table 8.1-2 refer to the textbook, *Chemical, Biochemical and Engineering Thermodynamics,* 4th ed., S. I. Sandler (John Wiley & Sons, Inc., 2006).]

TABLE 8.1-2 Recommended equation of state models

Vapor Mixtures

1. **Low pressures, no components that associate (such as carboxylic acids or HF):** Assume mixture is an ideal gas mixture. $\overline{f}_i^V(T, P, \underline{y}) = y_i P$
2. **Low pressure for a mixture that contains an associating component:** Use virial equation of state, retaining only second virial coefficient, and search for experimental pure component and cross virial coefficient data for all components in the mixture.
3. **Slightly elevated pressures:** Use choices 1 or 2 above. Alternatively use 4 below if there are no associating components, or 5 if associating components are present.
4. **Elevated pressures for a vapor mixture that contains hydrocarbons, nitrogen, oxygen, carbon dioxide and/or other inorganic gases (but not HF):** Use an equation of state, such as the Peng–Robinson or Soave–Redlich–Kwong equation with van der Waals one-fluid mixing rules (see Section 9.4).
5. **Elevated pressures for a vapor mixture that contains one or more polar and/or associating compounds:** Use an equation of state, such as the Peng–Robinson or Soave–Redlich–Kwong equation with the excess Gibbs energy based mixing rules (see Section 9.9) and the appropriate activity coefficient model (see Table 8.1-1).

The **Methods Assistant** is accessed from the **Properties>Methods>Start Page** by clicking as shown in Fig. 8.1-1.

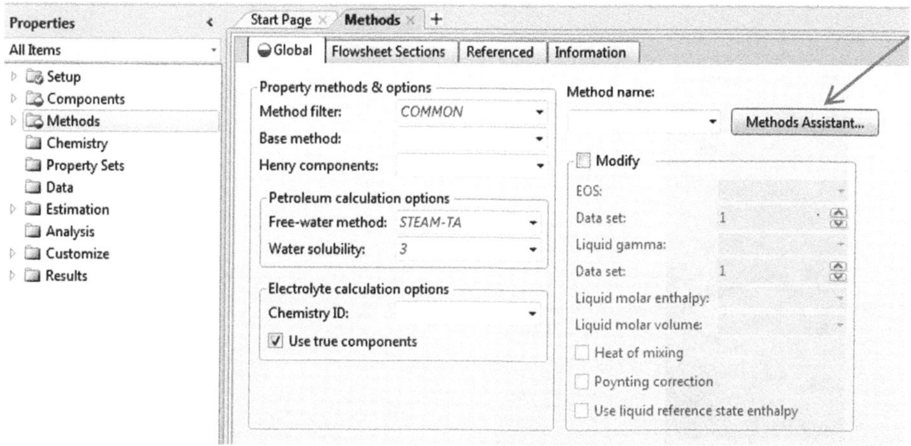

Figure 8.1-1

This will bring up in succession the windows in the following figures.

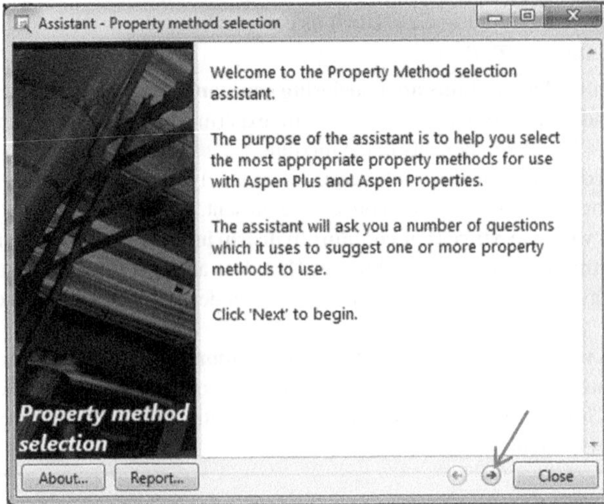

Figure 8.1-2

Note that the instructions are to “Click ‘Next’ to begin.” (Fig. 8.1-2) but there is no Next button. Click on the arrow at the bottom right, which brings up (Fig. 8.1-3):

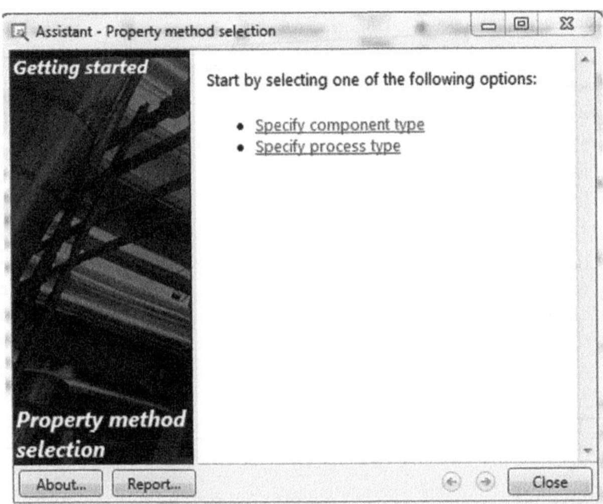

Figure 8.1-3

In the window of Fig. 8.1-3 one can choose to follow the path of specifying the component type (e.g., hydrocarbons, acids) or the process type (oil and gas, refrigerants, pharmaceuticals, etc.). With the latter choice the user is directed to the classes of models appropriate for the components in that type of process.

The choice of **Specify component type** leads to the window in Fig. 8.1-4.

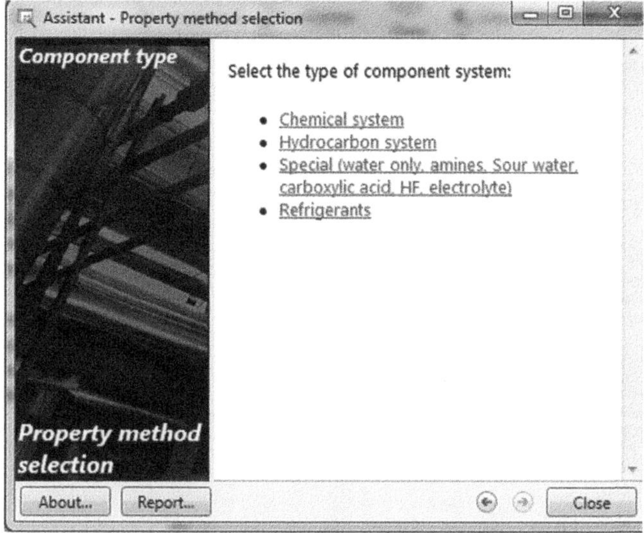

Figure 8.1-4

Choosing **Chemical system** then leads to the window in Fig. 8.1-5.

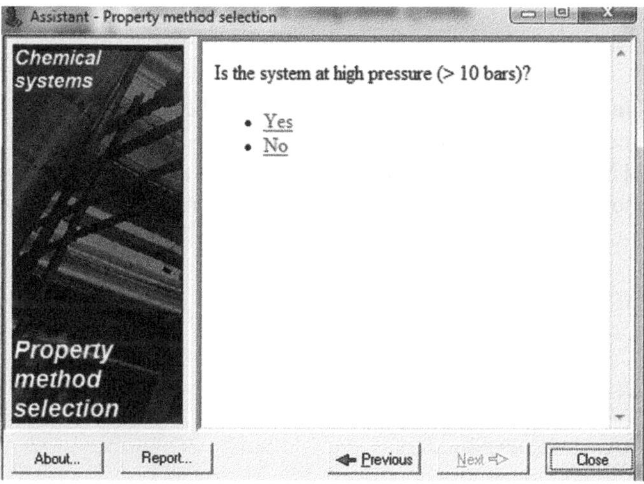

Figure 8.1-5

Choosing **Yes** (high pressure) then leads the user to the collection equations of state and mixing rule models that the logic in **Property Method Selection** indicates are appropriate. This is shown in Fig. 8.1-6.

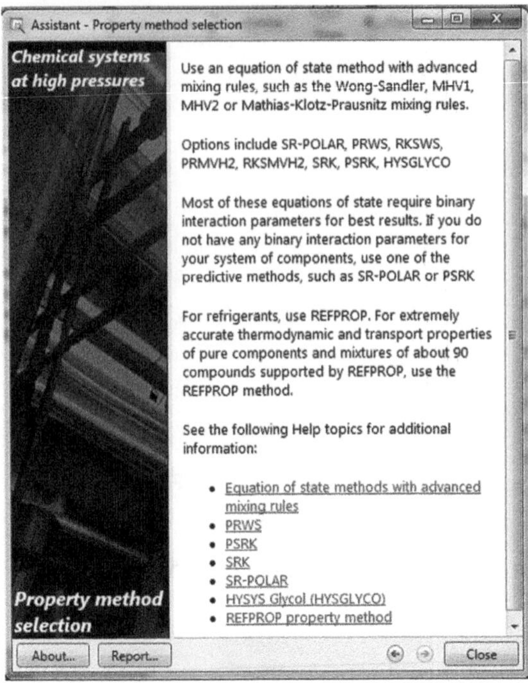

Figure 8.1-6

This screen provides links to further information about each of the equation of state models. However, it is left to the user to choose one of these models.

If the user entered **No** in response to the question of whether the pressure was > **10 bar** in Fig. 8.1-5, the window in Fig. 8.1-7 appears directing the user to use activity coefficient models.

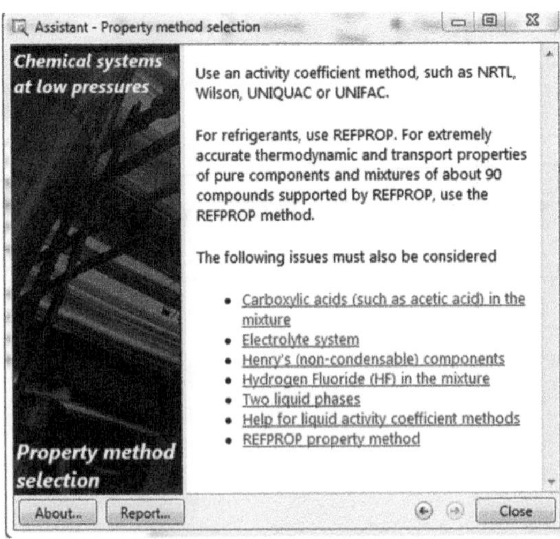

Figure 8.1-7

Further information is then provided on the choice of activity coefficient models by clicking on each of the items listed.

Going back to Fig. 8.1-3, if the user had chosen **Specify process type**, the window in Fig. 8.1-8 would have appeared.

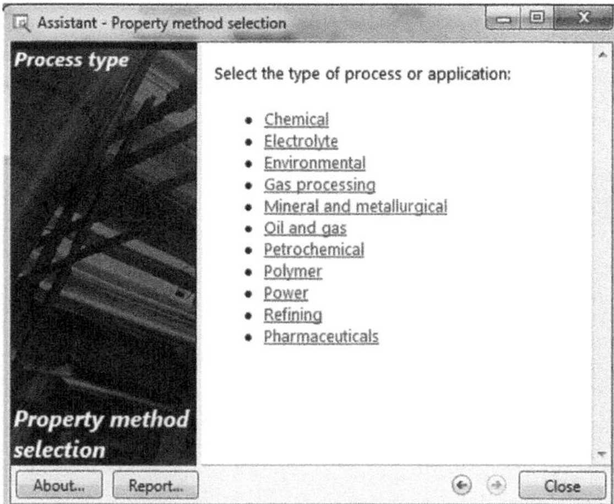

Figure 8.1-8

As an example, clicking on **Oil and gas** leads the user to equations of state models in Fig. 8.1-9.

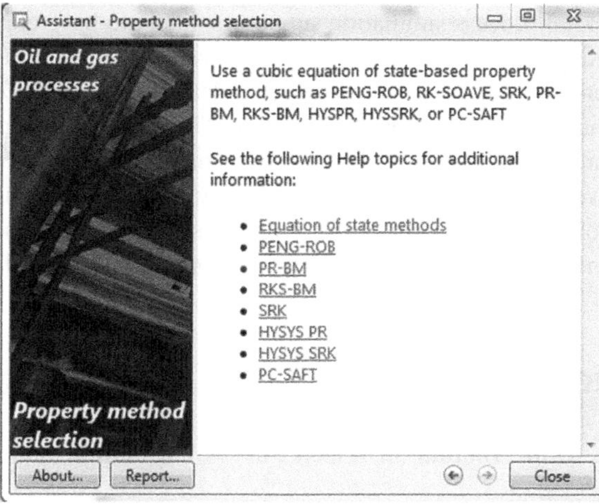

Figure 8.1-9

While choosing pharmaceuticals directs the user to a number of appropriate activity coefficient models shown in Fig. 8.1-10.

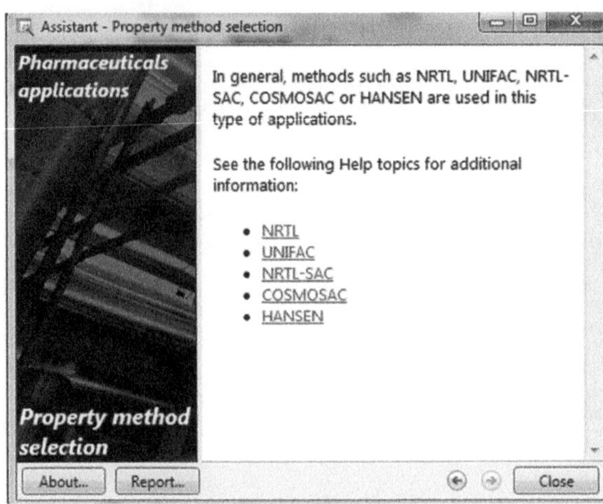

Figure 8.1-10

Again, clicking on one of these models provides further information about it.

As these examples show, the **Property>Methods Assistant** can be helpful in directing the user to the appropriate class of models. However, it does not choose the specific model to be used. That decision rests with the user.

There is one caveat that should be kept in mind when using commercial simulators, including Aspen Plus. Over the years new and improved thermodynamic models have appeared. However, a commercial process simulation supplier is likely to have companies and users who have designed processes many years ago. Now or in the future, those users may need to exactly reproduce those original design calculations to analyze plant performance, to make plant modifications or for other reasons. Therefore, while providers of process simulation software will add new thermodynamic models as they believe is appropriate, they rarely remove an old one. Consequently, in any new process simulation, the user should carefully choose the thermodynamic properties model to be used, and avoid outdated models that may no longer be useful but still appear in the simulator for historic reasons.

8.2 PROPERTY ESTIMATION

One of the options from the **Main Toolbar** in Aspen Plus is **Estimation** indicated in Fig. 8.2-1.

Figure 8.2-1

Aspen Plus describes its function as "Estimate missing properties from molecular structure and limited experimental data." This is a completely different function than the **Property Method Selection Assistant** or **Analysis>Property** on the main menu that has already been discussed. The purpose of an **Estimation** run is to generate values for parameters that are not in the Aspen Plus data bank, usually because the compound is a new or unusual one. This could be the case for a newly synthesized compound, for example, a complicated organic molecule or a new pharmaceutical. Consequently, you are unlikely to need to use the **Estimation** run type in the traditional thermodynamics course. Nonetheless, it will be briefly discussed here.

Among the many pure component properties that can be estimated in an **Estimation** run are the normal boiling point, the critical pressure, temperature and volume, the ideal gas, liquid, and solid heat capacities and Gibbs energies of formation, vapor pressure, volume, enthalpy of vaporization, the R and Q parameters for UNIQUAC and UNIFAC models, and many other parameters, including some transport properties such as thermal conductivity and viscosity.

In addition, mixture parameters can be predicted in an **Estimation** run. These include some equation of state parameters and binary parameters in the Wilson, NRTL, and UNIQUAC activity coefficient models. This is discussed later in this chapter.

Physical property estimates can be made by a number of different methods, even for a single parameter such as a boiling point, and the different methods require different types of input data. For a pure component property, the information needed to make a property estimate, depending on the method, may require some data for that property, data for other properties, the molecular structure, or some combination of all three. To be more specific, to estimate the normal boiling point, several methods require only the structure of the molecule, while another method requires the critical temperature and pressure and some vapor pressure data. To estimate the critical temperature of a species, two methods require only the structure, three require the structure and the normal boiling point, one requires some vapor pressure data, and the simplest (and least accurate method) requires the normal boiling point and the molecular weight (which **Estimation** can calculate from the structure).

One method of estimation of pure component parameters is by considering the molecule of interest to be composed of a collection of atoms and assigning a contribution of each atom to the total value of the property. However, methods of better accuracy consider not single atoms, but instead groups of atoms that are thought of as functional groups, for example, the $-CH_3$ methyl group, the $-COOH$ carboxylic acid group, the $-NH_2$ amine group. In this case the estimate of the total property is obtained as the sum of the contributions of the functional groups. This class of models is called Group Contribution methods, which are more accurate than atom-based models since they take into account the bonding states of each atom. However, there is no universal set of functional groups in use, so that different property estimation methods may use different definitions (combinations of atoms) for their functional groups. By entering the structure of the molecule, as will be described shortly, and the method of property estimation you want to use, the **Estimation** run will assign the appropriate groups.

We consider pure component property estimation first, and then move on to estimating binary parameters in various mixture models.

As an example of using the **Estimation** module, after going through the usual **Setup**, entering **Components**, here only *n*-butanol for this illustration, and then going to **Estimation>Input**, we see the window in Fig. 8.2-2

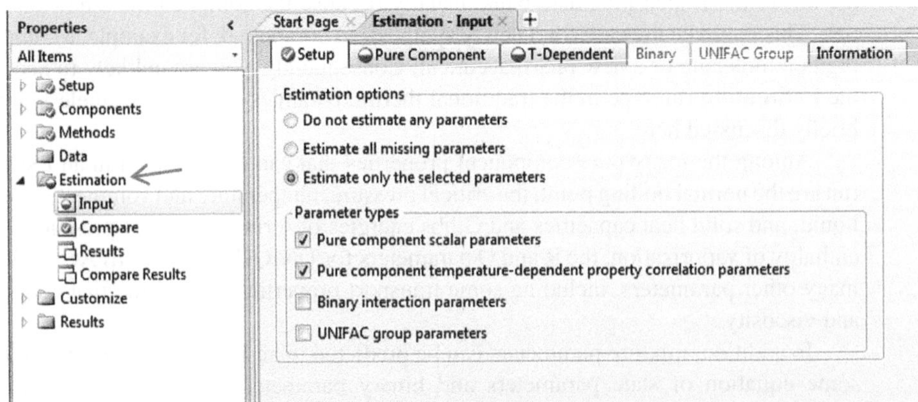

Figure 8.2-2

in which the following choices have been made: **Estimate only the selected parameters** has been chosen under Estimate options, and **Pure component scalar parameters** and **Pure component temperature-dependent property correlation parameters** have been chosen under **Parameter types**.

Next going to the **Pure Component** tab brings up the window in Fig. 8.2-3, now populated.

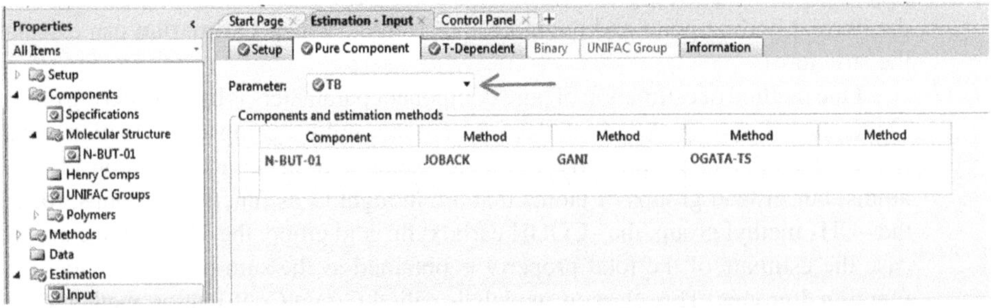

Figure 8.2-3

Clicking on the down arrow in that window by **Parameter** results in a long list of parameters that can be estimated as indicated by check marks; here the normal boiling temperature **TB** has been chosen. In the **Components and estimation method** box one choses the component and the one or more methods that can be used for the property (parameter) chosen. One can then go back to the **Parameter** drop-down menu and repeat the process to choose a number of other properties (the ones in which the component and method(s) have previously been chosen will now appear in bold on the parameter list).

One can also go to the **T-Dependent** tab (shown in Fig. 8.2-4) to estimate properties such as the ideal gas heat capacity **CPIG** (Fig. 8.2-4) and the vapor pressure **PL** (Fig. 8.2-5).

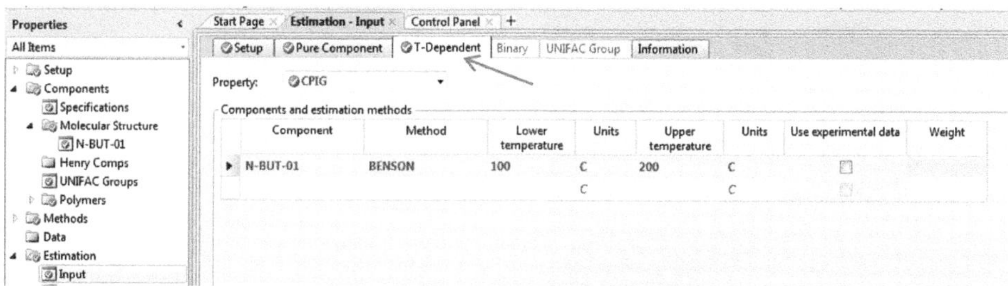

Figure 8.2-4

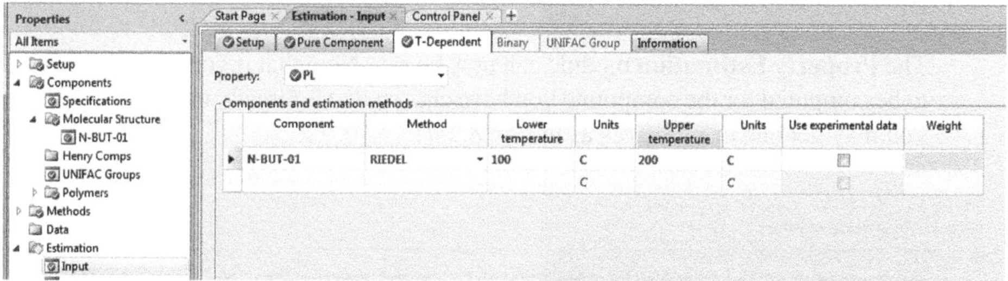

Figure 8.2-5

Some of the estimation methods require an abbreviated description of the structure of the molecule. This is entered in the **Properties>Components>Molecular Structure** window as will now be described. For *n*-butanol the structure is

$$CH_3 - CH_2 - CH_2 - CH_2 - OH$$

In the **Molecular Structure** window in Fig. 8.2-6 all the atoms, except hydrogens (which the Aspen Plus program adds internally) are entered one at a time. The procedure is that in the window under **Atom1** we designate a nonhydrogen atom in the molecule to be atom 1 and choose its atom type from the drop-down menu that appears under **Type**. For **Atom2** we choose the (nonhydrogen) atom to which Atom1 is chemically bonded, and designate the type of bond (single bond, double bond, etc.) from the **Bond type** drop-down menu. This is then repeated for Atom1 if it is also bonded to other nonhydrogen atoms. If not, then we move on to Atom2 and its bonds to other atoms (other than atom 1). We then continue in this way until all the nonhydrogen atoms are accounted for.

In the window in Fig. 8.2-6, starting from the left in the chemical formula above for *n*-butanol, we have entered the first carbon as **Atom1** and that it is singly bonded to the second carbon. The second carbon (now **Atom2**) on the second line below is singly bonded to the third carbon atom, the third carbon is singly bonded to the fourth carbon atom, which is also singly bonded to the oxygen of the alcohol group.

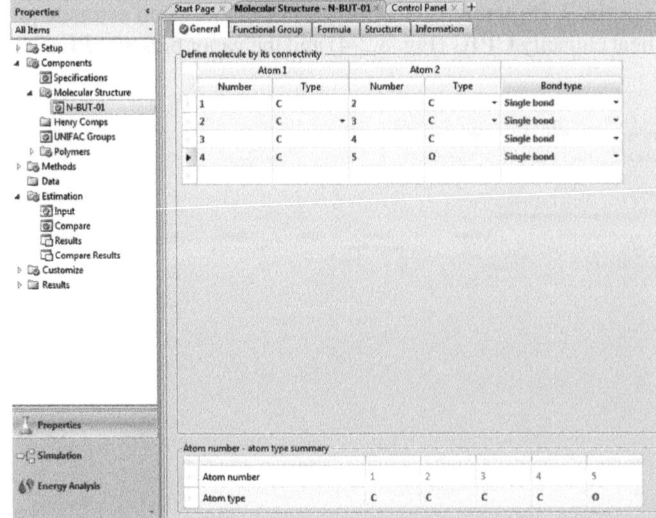

Figure 8.2-6

The **Property Estimation** module can now be run. Note that if your list includes a property to be computed for the compound you have chosen that is already in the Aspen Plus database you may get one or messages as in Fig. 8.2-7.

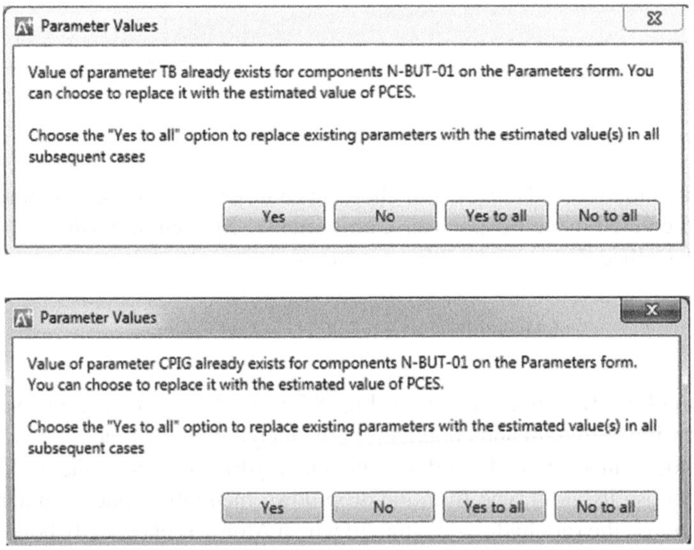

Parameter Values

Value of parameter TB already exists for components N-BUT-01 on the Parameters form. You can choose to replace it with the estimated value of PCES.

Choose the "Yes to all" option to replace existing parameters with the estimated value(s) in all subsequent cases

Yes | No | Yes to all | No to all

Parameter Values

Value of parameter CPIG already exists for components N-BUT-01 on the Parameters form. You can choose to replace it with the estimated value of PCES.

Choose the "Yes to all" option to replace existing parameters with the estimated value(s) in all subsequent cases

Yes | No | Yes to all | No to all

Figure 8.2-7

Choose **No to all** here. Now going to >**Estimation**>**Results** and then the **Pure Component** tab we see the results shown in Fig. 8.2-8.

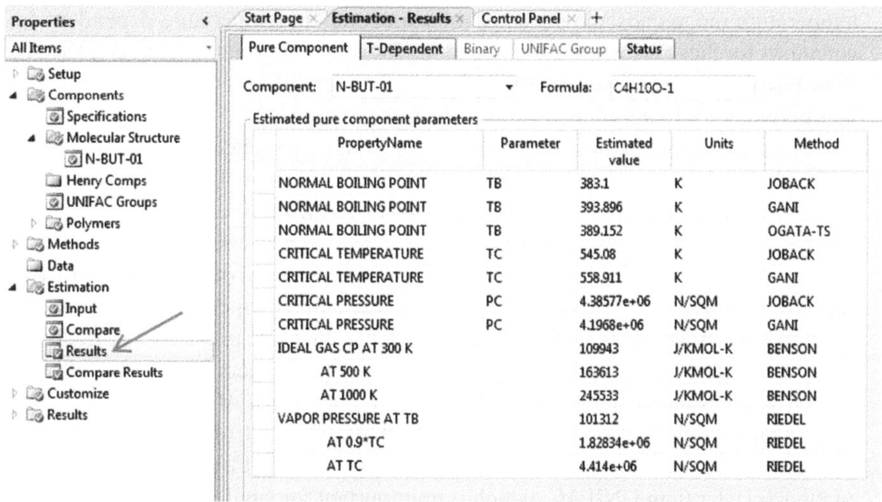

PropertyName	Parameter	Estimated value	Units	Method
NORMAL BOILING POINT	TB	383.1	K	JOBACK
NORMAL BOILING POINT	TB	393.896	K	GANI
NORMAL BOILING POINT	TB	389.152	K	OGATA-TS
CRITICAL TEMPERATURE	TC	545.08	K	JOBACK
CRITICAL TEMPERATURE	TC	558.911	K	GANI
CRITICAL PRESSURE	PC	4.38577e+06	N/SQM	JOBACK
CRITICAL PRESSURE	PC	4.1968e+06	N/SQM	GANI
IDEAL GAS CP AT 300 K		109943	J/KMOL-K	BENSON
AT 500 K		163613	J/KMOL-K	BENSON
AT 1000 K		245533	J/KMOL-K	BENSON
VAPOR PRESSURE AT TB		101312	N/SQM	RIEDEL
AT 0.9*TC		1.82834e+06	N/SQM	RIEDEL
AT TC		4.414e+06	N/SQM	RIEDEL

Figure 8.2-8

For comparison, using the **NIST TDE** described in Chapter 4, the reported normal boiling point is 390.79 K, compared to 383.1 (Joback method) and 393.9 K (Gani method), predicted here by the two different methods. Similarly, the measured critical temperature for *n*-butanol is 561.9 K compared to the two predicted values of 545.1 K and 558.9 K, the measured critical pressure is 4,412,553 N/m^2 compared to predictions of 4,385,773 and 4,196,802 N/m^2, and the predicted acentric factor is 0.6602 compared to the measured value of 0.6002. So we see that the estimates made using the various methods are not perfect, but they are reasonable.

Now clicking on the **T-Dependent** tab in the window shown in Fig. 8.2-8, the window in Fig. 8.2-9 appears that contains, not a list of the values of the property for the

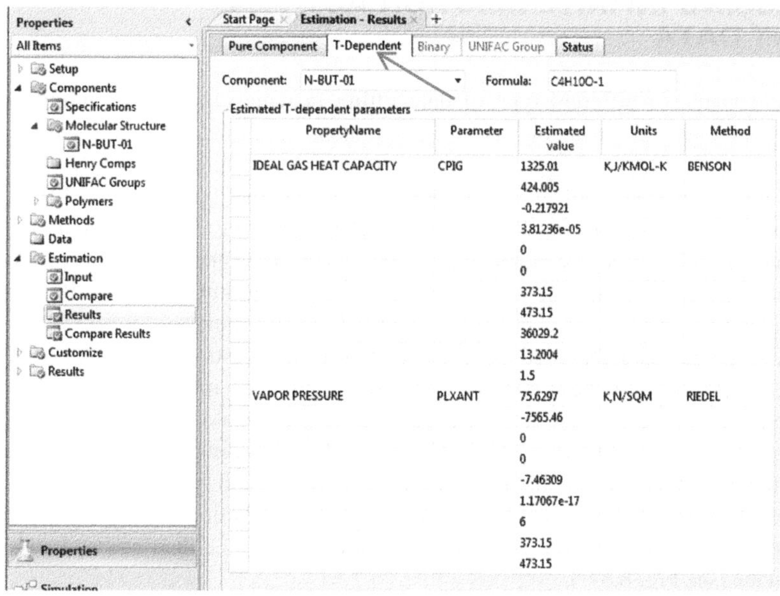

PropertyName	Parameter	Estimated value	Units	Method
IDEAL GAS HEAT CAPACITY	CPIG	1325.01	KJ/KMOL-K	BENSON
		424.005		
		-0.217921		
		3.81236e-05		
		0		
		0		
		373.15		
		473.15		
		36029.2		
		13.2004		
		1.5		
VAPOR PRESSURE	PLXANT	75.6297	K,N/SQM	RIEDEL
		-7565.46		
		0		
		0		
		-7.46309		
		1.17067e-17		
		6		
		373.15		
		473.15		

Figure 8.2-9

temperature range chosen, but rather the parameters in temperature-dependent correlative equations for these properties, the equations for which are found elsewhere. [See the Aspen Plus Help files when running the program or refer to a standard chemical engineering thermodynamics textbook.]

For mixtures, to estimate the values of the binary parameters in the Wilson, NRTL, or UNIQUAC models, generally one of two methods are used. The first is if infinite dilution activity coefficient data are available, **Property Estimation** can use these to compute the binary parameters in these models. This is considered in the next section. [Note that if activity coefficient data are available over a range of compositions, the **Data Regression** run type discussed in Chapter 5 should be used, not **Property Estimation**.] If infinite dilution activity coefficient data are not available, then the predictive UNIFAC method is used by **Property Estimation** to obtain the binary parameters in the other models.

Now a reasonable question is why use one model (UNIFAC) to estimate the parameters in another model (Wilson, NRTL, or UNIQUAC) instead of using the UNIFAC model directly in all calculations? There are two reasons for this. First, is that these other models run much faster than UNIFAC, which is unimportant for single calculations, but can have a noticeable effect when running a large simulation with multiple recycle streams that requires many iterations. Second, and more important, is for calculations involving multicomponent mixtures the user might have binary parameters in an activity coefficient model based on experimental data for some of the pairs of molecules in the mixture, but not all. Then there are two choices: do all the calculations with UNIFAC (and thereby ignore the experimental data upon which the parameters in the other model were based, which is not a good idea) or use UNIFAC to predict the values of the binary parameters in the model being used for the pairs of components in the mixture without experimental-based parameters. This latter method is preferable since it retains available experimental data and calculations using these other models are quicker.

8.3 REGRESSING INFINITE DILUTION ACTIVITY COEFFICIENT DATA

Here we consider a different type of data fitting in which the mutual infinite dilution activity coefficients are available for two components in a binary mixture and the goal is to compute the values of the parameters in an activity coefficient model. Such data may be available from static cell or ebulliometric measurements (see Figs. 10.2-9 and 10.2-10, respectively, in *Chemical, Biochemical and Engineering Thermodynamics*, 4th ed., S. I. Sandler, John Wiley & Sons, Inc., 2006). Once the binary parameters have been obtained from experiment, the complete vapor–liquid equilibrium, T-xy, P-xy, and x-y phase diagrams can be computed. This is what will be done here.

The example data that will be used here is from Illustration 10.2-6 of the above referenced textbook for the *n*-pentane and propionaldehyde mixture at 40°C, for which the *n*-pentane infinite dilution activity coefficient in propionaldehyde is 3.848 and that of propionaldehyde in *n*-pentane is 3.979. [Complete P-xy data are presented in the aforementioned illustration. However, only the infinite dilution data will be used here for regression. The P-xy data will then be used to compare with the predictions based on the regressed parameters.] The starting point is to go to **Components>Specifications** and add the two components (see Fig. 8.3-1).

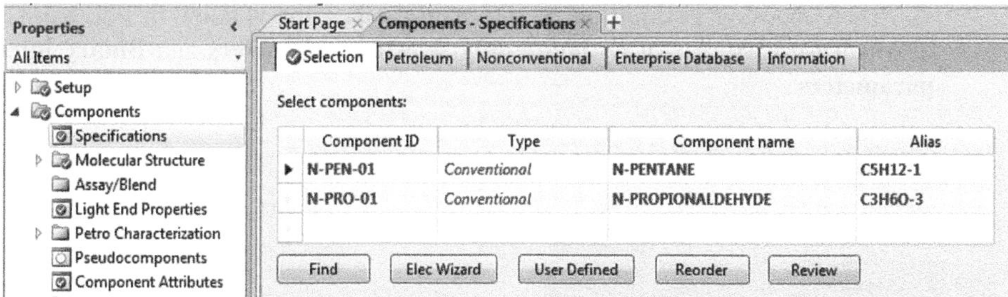

Figure 8.3-1

Then choose **Methods** and choose the NRTL model that will be examined first. Clicking on **Methods>Parameters>Binary Interaction>NRTL-1** brings up the following list of parameters shown in Fig. 8.3-2.

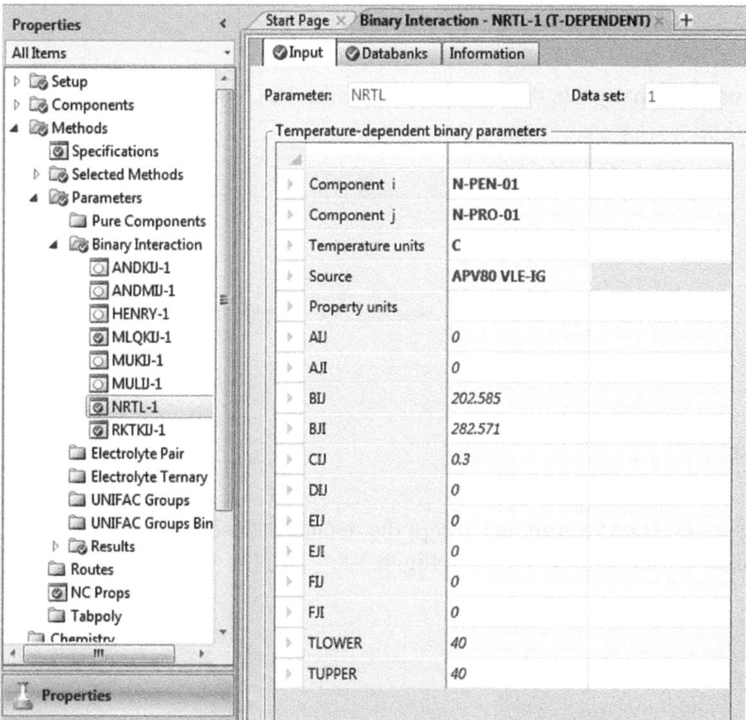

Figure 8.3-2

We need to do this even though a new set of parameters will be calculated here.

Then choosing **Estimation** on the main toolbar, the window in Fig. 8.3-3 appears in which we have chosen **Estimate only selected parameters** and **Binary interaction parameters**.

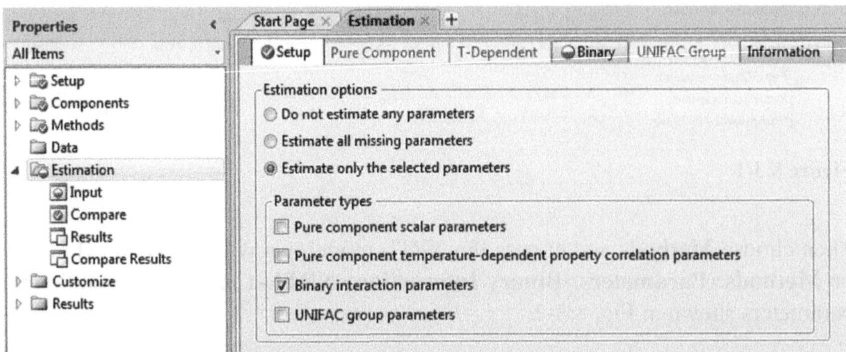

Figure 8.3-3

Next click on the **Binary** tab, then on **New** and make the choices in Fig. 8.3-4.

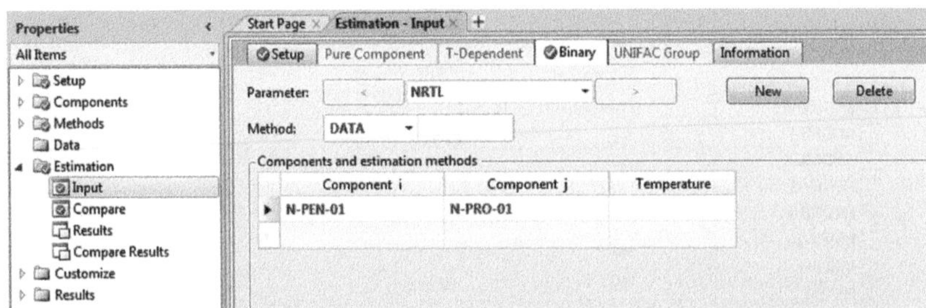

Figure 8.3-4

Then click on the **Data>Setup** and accept the default data set name of **D-1** (or choose a data set name of your own) as shown in Fig. 8.3-5.

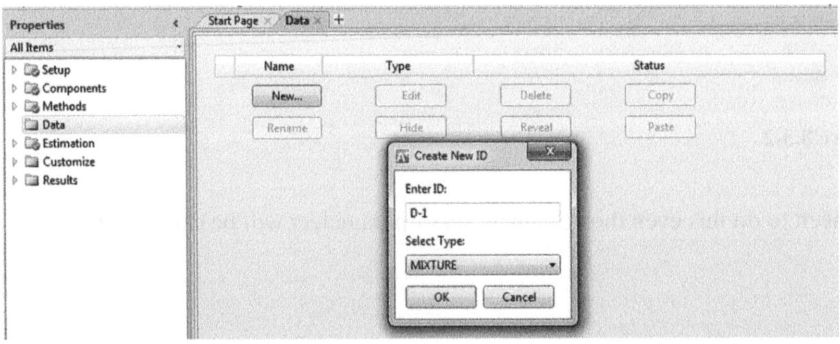

Figure 8.3-5

Click **OK**. Next go to **Data>Setup**, and choose **For estimation** under **Category** and then move the two **Available components** to **Selected components** by highlighting and clicking the > key. This is shown in Fig. 8.3-6.

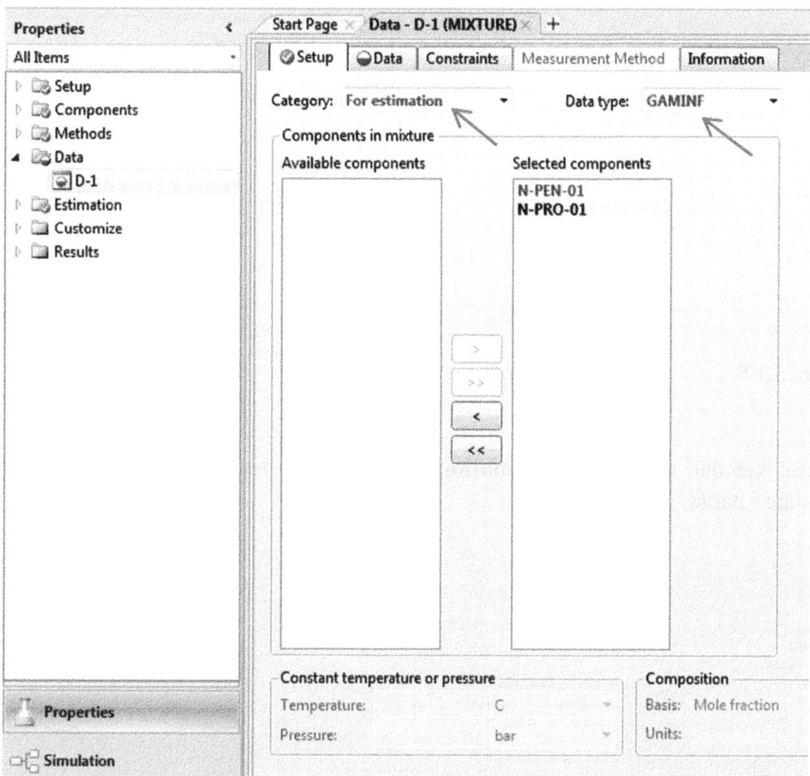

Figure 8.3-6

Then click on the **Data** tab and enter the infinite dilution data by choosing **GAMINF** for **Data type**, and then enter the temperature in appropriate units and the values of the infinite dilution activity coefficients as shown in Fig. 8.3-7.

Data type: GAMINF

Experimental data

Usage	TEMP1	GAMINF1	TEMP2	GAMINF2
	C	N-PEN-01	C	N-PRO-01
STD-DEV	0.1	1%	0.1	1%
DATA	40	3.848	40	3.979

Figure 8.3-7

Then click on > or press **F5** to run the regression, and the following message (Fig. 8.3-8) will appear.

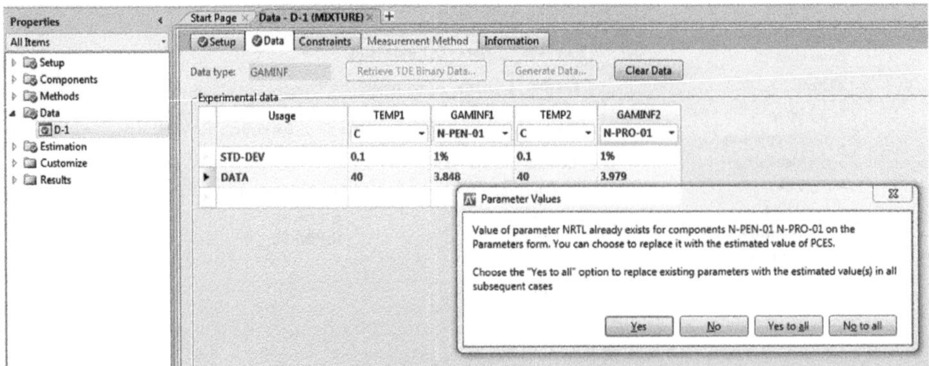

Figure 8.3-8

Choose **Yes** and then go to **Estimation>Results** and then to the **Binary** tab to see the calculated parameters (Fig. 8.3-9).

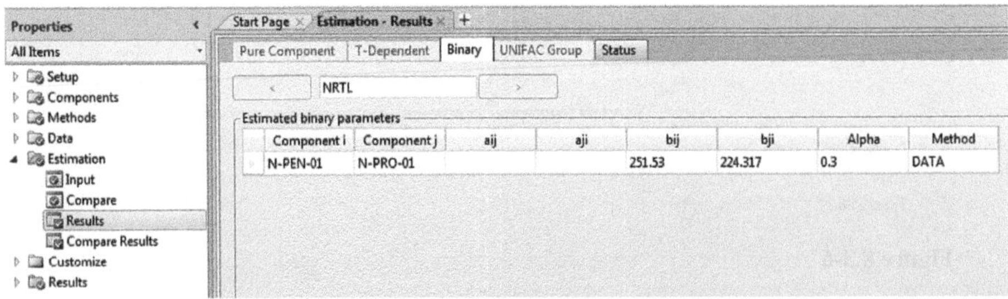

Figure 8.3-9

Then going to **Methods>Parameters>Binary Interaction>NRTL-1** you will see that the parameters just computed now appear as shown in Fig. 8.3-10. Also note that now the **Source** (of parameters) is listed as **R-PCES**.

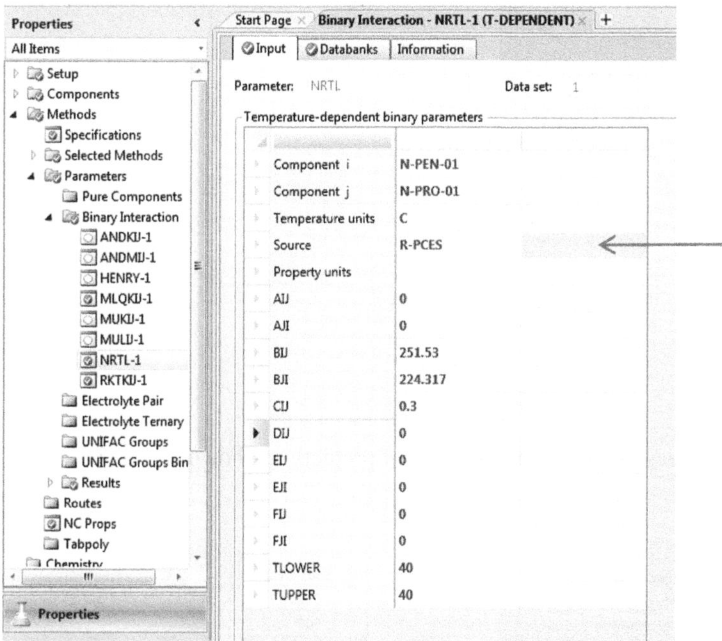

Figure 8.3-10

To then calculate the P-xy diagram for this system with these parameters, click **Binary** in the **Analysis** section of the main toolbar and populate the window as shown in Fig. 8.3-11.

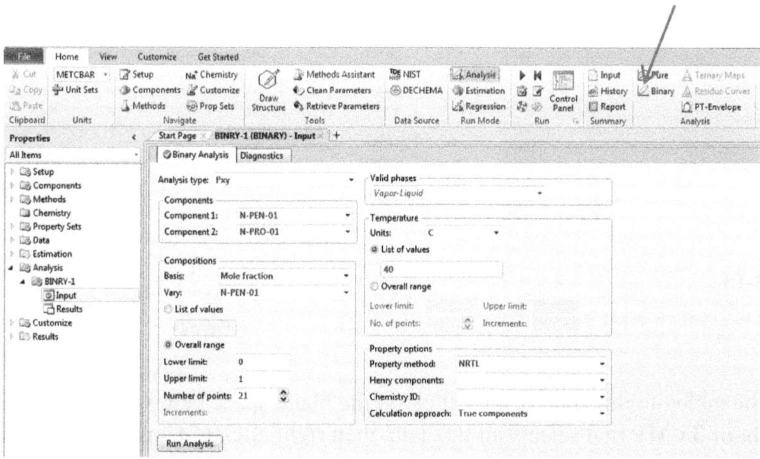

Figure 8.3-11

The result is shown in Fig. 8.3-12.

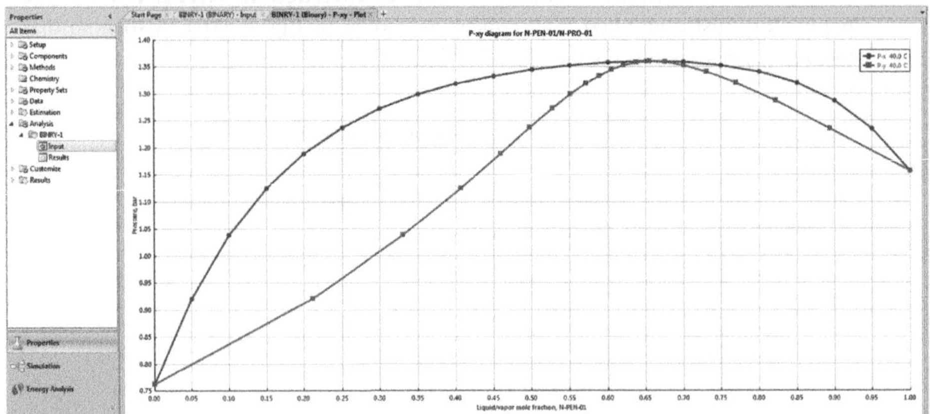

Figure 8.3-12

Then clicking on results brings up the table of predictions shown in Fig. 8.3-13.

Binary analysis results

TEMP	MOLEFRAC N-PEN-01	TOTAL PRES	TOTAL KVL N-PEN-01	TOTAL KVL N-PRO-01	LIQUID GAMMA N-PEN-01	LIQUID GAMMA N-PRO-01	VAPOR MOLEFRAC N-PEN-01	VAPOR MOLEFRAC N-PRO-01	LIQUID MOLEFRAC N-PEN-01	LIQUID MOLEFRAC N-PRO-01
C		bar								
40	0	0.761626	5.84658	1	3.848	1	0	1	0	1
40	0.05	0.920716	4.22892	0.830057	3.36471	1.00344	0.211446	0.788554	0.05	0.95
40	0.1	1.03814	3.30642	0.743731	2.96622	1.01374	0.330642	0.669358	0.1	0.9
40	0.15	1.12492	2.71112	0.698038	2.6355	1.031	0.406668	0.593332	0.15	0.85
40	0.2	1.18917	2.29596	0.676009	2.35939	1.05549	0.459193	0.540807	0.2	0.8
40	0.25	1.23681	1.9907	0.669767	2.12766	1.08764	0.497675	0.502326	0.25	0.75
40	0.3	1.27225	1.75755	0.675337	1.93229	1.12811	0.527264	0.472736	0.3	0.7
40	0.35	1.29871	1.57443	0.69069	1.76697	1.17775	0.551051	0.448949	0.35	0.65
40	0.4	1.31855	1.42762	0.714921	1.62667	1.23769	0.571047	0.428953	0.4	0.6
40	0.45	1.33344	1.30816	0.747866	1.5074	1.30935	0.588673	0.411327	0.45	0.55
40	0.5	1.34455	1.21005	0.789948	1.40595	1.39455	0.605026	0.394974	0.5	0.5
40	0.55	1.35255	1.12915	0.842145	1.31977	1.49554	0.621034	0.378966	0.55	0.45
40	0.6	1.35774	1.06263	0.906053	1.24679	1.61521	0.637579	0.362421	0.6	0.4
40	0.65	1.36	1.00859	0.984041	1.18535	1.75715	0.655586	0.344414	0.65	0.35
40	0.7	1.35876	0.965908	1.07935	1.13415	1.92594	0.676135	0.323865	0.7	0.3
40	0.75	1.35295	0.934138	1.19759	1.09216	2.1274	0.700604	0.299397	0.75	0.25
40	0.8	1.34086	0.913598	1.34561	1.05889	2.36896	0.730879	0.269121	0.8	0.2
40	0.85	1.31988	0.905573	1.53509	1.03288	2.66027	0.769737	0.230263	0.85	0.15
40	0.9	1.28627	0.912827	1.78456	1.01464	3.01385	0.821544	0.178456	0.9	0.1
40	0.95	1.23462	0.940742	2.1259	1.00368	3.44615	0.893705	0.106295	0.95	0.05
40	1	1.1572	1	2.61883	1	3.979	1	0	1	0

Figure 8.3-13

To copy the table of computed results click on the blank space below the B in **Binary** and to the right of **TEMP** that selects all the data, then right click to **Copy**, and then paste the table into a graphing software package. I then added the experimental data listed in the aforementioned Illustration 10.2-3 to produce the graph in Fig. 8.3-14 that compares the calculated results (lines) based on the measured infinite dilution activity coefficients with the reported experimental P-xy data (points).

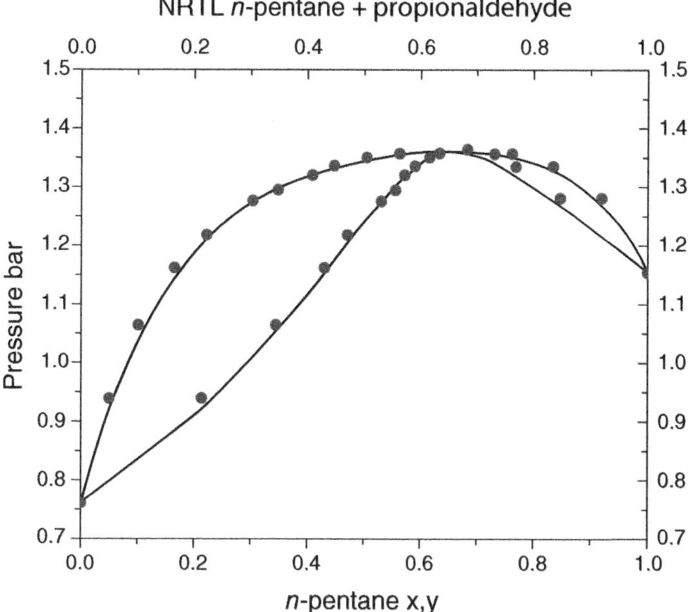

Figure 8.3-14

We see that the predictions of the complete P-xy diagram based only on the measured infinite dilution data and the NRTL model are in excellent agreement with the experimental data. [This will not be the case for all mixtures.]

Of course, the NRTL model is not the only model that could be used. Without much further discussion, the screenshots below follow the procedure just described, but now with the Wilson model. We start by going back to **Methods** and choosing **Wilson** (Fig. 8.3-15).

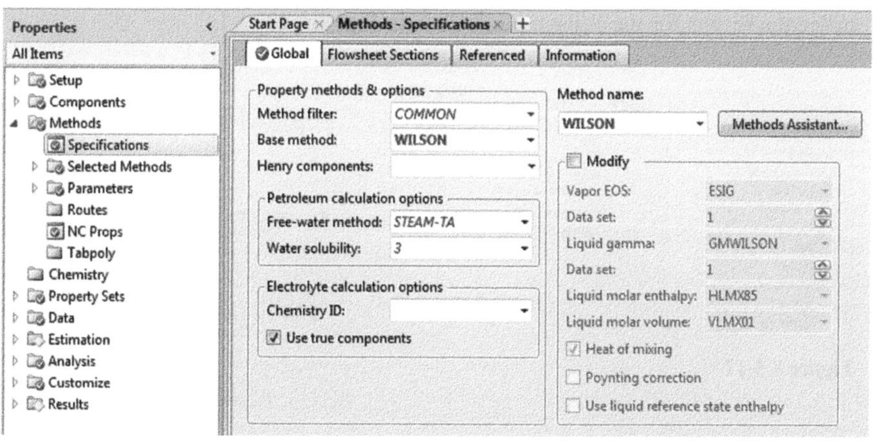

Figure 8.3-15

Then going to **Parameters>Binary Interaction>Wilson-1** (Fig. 8.3-16):

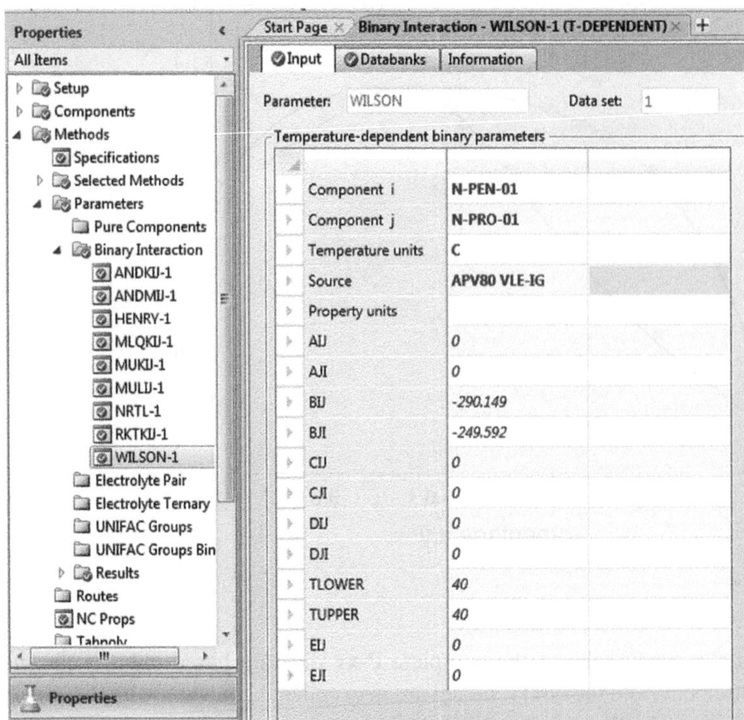

Figure 8.3-16

Repeating the same sequence as for the NRTL model, then clicking on > or pressing **F5** to do the calculation and, when completed, going to **Estimation>Results** and the **Binary** tab to see the results for the Wilson parameters in Fig. 8.3-17.

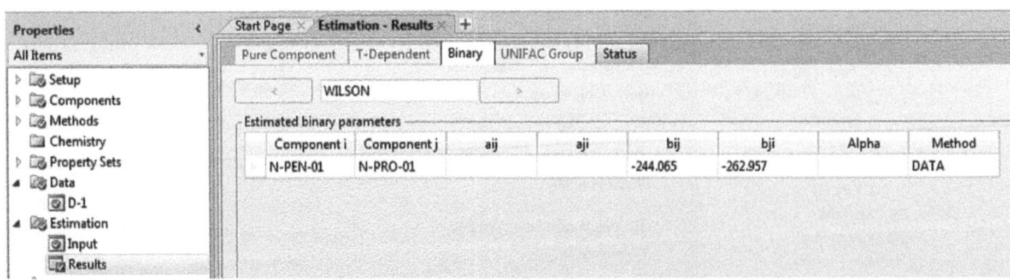

Figure 8.3-17

Then going to **Binary** in the **Analysis** section of the main toolbar, completing the window as for the NRTL model but now choosing **Wilson** as the **Property Method**, and then clicking on **Run Analysis** produces the graph and table in Fig. 8.3-18.

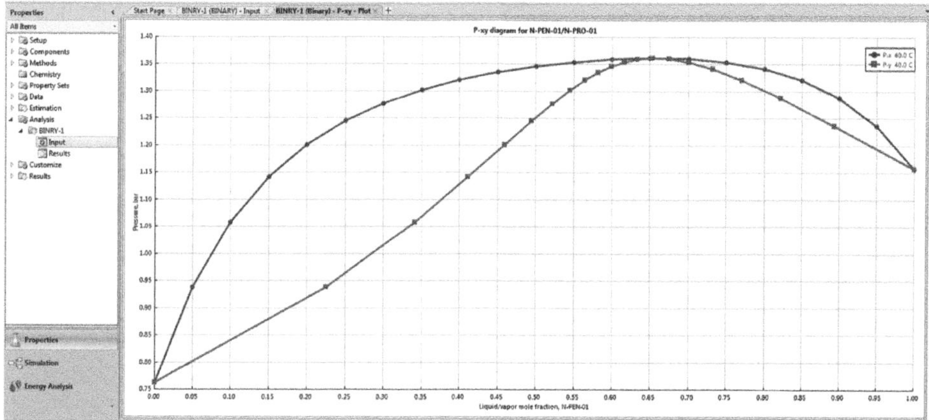

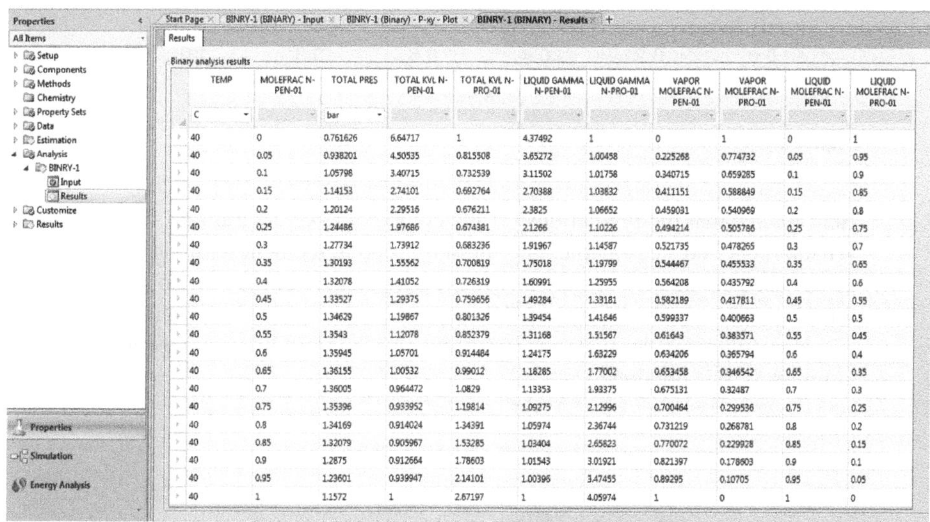

Binary analysis results

TEMP	MOLEFRAC N-PEN-01	TOTAL PRES	TOTAL KVL N-PEN-01	TOTAL KVL N-PRO-01	LIQUID GAMMA N-PEN-01	LIQUID GAMMA N-PRO-01	VAPOR MOLEFRAC N-PEN-01	VAPOR MOLEFRAC N-PRO-01	LIQUID MOLEFRAC N-PEN-01	LIQUID MOLEFRAC N-PRO-01
C		bar								
40	0	0.761626	6.64717	1	4.37492	1	0	1	0	1
40	0.05	0.938201	4.50535	0.815508	3.65272	1.00458	0.225268	0.774732	0.05	0.95
40	0.1	1.05798	3.40715	0.732539	3.11502	1.01758	0.340715	0.659285	0.1	0.9
40	0.15	1.14153	2.74101	0.692764	2.70388	1.03832	0.411151	0.588849	0.15	0.85
40	0.2	1.20124	2.29516	0.676211	2.3825	1.06652	0.459031	0.540969	0.2	0.8
40	0.25	1.24486	1.97686	0.674381	2.1266	1.10226	0.494214	0.505786	0.25	0.75
40	0.3	1.27734	1.73912	0.683236	1.91967	1.14587	0.521735	0.478265	0.3	0.7
40	0.35	1.30193	1.55562	0.700819	1.75018	1.19799	0.544467	0.455533	0.35	0.65
40	0.4	1.32078	1.41052	0.726319	1.60991	1.25955	0.564208	0.435792	0.4	0.6
40	0.45	1.33527	1.29375	0.759656	1.49284	1.33181	0.582189	0.417811	0.45	0.55
40	0.5	1.34629	1.19867	0.801326	1.39454	1.41646	0.599337	0.400663	0.5	0.5
40	0.55	1.3543	1.12078	0.852379	1.31168	1.51567	0.61643	0.383571	0.55	0.45
40	0.6	1.35945	1.05701	0.914484	1.24175	1.63229	0.634206	0.365794	0.6	0.4
40	0.65	1.36155	1.00532	0.99012	1.18285	1.77002	0.653458	0.346542	0.65	0.35
40	0.7	1.36005	0.964472	1.0829	1.13353	1.93375	0.675131	0.32487	0.7	0.3
40	0.75	1.35396	0.933952	1.19814	1.09275	2.12996	0.700464	0.299536	0.75	0.25
40	0.8	1.34169	0.914024	1.34391	1.05974	2.36744	0.731219	0.268781	0.8	0.2
40	0.85	1.32079	0.905967	1.53285	1.03404	2.65823	0.770072	0.229928	0.85	0.15
40	0.9	1.2875	0.912664	1.78603	1.01543	3.01921	0.821397	0.178603	0.9	0.1
40	0.95	1.23601	0.939947	2.14101	1.00396	3.47455	0.89295	0.10705	0.95	0.05
40	1	1.1572	1	2.67197	1	4.05974	1	0	1	0

Figure 8.3-18

After exporting the data and replotting with the predictions as line and the experimental data as points Fig. 8.3-19 is obtained.

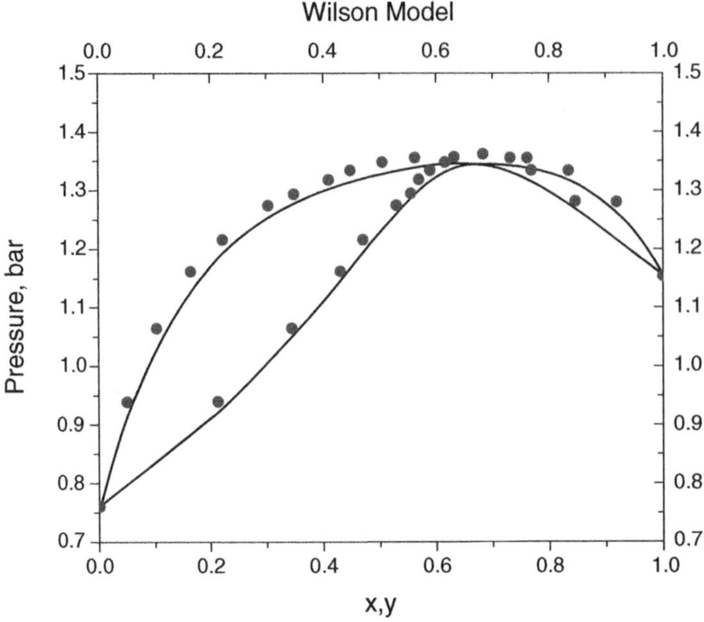

Figure 8.3-19

Here we see that the predictions for the complete P-xy diagram for this mixture using the Wilson model are reasonable, but not in as good agreement with the experimental data as those based on the NRTL model. However, we cannot take this to be a general result, since for other mixtures the Wilson model may lead to better results than the NRTL model.

Now following the analogous procedure for the UNIQUAC model the series of windows in Fig. 8.3-20 is obtained.

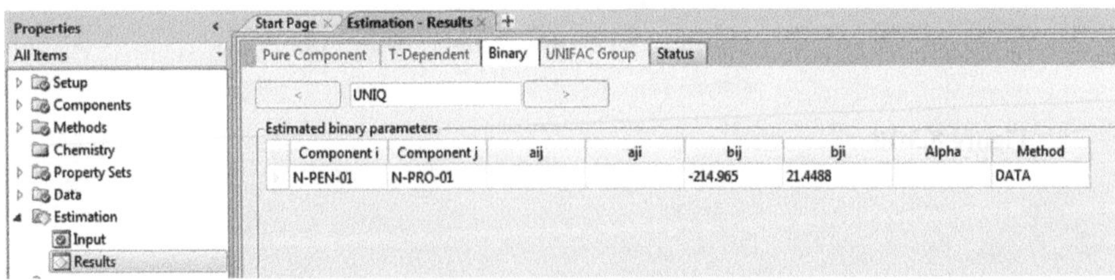

Figure 8.3-20

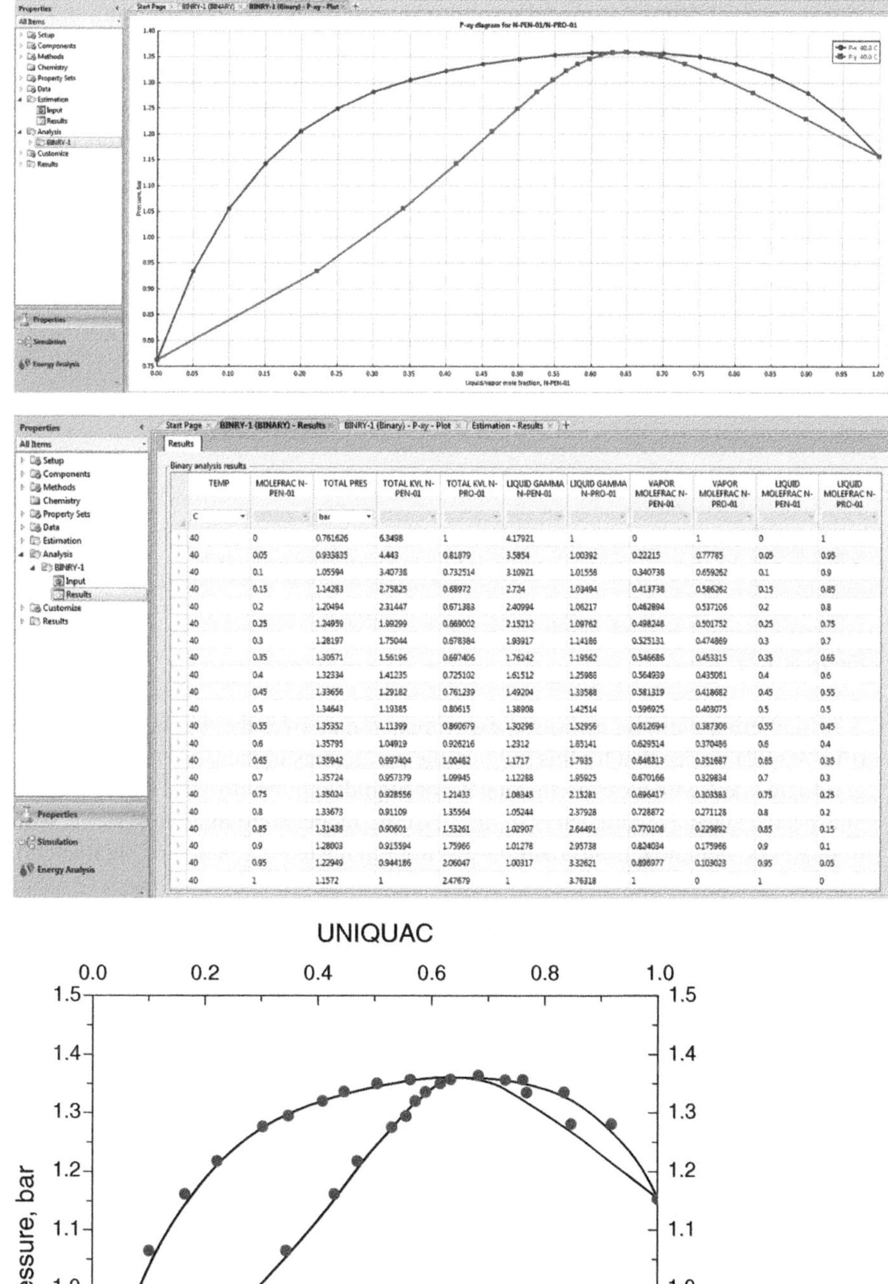

Binary analysis results

TEMP	MOLEFRAC N-PEN-01	TOTAL PRES	TOTAL KVL N-PEN-01	TOTAL KVL N-PRO-01	LIQUID GAMMA N-PEN-01	LIQUID GAMMA N-PRO-01	VAPOR MOLEFRAC N-PEN-01	VAPOR MOLEFRAC N-PRO-01	LIQUID MOLEFRAC N-PEN-01	LIQUID MOLEFRAC N-PRO-01
C		bar								
40	0	0.761626	6.3498	1	4.17921	1	0	1	0	1
40	0.05	0.933835	4.443	0.81879	3.5854	1.00392	0.22215	0.77785	0.05	0.95
40	0.1	1.05594	3.40738	0.732514	3.10921	1.01558	0.340738	0.659262	0.1	0.9
40	0.15	1.14283	2.75825	0.68972	2.724	1.03494	0.413738	0.586262	0.15	0.85
40	0.2	1.20494	2.31447	0.671383	2.40994	1.06217	0.462894	0.537106	0.2	0.8
40	0.25	1.24959	1.99299	0.669002	2.15212	1.09762	0.498248	0.501752	0.25	0.75
40	0.3	1.28197	1.75044	0.678384	1.93917	1.14186	0.525131	0.474869	0.3	0.7
40	0.35	1.30571	1.56196	0.697406	1.76242	1.19562	0.546686	0.453315	0.35	0.65
40	0.4	1.32334	1.41235	0.725102	1.61512	1.25988	0.564939	0.435061	0.4	0.6
40	0.45	1.33656	1.29182	0.761239	1.49204	1.33588	0.581319	0.418682	0.45	0.55
40	0.5	1.34643	1.19385	0.80615	1.38908	1.42514	0.596925	0.403075	0.5	0.5
40	0.55	1.35352	1.11399	0.860679	1.30296	1.52956	0.612694	0.387306	0.55	0.45
40	0.6	1.35795	1.04919	0.926216	1.2312	1.65141	0.629514	0.370486	0.6	0.4
40	0.65	1.35942	0.997404	1.00482	1.1717	1.7935	0.648313	0.351687	0.65	0.35
40	0.7	1.35724	0.957379	1.09945	1.12288	1.95925	0.670166	0.329834	0.7	0.3
40	0.75	1.35024	0.928556	1.21433	1.08345	2.15281	0.696417	0.303583	0.75	0.25
40	0.8	1.33673	0.91109	1.35564	1.05244	2.37928	0.728872	0.271128	0.8	0.2
40	0.85	1.31438	0.90601	1.53261	1.02907	2.64491	0.770108	0.229892	0.85	0.15
40	0.9	1.28003	0.915594	1.75966	1.01278	2.95738	0.824034	0.175966	0.9	0.1
40	0.95	1.22949	0.944186	2.06047	1.00317	3.32621	0.896977	0.103023	0.95	0.05
40	1	1.1572	1	2.47679	1	3.76318	1	0	1	0

Figure 8.3-20 *(Continued)*

So we see the UNIQUAC model also fits the experimental data quite well, indeed with a similar accuracy to the NRTL model.

Finally, following the analogous procedure, but now using the completely predictive UNIFAC model, that ignores the measured infinite dilution data (so that **Estimation** and the experimental infinite dilution activity coefficient data are not used to obtain the binary parameters), produces the results in Fig. 8.3-21.

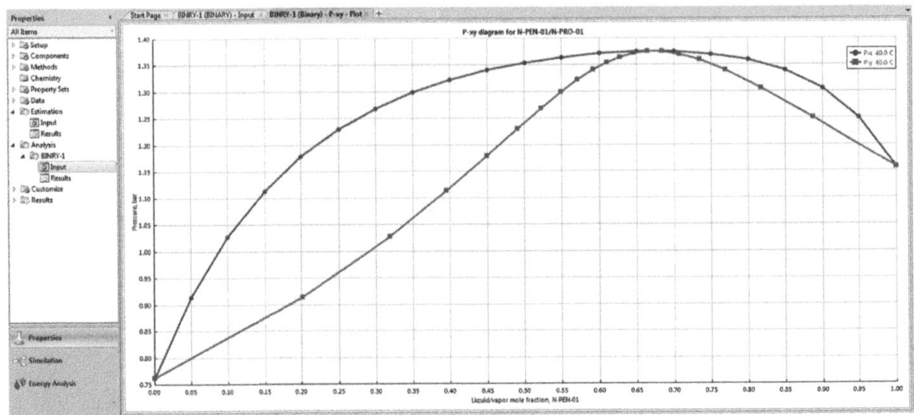

Figure 8.3-21

Now replotting with the measured experimental data over the whole concentration range (Fig. 8.3-22) we see that for this mixture the predictions of the UNIFAC model are in quite good agreement with the experimental vapor–liquid equilibrium data. This is not completely unexpected since the parameters in the UNIFAC model were fitted to systems such as this (though I do not know if this data set was used in the determination of parameters).

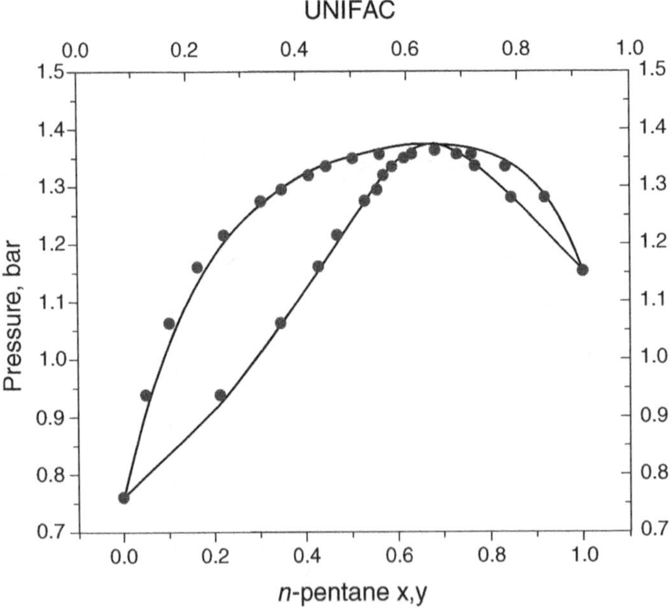

Figure 8.3-22

PROBLEMS

8.1. Pividal and Sandler, *J. Chem. Eng. Data*, **33**, 438 (1988) have reported the following infinite dilution data for 1,4 dioxane (1) + *n*-heptane (2) mixture at 1 atm

Temperature (°C)	γ_1^∞	γ_2^∞
40.2	3.59	5.47
60	3.21	4.83
80	2.63	3.92

[Do you have qualitative explanation of why the infinite dilution activity coefficients decrease in value as the temperature increases?]

(a) Fit the UNIQUAC model to these data simultaneously to find a set of four parameters that fit these data.

(b) Use the parameters in part (a) to estimate the vapor–liquid equilibrium for this system at each of the temperatures above.

8.2. Repeat Problem 8.1 with the NRTL model.

8.3. Repeat Problem 8.1 with the Wilson model.

8.4. Pividal and Sandler, *J. Chem. Eng. Data*, **33**, 438 (1988) have reported the following infinite dilution data for ethyl acetate (1) + cyclohexane (2) mixture at 1 atm

Temperature (°C)	γ_1^∞	γ_2^∞
40	3.33	2.82
60	2.95	2.56

(a) Fit the UNIQUAC model to these data simultaneously to find a set of four parameters that fit these data.

(b) Use the parameters in part (a) to estimate the vapor–liquid equilibrium for this system at each of the temperatures above.

8.5. Repeat Problem 8.4 with the NRTL model.

8.6. Repeat Problem 8.4 with the Wilson model.

8.7. Pividal and Sandler, *J. Chem. Eng. Data*, **33**, 438 (1988) have reported the following infinite dilution data for tetrahydrofuran (1) + cyclohexane (2) mixture at 1 atm

Temperature (°C)	γ_1^∞	γ_2^∞
40	1.72	1.76
60	1.63	1.65

(a) Fit the UNIQUAC model to these data simultaneously to find a set of four parameters that fit these data.

(b) Use the parameters in part (a) to estimate the vapor–liquid equilibrium for this system at each of the temperatures above.

8.8. Repeat Problem 8.7 with the NRTL model.

8.9. Repeat Problem 8.7 with the Wilson model.

8.10. Find the structure and properties of 1,4 dioxane, for example, from Wikipedia or other sources, and then use the **Properties>Estimation** procedure to estimate its critical properties, normal boiling point, and its vapor pressure and ideal gas heat capacity as function of temperature, and compare with the measured properties.

8.11. Find the structure and properties of tetrahydrofuran, for example, from Wikipedia or other sources, and then use the **Properties>Estimation** procedure to estimate its critical properties, normal boiling point, and its vapor pressure and ideal gas heat capacity as function of temperature, and compare with the measured properties.

Chapter 9

Chemical Reaction Equilibrium in Aspen Plus®

This chapter deals with the use of Aspen Plus® for the calculation of chemical reaction equilibrium. The starting point is to open a new simulation and start with a blank **Flowsheet**. Going to the **Model Palette** and then clicking on the **Reactors** tab brings up the collection of reactors shown in Fig. 9-1.

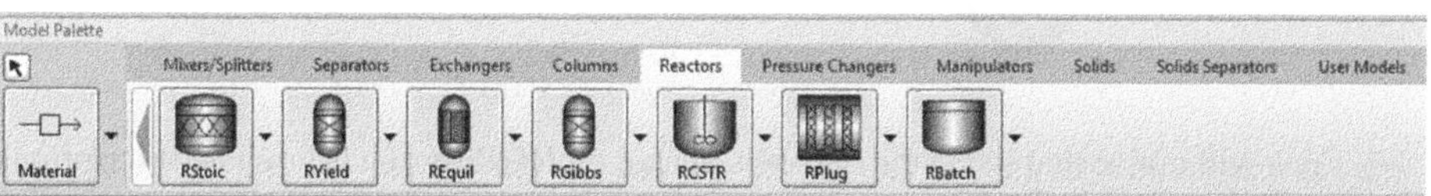

Figure 9-1

The stoichiometric reactor **RStoic** and **RYield** merely do mass balance calculations if you are specifying the extent of reaction or yield, respectively, in advance. These are not of interest here since what we are trying to do is find the equilibrium extent(s) of reaction that are not known in advance. The continuous flow stirred tank reactor (**RCSTR**), the plug flow reactor (**RPlug**), and the batch reactor (**RBatch**) will be considered in reaction kinetics/reactors course and also will not be dealt here. The reactor of interest here is **RGibbs** since it can efficiently calculate chemical equilibrium in multiphase, multireaction systems; **REquil** can also be used, but only for single-phase systems.

First, a short explanation about the **RGibbs** method of calculation. The way most chemical equilibrium problems are solved in thermodynamics textbooks is to first calculate a numerical value of the equilibrium constant (or constants in a multiple reaction system) in the standard states at the temperature of interest and then develop the equation(s) for the extent (or extents) of reaction using the stoichiometry of the reaction and departures from the standard states. The difficulty with this method of calculation in a general process simulation program is that since there is different stoichiometry for each reaction, the equation (or equations) to be solved is different in each case. This is not very useful in automated calculations in a process simulation program. What is needed instead is a single general method and algorithm applicable to all chemical equilibrium problems.

The concept of an equilibrium constant has been introduced in textbooks for computational convenience, especially for calculations by hand. The underlying fundamental principle is that at equilibrium at constant temperature and pressure, the Gibbs energy should

Using Aspen Plus® in Thermodynamics Instruction: A Step-by-Step Guide, First Edition. Stanley I. Sandler.

be a minimum. Therefore, in process simulation programs what is done is to develop a general expression for the Gibbs energy of the system in terms of the number of moles of all species present, reactants, products, and inert species, and in all phases. The calculation then is to vary the number of moles of each species in each phase subject to the stoichiometric constraints and find a solution that minimizes the total Gibbs energy of the system. In this way, the problem is transformed from solving a specific and frequently nonlinear algebraic equation (or set of equations for multiple reaction systems) to the problem of minimization of a function regardless of how many reactions or phases are involved. In this way, a general minimization algorithm can be used to solve all chemical reaction equilibrium problems. This is precisely what the **RGibbs** block does.

To solve a chemical equilibrium problem, we start by inserting a **RGibbs** reactor block into the flow sheet window and adding the inlet and exit streams as shown in Fig. 9-2.

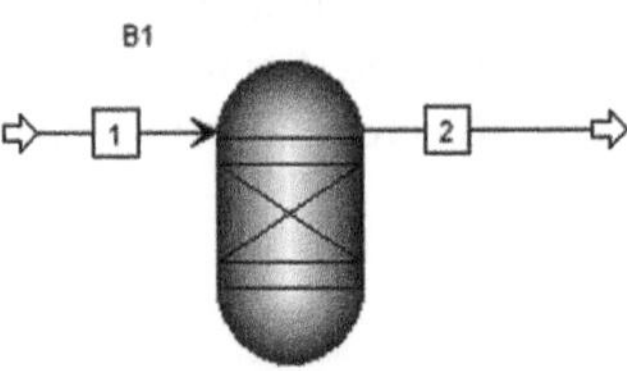

Figure 9-2

You will notice that when the cursor is placed on **RGibbs**, the following description appears at the bottom of the **Model Palette** "Rigorous reaction and/or multiphase equilibrium based on Gibbs free energy minimization."

We will use, as an example, the reaction of Illustration 13.1-4 in *Chemical, Biochemical and Engineering Thermodynamics*, 4th ed., S. I. Sandler (John Wiley & Sons, Inc., 2006), which is the reaction of nitrogen and hydrogen to form ammonia in the presence of a catalyst

$$\frac{1}{2}N_2 + \frac{3}{2}H_2 \leftrightarrow NH_3$$

The reaction is to take place at 450 K and 4 bar first (a) with stoichiometric amounts of nitrogen and hydrogen present, and then (b) with equal amounts of nitrogen, hydrogen, and an inert diluent (for which argon will be used here).

The next step is to go to **Components>Specifications** and add nitrogen, hydrogen, ammonia, and argon (Fig. 9-3).

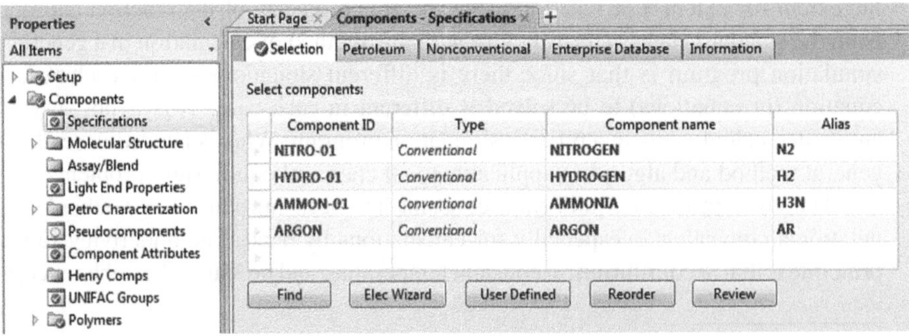

Component ID	Type	Component name	Alias
NITRO-01	Conventional	NITROGEN	N2
HYDRO-01	Conventional	HYDROGEN	H2
AMMON-01	Conventional	AMMONIA	H3N
ARGON	Conventional	ARGON	AR

Figure 9-3

Since the temperature is reasonably high, 450 K, the pressure is only 4 bar, and all the components are gases, in **Methods** we will use **IDEAL** for the **Base Method** (Fig. 9-4).

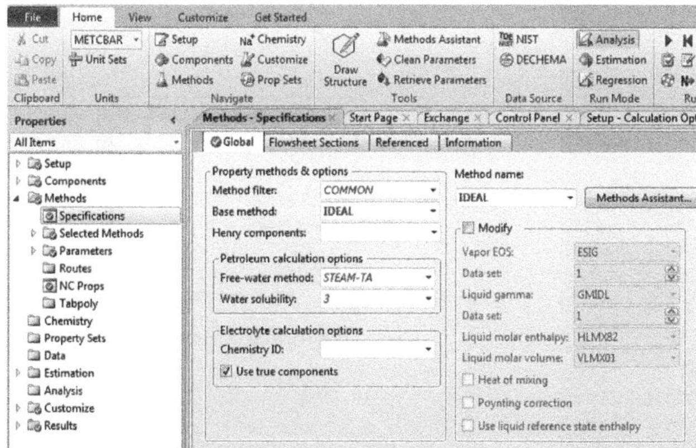

Figure 9-4

Next, in the **Simulation** window go to **Streams>1** (Fig. 9-5) and enter the stoichiometric amounts of nitrogen and hydrogen, and zero for the amount of the inert diluent for part (a).

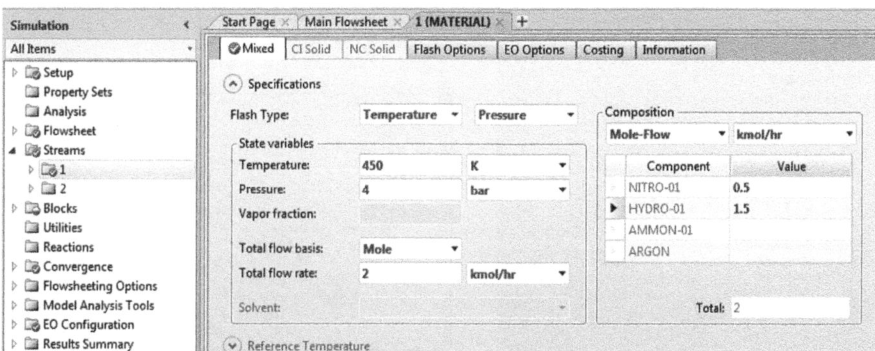

Figure 9-5

Go to **Blocks>B1** and enter the temperature and pressure of the reactor operation (Fig. 9-6)

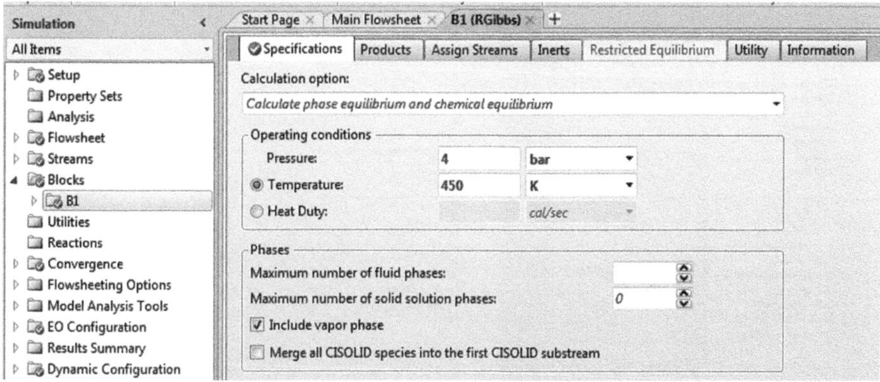

Figure 9-6

Next go to **Reactions** and click on **New** and then click on **New** again to bring up the window in Fig. 9-7, and choose **GENERAL** for the reaction type.

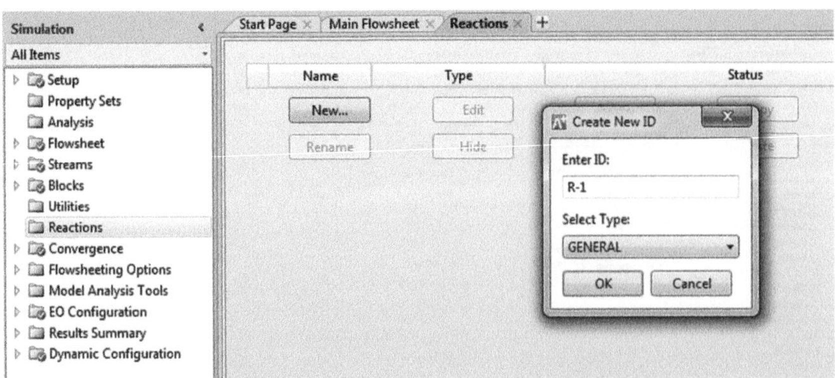

Figure 9-7

Accept the default reaction stoichiometry ID or create a new name of your choice, and then click on **OK,** which brings up the window in Fig. 9-8.

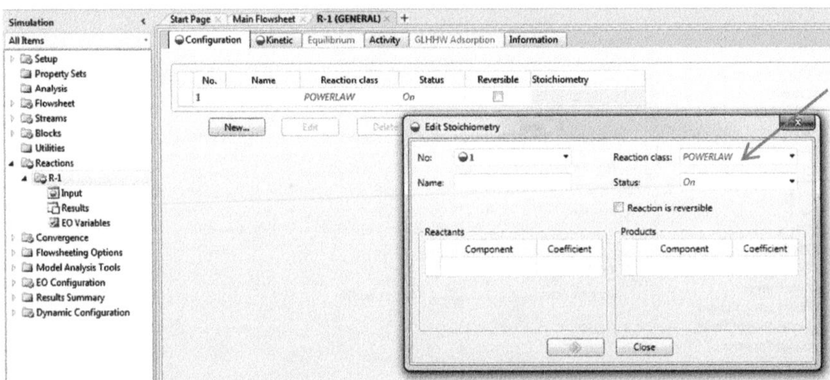

Figure 9-8

Before proceeding, change ***POWERLAW*** to **EQUILIBRIUM** here and throughout this chapter. Then populate the rest of the window as shown in Fig. 9-9.

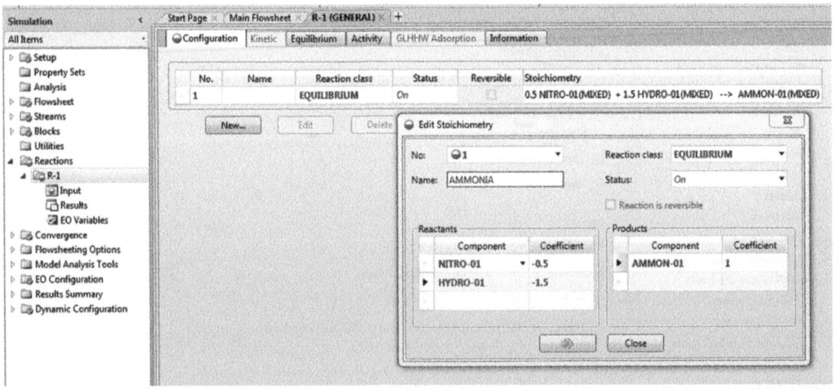

Figure 9-9

Click on **Close**. The simulation can now be run by pressing N>. The following results are obtained as seen by going to **Results Summary>Streams** (Fig. 9-10). [Note: Give the reaction a name, here AMMONIA, since Aspen Plus will ask for that when the simulation is run.]

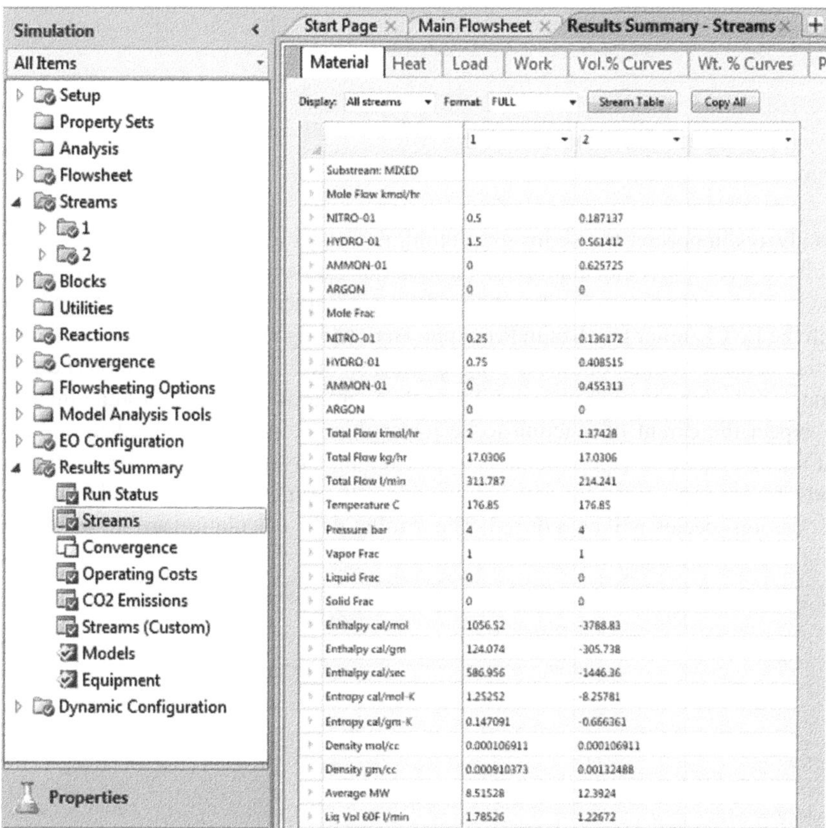

Figure 9-10

The result for the mole fraction of ammonia is 0.4553 (compared to 0.4605 in the textbook), the mole fraction of nitrogen is 0.1362 (compared to 0.1349) and 0.4085 for hydrogen (compared to 0.4046). So the agreement is quite good, but not exact. The small differences are the result of slightly different free energies of formation and heat capacities in the textbook and in the Aspen Plus database. The Aspen Plus result should be the more accurate one because of its extensive and frequently updated database. The extent of reaction X, using the notation in the textbook and mole numbers (not mole fractions) is

$$X = \frac{N_i - N_{i,0}}{\alpha_i} = \left(\frac{0.187137 - 0.5}{-0.5}\right)_{N_2} = \left(\frac{0.561412 - 1.5}{-1.5}\right)_{H_2} = \left(\frac{0.625725 - 0}{1}\right)_{NH_3}$$

$$= 0.6257$$

Before proceeding to part (b), consider the case in which 2 moles of the diluent argon are added (Change **Streams>1** in Fig. 9-11).

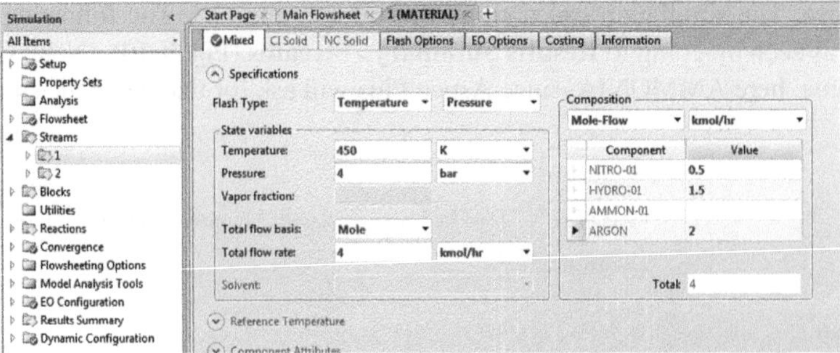

Figure 9-11

It is always important to use insight, intuition, or past experience to try to verify that the results of a computer calculation are (at least qualitatively) correct. So before using Aspen Plus to do the calculation for this changed case, consider the guidance that can be obtained from the Le Chatelier and Braun Principle. Here adding argon at fixed pressure reduces the partial pressure of each of the reacting species, so the effect of adding a inert diluent is the same as reducing the total pressure. Since there are fewer moles of product than reactants, we expect the extent of reaction to be lower than in the case without the diluent. Note that it is the extent of reaction that is to be examined using mole numbers of reactants and products, not mole fractions since by the addition of the diluent, the mole fractions of all the reaction species will be reduced. The results of the calculation are shown in Fig. 9-12.

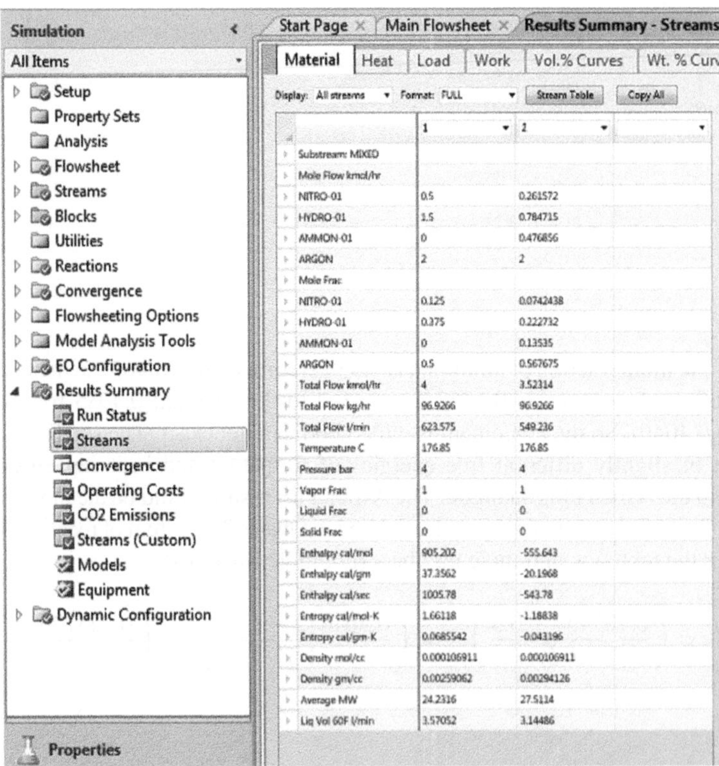

Figure 9-12

$$X = \frac{N_i - N_{i.0}}{\alpha_i} = \left(\frac{0.261572 - 0.5}{-0.5}\right)_{N_2} = \left(\frac{0.784715 - 1.5}{-1.5}\right)_{H_2} = \left(\frac{0.476856 - 0}{1}\right)_{NH_3}$$
$$= 0.4769$$

So, as expected, the extent of reaction is lower than previously. Conversely, if there is no diluent and the pressure is increased, say to 10 bar, by the principle of LeChatelier and Braun, the extent of reaction should increase since there are fewer moles of products than reactants. This is verified by changing **Stream>1** to eliminate the diluent and changing **Blocks>B1** to increase the pressure to 10 bar. The results are in Fig. 9-13

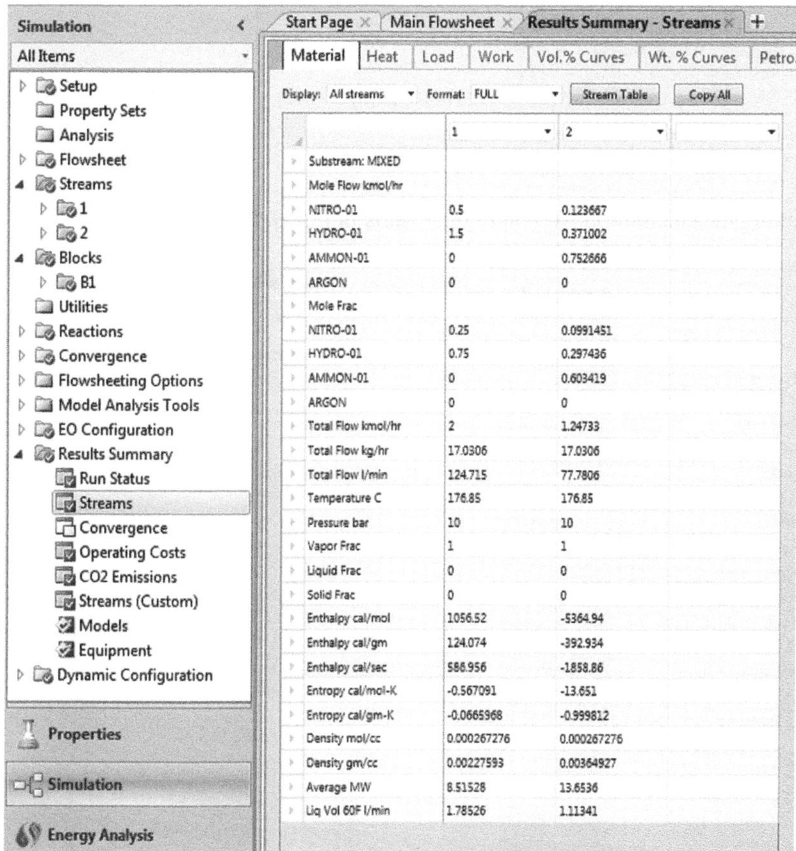

	1	2
Substream: MIXED		
Mole Flow kmol/hr		
NITRO-01	0.5	0.123667
HYDRO-01	1.5	0.371002
AMMON-01	0	0.752666
ARGON	0	0
Mole Frac		
NITRO-01	0.25	0.0991451
HYDRO-01	0.75	0.297436
AMMON-01	0	0.603419
ARGON	0	0
Total Flow kmol/hr	2	1.24733
Total Flow kg/hr	17.0306	17.0306
Total Flow l/min	124.715	77.7806
Temperature C	176.85	176.85
Pressure bar	10	10
Vapor Frac	1	1
Liquid Frac	0	0
Solid Frac	0	0
Enthalpy cal/mol	1056.52	-5364.94
Enthalpy cal/gm	124.074	-392.934
Enthalpy cal/sec	586.956	-1858.86
Entropy cal/mol-K	-0.567091	-13.651
Entropy cal/gm-K	-0.0665968	-0.999812
Density mol/cc	0.000267276	0.000267276
Density gm/cc	0.00227593	0.00364927
Average MW	8.51528	13.6536
Liq Vol 60F l/min	1.78526	1.11341

Figure 9-13

$$X = \frac{N_i - N_{i.0}}{\alpha_i} = \left(\frac{0.123667 - 0.5}{-0.5}\right)_{N_2} = \left(\frac{0.371002 - 1.5}{-1.5}\right)_{H_2} = \left(\frac{0.752666 - 0}{1}\right)_{NH_3}$$
$$= 0.7527$$

In this case, as expected, the extent of reaction increased as the pressure increased. Consider next the impact of using equal amounts, rather than stoichiometric amounts, of nitrogen and hydrogen, and also an equal amount of the diluent, the part (b) problem. Here we would expect the extent of reaction to be increased by the presence of the diluent, but reduced as a result of hydrogen being present in less than the stoichiometric amount; this latter effect should be dominant. The **Streams>1** input for this case is given in Fig. 9-14.

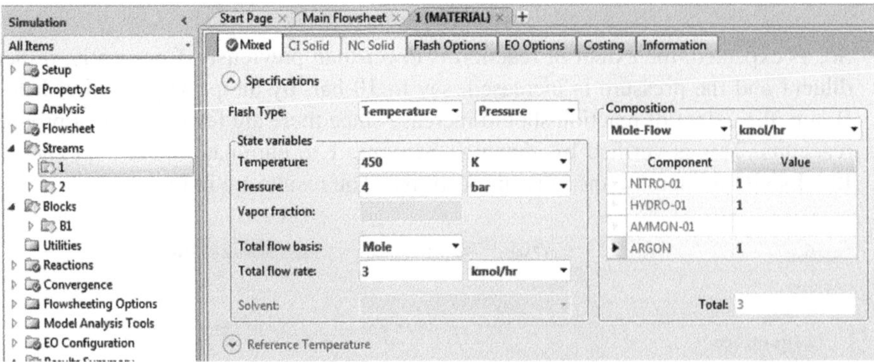

Figure 9-14

On running the calculation we obtain the results in Fig. 9-15.

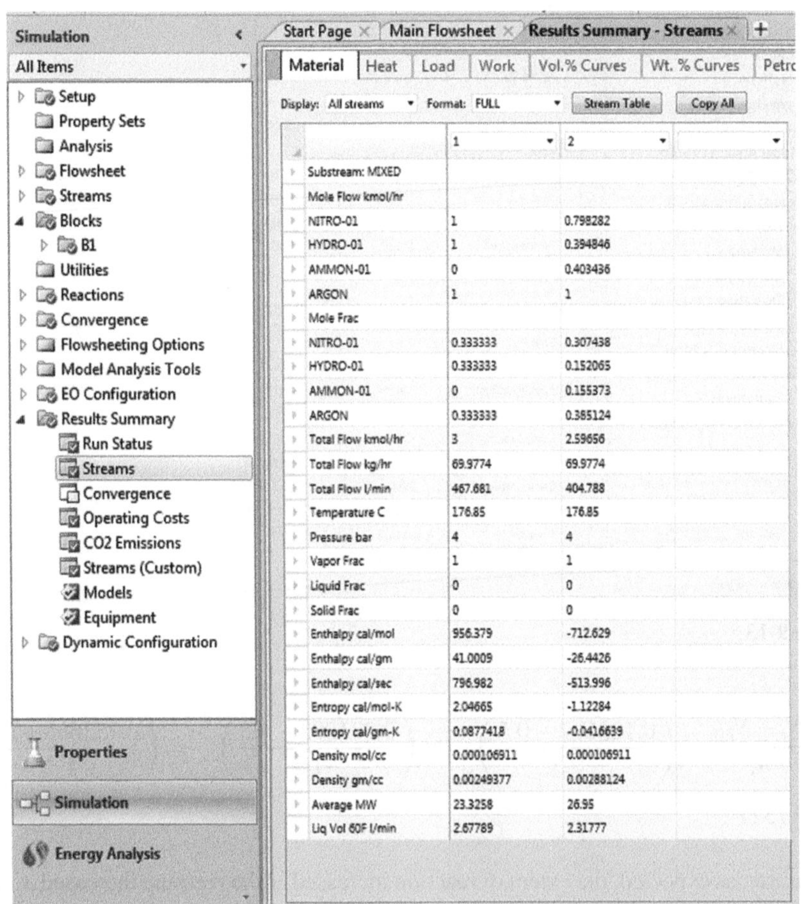

Figure 9-15

The calculated results for this case are a mole fraction of ammonia of 0.1554 (compared to 0.157 in the textbook), of nitrogen 0.3074 compared to 0.3072 (textbook), of hydrogen 0.1521 compared to 0.1501, and 0.3857 for the diluent compared to 0.3851. Again, the small differences are due to the difference in the databases. The extent of reaction in this case, calculated from mole numbers, is

$$X = \left(\frac{0.798282 - 1}{-0.5}\right)_{N_2} = \left(\frac{0.394846 - 1}{-1.5}\right)_{H_2} = \left(\frac{0.403436 - 0}{1.0}\right)_{NH_3} = 0.4034$$

This is considerably lower than when stoichiometric amounts of reactants and no diluent are used.

The important observation from these various calculations is that once the simulation is set up, to obtain the results for any change in operating conditions (e.g., flow rates, temperature, or pressure) takes very little additional effort. The only changes that need to be made are in the **Streams** and/or **Blocks** portion of the simulation input. In Chapter 13 we show how to do calculations for a number of different parameter values in a single run.

As the next example, consider Illustration 13.1-2 in *Chemical, Biochemical and Engineering Thermodynamics*, 4th ed., S. I. Sandler (John Wiley & Sons, Inc., 2006), the high temperature (700°C), low pressure (1 bar) dissociation of nitrogen tetroxide in the presence of an initial nitrogen

$$N_2O_4 \leftrightarrow 2NO_2$$

This reacting system at these conditions can be considered to be an ideal gas mixture. We proceed as in the previous example, as shown in the windows of Figs. 9-16–9-21, with little explanation.

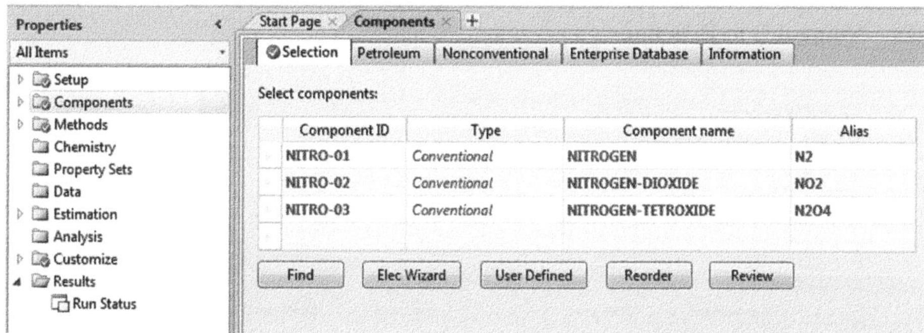

Figure 9-16

The Aspen Plus **Component IDs** are not very useful, so the user has to remember that NITRO-1 is nitrogen, NITRO-2 is nitrogen dioxide and NITRO-3 is nitrogen tetroxide, since it is NITRO-*x* that is used as identifiers in what follows. The **Streams>1 input** in the **Simulation** window is

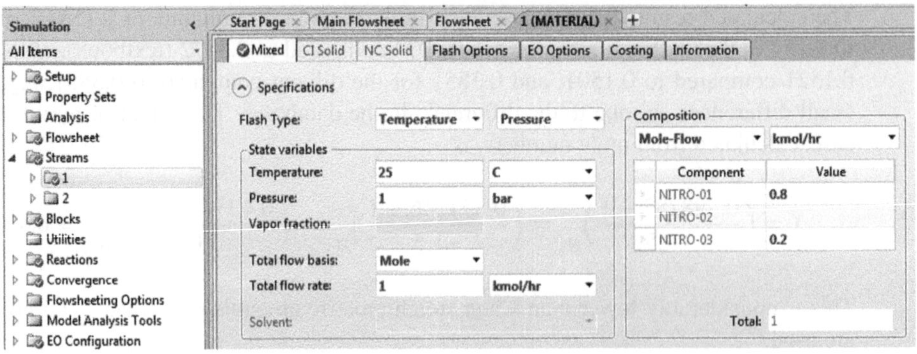

Figure 9-17

and the **Block>B1** input is

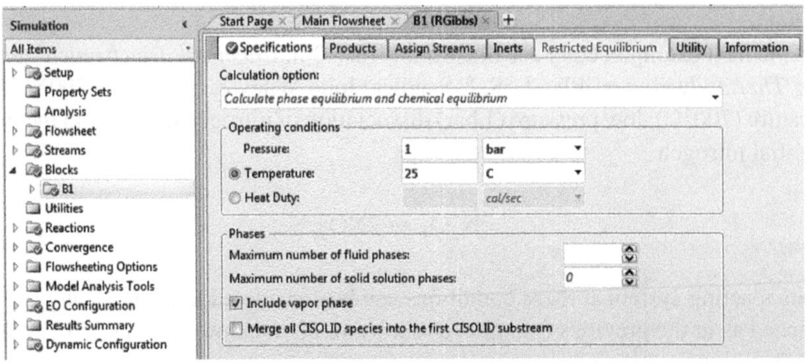

Figure 9-18

Now going to **Reactions**:

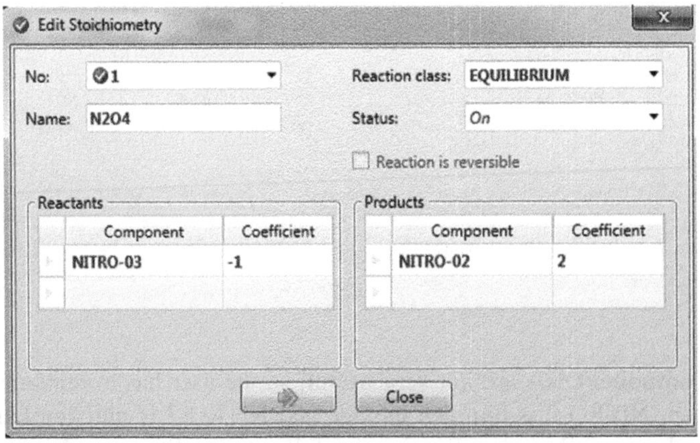

Figure 9-19

and

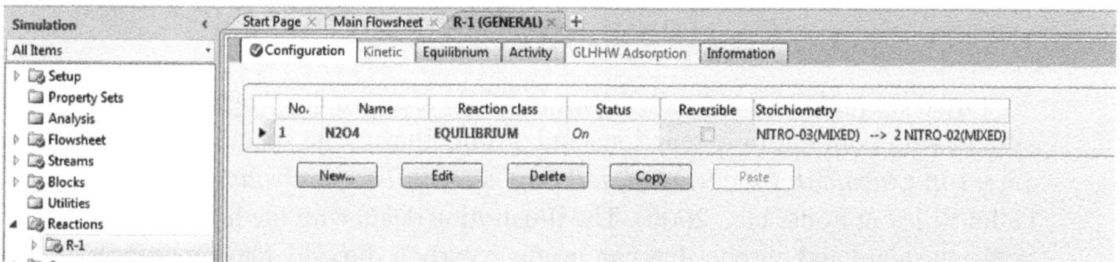

Figure 9-20

Running the simulation produces the results (seen in **Results Summary>Streams**) in Fig. 9-21

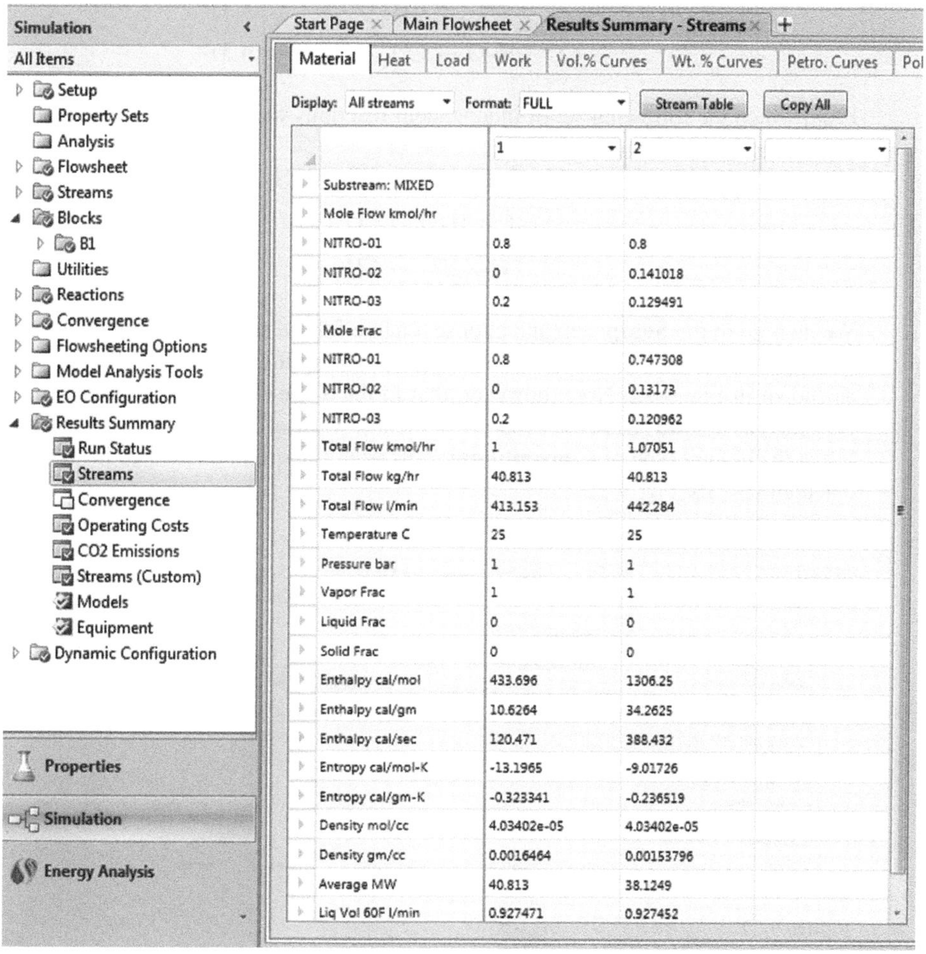

	1	2
Substream: MIXED		
Mole Flow kmol/hr		
NITRO-01	0.8	0.8
NITRO-02	0	0.141018
NITRO-03	0.2	0.129491
Mole Frac		
NITRO-01	0.8	0.747308
NITRO-02	0	0.13173
NITRO-03	0.2	0.120962
Total Flow kmol/hr	1	1.07051
Total Flow kg/hr	40.813	40.813
Total Flow l/min	413.153	442.284
Temperature C	25	25
Pressure bar	1	1
Vapor Frac	1	1
Liquid Frac	0	0
Solid Frac	0	0
Enthalpy cal/mol	433.696	1306.25
Enthalpy cal/gm	10.6264	34.2625
Enthalpy cal/sec	120.471	388.432
Entropy cal/mol-K	-13.1965	-9.01726
Entropy cal/gm-K	-0.323341	-0.236519
Density mol/cc	4.03402e-05	4.03402e-05
Density gm/cc	0.0016464	0.00153796
Average MW	40.813	38.1249
Liq Vol 60F l/min	0.927471	0.927452

Figure 9-21

and

$$X = \left(\frac{0.129491 - 0.2}{-1.0}\right)_{N_2O_4} = \left(\frac{0.141018 - 0}{2.0}\right)_{NO_2} = 0.07051$$

As a final example, we will consider a multiple reaction, heterogeneous system. That is, a system with both multiple reactions and a solid phase. The specific example is Illustration 13.3-1 in *Chemical, Biochemical and Engineering Thermodynamics*, 4th ed., S. I. Sandler (John Wiley & Sons, Inc., 2006). The illustration deals with the high temperature reactions between steam and carbon that can produce carbon dioxide, carbon monoxide, hydrogen, and methane, depending on the conditions. The following reactions may occur:

$$C + 2H_2O \rightarrow CO_2 + 2H_2 \quad \text{(rxn 1)}$$

$$C + H_2O \rightarrow CO + H_2 \quad \text{(rxn 2)}$$

$$C + CO_2 \rightarrow 2CO \quad \text{(rxn 3)}$$

$$C + 2H_2 \rightarrow CH_4 \quad \text{(rxn 4)}$$

$$CO + H_2O \rightarrow CO_2 + H_2 \quad \text{(rxn 5)}$$

From which the following set of independent reactions was identified in the textbook:

$$C + 2H_2O \rightarrow CO_2 + 2H_2$$

$$C + H_2O \rightarrow CO + H_2$$

$$C + 2H_2 \rightarrow CH_4$$

We then go to the **Setup** step and choose a name for the simulation (optional) and choose metric (METCBAR) for the units. Next, go to **Components** and add all the components as shown in Fig. 9-22. Note, however, that in the **Type** characterization carbon has been classified as a **Solid** (from the drop-down menu), while all the other compounds have been left at the default type of **Conventional**. This is an important difference from the previous calculations. See Fig. 9-22.

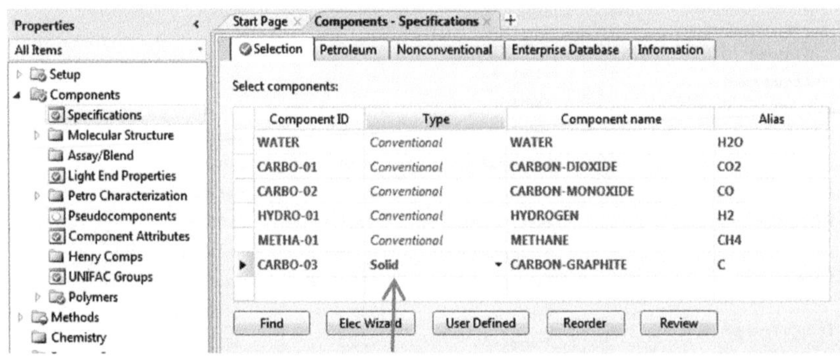

Figure 9-22

Because of the high temperatures and low pressures involved, **IDEAL** has been chosen in **Methods>Specifications**. Next going to **Simulation**, using the **RGIBBS** reactor, and then to **Reactions**, you will see the window in Fig. 9-23 after clicking on **New**.

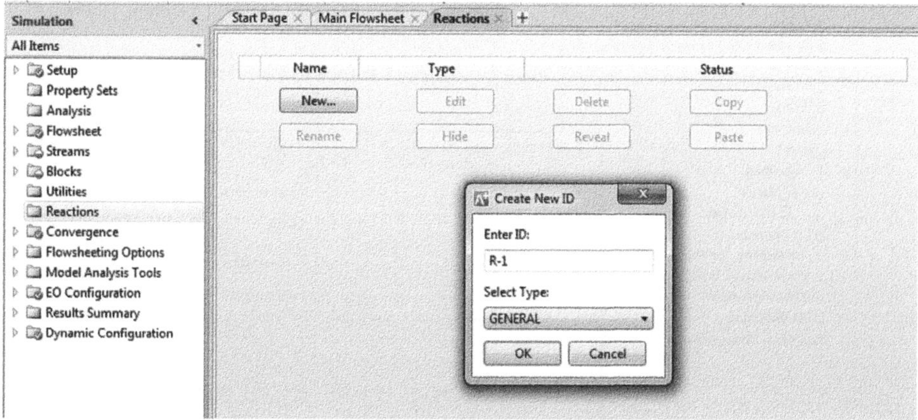

Figure 9-23

Clicking on **Reactions** as in the previous examples brings up the window in Fig. 9-24.

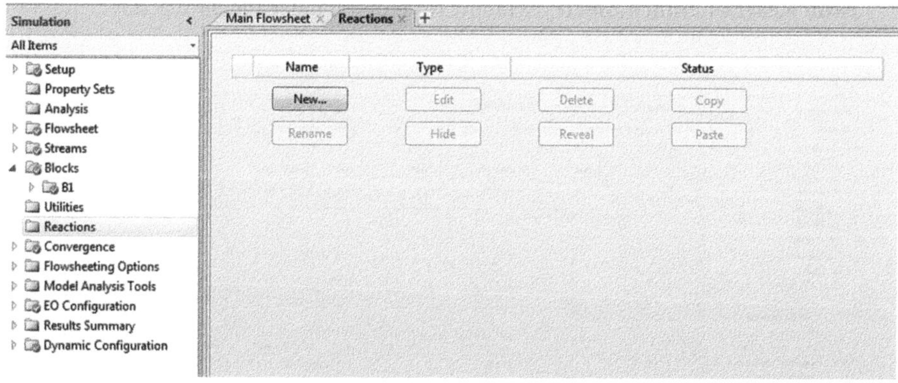

Figure 9-24

Then clicking on **New**, and then **Input** gives Fig. 9-25.

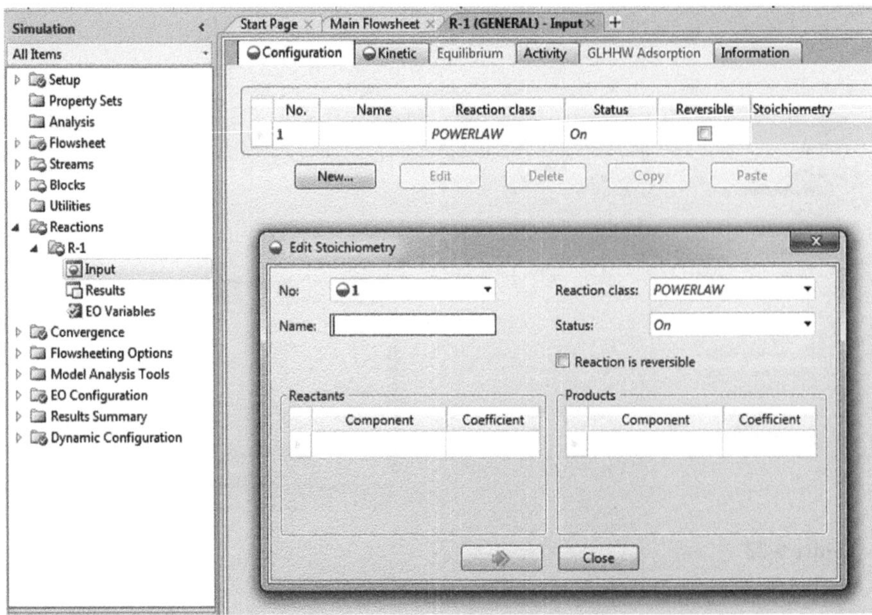

Figure 9-25

Complete the reaction stoichiometry as shown in Fig. 9-26, including changing **Reaction class** from the default of ***POWERLAW*** to **EQUILIBRIUM** and populating the **Equilibrium Reaction Stoichiometry** window for the first independent reaction.

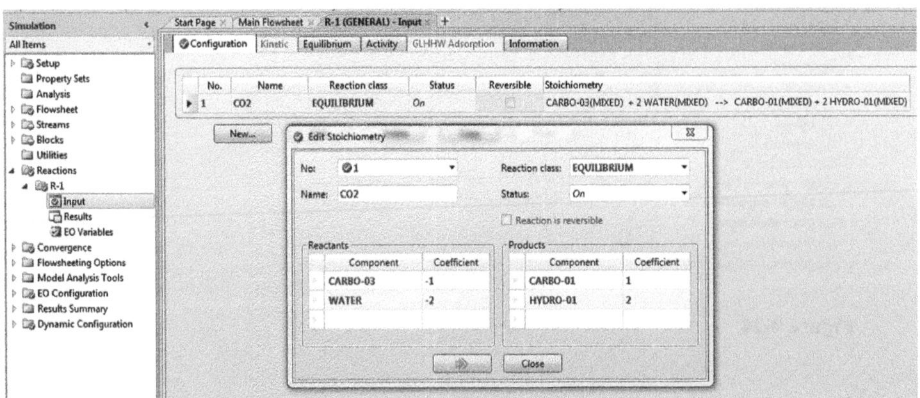

Figure 9-26

Clicking on **New** allows one to enter the second reaction following the same procedure as above, and then finally the third reaction can be entered, resulting in Fig. 9-27.

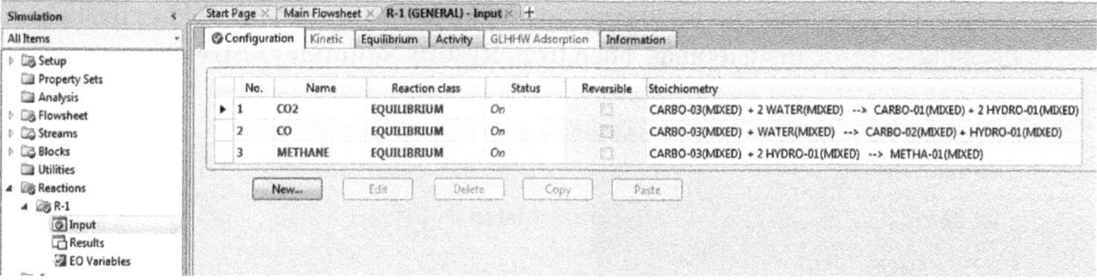

Figure 9-27

Next go to **Streams>1** and enter one mole (actually 1 kmol/hr) of each of carbon and steam at 800 K for this calculation (Fig. 9-28).

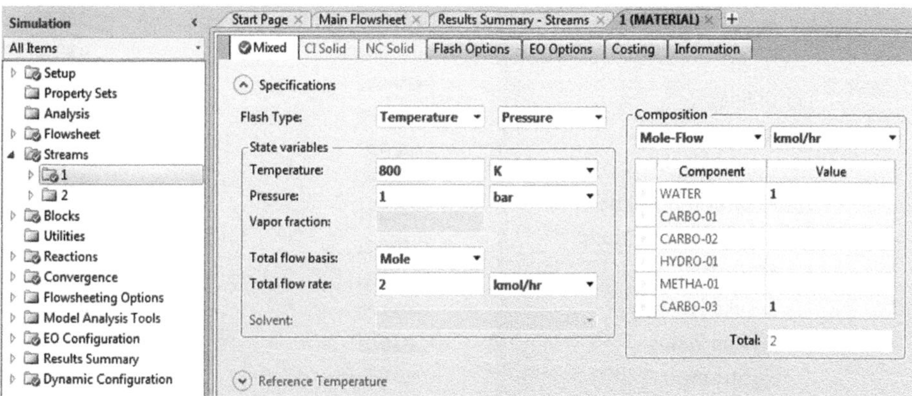

Figure 9-28

Next, go to **Blocks>B1>Specifications** and choose the operating conditions (here 1 bar and 800 K), as shown in Fig. 9-29.

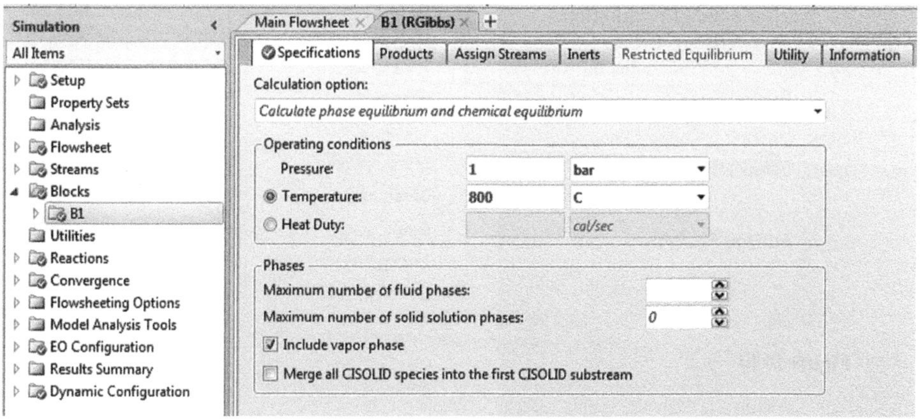

Figure 9-29

Notice that now there are no longer any red boxes, so we can proceed to the calculation by pressing > on the main toolbar. Then from **Results Summary>Streams** we see (Fig. 9-30).

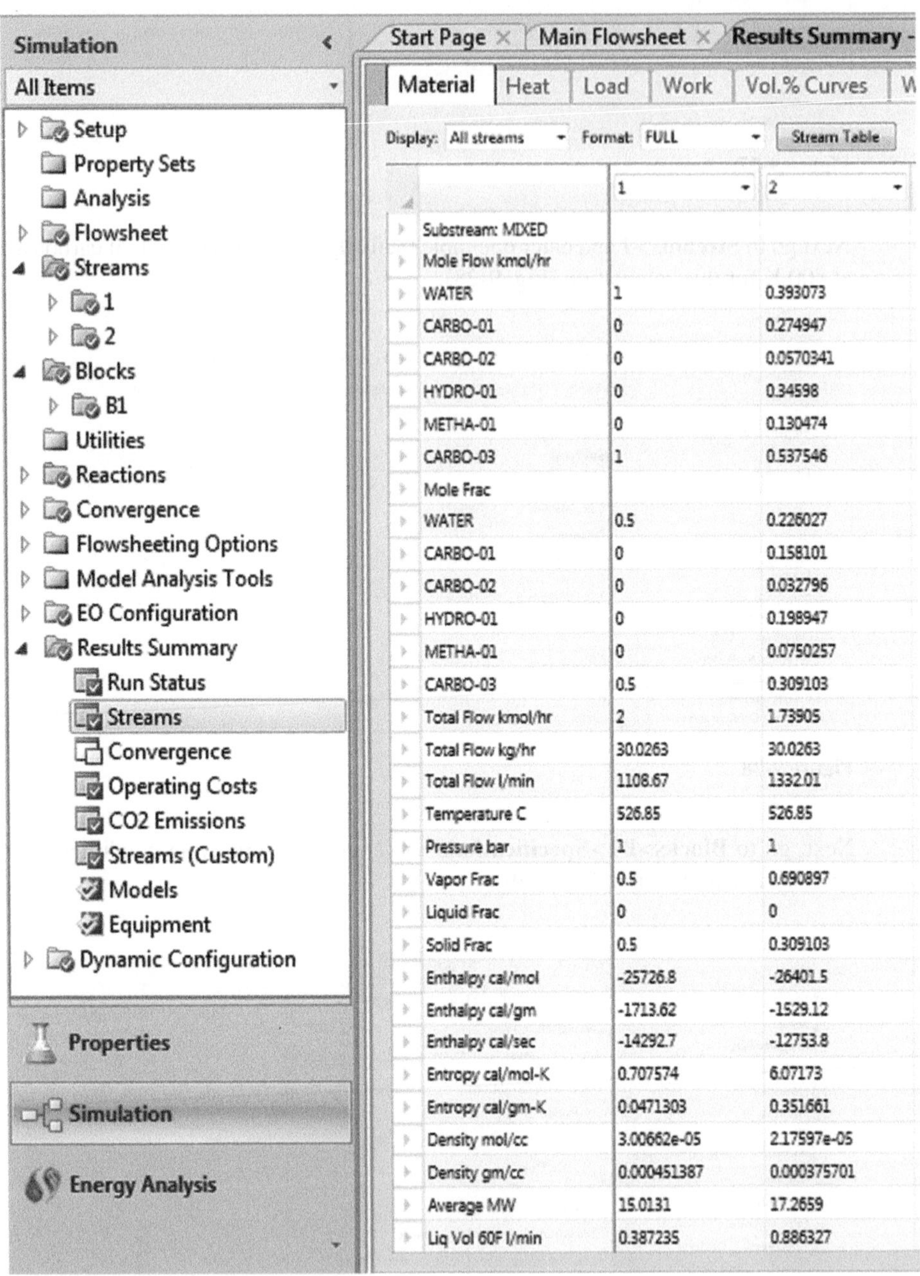

	1	2
Substream: MIXED		
Mole Flow kmol/hr		
WATER	1	0.393073
CARBO-01	0	0.274947
CARBO-02	0	0.0570341
HYDRO-01	0	0.34598
METHA-01	0	0.130474
CARBO-03	1	0.537546
Mole Frac		
WATER	0.5	0.226027
CARBO-01	0	0.158101
CARBO-02	0	0.032796
HYDRO-01	0	0.198947
METHA-01	0	0.0750257
CARBO-03	0.5	0.309103
Total Flow kmol/hr	2	1.73905
Total Flow kg/hr	30.0263	30.0263
Total Flow l/min	1108.67	1332.01
Temperature C	526.85	526.85
Pressure bar	1	1
Vapor Frac	0.5	0.690897
Liquid Frac	0	0
Solid Frac	0.5	0.309103
Enthalpy cal/mol	-25726.8	-26401.5
Enthalpy cal/gm	-1713.62	-1529.12
Enthalpy cal/sec	-14292.7	-12753.8
Entropy cal/mol-K	0.707574	6.07173
Entropy cal/gm-K	0.0471303	0.351661
Density mol/cc	3.00662e-05	2.17597e-05
Density gm/cc	0.000451387	0.000375701
Average MW	15.0131	17.2659
Liq Vol 60F l/min	0.387235	0.886327

Figure 9-30

Note that while carbon (here CARBO-3) is a solid, Aspen Plus has included it in the mole fraction calculation. Therefore, the mole fractions reported are of the total reaction product, and not only of the gas phase, so that the gas-phase mole fractions are incorrect. One way to correct the gas-phase mole fractions is to copy the data into, for example, an EXCEL spreadsheet and manually correct the entries.

However, a simple automatic way to get the correct gas-phase mole fractions is to modify the process **Flowsheet** and add a **Flash2** as a separator (as was done for vapor–liquid separations after the adiabatic flash in the liquefaction of propane in Chapter 2, but here an isothermal flash is done) as shown in Fig. 9-31

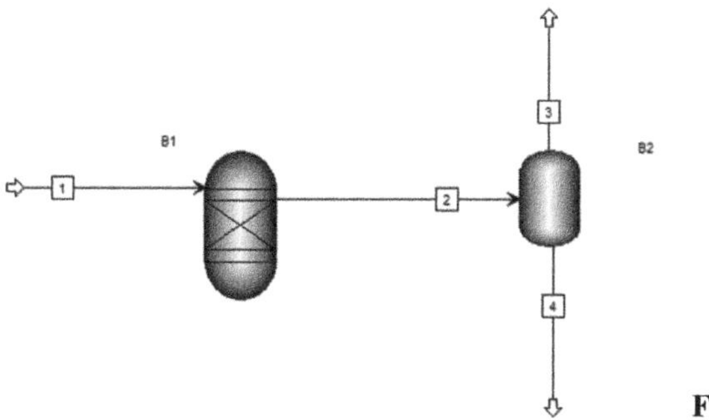

Figure 9-31

with the following input for the new **Flash2** (separator) block (Fig. 9-32).

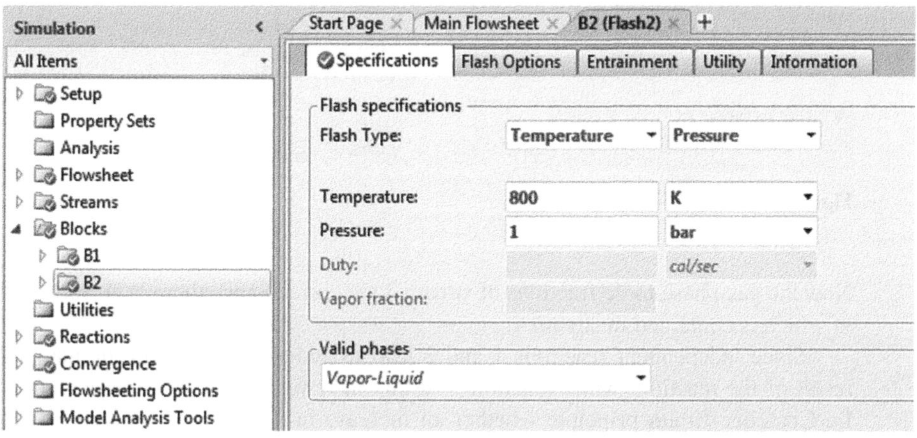

Figure 9-32

The results obtained after running this new simulation at 800 K are given in Fig. 9-33.

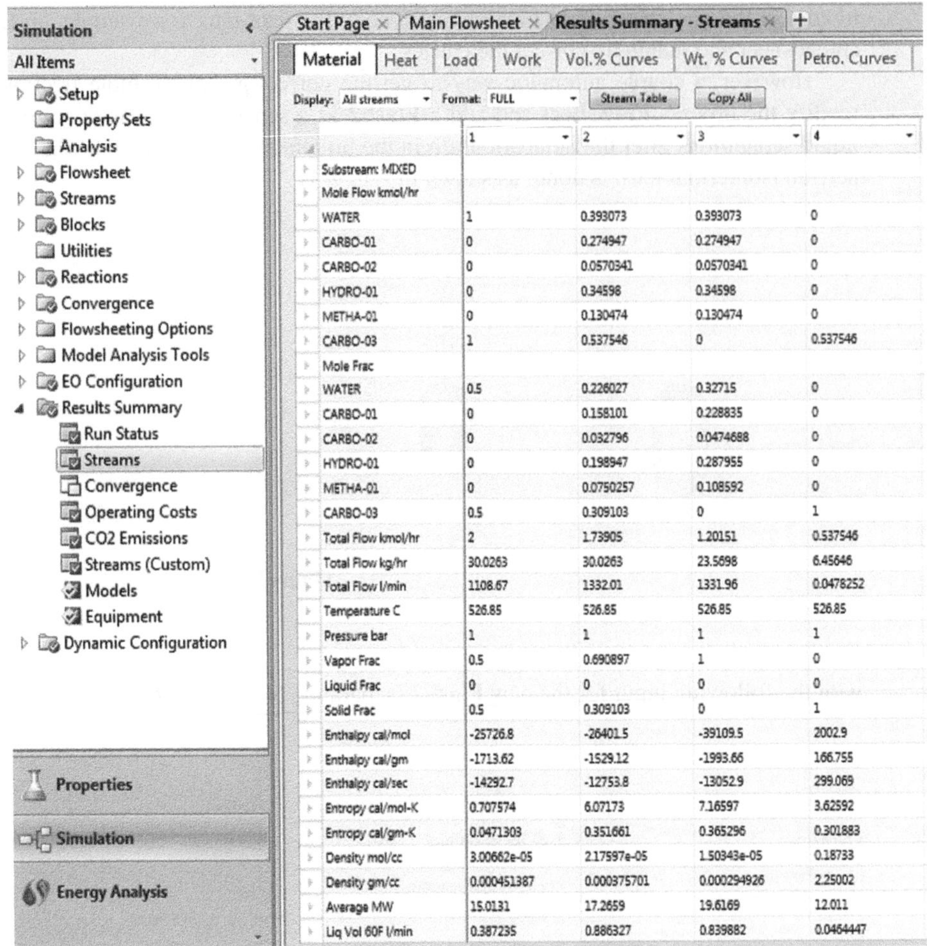

	1	2	3	4
Substream: MIXED				
Mole Flow kmol/hr				
WATER	1	0.393073	0.393073	0
CARBO-01	0	0.274947	0.274947	0
CARBO-02	0	0.0570341	0.0570341	0
HYDRO-01	0	0.34598	0.34598	0
METHA-01	0	0.130474	0.130474	0
CARBO-03	1	0.537546	0	0.537546
Mole Frac				
WATER	0.5	0.226027	0.32715	0
CARBO-01	0	0.158101	0.228835	0
CARBO-02	0	0.032796	0.0474688	0
HYDRO-01	0	0.198947	0.287955	0
METHA-01	0	0.0750257	0.108592	0
CARBO-03	0.5	0.309103	0	1
Total Flow kmol/hr	2	1.73905	1.20151	0.537546
Total Flow kg/hr	30.0263	30.0263	23.5698	6.45646
Total Flow l/min	1108.67	1332.01	1331.96	0.0478252
Temperature C	526.85	526.85	526.85	526.85
Pressure bar	1	1	1	1
Vapor Frac	0.5	0.690897	1	0
Liquid Frac	0	0	0	0
Solid Frac	0.5	0.309103	0	1
Enthalpy cal/mol	-25726.8	-26401.5	-39109.5	2002.9
Enthalpy cal/gm	-1713.62	-1529.12	-1993.66	166.755
Enthalpy cal/sec	-14292.7	-12753.8	-13052.9	299.069
Entropy cal/mol-K	0.707574	6.07173	7.16597	3.62592
Entropy cal/gm-K	0.0471303	0.351661	0.365296	0.301883
Density mol/cc	3.00662e-05	2.17597e-05	1.50343e-05	0.18733
Density gm/cc	0.000451387	0.000375701	0.000294926	2.25002
Average MW	15.0131	17.2659	19.6169	12.011
Liq Vol 60F l/min	0.387235	0.886327	0.839882	0.0464447

Figure 9-33

Now the gas-phase mole fractions of stream 3 are correct since there is no solid carbon, all of which is contained in stream 4.

Since independent reactions 1 and 2 lead to additional moles in the gas phase as a result of the reaction, while reaction 3 results in fewer moles, it is not evident from the Le Chatelier–Braun principle whether an increase in pressure will increase or decrease the extent of reaction of carbon and steam. This is because a pressure increase will favor some reactions at the expense of others. This is especially so at 800 K, a temperature at which the equilibrium constants of the three reactions are comparable (see Fig. 13.3-1 in *Chemical, Biochemical and Engineering Thermodynamics*, 4th ed., S. I. Sandler, John Wiley & Sons, Inc., 2006). The easiest way to determine this is to change the pressure to, say, 10 bar and rerun the simulation. The results are shown in Fig. 9-34.

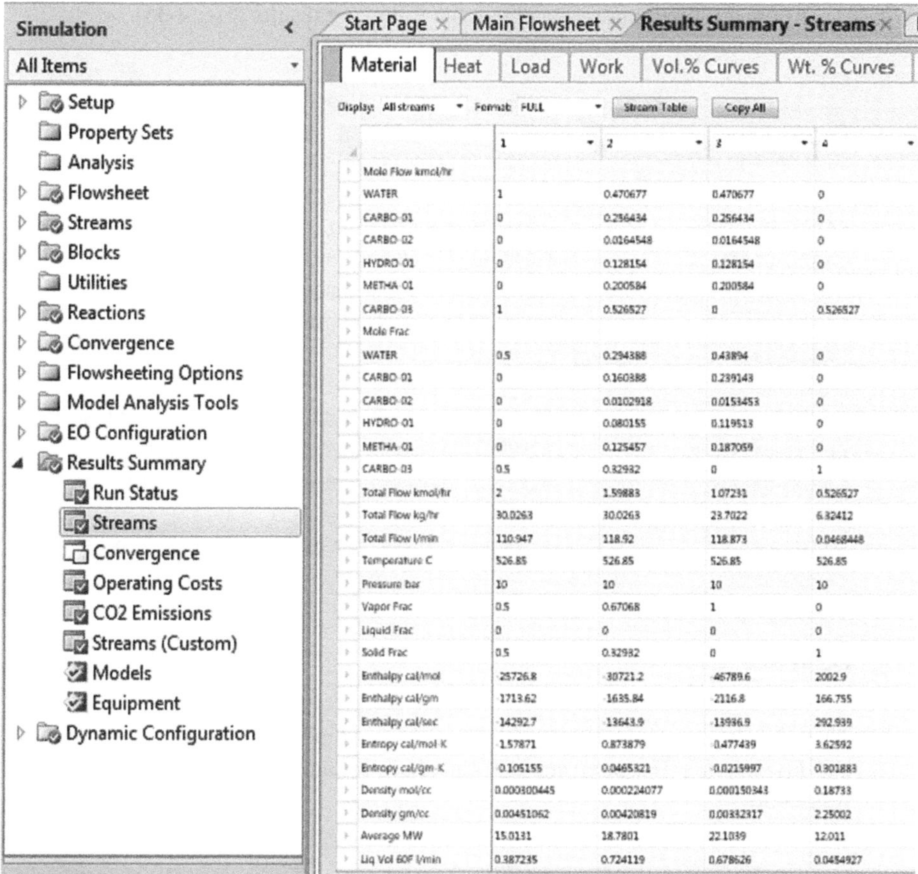

	1	2	3	4
Mole Flow kmol/hr				
WATER	1	0.470677	0.470677	0
CARBO-01	0	0.256434	0.256434	0
CARBO-02	0	0.0164548	0.0164548	0
HYDRO-01	0	0.128154	0.128154	0
METHA-01	0	0.200584	0.200584	0
CARBO-03	1	0.526527	0	0.526527
Mole Frac				
WATER	0.5	0.294388	0.43894	0
CARBO-01	0	0.160388	0.239143	0
CARBO-02	0	0.0102918	0.0153453	0
HYDRO-01	0	0.080155	0.119513	0
METHA-01	0	0.125457	0.187059	0
CARBO-03	0.5	0.32932	0	1
Total Flow kmol/hr	2	1.59883	1.07231	0.526527
Total Flow kg/hr	30.0263	30.0263	23.7022	6.32412
Total Flow l/min	110.947	118.92	118.873	0.0468448
Temperature C	526.85	526.85	526.85	526.85
Pressure bar	10	10	10	10
Vapor Frac	0.5	0.67068	1	0
Liquid Frac	0	0	0	0
Solid Frac	0.5	0.32932	0	1
Enthalpy cal/mol	-25726.8	-30721.2	-46789.6	2002.9
Enthalpy cal/gm	-1713.62	-1635.84	-2116.8	166.755
Enthalpy cal/sec	-14292.7	-13643.9	-13936.9	292.939
Entropy cal/mol-K	-1.57871	0.873879	-0.477439	3.62592
Entropy cal/gm-K	-0.105155	0.0465321	-0.0215997	0.301883
Density mol/cc	0.000300445	0.000224077	0.000150343	0.18733
Density gm/cc	0.00451062	0.00420819	0.00332317	2.25002
Average MW	15.0131	18.7801	22.1039	12.011
Liq Vol 60F l/min	0.387235	0.724119	0.678526	0.0454927

Figure 9-34

Comparing the extent of reaction of the carbon, which is 0.4625 at 1 bar and 0.4735 at 10 bar, we see that there is very little change with pressure. However, there are significant changes in the product distribution as seen below:

	Mole fractions	
	1 bar	10 bar
H_2O	0.3272	0.4389
CO_2	0.2288	0.2391
CO	0.0475	0.0153
H_2	0.2880	0.1195
CH_4	0.1086	0.1871

Clearly the reactions that produced additional moles of product than reactant have been suppressed by the higher pressure resulting in lower mole fractions of carbon dioxide, carbon monoxide, and hydrogen, while the one reaction that produced fewer moles of product than reactant has been enhanced resulting in a larger conversion to methane.

At other temperatures, we obtain the following results (Fig. 9-35).

	FEED	Vapor	Vapor	Vapor	Vapor	Vapor	Vapor
Temperature K		600	800	1000	1200	1400	1600
Mole Flow kmol/hr							
WATER	1	0.567714	0.393073	0.122006	0.014099	0.002305	0.000609
CARBO-01 CO2	0	0.215835	0.274947	0.132955	0.010223	0.001067	0.000204
CARBO-02 CO	0	0.000616	0.057034	0.612084	0.965455	0.995562	0.998984
HYDRO-01	0	0.046038	0.34598	0.809629	0.97267	0.994301	0.998244
METHA-01	0	0.193124	0.130474	0.034183	0.006615	0.001697	0.000573
CARBO-03 C	1	0	0	0	0	0	0
Mole Frac							
WATER	0.5	0.554773	0.32715	0.071313	0.00716	0.001156	0.000305
CARBO-01 CO2	0	0.210915	0.228835	0.077713	0.005192	0.000535	0.000102
CARBO-02 CO	0	0.000602	0.047469	0.357765	0.490312	0.499046	0.499838
HYDRO-01	0	0.044988	0.287955	0.47323	0.493976	0.498414	0.499468
METHA-01	0	0.188722	0.108592	0.01998	0.00336	0.000851	0.000287
CARBO-03 C	0.5	0	0	0	0	0	0
Pressure bar	1	1	1	1	1	1	1

Figure 9-35

These results are in agreement with those reported in Fig. 13.3-2 in *Chemical, Biochemical and Engineering Thermodynamics*, 4th ed., S. I. Sandler (John Wiley & Sons, Inc., 2006) (Fig. 9-36).

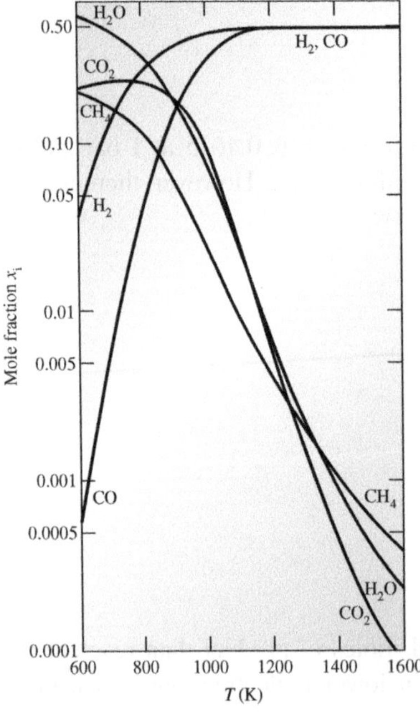

Figure 9-36

The results Aspen Plus provides after each simulation are rather detailed, but sometimes the user wants additional information. This additional information can be obtained by adding the properties that are desired using the **Property Sets** option as was done in Chapter 3 and in Chapter 5. This is illustrated here for the equilibrium chemical reaction

$$CO_2 + H_2 \leftrightarrow CO + H_2O$$

at 1000 K and 500 bar starting with an equimolar mixture of carbon dioxide and hydrogen using the **RGibbs** reactor as in Fig. 9-2. Use the usual **Setup** procedure and choose the Peng–Robinson (**PENG-ROB**) or another equation of state in the **Properties> Methods>Specifications**, and then choose **Parameters>Binary Interaction>PRKBV-1.** Next, go to **Simulation** and to **Streams** and enter the starting equimolar composition as shown in Fig. 9-37 for **Stream 1**.

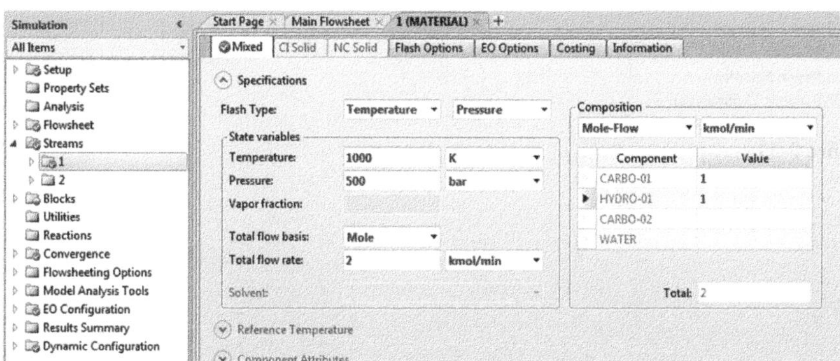

Figure 9-37

Next go to **Blocks B1** and enter the reaction conditions (Fig. 9-38).

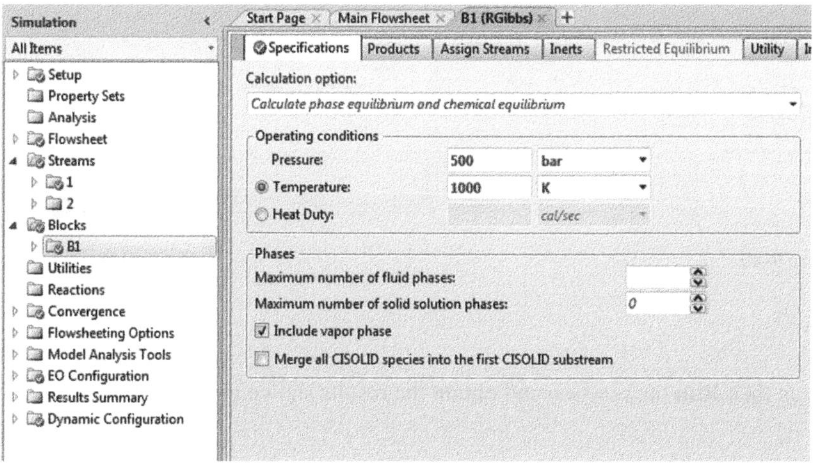

Figure 9-38

Then go to **Reactions>Chemistry** and enter a **New** reaction (Fig. 9-39) **R-1** following the procedure described earlier (Fig. 9-40).

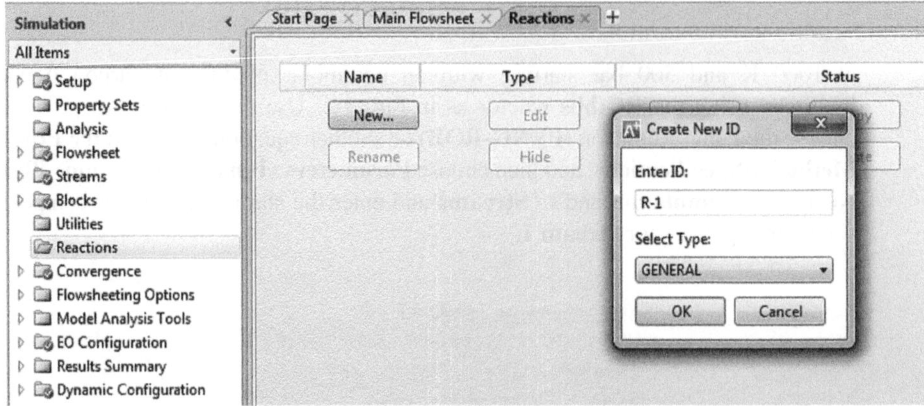

Figure 9-39

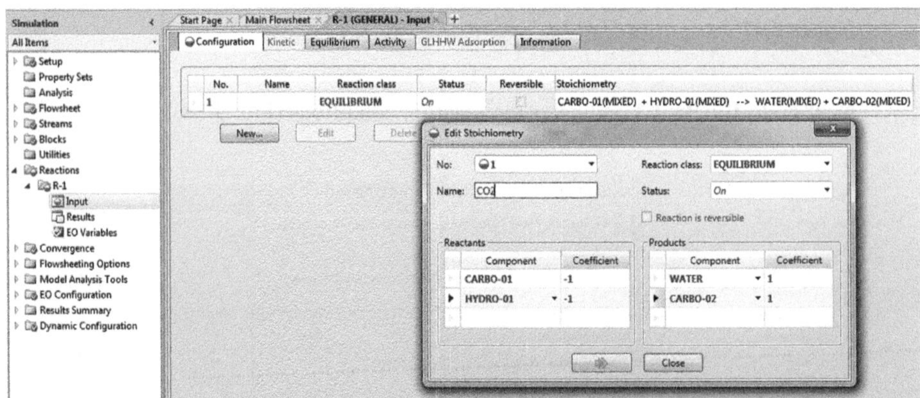

Figure 9-40

We can then **Run** the reaction and obtain the results shown in Fig. 9-41.

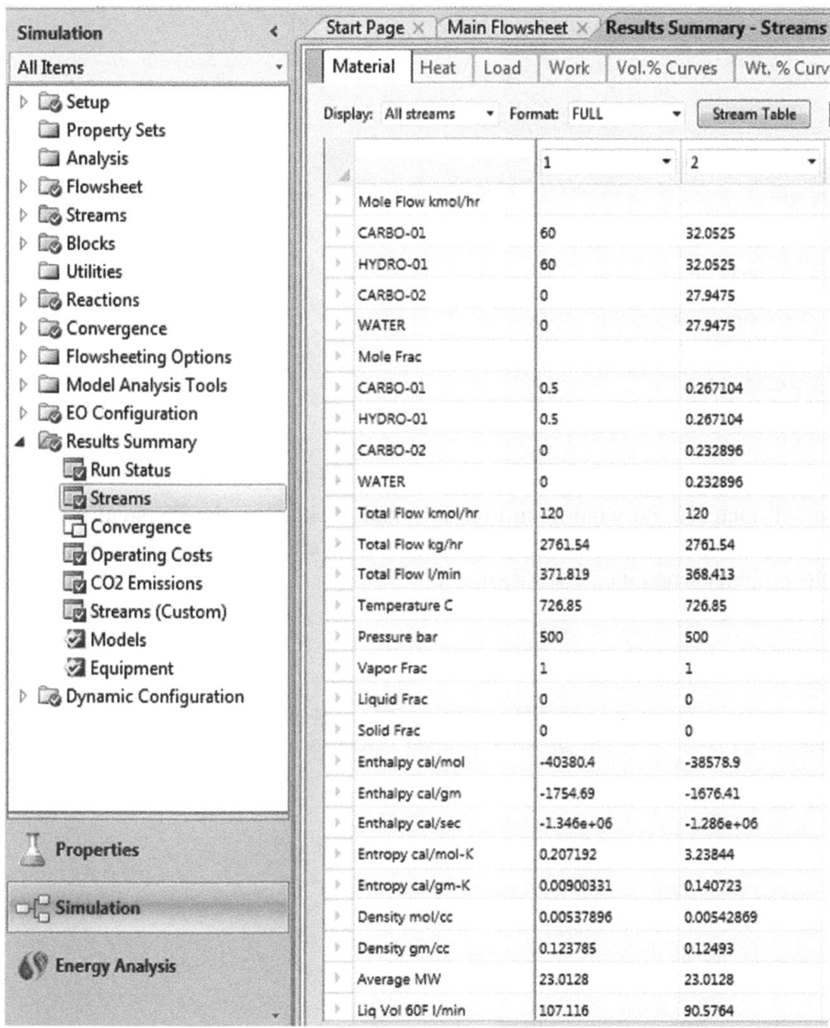

Figure 9-41

But suppose that we also want additional information, for example, the fugacity coefficient of each species in the mixture and the difference between the enthalpy of the mixture and that of an ideal gas mixture at the same temperature, pressure, and composition. We can get such information in the course of a simulation by adding a **Property Sets** to the report specifications in **Setup**. This is done in the following way. First go to **Properties>Property Sets,** then click on **New**, which brings up the **Create New ID** pop-up box, choose the default ID name of **PS-1** or a name of your own choice (Fig. 9-42).

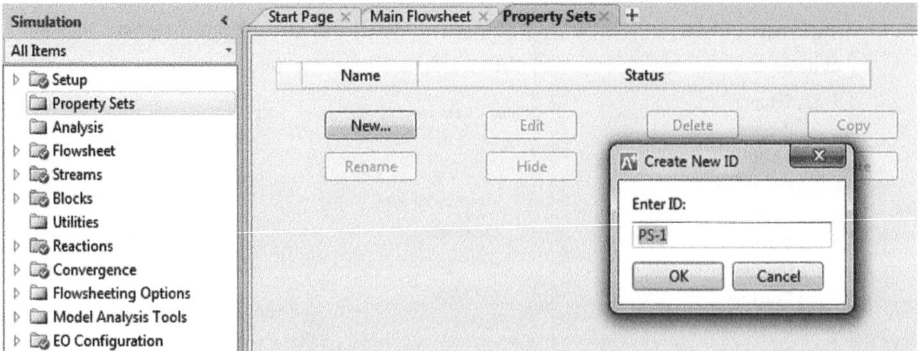

Figure 9-42

You will then see the window in Fig. 9-43 and clicking on the **Physical properties** box will lead a drop-down menu with a large number of choices, each with a brief explanation of the property when the cursor is over it.

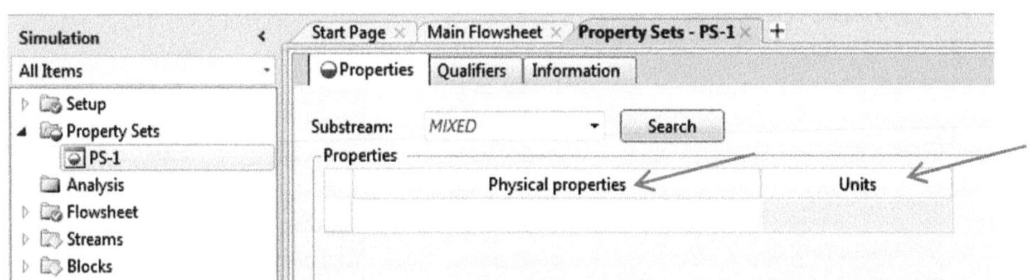

Figure 9-43

For the example here, **PHIMX**, the fugacity coefficient for each species in the mixture, and **DHMX,** the enthalpy departure of the mixture, that is, the difference between the enthalpy of the mixture and that of an ideal gas mixture at the same temperature, pressure, and composition have been chosen. Next, one has to specify the **Units** from the drop-down menu of appropriate units that will appear (Fig. 9-43). Next click on the **Qualifiers** tab (9-44) to specify the **Phase** (several may be specified for multiphase systems), and then each **Component** (from the drop-down menu that will appear) for which the **Property Sets** information is desired. Finally, either leave the **System** boxes checked if you want information at the stream conditions, or uncheck the boxes and set your own temperatures and pressures. We will leave the system boxes checked as shown in Fig. 9-45.

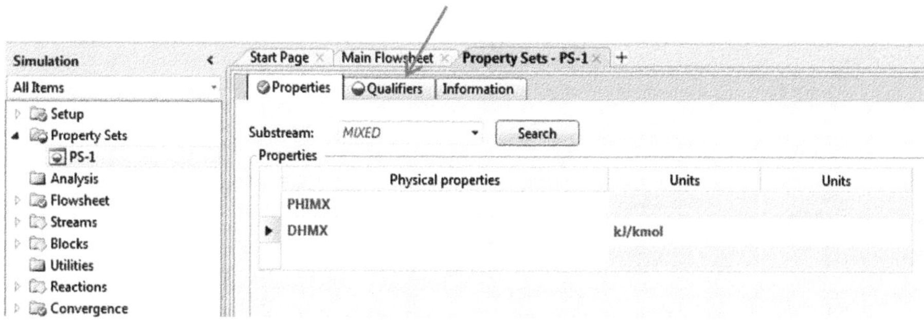

Figure 9-44

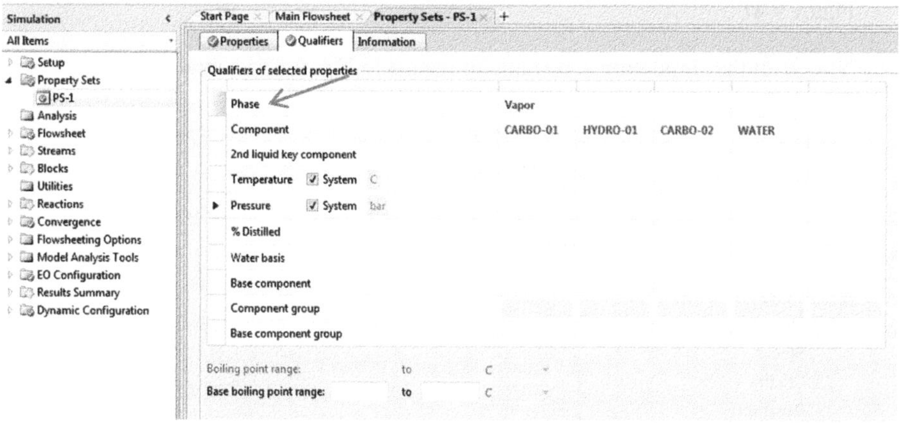

Figure 9-45

Now go to Setup>Report Options>Stream (Fig. 9-46).

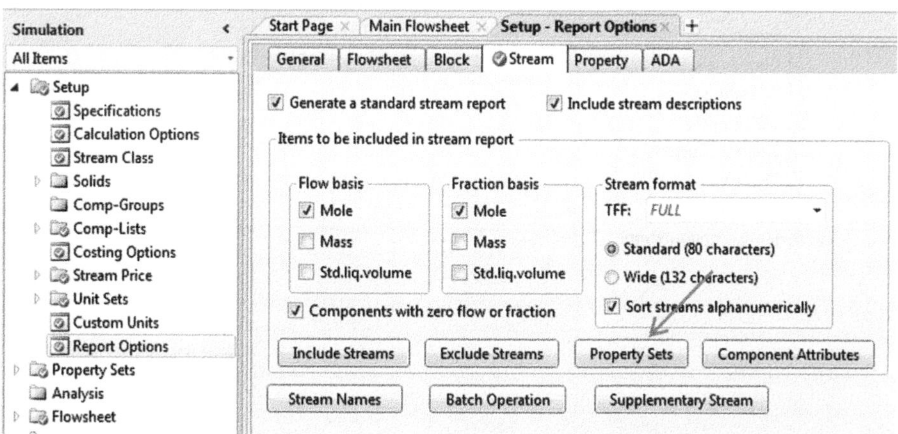

Figure 9-46

Then click on **Property Sets** to bring up the **Property Sets** box, select **PS-1** and move it to **Selected property sets,** then click on **Close** (see Fig. 9-47).

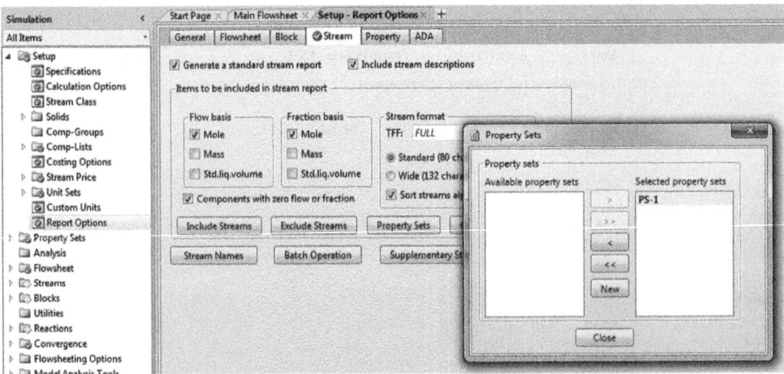

Figure 9-47

Now **Run** the simulation. The table of results in Fig. 9-48 is obtained.

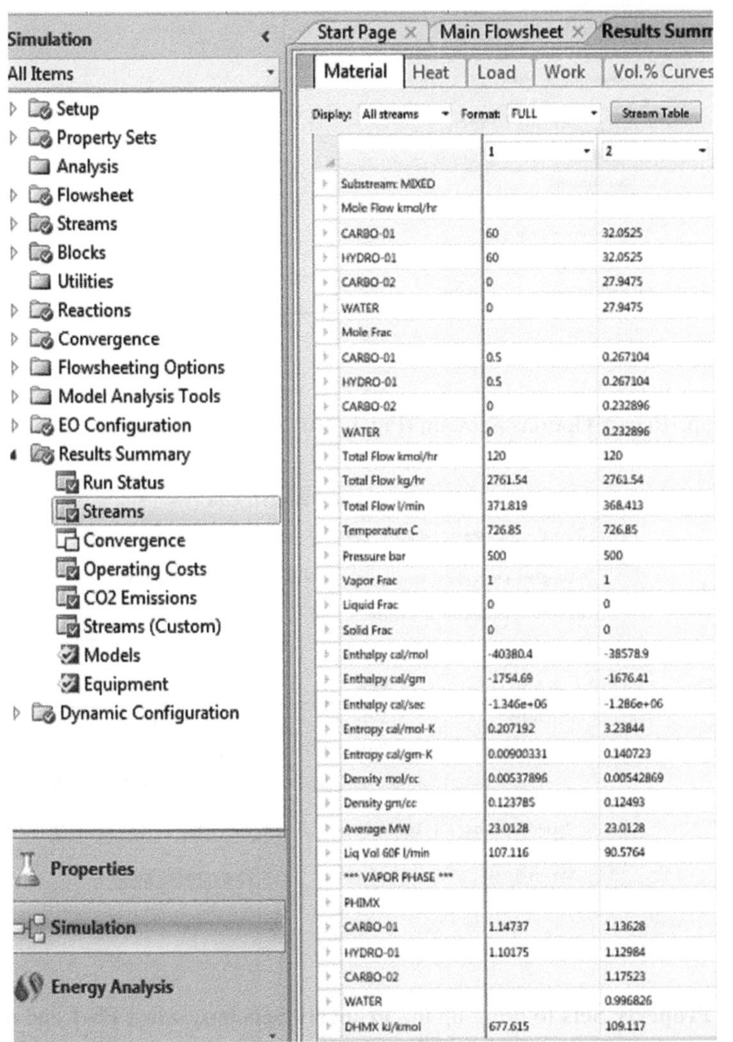

	1	2
Substream: MIXED		
Mole Flow kmol/hr		
CARBO-01	60	32.0525
HYDRO-01	60	32.0525
CARBO-02	0	27.9475
WATER	0	27.9475
Mole Frac		
CARBO-01	0.5	0.267104
HYDRO-01	0.5	0.267104
CARBO-02	0	0.232896
WATER	0	0.232896
Total Flow kmol/hr	120	120
Total Flow kg/hr	2761.54	2761.54
Total Flow l/min	371.819	368.413
Temperature C	726.85	726.85
Pressure bar	500	500
Vapor Frac	1	1
Liquid Frac	0	0
Solid Frac	0	0
Enthalpy cal/mol	-40380.4	-38578.9
Enthalpy cal/gm	-1754.69	-1676.41
Enthalpy cal/sec	-1.346e+06	-1.286e+06
Entropy cal/mol-K	0.207192	3.23844
Entropy cal/gm-K	0.00900331	0.140723
Density mol/cc	0.00537896	0.00542869
Density gm/cc	0.123785	0.12493
Average MW	23.0128	23.0128
Liq Vol 60F l/min	107.116	90.5764
*** VAPOR PHASE ***		
PHIMX		
CARBO-01	1.14737	1.13628
HYDRO-01	1.10175	1.12984
CARBO-02		1.17523
WATER		0.996826
DHMX kJ/kmol	677.615	109.117

Figure 9-48

Note the additional information on the species fugacities and enthalpy departure appear in the results beyond that in the default **Streams>Results** output shown in Fig. 9-41.

PROBLEMS

9.1. Calculate the equilibrium extent of decomposition of nitrogen tetroxide as a result of the chemical reaction $N_2O_4(g) \leftrightarrow 2NO_2(g)$ at 25°C and 1 bar.

9.2. Compute the equilibrium extent of decomposition of pure nitrogen tetroxide in Problem 9.1 over the temperature range of 200 K to 400 K, at pressures of 0.1, 1, and 10 bar.

9.3. Nitrogen and hydrogen react to form ammonia in the presence of a catalyst, $N_2 + 3H_2 \leftrightarrow 2NH_3$. The reactor in which this reaction is to be run is maintained at 450 K and has a sufficiently long residence time that equilibrium is achieved at the reactor exit.

(a) What will be the mole fractions of nitrogen, hydrogen, and ammonia exiting the reactor if stoichiometric amounts of nitrogen and hydrogen enter the reactor, which is kept at 10 bar?

(b) What will be the exit mole fractions if the reactor operates at 10 bar and the feed consists of equal amounts of nitrogen, hydrogen, and an inert diluent?

9.4. At high temperatures, hydrogen sulfide dissociates into molecular hydrogen and sulfur:

$$2H_2S \leftrightarrow 2H_2 + S_2$$

Estimate the extent of dissociation of pure hydrogen sulfide at 700°C and $P = 10$ bar.

9.5. Starting with a 2:1 molar ratio of carbon dioxide-to-hydrogen, compute the equilibrium mole fraction of each of the species in the gas-phase reaction

$$CO_2 + H_2 \leftrightarrow CO + H_2O$$

at 1000 K and (a) 1 bar total pressure and (b) 500 bar total pressure.

9.6. The nitrogen fixation reaction to form ammonia, considered in Problem 9.3, is run at higher temperatures in commercial reactors to take advantage of the faster reaction rates. However, since the reaction is exothermic, at a fixed pressure the equilibrium conversion (extent of reaction) decreases with increasing temperature. To overcome this, commercial reactors are operated at high pressures. The operating range of commercial reactors is pressures of about 350 bar and temperatures from 350°C to 600°C. In trying to find the economically optimal operating conditions for design, it is frequently necessary to consider a wide range of conditions. Therefore, determine the equilibrium extent of reaction over the temperature range of 500 K to 800 K and the pressure range from 1 to 1000 bar.

9.7. Carbon black is to be produced from methane in a reactor maintained at a pressure of 1 bar and a temperature of 700°C according to the reaction $CH_4 \leftrightarrow C(s) + 2\ H_2$. Compute the equilibrium fraction of pure methane charged that is reacted.

9.8. Compute the equilibrium mole fractions of H_2O, CO, CO_2, H_2, and CH_4 in the steam–carbon system at a total pressure of 1 bar and over the temperature range of 600 K to 1600 K.

A set of independent reactions is

$$C + 2H_2O \leftrightarrow CO_2 + 2H_2$$

$$C + H_2O \leftrightarrow CO + H_2$$

$$C + 2H_2 \leftrightarrow CH_4$$

9.9. One mole of nitrogen, 3 moles of hydrogen, and 5 moles of water react at 25°C and kPa and, using the appropriate catalyst and stirring, are allowed to attain phase and chemical equilibrium. Compute the amount and composition of each phase at equilibrium (neglecting the aqueous-phase reaction of ammonia to form NH_4OH, and its subsequent ionization).

9.10. Repeat the calculations of the previous problem for a range of pressures.

9.11. Isopropyl alcohol is to be dehydrogenated in the gas phase to form propionaldehyde according to the reaction

$$(CH_3)_2CHOH(g) \leftrightarrow CH_3CH_2CHO(g) + H_2(g)$$

Compute the equilibrium fraction of isopropyl alcohol that would be dehydrogenated at 500 K and 1.013 bar.

9.12. Carbon dioxide can react with graphite to form carbon monoxide,

$$C(graphite) + CO_2(g) \leftrightarrow 2CO(g)$$

and the carbon monoxide formed can further react to form carbon and oxygen:

$$2CO(g) \leftrightarrow 2C(s) + O_2(g)$$

Determine the equilibrium composition when pure carbon dioxide is passed over a hot carbon bed maintained at 1 bar and (a) 2000 K and (b) 1000 K.

9.13. The production of NO by the direct oxidation of nitrogen from air,

$$N_2 + O_2 \leftrightarrow 2NO$$

occurs naturally in internal combustion engines. This reaction is also used to commercially produce nitric oxide in electric arcs in the Berkeland–Eyde process. If air is used as the feed, compute the equilibrium conversion of oxygen at 1atm total pressure over the temperature range of 1500–3000°C. Air contains 21 mol% oxygen and 79 mol% nitrogen.

9.14. Hydrogen gas can be produced by the following reactions between propane and steam in the presence of a nickel catalyst:

$$C_3H_8 + 3H_2O \leftrightarrow 3CO + 7H_2 \text{ and}$$

$$C_3H_8 + 6H_2O \leftrightarrow 3CO_2 + 10H_2$$

(a) What is the equilibrium composition of the product gas at 1000 K and 1 bar if the inlet to the catalytic reactor is pure propane and steam in a 1:10 ratio?
(b) Repeat calculation of part (a) at 1100 K.

9.15. An important step in the manufacture of sulfuric acid is the gas-phase oxidation reaction

$$SO_2 + \tfrac{1}{2}O_2 \leftrightarrow SO_3$$

Compute the equilibrium conversion of sulfur dioxide to sulfur trioxide over the temperature range of 0 to 1400°C for a reactant mixture consisting of initially pure sulfur dioxide and a stoichiometric amount of air at a total pressure of 1.013 bar. (Air contains 21 mol% oxygen and 79 mol% nitrogen.)

9.16. Ethylene dichloride is produced by the direct chlorination of ethylene using small amounts of ethylene dibromide as a catalyst:

$$C_2H_4 + Cl_2 \leftrightarrow C_2H_4Cl_2$$

If stoichiometric amounts of ethylene and chlorine are used, and the reaction is carried out at 50°C and 1 bar, what is the equilibrium conversion of ethylene?

9.17. Acetaldehyde is produced from ethanol by the following gas-phase reactions:

$$C_2H_5OH + {}^1\!/\!_2O_2 \leftrightarrow CH_3CHO + H_2O \text{ and}$$
$$C_2H_5OH \leftrightarrow CH_3CHO + H_2$$

The reactions are carried out at 540°C and 1 bar pressure using a silver gauze catalyst and air as an oxidant. If 50% excess air (sufficient air that 50% more oxygen is present) than is needed for all the ethanol to react by the first reaction is used, calculate the equilibrium composition of the reactor effluent.

9.18. When propane is heated to high temperatures, it pyrolyzes or decomposes. Assume that the only reactions that occur are

$$C_3H_8 \leftrightarrow C_3H_6 + H_2$$
$$C_3H_8 \leftrightarrow C_2H_4 + CH_4$$

and that these reactions take place in the gas phase. Calculate the composition of the equilibrium mixture of propane and its pyrolysis products at a pressure of 1 bar and over a temperature range of 1000 K to 2000 K.

9.19. The simple statement of the Le Chatelier–Braun principle leads one to expect that if the concentration of a reactant were increased, the reaction would proceed so as to consume the added reactant. This, however, is not always true. Consider the gas-phase reaction

$$N_2 + 3H_2 \leftrightarrow 2NH_3$$

Show that if the mole fraction of nitrogen is less than 0.5, the addition of a small amount of nitrogen to the system at constant temperature and pressure results in the reaction of nitrogen and hydrogen to form ammonia, whereas if the mole fraction of nitrogen is greater than 0.5, the addition of a small amount of nitrogen leads to the dissociation of some ammonia to form more nitrogen and hydrogen. Why does this occur?

9.20. By catalytic dehydrogenation, 1-butene can be produced from *n*-butane,

$$C_4H_{10} \leftrightarrow C_4H_8 + H_2$$

However, 1-butene may also be dehydrogenated to form 1,3-butadiene,

$$C_4H_8 \leftrightarrow C_4H_6 + H_2$$

Compute the equilibrium conversion of *n*-butane to 1-butene and 1,3-butadiene at 1 bar and (a) 900 K and (b) 1000 K

9.21. Styrene can be hydrogenated to ethyl benzene at moderate conditions in both the liquid phase and the gas phase. Calculate the equilibrium compositions in the vapor and liquid phases of hydrogen, styrene, and ethyl benzene at each of the following conditions:

(a) 3 bar pressure and 25°C with a starting mole ratio of hydrogen-to-styrene of 2:1

(b) 3 bar pressure and 150°C, with a starting mole ratio of hydrogen-to-styrene of 2:1
Reaction stoichiometry: $C_6H_5\ HC = CH_2 + H_2 \leftrightarrow C_6H_5CH_2CH_3$

9.22. Formamide is an important industrial solvent and also a raw material in chemical manufacture. At elevated temperatures, it dissociates into ammonia and carbon monoxide in the following gas-phase reaction:

$$HCONH_2 \leftrightarrow NH_3 + CO$$

(a) Compute the equilibrium dissociation of formamide at 1 atm over the temperature range of 400 to 500 K.

(b) Repeat the calculation at 10 atm.

9.23. The industrial solvent carbon disulfide can be made from the reaction between methane and hydrogen sulfide at elevated temperatures using the following gas-phase reaction:

$$2H_2S + CH_4 \leftrightarrow CS2 + 4H_2$$

(a) Starting with equal number of moles of H_2S and CH_4, compute the equilibrium compositions of all components in the reaction at 1 bar and over the temperature range of 500 to 800°C;

(b) Repeat the calculation of part (a) for a pressure of 10 bar.

Chapter 10

Shortcut Distillation Calculations

Distillation is the most used separation/purification method in the oil, chemical, and petrochemical industries, and the distillation process is based on vapor–liquid equilibrium. The design of a distillation column is discussed in detail in courses on mass transfer, stage-wise operations, and in design. Here, just to introduce the ideas of distillation and its relation to thermodynamics, we consider only a very simple description paralleling the one in the textbook, *Chemical, Biochemical and Engineering Thermodynamics*, 4th ed., S. I. Sandler (John Wiley & Sons, Inc., 2006); more complete discussions can be found in specialized textbooks. A very simplified picture of a distillation column is shown in the Fig. 10-1.

In its most basic form, the internals of a distillation column consists of a number of stages, which are individual trays or packings of various kinds that promote vapor–liquid contact and mass transfer between the two phases; trays are used in the diagram below to make the process easier to visualize. Each tray has a layer of liquid on it, and contains holes (sieve tray) or other devices (e.g., bubble caps) that allow some of the liquid to fall

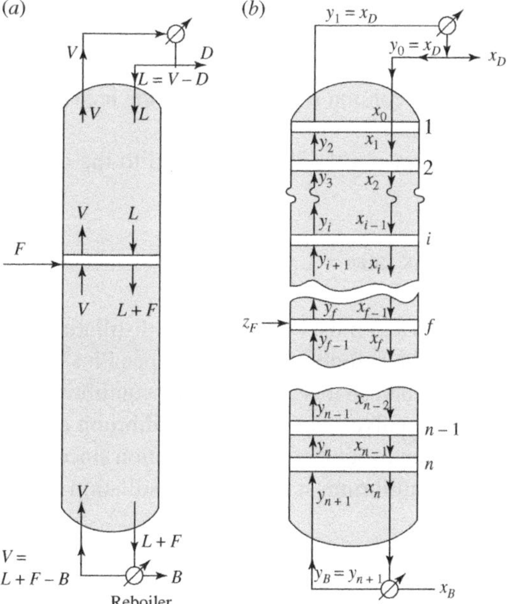

Figure 10-1 (a) Schematic diagram of a distillation column, and (b) showing the tray-to-tray flows and compositions.

Using Aspen Plus® in Thermodynamics Instruction: A Step-by-Step Guide, First Edition. Stanley I. Sandler.

to the tray below and for the vapor from the tray below to rise into the liquid. Therefore, each tray receives liquid flowing down from the tray above it and vapor rising from the tray below it, while some of its liquid flows to the tray below it and its vapor flows to the tray immediately above it. An important assumption in modeling distillation is that the liquid on each tray and the vapor traveling through it are sufficiently in contact that they achieve thermodynamic equilibrium. That is, the vapor and liquid leaving a tray are in equilibrium; though they are not in equilibrium with the liquid entering from the tray above or with the vapor flowing into it from the tray below it.

The simple distillation column shown receives the feed mixture F at an appropriately chosen feed stage (to be discussed later), produces the distillate D from the condenser at the top of the column and a bottoms product B from the heat exchanger (called a reboiler) at the bottom of the column. For simplicity the diagram has been drawn as if the tray-to-tray vapor V and liquid L molar flow rates (usually referred to as the column traffic) above and below the feed tray are constant (we discuss this more later), that the feed F is a saturated liquid (it could also be also be a subcooled liquid, a vapor–liquid mixture, or all vapor, or a superheated vapor; we leave those cases to the mass transfer or stage-wise operations courses). In this simple picture, the liquid and vapor molar flow rates above the feed tray are L and V, respectively, while those below the feed tray are L + F and V.

The reboiler receives the liquid flow L + F leaving the bottom of the column, vaporizes part of part of it sending it back up the column, while an amount of liquid B is withdrawn as the bottoms product. Note that if no liquid were vaporized in the reboiler, there would not be any vapor and therefore only liquid in the column, so there would no vapor-liquid equilibrium and consequently no purification. Similarly, the column has a condenser at the top that, depending on the process, may condense all (total condenser) or some (partial condenser) of the vapor V coming from the topmost tray of the column. In normal operations some distillate D product will be removed, and the remaining condensed distillate (V−D = L) is returned to the column as a liquid. The amount of liquid returned to the column is referred to as the reflux. If there were no reflux, there would be no liquid in the column above the feed tray, only vapor, and again no vapor–liquid equilibrium (except perhaps below the feed tray at which the feed to the column F is supplied, if the feed is liquid), and no separations in that top part of the column.

The reflux ratio R is defined as the ratio of the liquid returned to the column to the amount of distillate withdrawn. That is

$$R = \text{reflux ratio} = L/D$$

This is one of the key operating parameters that must be specified in distillation. Another is the number of trays or stages. In specifying the number of trays in Aspen Plus® note that the reboiler involves vapor–liquid equilibrium and so it is counted as an equilibrium stage, and that if a partial condenser is used, it also involves vapor–liquid equilibrium and is another equilibrium stage. A total condenser is a stage, but there is no separation since all the vapor is condensed. Therefore, the number of equilibrium stages S in a distillation column is

$$S = \begin{pmatrix} \text{number of stages} + 1 & \text{for a total condenser} \\ \text{number of stages} + 2 & \text{for a partial condenser} \end{pmatrix}$$

An efficiency factor is sometimes introduced to account for the fact that complete vapor–liquid equilibrium may not be achieved on each tray or stage, so in an actual design it is common to distinguish between the calculated number of theoretical equilibrium stages and actual number of stages needed to achieve the desired separation. This will not be done here.

There are a number of variables to be considered in the design of a distillation column:

- The desired composition of the distillate product D
- The desired composition of the bottoms product B
- The reflux ratio R
- The stage at which the feed will be added
- Whether a total or partial condenser will be used
- The condition of the feed (a saturated or subcooled liquid, a saturated or superheated vapor, or a vapor–liquid mixture)
- Pressure at which the column will operate

As can be seen by looking at the diagram of a simple distillation column in Fig. 10-1, there are only two product streams, the distillate D overhead and the bottoms product B from the reboiler at the bottom of the column. Before discussing the design process, it is useful to consider the insight that can be obtained from classical thermodynamics. For a binary feed mixture, since there are two product streams from the distillation column, it might be tempting to think one could produce two extremely pure product streams, one for each of the two components. However, consider the Gibbs free energy change that would be required for a process starting with 1 mole of a binary mixture feed of species 1 and 2, (for simplicity assuming ideal solutions; the conclusion would be the same for nonideal solutions) is

$$\begin{aligned}\Delta G &= x_{\rm D}\underline{G}_{\rm D} + x_{\rm B}\underline{G}_{\rm B} - \underline{G}_{\rm F} = \\ &= D[x_{1,\rm D}(\underline{G}_{1,\rm D} - RT\ln x_{1,\rm D}) + x_{2,\rm D}(\underline{G}_{2,\rm D} - RT\ln x_{2,\rm D})] \\ &\quad + B[x_{1,\rm B}(\underline{G}_{1,\rm B} - RT\ln x_{1,\rm B}) + x_{2,\rm B}(\underline{G}_{2,\rm B} - RT\ln x_{2,\rm B})] \\ &\quad - F[x_{1,\rm F}(\underline{G}_{1,\rm F} - RT\ln x_{1,\rm F}) + x_{2,\rm F}(\underline{G}_{2,\rm F} - RT\ln x_{2,\rm F})]\end{aligned}$$

Now from the mass (mole balances) on each component we have

$$x_{1,\rm D}D + x_{1,\rm B}B = x_{1,\rm F}F \quad \text{and} \quad x_{2,\rm D}D + x_{2,\rm B}B = x_{2,\rm F}F$$

Combining this with the Gibbs energy change above gives

$$\begin{aligned}\Delta G/RT &= -D[x_{1,\rm D}\ln x_{1,\rm D} + x_{2,\rm D}\ln x_{2,\rm D}] - B[x_{1,\rm B}\ln x_{1,\rm B} + x_{2,\rm B}\ln x_{2,\rm B}] \\ &\quad + F[x_{1,\rm F}\ln x_{1,\rm F} + x_{2,\rm F}\ln x_{2,\rm F}]\end{aligned}$$

To produce a distillate of pure component 1 requires $x_{1,\rm D} \to 1$ and $x_{2,\rm D} \to 0$. Likewise to obtain a bottoms product of pure component 2 requires $x_{2,\rm B} \to 1$ and $x_{1,\rm B} \to 0$. However, since $\lim_{x\to 0} x\ln x \to -\infty$, if either or both of the distillate and bottoms streams could be purified to the extent of containing only a single pure component, the Gibbs free energy change required to accomplish this would be infinite. That is, $\Delta G \to \infty$ if either $x_{2,\rm D} \to 0$ and/or $x_{1,\rm B} \to 0$. Therefore, classical thermodynamics tells us that we cannot produce pure streams in distillation. Further, we can see from the equation above that the lower the concentration of the impurities in either of the product streams, the more Gibbs energy, and therefore more work (actually heat in a distillation column), will be required to produce such streams. Also, since the Gibbs energy requirement will increase with increasing purity, one can expect that the separation will become more difficult, more energy intensive, and more expensive to accomplish as the purity specification increases. Therefore, the goal in distillation, or any other purification/separation process, is usually not to produce streams of extreme purity, but rather to meet the realistic specifications on the desired product. So, depending on the need, one might aim for 95% purity, 99% purity, 99.9% purity, but never for an unrealistic 100% purity.

The discussion so far has been limited to binary distillation. In industry, there are usually more than two components in the feed stream. As the distillation columns considered here produce only two product streams, a single simple distillation column cannot produce three, four, or more streams each concentrated in one of the feed components. However, this can be accomplished in a train of several distillation columns. So what can be done in a single distillation column with a multicomponent feed? The usual procedure would be to rank all the components according to their volatility (essentially their vapor pressures, unless the values of activity coefficients are very large or very small), and then place a virtual division line between two adjacent components in the list that you wish to separate. The more volatile components above the line will appear predominantly in the distillate, and components below the line will appear predominantly in the bottoms product. The component immediately above the division line is referred to as the light key component and the component just below the line is the heavy key component. The procedure then is to design the distillation column based on the separation of the light key and the heavy key components. Components more volatile than the light key will largely (but not exclusively) appear in the distillate, and components less volatile than the heavy key will then largely appear in the bottoms product. For generality, we will use the designation light key and heavy key components in what follows. In a binary mixture, the light key component is the more volatile of the two components, and the less volatile component is the heavy key.

In principle, there are two ways that the design of a distillation column is done. The first, and most difficult but would be most useful, is the design mode wherein one specifies the desired purity of the distillate and bottoms product, or alternatively, the recovery, that is, the percentage of the component in the original feed that is to appear in the desired product, and directly designs a distillation column for this specification. The difficulty with such a procedure is that meeting the specification requires the search of the multidimensional space of all of the variables listed previously, and there is not a single solution since the impact of changing one operational variable can be compensated by changing another variable. For example, within limits, reducing the reflux ratio can be compensated for by increasing the number of stages. The final column design among the many possibilities that meet the composition requirements would then be based on other considerations such as keeping energy costs low, or keeping capital costs low.

The other type of distillation calculation is referred to as a rating calculation. In this case, one specifies all the distillation column parameters (number of stages, reflux ratio, pressure, feed location, condition of the feed, and type of condenser), and then does a calculation to determine the resulting product streams (distillate and bottoms compositions or recoveries, and flow rates) that can be obtained. One can then change the column specifications and determine how the product streams change. By trial-and-error, or repeated guesses, one can identify a column specification to meet all the separation requirements. This is a simpler calculation than the design calculation and more likely to converge to a solution, but requires much manual intervention in changing the column characteristics to meet the distillate and bottoms composition specifications.

As mentioned above, there is an interplay between the operating parameters of a distillation column. The most important is between the reflux ratio and the number of stages in the column. As the reflux ratio increases, that is, more of the liquid leaving the condenser is returned to the column, the number of equilibrium stages needed to accomplish a given separation decreases. Conversely, as the reflux ratio is reduced, more stages are needed. There are two useful extremes to consider. The first is the infinite reflux ratio when no product is taken from the column; all the vapor coming from the top stage of the column is condensed and returned to the column. In this case there is no distillate or bottoms product and no added feed, just constant internal circulation within the column. The number of stages to meet the product specification in this case, which is the minimum number of stages needed, is referred to as N_{min}. The other limiting case is the minimum reflux ratio R_{min} to accomplish the separation, which occurs when an infinite number of stages is used to accomplish the desired separation. There will be a large number of pairs of the reflux ratio R and number of stages N for which $R > R_{\text{min}}$ and $N > N_{\text{min}}$ that can be used to accomplish the specified separation, which is why there is not a single solution.

You will learn how to do rigorous distillation column calculations in later courses; for completeness, we briefly outline the equations that are used here. First, consider the total column, the leftmost diagram in Fig. 10-1. The overall and species mass balances, and the energy balance over the whole column are, respectively,

$$F = D + B \quad \text{and} \quad x_{i,\text{F}}F = x_{i,\text{D}}D + x_{i,\text{B}}B$$

and

$$\underline{H}_{\text{F}}F + Q_{\text{B}} = \underline{H}_{\text{D}}D + \underline{H}_{\text{B}}B + Q_{\text{D}} \quad \text{or}$$

$$F\sum_i x_{i,\text{F}}\overline{H}_{i,\text{F}} + Q_{\text{B}} = D\sum_i x_{i,\text{D}}\overline{H}_{i,\text{D}} + B\sum_i x_{i,\text{B}}\overline{H}_{i,\text{B}} + Q_{\text{D}}$$

In these equations, Q_{D} is the heat removed in the condenser and Q_{B} is the heat added in the reboiler.

Numbering stages from the top of the column (starting with 0 for the condenser in the diagram, but 1 in the Aspen Plus distillation blocks including **DSTWU** that will be used here), consider the balance equations for stage n. This stage receives a liquid flow from stage $n - 1$ and delivers vapor to this stage. It also delivers liquid to stage $n + 1$, from which it receives a vapor flow. The overall and species mass balances for this stage are

$$V_{n+1} + L_{n-1} + F_n = V_n + L_n \quad \text{and}$$

$$y_{i,n+1}V_{n+1} + x_{i,n-1}L_{n-1} + x_{i,n}F_n = y_{i,n}V_n + x_{i,n}L_n$$

For generality, we have written this equation allowing for stage n to be the feed stage. If it is not, then $F_n = 0$. The energy balance for the stage is

$$\underline{H}^V_{n+1}V_{n+1} + \underline{H}^L_{n-1}L_{n-1} + \underline{H}^F_nF_n + Q_n = \underline{H}^V_nV_n + \underline{H}^L_nL_n \quad \text{or}$$

$$V_{n+1}\sum_i y_{i,n+1}\overline{H}^V_{i,n+1} + L_{n-1}\sum_i x_{i,n+1}\overline{H}^L_{i,n-1} + F_n\sum_i x_{i,n+1}\overline{H}^L_{i,Fn} + Q_n$$

$$= V_n\sum_i y_{i,n}\overline{H}^V_{i,n} + L_n\sum_i x_{i,n}\overline{H}^L_{i,n}$$

There are several points to note about this equation. First, for simplicity, a single phase (typically liquid) feed has been assumed. Second, since real distillation columns are not perfectly insulated, a heat loss Q_n has been allowed on each tray. Perhaps, most important is that the stage-to-stage molar flow rates (the column traffic), that is L_n and V_n have not been assumed to be the same on all stages, though they are shown as such in Fig. 10-1. The most obvious reason is that at least one stage will be the feed stage, so that for a liquid feed, there will be a greater liquid flow leaving the stage for the stage below than flowing to the stage from the stage above. Another reason for the changes in molar flow is that there will be some partial condensation if there are heat losses along the column. There will also be changes in molar flows to satisfy the energy balance on each stage as a result of the temperature differences across the column (from the hot bottom section to the cooler top section) and differences in species' partial molar enthalpies and concentrations. Finally, there is also a transfer of species between the vapor and liquid on each stage to achieve thermodynamic equilibrium. [In some distillation calculations a simplification referred to as "constant molar overflow" is made in which it is assumed that the only variation in the stage-to-stage molar flows of vapor and liquid throughout the column are the result of the feed addition. This assumption is not used in rigorous calculations.]

The final equation to be satisfied is that of phase equilibrium for each component on each stage. That is, that the vapor and liquid leaving each stage are in equilibrium

$$\bar{f}^{L}_{i,n}(T_n, P_n, x_{1,n}, x_{2,n}, \ldots) = \bar{f}^{V}_{i,n}(T_n, P_n, y_{1,n}, y_{2,n}, \ldots) \quad \text{or}$$
$$x_{i,n}\gamma_{i,n}(T_n, P_n, x_{1,n}, x_{2,n}, \ldots)P_i^{\text{vap}}(T_n) = y_{i,n}P_n$$

Written in this way, the equations allow for temperature, composition, and pressure changes (due to the hydrostatic head resulting from the liquid holdup on each stage). In a rigorous equilibrium calculation, all the above equations must be solved simultaneously. This is done, for example, in Aspen Plus in the **RadFrac** block discussed in the next chapter.

For both computer-based design and rating calculations there is the problem of deciding upon a meaningful set of initial column specifications. In the past, a number of shortcut recipes were developed to make such *initial* guesses. The idea being to get a rough estimate of the column parameters from simple hand calculations that would then be the starting point for rigorous distillation calculations (that were extremely time-consuming when done before computers). We consider only the simplest of these simple shortcut methods here.

Before starting this discussion, we need to introduce the term relative volatility α_{ij} for any two species i and j in a mixture that is defined to be

$$\alpha_{\text{ij}} = \frac{y_{\text{i}}/x_{\text{i}}}{y_{\text{j}}/x_{\text{j}}}$$

The more volatile component in a mixture will have a relative volatility greater than one, while at a mixture azeotropic point $\alpha_{\text{ij}} = 1$. The greater the difference in the relative volatilities of the two components in a binary mixture, the easier it is to accomplish the separation by vapor–liquid equilibrium, and the fewer stages that are required in the distillation column. Consequently, the relative volatility is an important parameter in distillation column design.

The relative volatility of the two components is a function of composition (as a result of the dependence of activity coefficients of composition in nonideal mixtures) and temperature (as a result of the different vapor pressure changes of the components with temperature), and therefore varies throughout the distillation column. Also, the numeric value depends on the thermodynamic model or data used in the calculation. Consequently, the relative volatility of the two components to be separated will be different at the bottoms conditions (temperature and composition) $\alpha_{ij,B}$, at the feed $\alpha_{ij,F}$, and in the distillate $\alpha_{ij,D}$.

In a binary mixture distillation, there is one relative volatility α_{12} at each of these conditions. In the analysis of a multicomponent distillation, the relative volatility of the light key to heavy key components $\alpha_{LK,HK}$ is most important. It is common to refer the relative volatility of each component with reference to one of the components, typically the heavy key. That is, the relative volatility for any component j is $\alpha_{j,HK}$.

The first step in simple shortcut calculations for a desired separation is to estimate the minimum reflux ratio R_{min} and the minimum number of stages N_{min} that will be needed for the separation. The minimum number of stages, N_{min}, required to meet the distillate and bottoms composition specifications occurs at total reflux. The empirical Fenske equation

$$N_{min} = \frac{\log\left(\frac{x_{LK,D}\, x_{HK,B}}{x_{LK,B}\, x_{HK,D}}\right)}{\log(\bar{\alpha}_{LK,HK})}$$

or one of its more complicated later versions, can be used to make a rough estimate of the minimum number of stages. In this equation the relative volatility of the light key to the heavy key is generally taken to be the geometric mean over the column or the geometric mean of values at the bottoms, feed, and distillate conditions, that is, $\bar{\alpha} = \sqrt[3]{\alpha_B \alpha_F \alpha_D}$. This will be used here.

The other extreme is the minimum reflux ratio R_{min} required to obtain the desired separation, even though this would require an infinite number of stages. The Underwood equations (only the original simplest forms are presented here) are used to compute the minimum reflux ratio. The starting point is solving the following equation for the single variable ϕ

$$\frac{\Delta V}{F} = \sum_i \frac{\bar{\alpha}_{i,HK} x_{i,F}}{\bar{\alpha}_{i,HK} - \phi}$$

In this equation F is the molar flow rate of the feed, ΔV is the change in vapor flow rate as a result of the feed (0 for a saturated liquid feed, F for a saturated vapor feed, between these two values for a mixed feed, negative for a subcooled liquid feed, and greater than F for a superheated vapor feed). The sum is over all the components in the mixture so that the order of this algebraic equation with respect to ϕ , and consequently the difficulty in solving it, increases with the number of components. The value of ϕ so obtained is then used in the second Underwood equation

$$R_{min} = \sum_i \frac{\bar{\alpha}_{i,HK} x_{i,D}}{\bar{\alpha}_{i,HK} - \phi} - 1$$

The shortcut distillation block **DSTWU** uses somewhat more sophisticated versions of these equations, and then relates the minimum number of stages N_{min} and the minimum reflux ratio R_{min} to the actual number of stages N and the actual reflux ratio R using the empirical correlation

$$\frac{N - N_{min}}{N+1} = 1 - \exp\left(\frac{1+54X}{11+117.2X} \cdot \frac{X-1}{X^{0.5}}\right) \text{ where } X = \frac{R - R_{min}}{R+1}$$

So once the N_{min} and R_{min} are computed and the actual reflux ratio R is chosen, the number of (theoretical) stages can be computed. Depending on energy and capital costs, the operating reflux ratio R is frequently taken to be in the range of $1.15R_{min}$ to $1.25R_{min}$

To complete this very preliminary shortcut design procedure, the location of the feed stage has to be specified. This is done by examining the concentrations on each stage, and choosing the feed location to be the stage whose composition is closest to that of the feed. [Note that in shortcut calculations the composition on each stage is obtained by solving the mass balance equations assuming constant relative volatility and constant molal overflow.]

Once these steps have been completed, one has a (very) preliminary design for a distillation column. This would then only be used as a starting point or initial guess for a rigorous distillation calculation in which the mass and energy balances and phase equilibria are solved simultaneously for a more accurate column rating or design. The Aspen Plus block **DSTWU** does the shortcut calculations discussed above.

To see how all this comes together we start by solving the following example problem by hand using Aspen Plus for thermodynamic modeling, and then using the **DSTWU** block. The problem to be considered is the design of a distillation column operating at 1 atm pressure with a saturated liquid equimolar mixture of benzene and toluene feed to produce a distillate that contains 90 mol% benzene and a bottoms product that contains 98 mol% toluene.

The Aspen Plus block **DSTWU** requires the separation specification to be presented as percent recovery of key components rather than compositions, so the starting point is to do the mass balances to relate the specified compositions to recoveries.

Using B for the bottoms product, D for the distillate, and F (= 1) for the feed, the mass balances are

Overall: $B + D = F = 1$

Species i: $z_i F = y_i D + x_i B$

So that for benzene: $0.5 = 0.90D + 0.02B$ and for toluene: $0.5 = 0.10D + 0.98B$

Subtracting the two equations: $0 = 0.80D - 0.96B$; so that $B = 0.80D/0.96 = 0.8333D$

Using the overall mass balance: $0.8333D + D = 1$

$D = 1/1.8333 = 0.5455$, and $B = 1 - 0.5455 = 0.4545$

Therefore, the fractional recovery of benzene in the distillate = $(0.9 * 0.5455)/(0.5 * 1) = 0.981$ and the fractional recovery of toluene in the distillate = $(0.1 * 0.5455)/(0.5 * 1) = 0.109$

The next (or perhaps even the first) useful step is to see whether the desired separation can be obtained by distillation, that is, to check to see whether this mixture has an azeotrope within the benzene composition range of 2 mol% (bottoms composition) to 90 mol% (distillate). Using the **NIST TDE** and choosing one of the available data sets at about 1 atm show that the system does not have an azeotrope (data in Fig. 10-2 and is plotted by clicking on **T-xy** in **Plot** on the **Main Toolbar** producing Fig. 10-3.) Therefore, simple distillation can be used to accomplish the desired separation. We could have guessed this for this system, since benzene and toluene are so chemically similar that they essentially form an ideal mixture. Nonetheless, a check on the possibility of azeotrope formation for any new mixture is always a useful first step in a distillation calculation. Also, the data provide information on the temperature range of the distillation.

Start Page | Components - Specifications | TDE Binary Results

Experimental Data | Consistency Test

- Data for BENZE-01(1) and TOLUE-01(2)
  - Binary diffusion coefficient
    - (Liquid) (X1(L), T, P)
      - Binary diffusion coefficient 0(
      - Binary diffusion coefficient 0(
  - Binary VLE
    - Isobaric
      - Binary VLE 017
      - Binary VLE 025
      - Binary VLE 027
      - Binary VLE 029
      - Binary VLE 031
      - Binary VLE 033
      - Binary VLE 035
      - Binary VLE 037
      - Binary VLE 039
      - Binary VLE 041
      - Binary VLE 052
      - Binary VLE 054
      - Binary VLE 056
      - Binary VLE 058
      - Binary VLE 061
    - Isothermal
      - Binary VLE 002
      - Binary VLE 004
      - Binary VLE 005
      - Binary VLE 006
      - Binary VLE 007
      - Binary VLE 008
      - Binary VLE 009
      - Binary VLE 010
      - Binary VLE 011
      - Binary VLE 012

No.	Liquid Mole fraction BENZE-01	Temperature (K)	Vapor Mole fraction BENZE-01	Total pressure(N/sqm)
1	0.0374	382.25	0.0821	101000
2	0.0703	380.85	0.149	101000
3	0.1061	379.35	0.2167	101000
4	0.1347	378.15	0.2678	101000
5	0.185	376.18	0.3495	101000
6	0.2047	375.45	0.3795	101000
7	0.229	374.55	0.4148	101000
8	0.2935	372.2	0.5007	101000
9	0.3763	369.41	0.5959	101000
10	0.3841	369.16	0.6042	101000
11	0.4553	366.95	0.6739	101000
12	0.5432	364.35	0.7486	101000
13	0.6391	361.75	0.8181	101000
14	0.731	359.45	0.8749	101000
15	0.8149	357.45	0.9197	101000
16	0.9148	355.25	0.9658	101000

Huang, X.; Xia, S.; Ma, P.; Song, S.; Ma, B. J. Chem. Eng. Data, 2008, 53, 252-255 Vapor Liquid Equilibrium of N-Formylmorpholine with Toluene and Xylene at 101.33 kPa

Figure 10-2

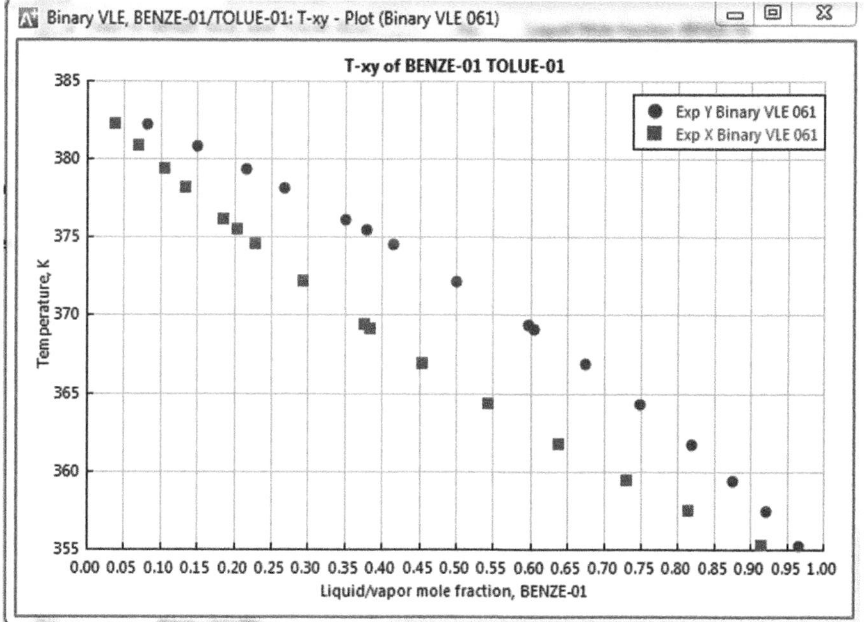

Figure 10-3

Based on the data in the figures, we will consider the benzene–toluene mixture to be ideal in what follows. Next, we need information on the benzene–toluene relative volatility. We could compute this using the experimental data already obtained. However, since the system is essentially ideal, we will instead compute this using **Properties>Analysis>Binary** and populate the window as shown in Fig. 10-4.

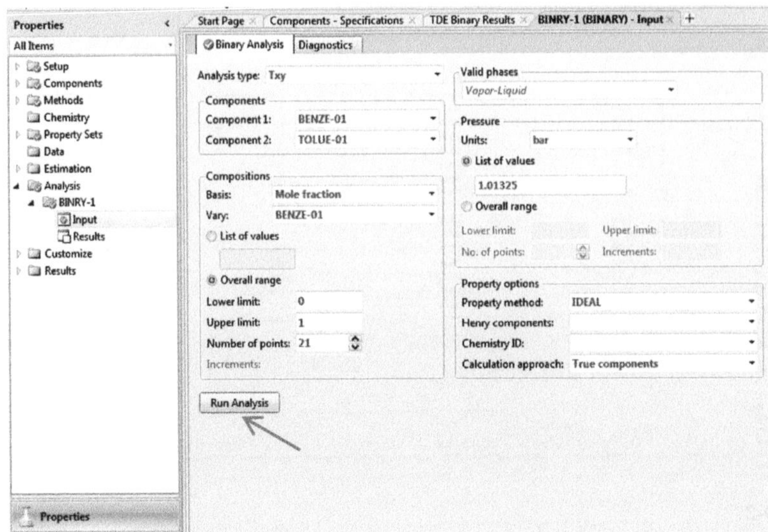

Figure 10-4

Clicking on **Run Analysis** produces the results in Fig. 10-5 that are in very good agreement with available experimental data, as would be expected for a mixture of very similar components.

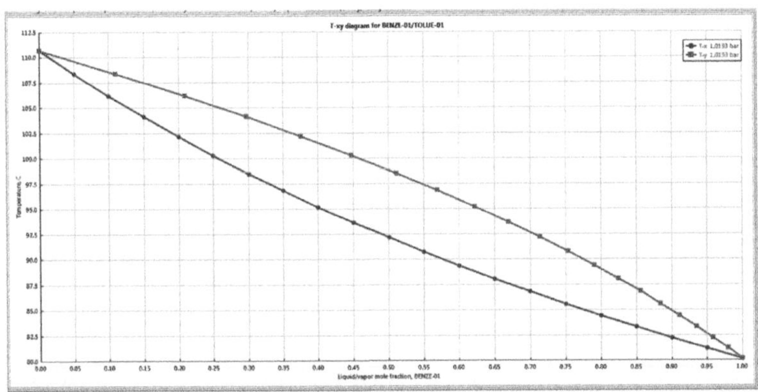

Figure 10-5

Now going to **Analysis>Binary-1>Results** we obtain Fig. 10-6.

Binary analysis results

PRES	MOLEFRAC BENZE-01	TOTAL TEMP	TOTAL KVL BENZE-01	TOTAL KVL TOLUE-01	LIQUID GAMMA BENZE-01	LIQUID GAMMA TOLUE-01	VAPOR MOLEFRAC BENZE-01	VAPOR MOLEFRAC TOLUE-01	LIQUID MOLEFRAC BENZE-01	LIQUID MOLEFRAC TOLUE-01
bar		C								
1.01325	0	110.679	2.34897	1	1	1	0	1	0	1
1.01325	0.05	108.373	2.21456	0.936076	1	1	0.110728	0.889272	0.05	0.95
1.01325	0.1	106.189	2.09282	0.878576	1	1	0.209282	0.790719	0.1	0.9
1.01325	0.15	104.119	1.98218	0.826674	1	1	0.297327	0.702673	0.15	0.85
1.01325	0.2	102.154	1.88134	0.779666	1	1	0.376267	0.623733	0.2	0.8
1.01325	0.25	100.285	1.78913	0.736955	1	1	0.447284	0.552716	0.25	0.75
1.01325	0.3	98.5044	1.7046	0.698029	1	1	0.51138	0.488621	0.3	0.7
1.01325	0.35	96.8069	1.62688	0.662449	1	1	0.569408	0.430592	0.35	0.65
1.01325	0.4	95.1859	1.55525	0.629837	1	1	0.622098	0.377902	0.4	0.6
1.01325	0.45	93.6361	1.48905	0.599865	1	1	0.670075	0.329926	0.45	0.55
1.01325	0.5	92.1523	1.42775	0.57225	1	1	0.713875	0.286125	0.5	0.5
1.01325	0.55	90.7303	1.37085	0.546746	1	1	0.753965	0.246036	0.55	0.45
1.01325	0.6	89.3657	1.31791	0.523136	1	1	0.790745	0.209255	0.6	0.4
1.01325	0.65	88.055	1.26857	0.501234	1	1	0.824568	0.175432	0.65	0.35
1.01325	0.7	86.7945	1.22248	0.480872	1	1	0.855739	0.144262	0.7	0.3
1.01325	0.75	85.5812	1.17937	0.461904	1	1	0.884524	0.115476	0.75	0.25
1.01325	0.8	84.4122	1.13895	0.444202	1	1	0.91116	0.0888404	0.8	0.2
1.01325	0.85	83.2848	1.101	0.427652	1	1	0.935852	0.0641477	0.85	0.15
1.01325	0.9	82.1965	1.06532	0.412151	1	1	0.958785	0.041215	0.9	0.1
1.01325	0.95	81.1451	1.03171	0.397609	1	1	0.98012	0.0198804	0.95	0.05
1.01325	1	80.1285	1	0.383944	1	1	1	0	1	0

Figure 10-6

After copying the results into Excel, eliminating the K-value = y_i/x_i and activity coefficient columns and computing the relative volatility $\alpha = \frac{y_B/x_B}{y_T/x_T}$, we obtain Fig. 10-7.

PRES	x BENZE	TOTAL TEMP	y BENZE	y TOLUE	x TOLUE	Relative
bar		C				volatility
1.01325	0	110.6793	0	1	1	
1.01325	0.05	108.3728	0.1107279	0.8892721	0.95	2.365788941
1.01325	0.1	106.1894	0.2092815	0.7907185	0.9	2.382053158
1.01325	0.15	104.1193	0.2973274	0.7026726	0.85	2.397781366
1.01325	0.2	102.1537	0.3762671	0.6237329	0.8	2.413001463
1.01325	0.25	100.2845	0.4472836	0.5527164	0.75	2.427738348
1.01325	0.3	98.5044	0.5113795	0.4886205	0.7	2.442015497
1.01325	0.35	96.80687	0.5694081	0.4305919	0.65	2.455857125
1.01325	0.4	95.1859	0.6220981	0.3779019	0.6	2.469284092
1.01325	0.45	93.63605	0.6700745	0.3299255	0.55	2.482317809
1.01325	0.5	92.15234	0.7138751	0.2861249	0.5	2.494977194
1.01325	0.55	90.73028	0.7539645	0.2460355	0.45	2.507280638
1.01325	0.6	89.36574	0.7907454	0.2092546	0.4	2.519244977
1.01325	0.65	88.05497	0.8245682	0.1754318	0.35	2.530888137
1.01325	0.7	86.79451	0.8557385	0.1442615	0.3	2.542224165
1.01325	0.75	85.58121	0.8845239	0.1154761	0.25	2.553266866
1.01325	0.8	84.41217	0.9111595	0.0888404	0.2	2.564034775
1.01325	0.85	83.28475	0.9358522	0.0641477	0.15	2.574533276
1.01325	0.9	82.19647	0.9587849	0.041215	0.1	2.584778735
1.01325	0.95	81.1451	0.9801196	0.0198804	0.05	2.594778883
1.01325	1	80.12853	1	0	0	

Figure 10-7

We find that benzene–toluene therelative volatility varies from approximately 2.35 at the bottoms composition (2 mol% benzene), 2.495 at the feed composition to 2.585 at the distillate composition. The geometric mean of these three values is 2.4748.

The minimum number of stages can now be estimated using the Fenske equation

$$N_{\min} = \frac{\log\left(\dfrac{x_{\mathrm{LK,D}}}{x_{\mathrm{LK,B}}}\dfrac{x_{\mathrm{HK,B}}}{x_{\mathrm{HK,D}}}\right)}{\log(\bar{\alpha}_{\mathrm{LK,HK}})} = \frac{\log\left(\dfrac{0.9}{0.02}\dfrac{0.98}{0.1}\right)}{\log(2.4748)} = 6.72 \text{ equilibrium stages}$$

So based on the Fenske equation, if a total condenser is used, a reboiler and 6.72–1 = 5.72 equilibrium stages (or trays) is the minimum required for the separation.

To compute the minimum reflux ratio using the Underwood method for a saturated liquid feed (so that $\Delta V = 0$) we start with the first Underwood equation

$$\frac{\Delta V}{F} = 0 = \sum_i \frac{\bar{\alpha}_{\mathrm{i,HK}} x_{\mathrm{i,F}}}{\bar{\alpha}_{\mathrm{i,HK}} - \phi} = \frac{\bar{\alpha}_{\mathrm{LK,HK}} x_{\mathrm{HK,F}}}{\bar{\alpha}_{\mathrm{LK,HK}} - \phi} + \frac{\bar{\alpha}_{\mathrm{HK,HK}} x_{\mathrm{HK,F}}}{\bar{\alpha}_{\mathrm{HK,HK}} - \phi} = \frac{2.4748 \times 0.5}{2.4748 - \phi} + \frac{1 \times x_{\mathrm{HK,F}}}{1 - \phi}$$

which has the solution $\phi = 1.4244$. Using this result in the second Underwood equation gives

$$R_{\min} = \sum_i \frac{\bar{\alpha}_{\mathrm{i,HK}} x_{\mathrm{i,D}}}{\bar{\alpha}_{\mathrm{i,HK}} - \phi} - 1 = \frac{\bar{\alpha}_{\mathrm{LK,HK}} x_{\mathrm{LK,D}}}{\bar{\alpha}_{\mathrm{LH,HK}} - \phi} + \frac{\bar{\alpha}_{\mathrm{HK,HK}} x_{\mathrm{HK,D}}}{\bar{\alpha}_{\mathrm{HK,HK}} - \phi} - 1$$

$$= \frac{2.4748 \times 0.9}{2.4748 - 1.4244} + \frac{1 \times 0.1}{2.4748 - 1.4244} - 1 = 0.8848 = (L/D)_{\min}$$

This result indicates that to accomplish the specified separation, as a minimum, 0.8848 mols of liquid reflux must be returned to the column for each mol of distillate (containing 0.9 mols of benzene and 0.1 mols of toluene) withdrawn from the column.

Now, for comparison, Aspen Plus is used to solve the shortcut distillation problem. We start, after the usual initial setup, by going to **Simulation**, then to the **Model Palette** and choosing the **DSTWU** block from the **Columns** tab as shown in Fig. 10-8.

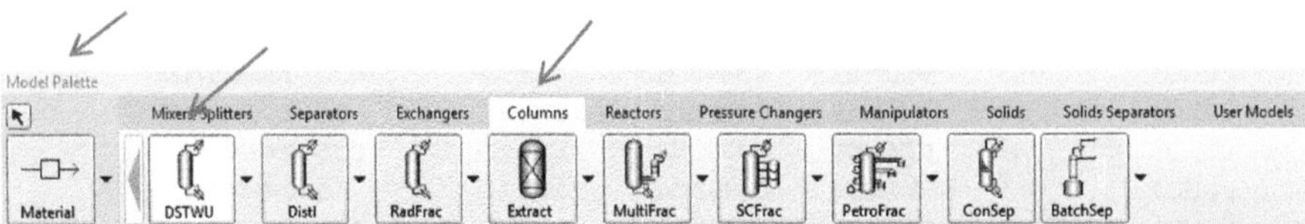

Figure 10-8

The flow sheet then obtained is shown in Fig. 10-9.

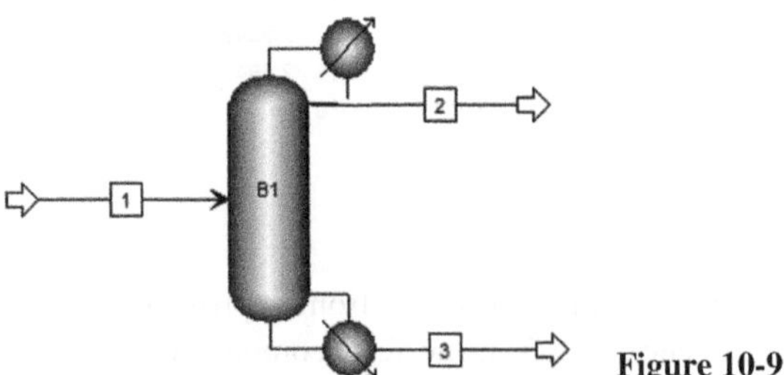

Figure 10-9

The feed stream specification used in **Streams>1** input is shown in Fig. 10-10.

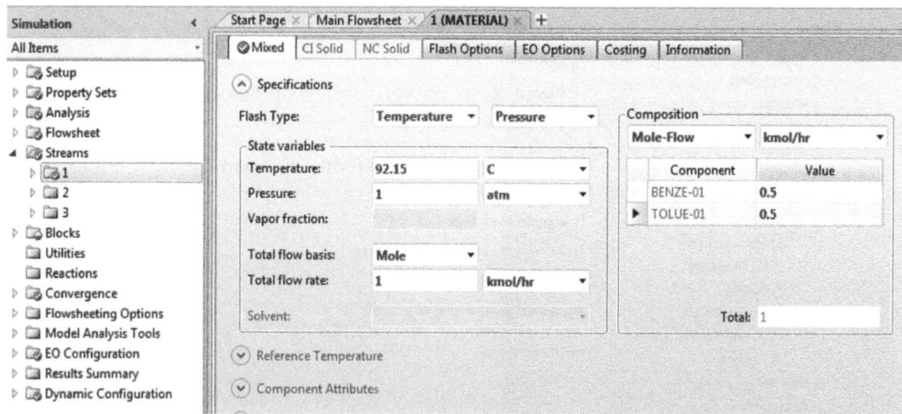

Figure 10-10

The temperature of the feed stream has been set at 92.15°C that, from a previous bubble point calculation, is the saturation temperature of the liquid feed. The Block specification **Blocks>B1** input is, based on the mass balance calculations done earlier, shown in Fig. 10-11.

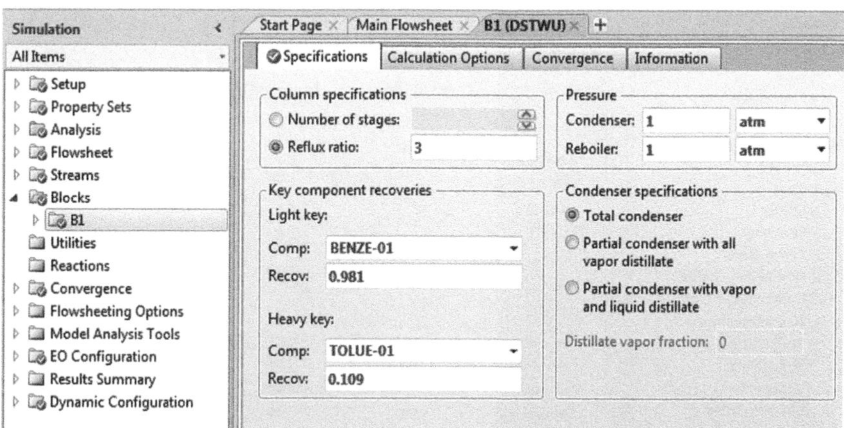

Figure 10-11

In this input, the benzene and toluene recoveries in the distillate are the values computed earlier from the mass balances. To initiate the calculation, a reflux ratio of 3 has been set, which is greater than the R_{min} value of 0.8848, a total condenser has been chosen, and the column pressure has been set at 1 atm throughout. In an actual column design, one would allow for a small pressure drop on each stage as a result of the hydrostatic pressure difference due to the liquid holdup on the stage.

It is also useful to go to the **Blocks>B1>Input>Calculation Options** tab and click on the option to **Generate table of reflux ratio vs number of theoretical trays**. That window has been populated as shown in Fig. 10-12.

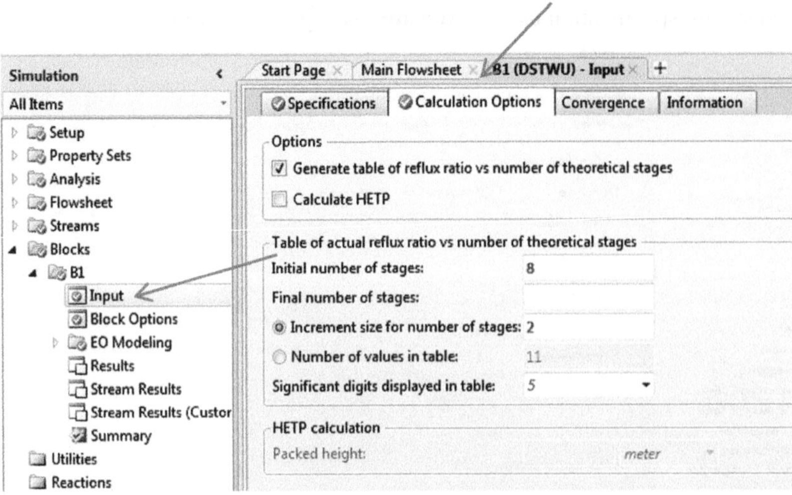

Figure 10-12

With this input complete, the simulation can now be run. Looking at the **Results Summary>Streams** in Fig. 10-13 we see that the composition specifications have been met (well almost, we specified a benzene mole fraction of 0.02 in the bottoms stream while 0.0209 was obtained).

	1	2	3
Substream: MIXED			
Mole Flow kmol/hr			
BENZE-01	0.5	0.4905	0.0095
TOLUE-01	0.5	0.0545	0.4455
Mole Frac			
BENZE-01	0.5	0.9	0.0208791
TOLUE-01	0.5	0.1	0.979121
Total Flow kmol/hr	1	0.545	0.455
Total Flow kg/hr	85.1271	43.3364	41.7907
Total Flow l/min	1.77575	0.890251	0.891153
Temperature C	92.15	82.1965	109.701
Pressure bar	1.01325	1.01325	1.01325
Vapor Frac	0	0	0
Liquid Frac	1	1	1
Solid Frac	0	0	0
Enthalpy cal/mol	9864.84	12861.7	6612.02
Enthalpy cal/gm	115.884	161.75	71.989
Enthalpy cal/sec	2740.24	1947.12	835.686
Entropy cal/mol-K	-62.0137	-55.7461	-70.7258
Entropy cal/gm-K	-0.728484	-0.701065	-0.770034
Density mol/cc	0.00938571	0.0102031	0.00850957
Density gm/cc	0.798978	0.811314	0.781584
Average MW	85.1271	79.5163	91.8477
Liq Vol 60F l/min	1.62023	0.819771	0.800455

Figure 10-13

Looking at the relevant parts of the **Blocks>B1> Results** (Fig. 10-14) we see that the minimum reflux ratio for the separation computed in the **DSTWU** block is 0.8901 (compared to 0.8848 computed in the previous hand calculation), and the minimum number of stages is 6.7905 (compared to the hand calculation result of 6.72). These differences are the result of the slightly different methods used in **DSTWU** and using slightly different averages of the relative volatility. Nonetheless, the conclusion is that the simple hand calculation results are reasonable. Also, we see for the chosen reflux ratio of 3, 8.71 actual (or rather equilibrium) stages are needed, that there should 3.03778 stages above the feed. Since there cannot be fractional stages, the interpretation of these results is that there should be 9 stages, that is 8 stages or trays and a reboiler in the distillation column, and that there should be 3 stages above the feed. However, the total condenser is counted as one of those stages in the Aspen Plus output. Therefore, the feed stage should be the third stage from the top of the column. [Note, all of this is based on the assumption that each stage or tray operates as a perfect, equilibrium stage. It is not uncommon to use a stage efficiency, either computed from mass transfer considerations or using an empirical factor, for example, 0.7, and scale up all the numbers of stages above accordingly.] The model summary also provides information on the heating rate required in the reboiler and the cooling rate in the partial condenser.

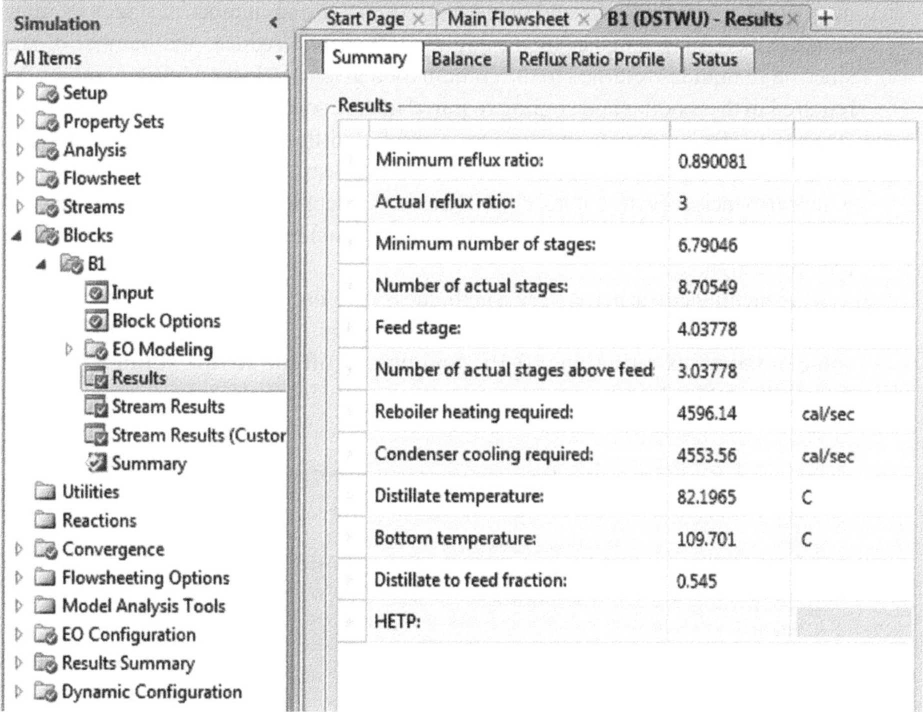

Figure 10-14

Next, clicking on the **Blocks>B1>Results>Reflux Ratio Profile**, we obtain estimates of how the number of theoretical stages required for the specified separation changes with the reflux ratio, as shown in Fig. 10-15.

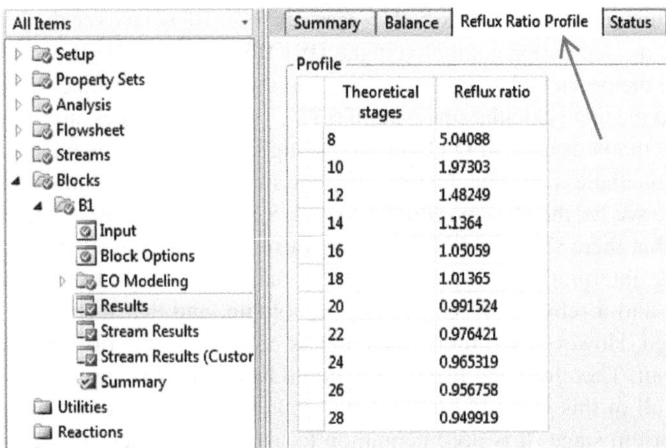

Figure 10-15

This information can then be used to make an initial guess for a more rigorous distillation calculation.

It is interesting to consider the implications of the results above. The capital cost of a distillation column will depend on its size. An increasing number of stages requires a taller column, increasing its cost. Increasing the reflux ratio reduces the number of stages, but since more liquid is returned to the column for a given product rate this increases the energy required in the reboiler and cooling required in the condenser, and therefore increasing their size. Also, since there are greater vapor and liquid flows in the column at a higher reflux ratio for a fixed product rate, the diameter of the distillation column will have to increase, which also increases the capital cost. Therefore, a detailed design and economic study (after doing rigorous, not shortcut, distillation calculations) must be done to choose the optimum column design.

As mentioned earlier, a very rough guess is to use a reflux ratio of about 20% greater than R_{min}. In this case, that would be $1.2 \times 0.8901 = 1.068$. Rerunning **DSTWU** using that value of the reflux ratio (Fig. 10-16) (and turning off the **Reflux Ratio Profile**) produces the results in Fig. 10-17.

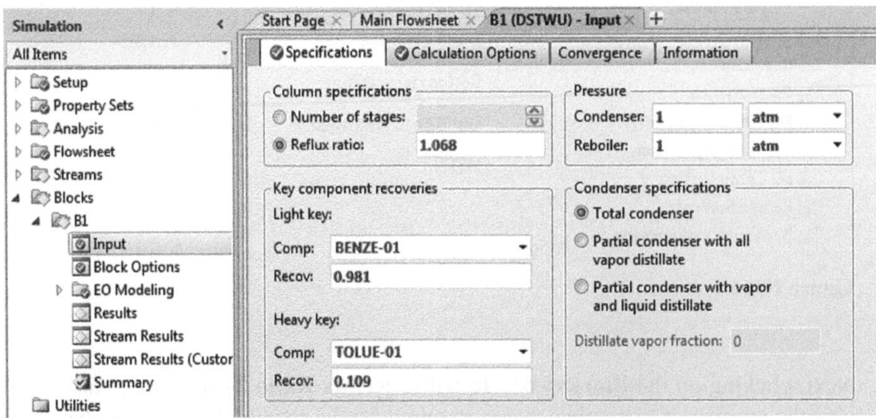

Figure 10-16

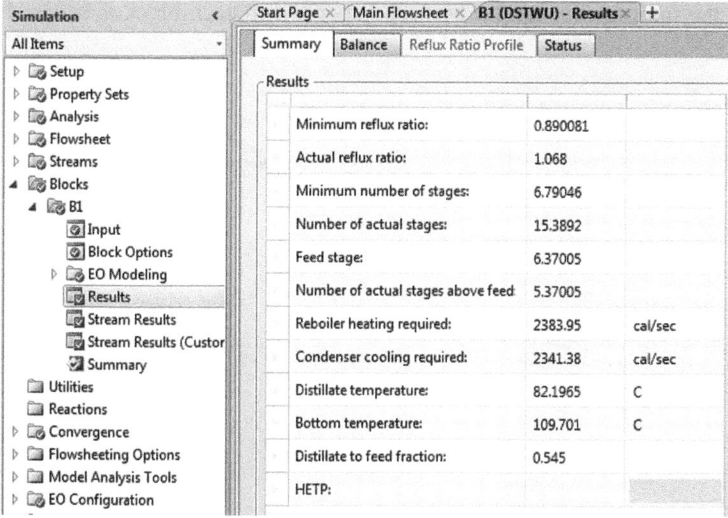

Results		
Minimum reflux ratio:	0.890081	
Actual reflux ratio:	1.068	
Minimum number of stages:	6.79046	
Number of actual stages:	15.3892	
Feed stage:	6.37005	
Number of actual stages above feed:	5.37005	
Reboiler heating required:	2383.95	cal/sec
Condenser cooling required:	2341.38	cal/sec
Distillate temperature:	82.1965	C
Bottom temperature:	109.701	C
Distillate to feed fraction:	0.545	
HETP:		

Figure 10-17

This shows that at $1.2 \times R_{\min}$, the number of theoretical stages to accomplish the separation has increased to 15.4 from the $N_{\min}$ value of 6.79.

To close this chapter, we briefly consider repeated flash calculations as another method of producing a product stream of 90 mol% benzene from the feed containing only 50 mol% at 100°C and 1 atm. Fig. 10-18 is a flow sheet for four flash separations in series in which each flash results in half of its inlet stream being vaporized at 1 atm.

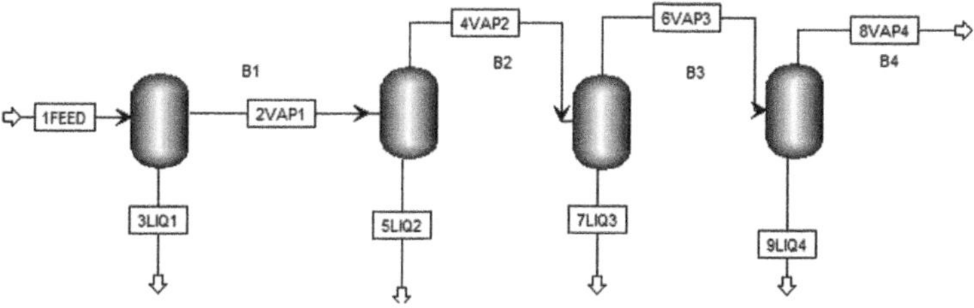

Figure 10-18

The Feed is specified in Fig. 10-19.

Figure 10-19

That half of the inlet feed stream to be flash vaporized in each block is set as shown in Fig. 10-20.

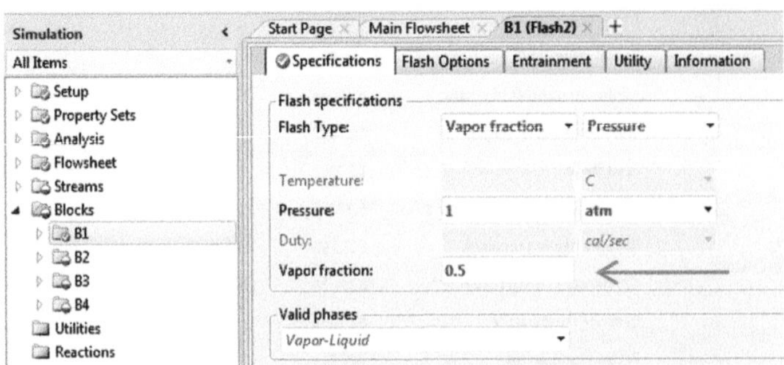

Figure 10-20

This is done from the pull-down menu under **Specifications>Flash Type** and setting the **Vapor fraction** instead of **Temperature**. The same is done for the other flash drums. An abbreviated list of the results is given in Fig. 10-21.

	1FEED	2VAP1	3LIQ1	4VAP2	5LIQ2	6VAP3	7LIQ3	8VAP4	9LIQ4
Substream: MIXED									
Mole Flow kmol/hr									
BENZE-01	0.5	0.305483	0.194517	0.179385	0.126098	0.101105	0.0782796	0.0549057	0.0461995
TOLUE-01	0.5	0.194517	0.305483	0.0706151	0.123902	0.0238947	0.0467204	0.00759427	0.0163004
Mole Frac									
BENZE-01	0.5	0.610966	0.389034	0.717539	0.504393	0.808842	0.626237	0.878492	0.739192
TOLUE-01	0.5	0.389034	0.610966	0.282461	0.495608	0.191158	0.373763	0.121508	0.260808
Total Flow kmol/hr	1	0.5	0.5	0.25	0.25	0.125	0.125	0.0625	0.0625
Total Flow kg/hr	85.1271	41.7853	43.3418	20.5189	21.2664	10.0994	10.4196	4.98863	5.11075
Total Flow l/min	510.32	252.107	0.908201	124.853	0.44354	61.8534	0.216284	30.6847	0.105631
Temperature C	100	95.5351	95.5351	92.025	92.025	88.6715	88.6715	85.8396	85.8396
Pressure bar	1.01325	1.01325	1.01325	1.01325	1.01325	1.01325	1.01325	1.01325	1.01325
Vapor Frac	1	1	0	1	0	1	0	1	0
Liquid Frac	0	0	1	0	1	0	1	0	1

Figure 10-21

What we see in the table is that the vapor from the fourth flash has a benzene mole fraction of 0.878, almost up to the specified 0.9 mole fraction. However, in this successive flash sequence we are recovering in the product stream only $0.046 \times 100/0.5 = 9.2\%$ of the benzene in the feed. While in the distillation design, 90% of the benzene is recovered. The conclusion then is that repeated flash separations are very inefficient for this separation leading to poor fractional recovery of the desired product from the feed, and a large waste stream.

The next chapter deals with rigorous distillation calculations using the **RadFrac** block.

PROBLEMS

10.1. An equimolar mixture of benzene and 2,2,4-trimethylpentane is to be purified by distillation to produce a distillate containing 0.90 mole fraction of benzene and a bottoms product that contains 0.02 mole fraction of benzene. The distillation column is to operate at 1 atm pressure and the feed to the column is a saturated liquid. Use the DSTWU program and the NRTL model (with Aspen default parameters) to estimate

(a) the minimum reflux ratio and the minimum number of stages to accomplish the separation;
(b) the actual number of stages and the location of the feed stage if the column is operated at 1.2 times the minimum reflux ratio.

10.2. Repeat the calculation above using the Wilson equation with the Aspen default parameters.

10.3. Repeat the calculation above using the UNIQUAC model with the Aspen default parameters.

10.4. An equimolar mixture of *n*-pentane and *n*-heptane is to be purified by distillation to produce a distillate that contains 98% of the pentane and 1% of the heptane in the initial feed. The column is to operate at 1 bar and the feed is at 20°C and 1 bar. Use the DSTWU program and the NRTL model (with Aspen default parameters) to estimate

(a) the minimum reflux ratio and the minimum number of stages to accomplish the separation;
(b) the actual number of stages and the location of the feed stage if the column is operated at 1.2 times the minimum reflux ratio.

10.5. Repeat the calculation above using the Wilson equation with the Aspen default parameters.

10.6. Repeat the calculation above using the UNIQUAC model with the Aspen default parameters.

Chapter 11

A Rigorous Distillation Calculation: RadFrac

The rigorous design of distillation columns is a very complicated and time-intensive process because of the complexity of the calculation and the many degrees of freedom in the design. These include the reflux ratio, operating temperature and pressure, choice of partial or total condenser, use of a single or multiple feeds and their location, possibility of withdrawing some liquid streams (called side streams) from stages within the column (e.g., in petroleum refining it may be advantageous to take a stream rich in mid-range components that are less volatile than the condenser distillate product but more volatile than the bottoms product), the type of reboiler, etc. Here we will consider only a simple distillation column with a single feed, the distillate, and the bottoms as the only product streams, and the use of a total condenser. Even in this case there are many choices and it might seem difficult to know where to begin. The shortcut analysis of the previous chapter is especially useful in providing initial guesses for the number of stages, the reflux ratio, and the feed location, and will be used in that way here.

We will do rigorous distillation calculation using the **RadFrac** block selected from the **Columns** tab in the **Model Palette** (Fig. 11-1).

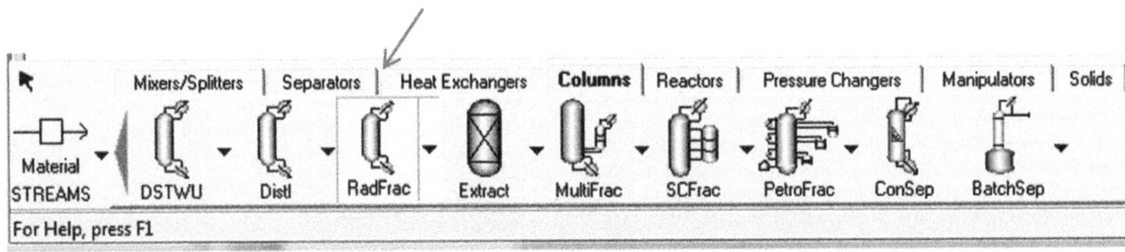

Figure 11-1

Aspen Plus® documentation describes **RadFrac** as "Rigorous 2- or 3-phase fractionation for single columns." It is a rating program for distillation, not a design program to meet composition specifications. That is, the user must supply all the column parameters such as the number of stages, the reflux ratio, the location of the feed stage, the type of condenser, and reboiler, and then **RadFrac** is used to compute the distillate and bottoms products for that distillation configuration. If the goal is to meet certain purity specifications, then repeated

Using Aspen Plus® in Thermodynamics Instruction: A Step-by-Step Guide, First Edition. Stanley I. Sandler.

calculations with varying column parameters will likely be needed until the composition specifications are met.

The flow sheet for a single distillation column using **RadFrac** is shown in Fig. 11-2.

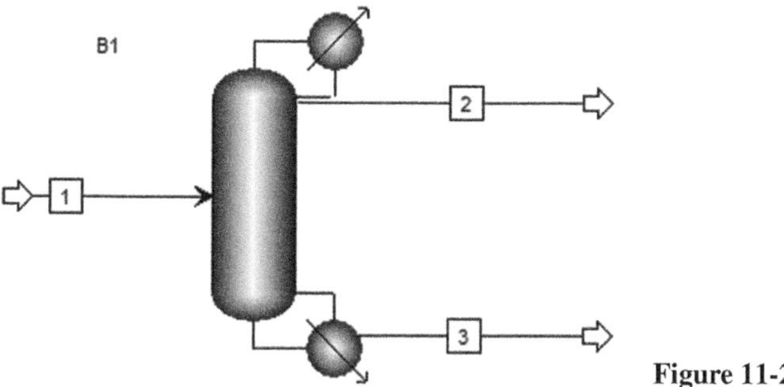

Figure 11-2

Here, as an example, we will consider the same problem as in the previous chapter of starting with an equimolar benzene–toluene mixture to produce a distillate containing 90 mol% benzene and a bottoms product containing 98 mol% toluene. From the mass balances of the previous chapter, the distillate flow D will be 0.5455 mols/mol of feed, and the flow of the bottoms product B will be 0.4545 mols/mol of feed.

The initial steps are to go through the usual **Setup** procedure in **Properties**, entering the components by adding benzene and toluene in **Components**, and then choosing **Ideal** under **Methods**. Next go to **Simulation.** Since **RADFRAC** has already been chosen above go to **Streams>1** to specify the feed composition and state (Fig. 11-3).

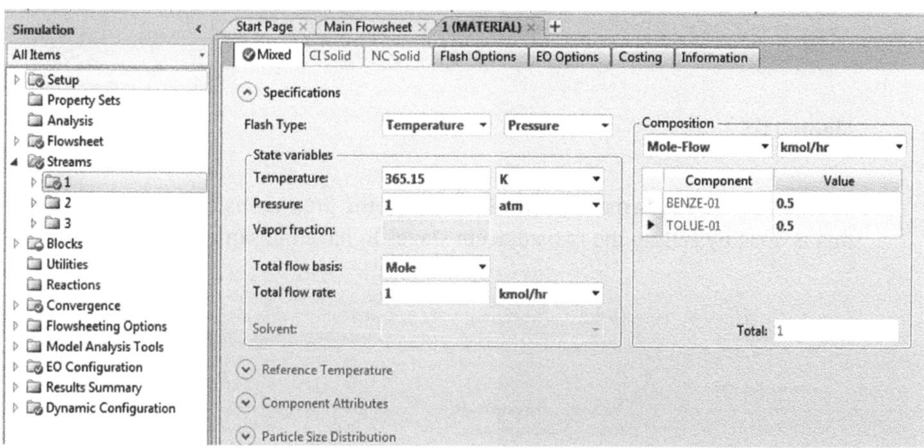

Figure 11-3

Note that the temperature entered is just below the saturation temperature to ensure that the feed is all liquid.

The next step is to set the initial guess specifications for the column, **Block B1**. The shortcut estimates (of the previous chapter) using the block **DSTWU** were $R_{\text{min}} = 0.8848$, $N_{\text{nim}} = 6.79$. For $R = 1.2R_{\text{min}} = 1.06175$, from the **DSTWU** results, the initial guess for the column configuration is 16 stages with the feed at stage 8, and these estimates are used in specifying the block parameters as shown in Fig. 11-4.

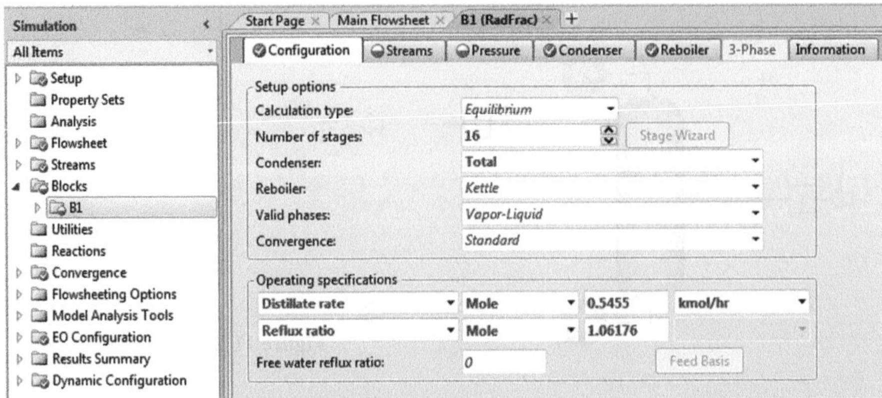

Figure 11-4

Next, go to the **Streams** tab and populate the window as in Fig. 11-5.

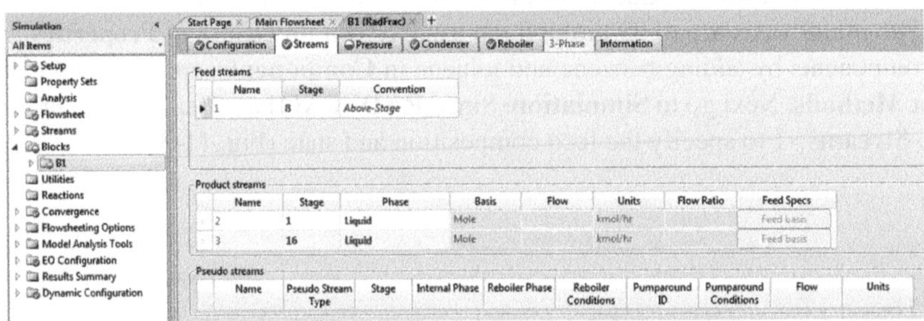

Figure 11-5

Then go to the **Pressure** tab and indicate that the pressure is constant in the column (i.e., this is done by setting the two adjacent stages to have the same pressure) as in Fig. 11-6.

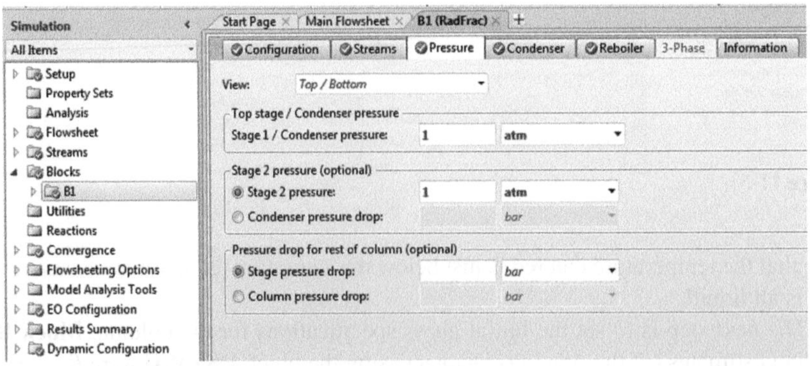

View: Top / Bottom
Top stage / Condenser pressure
Stage 1 / Condenser pressure: 1 atm
Stage 2 pressure (optional)
Stage 2 pressure: 1 atm
Condenser pressure drop: bar
Pressure drop for rest of column (optional)
Stage pressure drop: bar
Column pressure drop: bar

Figure 11-6

For simplicity, here we have not considered a pressure change across the column. In fact, because of the hydrostatic head due to the liquid hold up on each stage, a more realistic design might include perhaps a 0.5 psia (or less) change in pressure per stage.

Finally, still in the **Simulation** mode, go to **Setup>Report Options>Stream** and click on **Mole** under **Fraction basis** (Fig. 11-7) so that mole fractions appear in the output.

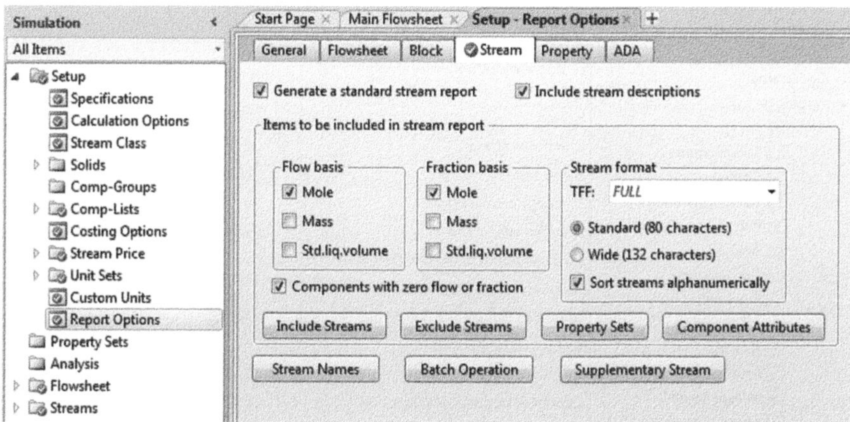

Figure 11-7

Since there are no remaining red flags (unspecified parameters), run the calculation leading to the following results, only some of which are shown are shown in Fig. 11-8.

Simulation — All Items: Setup, Property Sets, Analysis, Flowsheet, Streams, Blocks, Utilities, Reactions, Convergence, Flowsheeting Options, Model Analysis Tools, EO Configuration, Results Summary (Run Status, Streams, Convergence, Operating Costs, CO2 Emissions, Streams (Custom), Models, Equipment), Dynamic Configuration

Start Page | Main Flowsheet | Results Summary - Streams

Material | Heat | Load | Work | Vol.% Curves | Wt. % Curves | Petro. Curves | Poly. C

Display: All streams Format: FULL Stream Table Copy All

	1	2	3
Substream: MIXED			
Mole Flow kmol/hr			
BENZE-01	0.5	0.486316	0.0136843
TOLUE-01	0.5	0.0591843	0.440816
Mole Frac			
BENZE-01	0.5	0.891504	0.0301086
TOLUE-01	0.5	0.108496	0.969891
Total Flow kmol/hr	1	0.5455	0.4545
Total Flow kg/hr	85.1271	43.4412	41.6859
Total Flow l/min	1.77539	0.892692	0.888465
Temperature C	92	82.3787	109.275
Pressure bar	1.01325	1.01325	1.01325
Vapor Frac	0	0	0
Liquid Frac	1	1	1
Solid Frac	0	0	0
Enthalpy cal/mol	9858.58	12796.4	6669.33

Figure 11-8

From the **Blocks>B1>Summary**, a portion of which is shown in Fig. 11-9, we see that for a flow of 1 kmol/hr, the reboiler heat duty is 2388.4 cal/sec and the condenser cooling duty is 2345.8 cal/sec. This information is needed to design these two pieces of heat exchange equipment if this column configuration is used.

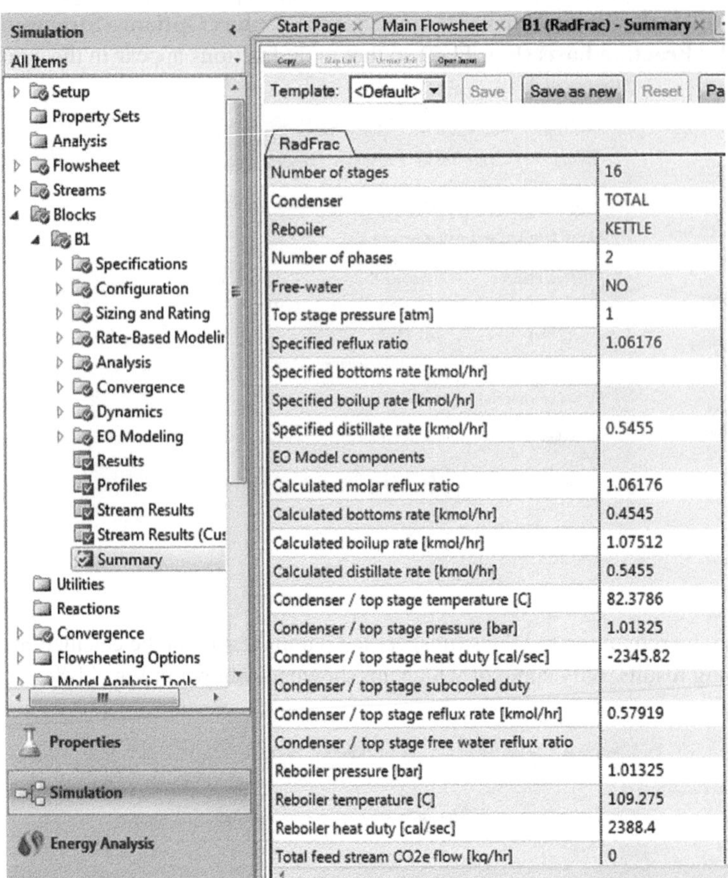

Figure 11-9

Now going to **Blocks>B1>Profiles>TPFQ** the column stage-by-stage temperature, pressure, liquid, and vapor flow rate profiles are obtained and displayed in Fig. 11-10.

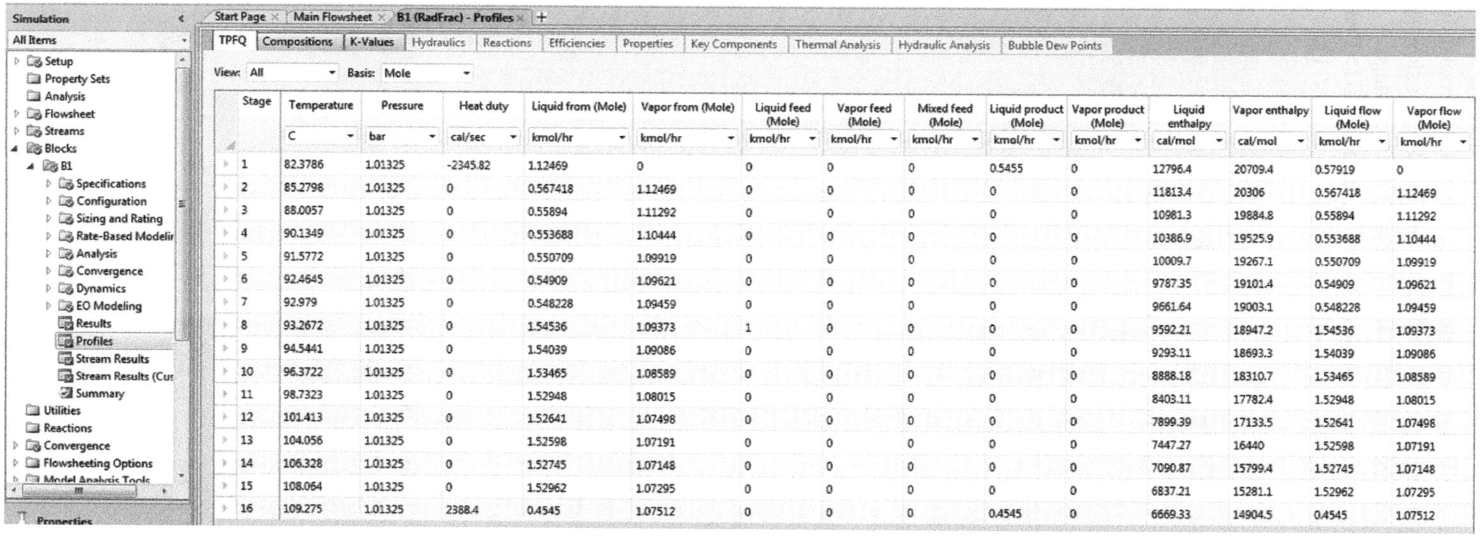

View: All Basis: Mole

Stage	Temperature	Pressure	Heat duty	Liquid from (Mole)	Vapor from (Mole)	Liquid feed (Mole)	Vapor feed (Mole)	Mixed feed (Mole)	Liquid product (Mole)	Vapor product (Mole)	Liquid enthalpy	Vapor enthalpy	Liquid flow (Mole)	Vapor flow (Mole)
	C	bar	cal/sec	kmol/hr	kmol/hr	kmol/hr	kmol/hr	kmol/hr	kmol/hr	kmol/hr	cal/mol	cal/mol	kmol/hr	kmol/hr
1	82.3786	1.01325	-2345.82	1.12469	0	0	0	0	0.5455	0	12796.4	20709.4	0.57919	0
2	85.2798	1.01325	0	0.567418	1.12469	0	0	0	0	0	11813.4	20306	0.567418	1.12469
3	88.0057	1.01325	0	0.55894	1.11292	0	0	0	0	0	10981.3	19884.8	0.55894	1.11292
4	90.1349	1.01325	0	0.553688	1.10444	0	0	0	0	0	10386.9	19525.9	0.553688	1.10444
5	91.5772	1.01325	0	0.550709	1.09919	0	0	0	0	0	10009.7	19267.1	0.550709	1.09919
6	92.4645	1.01325	0	0.54909	1.09621	0	0	0	0	0	9787.35	19101.4	0.54909	1.09621
7	92.979	1.01325	0	0.548228	1.09459	0	0	0	0	0	9661.64	19003.1	0.548228	1.09459
8	93.2672	1.01325	0	1.54536	1.09373	1	0	0	0	0	9592.21	18947.2	1.54536	1.09373
9	94.5441	1.01325	0	1.54039	1.09086	0	0	0	0	0	9293.11	18693.3	1.54039	1.09086
10	96.3722	1.01325	0	1.53465	1.08589	0	0	0	0	0	8888.18	18310.7	1.53465	1.08589
11	98.7323	1.01325	0	1.52948	1.08015	0	0	0	0	0	8403.11	17782.4	1.52948	1.08015
12	101.413	1.01325	0	1.52641	1.07498	0	0	0	0	0	7899.39	17133.5	1.52641	1.07498
13	104.056	1.01325	0	1.52598	1.07191	0	0	0	0	0	7447.27	16440	1.52598	1.07191
14	106.328	1.01325	0	1.52745	1.07148	0	0	0	0	0	7090.87	15799.4	1.52745	1.07148
15	108.064	1.01325	0	1.52962	1.07295	0	0	0	0	0	6837.21	15281.1	1.52962	1.07295
16	109.275	1.01325	2388.4	0.4545	1.07512	0	0	0	0.4545	0	6669.33	14904.5	0.4545	1.07512

Figure 11-10

Note the stages are numbered starting from the condenser as stage 1. Of the 16 stages, the total condenser of stage 1 is not an equilibrium stage, so there are 15 equilibrium stages. Looking at the liquid flow rates we see that that feed enters at stage 8 (actually the seventh stage below the condenser), and that unlike the constant molal overflow assumption, there is a small stage-to-stage variation in the liquid and vapor flow rates across the column in addition to the step change in the liquid flow at the feed stage.

By clicking on the **Compositions** tab (i.e., **Blocks>B1>Profiles>Compositions**) the vapor and liquid composition profiles are shown in Fig. 11-11.

TPFQ | Compositions | K-Values | Hydraulics

View: Vapor Basis: Mole

Stage	BENZE-01	TOLUE-01
1	0.955009	0.0449912
2	0.891504	0.108496
3	0.825834	0.174166
4	0.770239	0.229761
5	0.730328	0.269672
6	0.704843	0.295157
7	0.689736	0.310264
8	0.681163	0.318837
9	0.642256	0.357744
10	0.583796	0.416204
11	0.503338	0.496662
12	0.404831	0.595169
13	0.299876	0.700124
14	0.203128	0.796872
15	0.124974	0.875026
16	0.0682259	0.931774

TPFQ | Compositions | K-Values | Hydraulics

View: Liquid Basis: Mole

Stage	BENZE-01	TOLUE-01
1	0.891504	0.108496
2	0.762699	0.237301
3	0.651888	0.348112
4	0.571534	0.428466
5	0.519945	0.480055
6	0.489283	0.510717
7	0.471866	0.528134
8	0.462218	0.537782
9	0.420425	0.579575
10	0.363184	0.636816
11	0.293477	0.706523
12	0.21955	0.78045
13	0.151595	0.848405
14	0.0967461	0.903254
15	0.0569	0.9431
16	0.0301086	0.969891

Figure 11-11

These results can easily be plotted by going to **Plot** on the main toolbar and clicking first on **Temp** for temperature (Fig. 11-12).

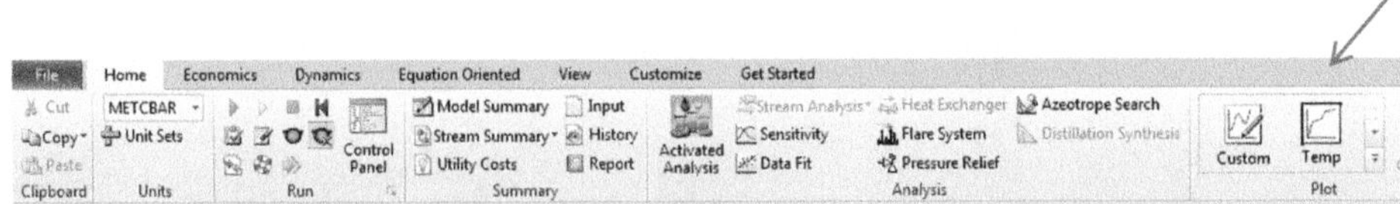

Figure 11-12

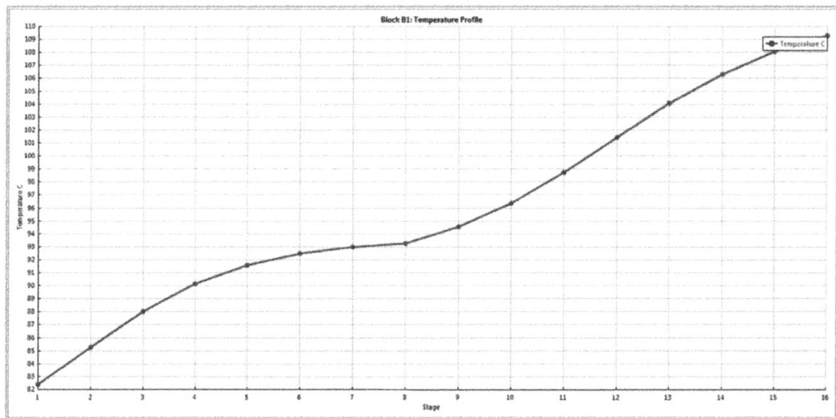

Figure 11-12 (*Continued*)

Now putting the cursor in the **Plot** area of the main toolbar, clicking on the down arrow, and choosing the **Comp** (composition) template brings up the menu in Fig. 11-13 in which the choices that have been made are to plot the liquid compositions of both components that produces the plot in Fig. 11-14.

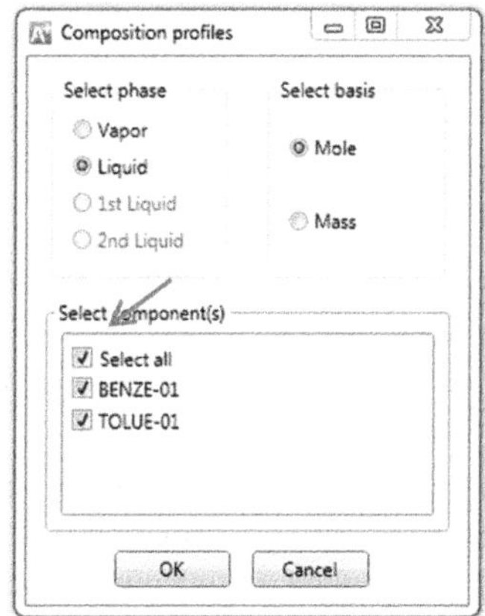

Figure 11-13

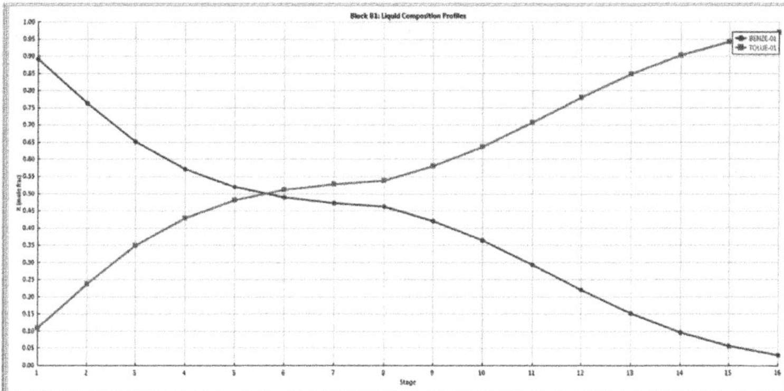

Figure 11-14

Repeat for the vapor compositions by clicking on the **vapor** button in the window shown in Fig. 11-15.

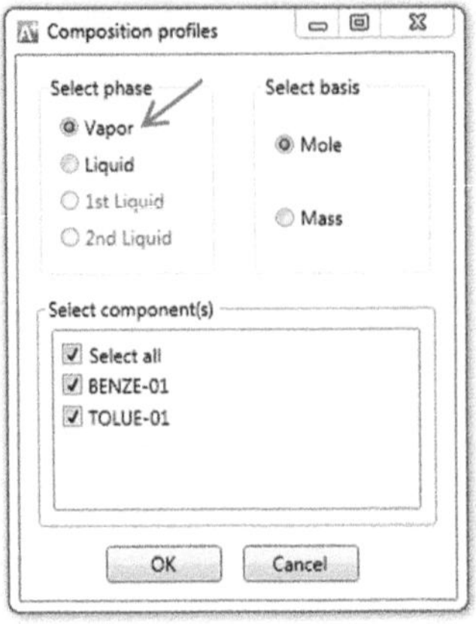

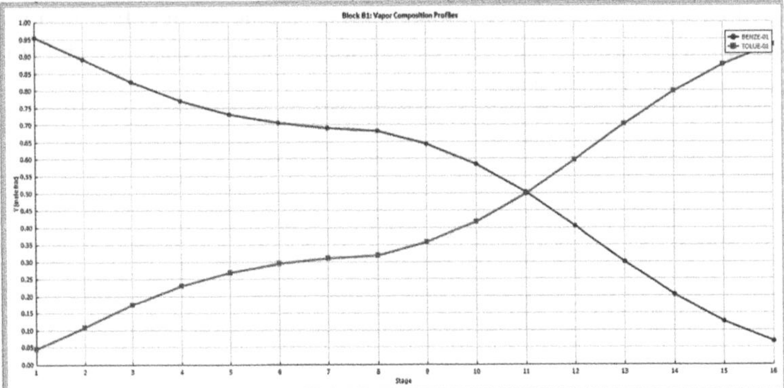

Figure 11-15

We see that the **RadFrac** block in Aspen Plus gives a lot of information about the distillation column. However, what is also seen from the results of this rating calculation is that while the shortcut design column parameters provided a very good initial guess, the results do not quite meet the required compositions. In particular, the benzene concentration is a bit low in the distillate (0.8915 instead of 0.9), and too high in the bottoms product (0.0301 instead of 0.02). So now we have to guess how to change the column parameters, for example, the number of stages, the reflux ratio, and/or the feed location to meet the desired concentrations.

We will try increasing the reflux ratio to 1.2 and reducing the number of stages to 15. However, before running the new calculation, it is advisable to first click on **Reinitialize** symbol in the Run area of the main toolbar as indicated in Fig. 11-16.

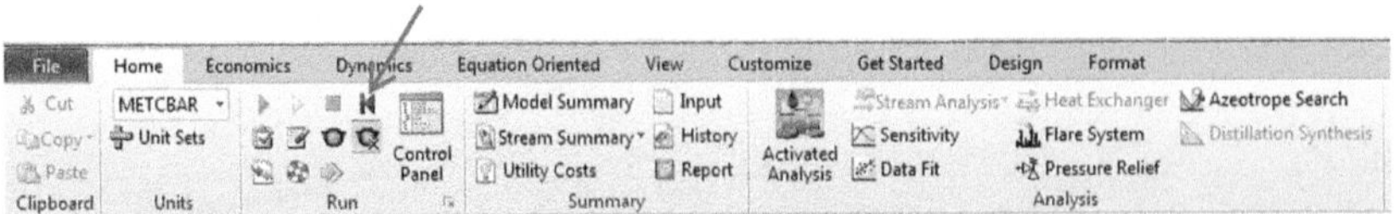

Figure 11-16

The reason is that this and other Aspen Plus calculations are usually done by iteration because of the complexity and nonlinearity of the coupled equations, and the program generally uses the previous solution (even though it was for a different set of conditions) as the starting guess for the new solution. This may be a particularly bad starting guess for the new problem specifications. Therefore, it is best to first reinitialize Aspen Plus calculations, which discards the previous solution, before attempting a new solution.

The new specification for the distillation block **B1** is as shown in Fig. 11-17.

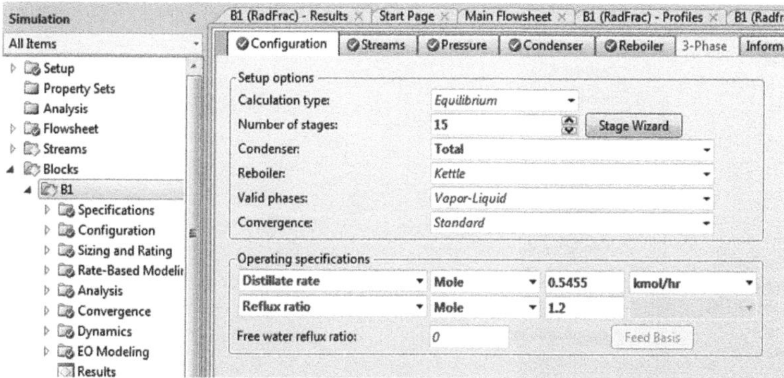

Figure 11-17

The results for this new column configuration are given in Fig. 11-18.

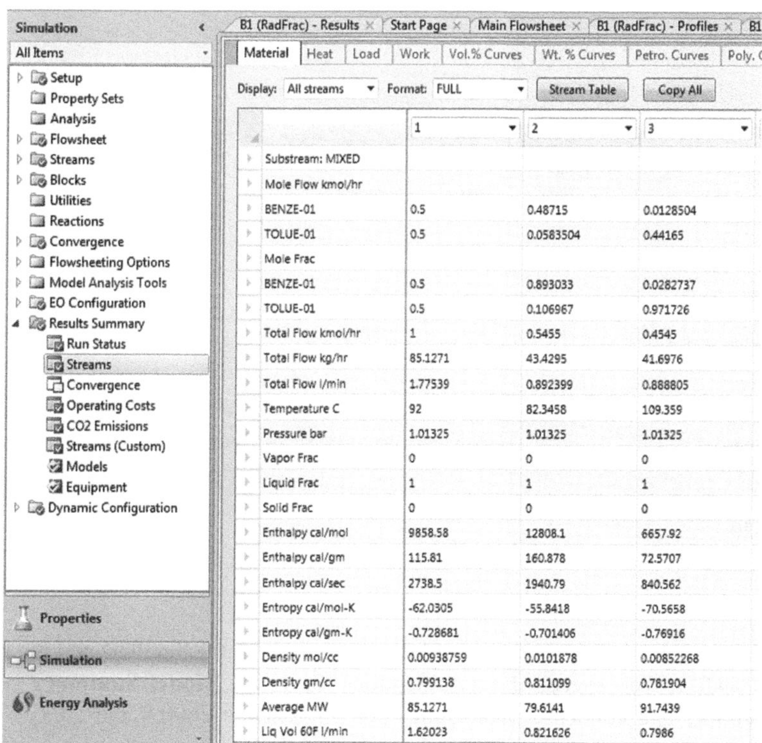

	1	2	3
Substream: MIXED			
Mole Flow kmol/hr			
BENZE-01	0.5	0.48715	0.0128504
TOLUE-01	0.5	0.0583504	0.44165
Mole Frac			
BENZE-01	0.5	0.893033	0.0282737
TOLUE-01	0.5	0.106967	0.971726
Total Flow kmol/hr	1	0.5455	0.4545
Total Flow kg/hr	85.1271	43.4295	41.6976
Total Flow l/min	1.77539	0.892399	0.888805
Temperature C	92	82.3458	109.359
Pressure bar	1.01325	1.01325	1.01325
Vapor Frac	0	0	0
Liquid Frac	1	1	1
Solid Frac	0	0	0
Enthalpy cal/mol	9858.58	12808.1	6657.92
Enthalpy cal/gm	115.81	160.878	72.5707
Enthalpy cal/sec	2738.5	1940.79	840.562
Entropy cal/mol-K	-62.0305	-55.8418	-70.5658
Entropy cal/gm-K	-0.728681	-0.701406	-0.76916
Density mol/cc	0.00938759	0.0101878	0.00852268
Density gm/cc	0.799138	0.811099	0.781904
Average MW	85.1271	79.6141	91.7439
Liq Vol 60F l/min	1.62023	0.821626	0.7986

Figure 11-18

Benzene mole fraction in the distillate has improved, but again is a little too low in the distillate, and a little too high in the bottoms product. The next guess is to increase the reflux ratio to 1.25 while keeping the number of stages unchanged, and moving the feed to stage 7 (see Figs. 11-19 and 11-20).

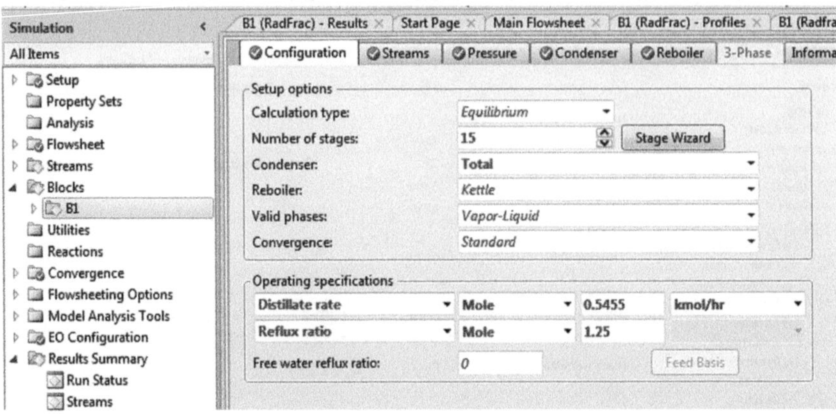

Figure 11-19

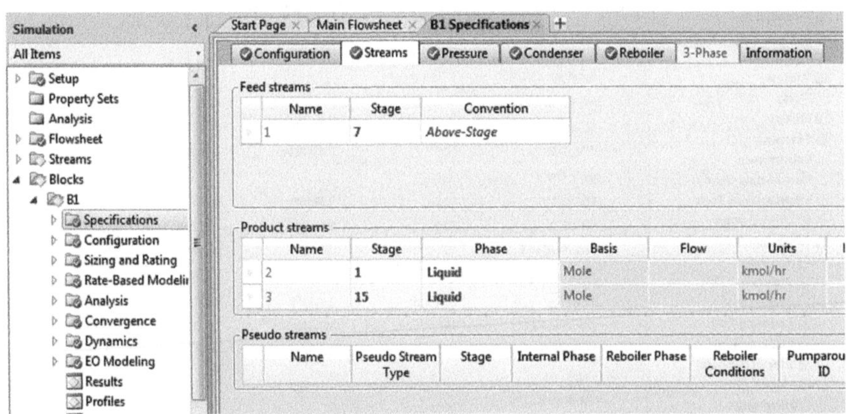

Figure 11-20

We see in the results that the benzene mole fractions in both the distillate and the bottoms now meet and, in fact, exceed the specifications. From the **Model Summary** (Fig. 11-21), the reboiler and condenser heat duties per kmol/hr are 2600.8 and 2556.3 cal/sec, respectively, which are higher than for the column with a lower reflux ratio. [Can you explain why?]

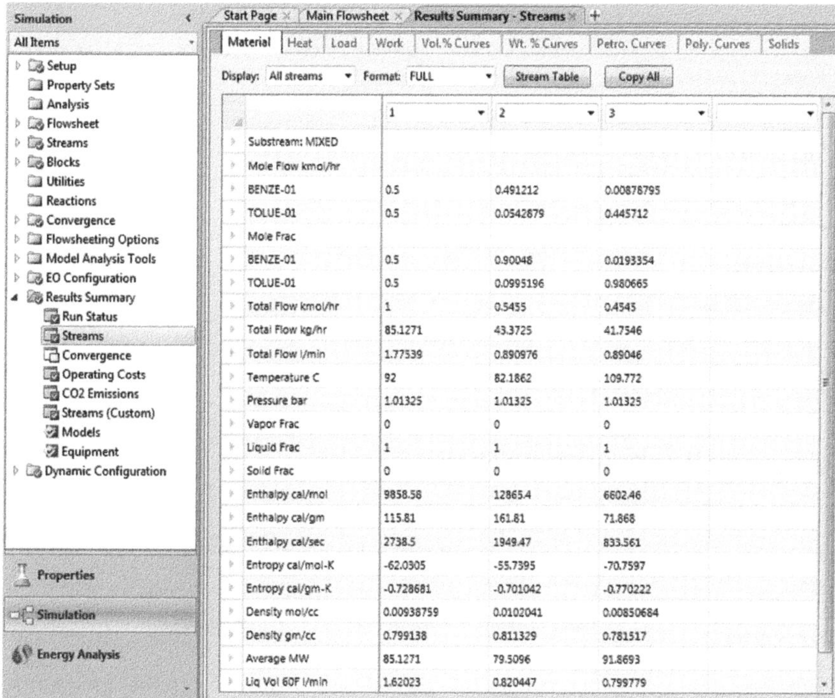

	1	2	3
Substream: MIXED			
Mole Flow kmol/hr			
BENZE-01	0.5	0.491212	0.00878795
TOLUE-01	0.5	0.0542879	0.445712
Mole Frac			
BENZE-01	0.5	0.90048	0.0193354
TOLUE-01	0.5	0.0995196	0.980665
Total Flow kmol/hr	1	0.5455	0.4545
Total Flow kg/hr	85.1271	43.3725	41.7546
Total Flow l/min	1.77539	0.890976	0.89046
Temperature C	92	82.1862	109.772
Pressure bar	1.01325	1.01325	1.01325
Vapor Frac	0	0	0
Liquid Frac	1	1	1
Solid Frac	0	0	0
Enthalpy cal/mol	9858.58	12865.4	6602.46
Enthalpy cal/gm	115.81	161.81	71.868
Enthalpy cal/sec	2738.5	1949.47	833.561
Entropy cal/mol-K	-62.0305	-55.7395	-70.7597
Entropy cal/gm-K	-0.728681	-0.701042	-0.770222
Density mol/cc	0.00938759	0.0102041	0.00850684
Density gm/cc	0.799138	0.811329	0.781517
Average MW	85.1271	79.5096	91.8693
Liq Vol 60F l/min	1.62023	0.820447	0.799779

Figure 11-21

However, this column design (number of stages, reflux ratio, and feed stage location) is not the only one that will meet the specifications. For example, the different set of column specifications (for example lower reflux ratio and increased number of stages) in Fig. 11-22 also meets the requirements, as can be seen in the results in Fig. 11-23.

Setup options
Calculation type: Equilibrium
Number of stages: 18
Condenser: Total
Reboiler: Kettle
Valid phases: Vapor-Liquid
Convergence: Standard
Operating specifications
Distillate rate Mole 0.5455 kmol/hr
Reflux ratio Mole 1.115
Free water reflux ratio: 0

Figure 11-22

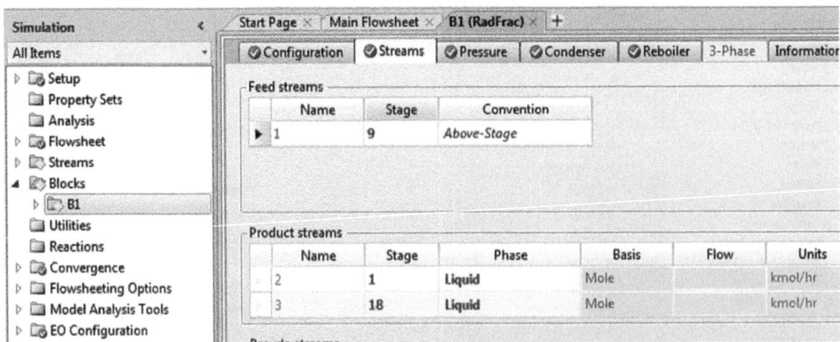

Figure 11-22 *(Continued)*

Looking at the Results Summary>Streams

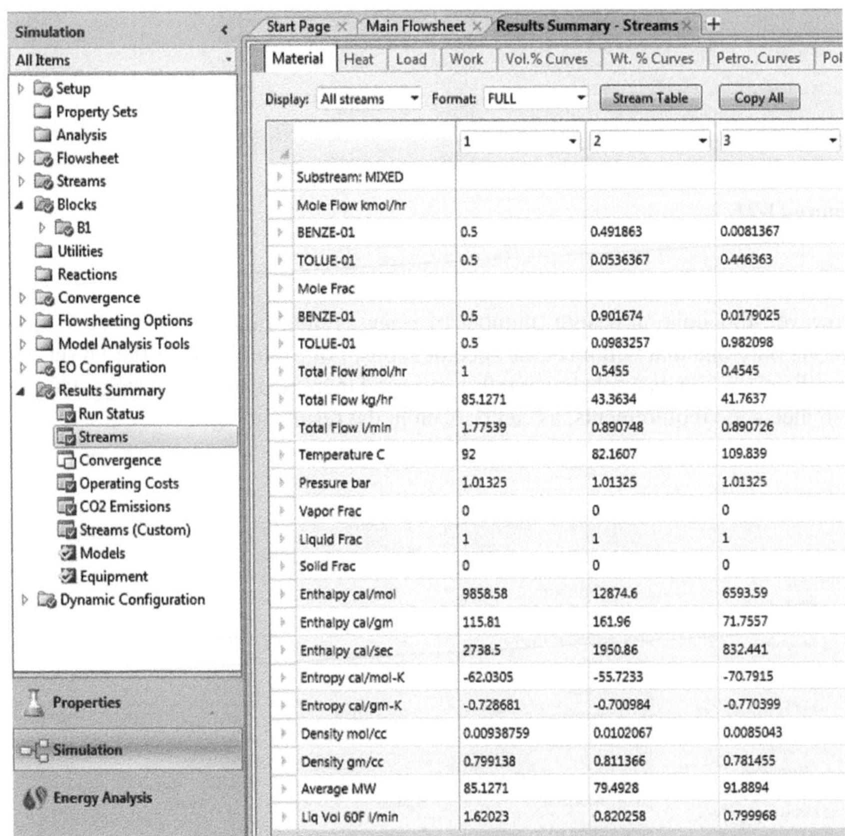

	1	2	3
Substream: MIXED			
Mole Flow kmol/hr			
BENZE-01	0.5	0.491863	0.0081367
TOLUE-01	0.5	0.0536367	0.446363
Mole Frac			
BENZE-01	0.5	0.901674	0.0179025
TOLUE-01	0.5	0.0983257	0.982098
Total Flow kmol/hr	1	0.5455	0.4545
Total Flow kg/hr	85.1271	43.3634	41.7637
Total Flow l/min	1.77539	0.890748	0.890726
Temperature C	92	82.1607	109.839
Pressure bar	1.01325	1.01325	1.01325
Vapor Frac	0	0	0
Liquid Frac	1	1	1
Solid Frac	0	0	0
Enthalpy cal/mol	9858.58	12874.6	6593.59
Enthalpy cal/gm	115.81	161.96	71.7557
Enthalpy cal/sec	2738.5	1950.86	832.441
Entropy cal/mol-K	-62.0305	-55.7233	-70.7915
Entropy cal/gm-K	-0.728681	-0.700984	-0.770399
Density mol/cc	0.00938759	0.0102067	0.0085043
Density gm/cc	0.799138	0.811366	0.781455
Average MW	85.1271	79.4928	91.8894
Liq Vol 60F l/min	1.62023	0.820258	0.799968

Figure 11-23

and then **Blocks>B1>Profiles>TPFQ** and **Summary** provides the table in Fig. 11-24 that includes the temperature, pressure, and liquid and vapor compositions that are in equilibrium and leaving each stage.

View: Summary Basis: Mole

Stage	Temperature C	Pressure bar	Heat duty cal/sec	Liquid from (Mole) kmol/hr	Vapor from (Mole) kmol/hr
1	82.1607	1.01325	-2402.49	1.15373	0
2	84.8337	1.01325	0	0.59637	1.15373
3	87.4756	1.01325	0	0.58723	1.14187
4	89.6575	1.01325	0	0.581251	1.13273
5	91.2159	1.01325	0	0.577795	1.12675
6	92.2205	1.01325	0	0.575854	1.1233
7	92.8267	1.01325	0	0.574811	1.12135
8	93.1784	1.01325	0	0.574283	1.12031
9	93.3783	1.01325	0	1.57119	1.11978
10	94.6845	1.01325	0	1.5662	1.11669
11	96.5285	1.01325	0	1.56074	1.1117
12	98.8761	1.01325	0	1.55604	1.10624
13	101.507	1.01325	0	1.5532	1.10154
14	104.072	1.01325	0	1.55271	1.0987
15	106.258	1.01325	0	1.554	1.09821
16	107.921	1.01325	0	1.5559	1.0995
17	109.08	1.01325	0	1.55767	1.1014
18	109.839	1.01325	2447.28	0.4545	1.10317

Figure 11-24

Now plotting the liquid compositions across the column as done described earlier results in Fig. 11-25.

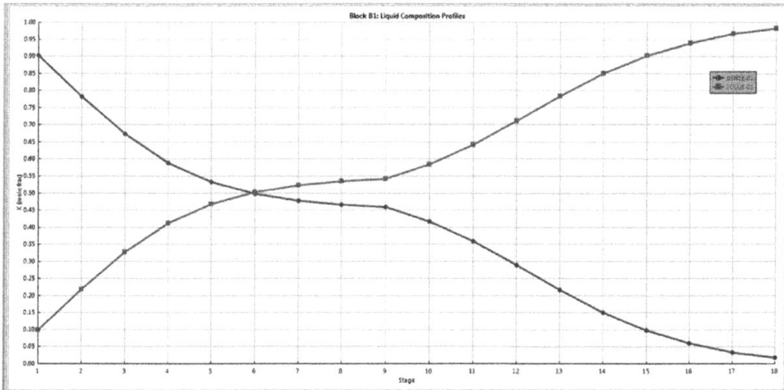

Figure 11-25

Then choosing instead **Temp** produces the following temperature profile plot across the column in Fig. 11-26.

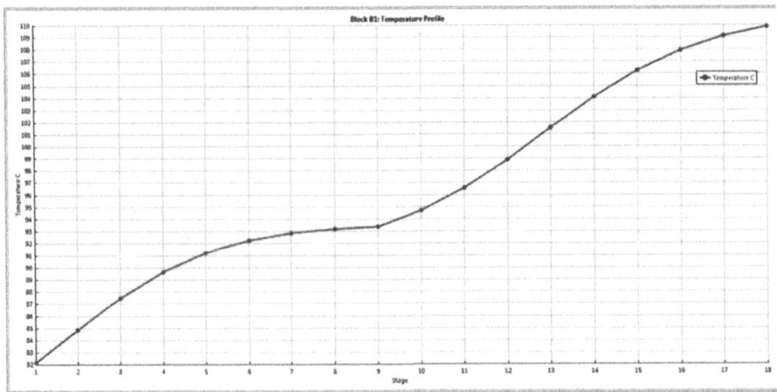

Figure 11-26

Of course, other quite different column parameters can also meet the specified compositions. For example, for a reflux ratio of $R = 3.0$, the shortcut estimate from the block **DSTWU** suggests there should be nine stages (one of which is the total condenser) and the fourth stage (third stage below the total condenser) should be the feed stage. [Note that we have rounded the estimates to the nearest whole number. For example, the **DSTWU** estimate is that the feed stage is 4.038, which we have taken to be 4. This was also done for the number of stages.] See Figs. 11-27 to 11-30 for the results of this column configuration.

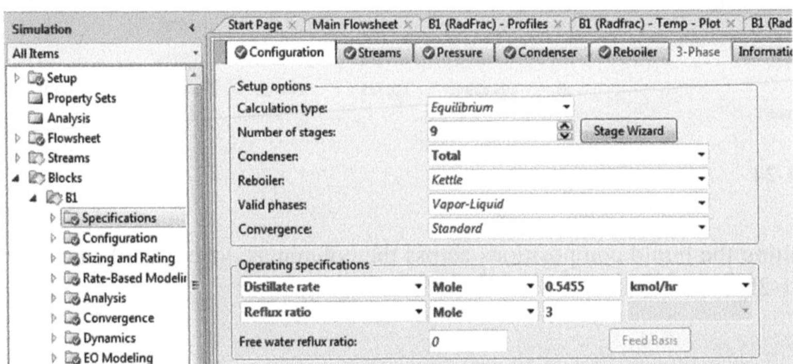

Figure 11-27

Next, go to the **Streams** tab and set the feed stage

Simulation | All Items | Start Page | Main Flowsheet | B1 (RadFrac) - Profiles | B1 (Radfrac) - Temp - Plot | B1 (Radfrac) -

Configuration | Streams | Pressure | Condenser | Reboiler | 3-Phase | Information

Feed streams

Name	Stage	Convention
1	3	Above-Stage

Product streams

Name	Stage	Phase	Basis	Flow	Units
2	1	Liquid	Mole		kmol/hr
3	9	Liquid	Mole		kmol/hr

Pseudo streams

Name	Pseudo Stream Type	Stage	Internal Phase	Reboiler Phase	Reboiler Conditions	Pumparo ID

Figure 11-28

then to the **Pressure** tab,

Figure 11-29

and finally, running the simulation produces the following results.

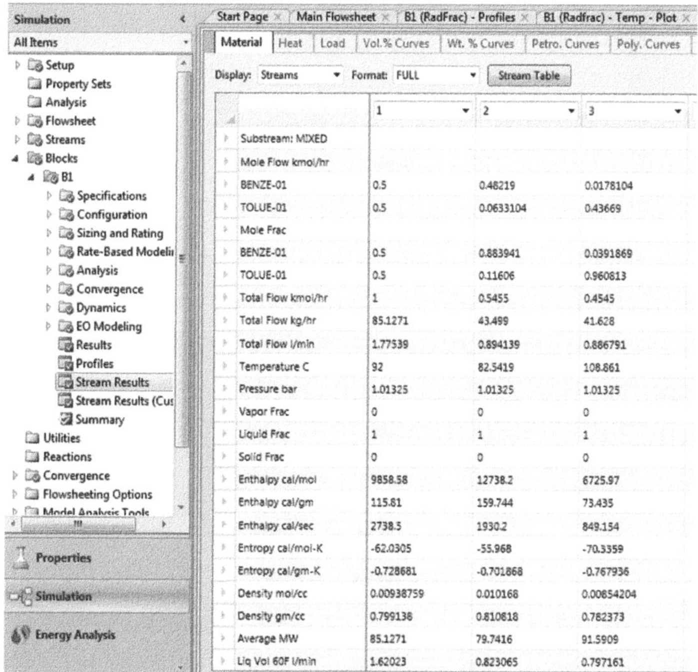

	1	2	3
Substream: MIXED			
Mole Flow kmol/hr			
BENZE-01	0.5	0.48219	0.0178104
TOLUE-01	0.5	0.0633104	0.43669
Mole Frac			
BENZE-01	0.5	0.883941	0.0391869
TOLUE-01	0.5	0.11606	0.960813
Total Flow kmol/hr	1	0.5455	0.4545
Total Flow kg/hr	85.1271	43.499	41.628
Total Flow l/min	1.77539	0.894139	0.886791
Temperature C	92	82.5419	108.861
Pressure bar	1.01325	1.01325	1.01325
Vapor Frac	0	0	0
Liquid Frac	1	1	1
Solid Frac	0	0	0
Enthalpy cal/mol	9858.58	12738.2	6725.97
Enthalpy cal/gm	115.81	159.744	73.435
Enthalpy cal/sec	2738.5	1930.2	849.154
Entropy cal/mol-K	-62.0305	-55.968	-70.3359
Entropy cal/gm-K	-0.728681	-0.701868	-0.767936
Density mol/cc	0.00938759	0.010168	0.00854204
Density gm/cc	0.799138	0.810818	0.782373
Average MW	85.1271	79.7416	91.5909
Liq Vol 60F l/min	1.62023	0.823065	0.797161

Figure 11-30

What is seen from the results of this rating calculation is that while the shortcut design column specification again provided a reasonable initial guess, it does not quite meet the required product compositions. In particular, the benzene concentration is a bit low in the distillate (0.889 instead of 0.9), and is too high in the bottoms product (0.03277 instead of 0.02). So again one has to guess how to change the specifications, for example, using additional trays, a higher reflux ratio, and/or a change in feed location to meet the desired concentrations. Adding two stages and moving the feed down one stage (Figs. 11-31 and 11-32) gives the results in Fig. 11-33.

Setup options

Calculation type:	Equilibrium
Number of stages:	11
Condenser:	Total
Reboiler:	Kettle
Valid phases:	Vapor-Liquid
Convergence:	Standard

Operating specifications

Distillate rate	Mole	0.5455	kmol/hr
Reflux ratio	Mole	3	
Free water reflux ratio:	0		

Figure 11-31

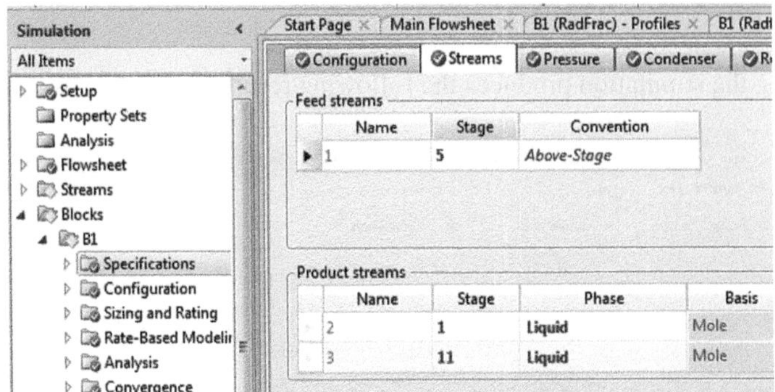

Feed streams

Name	Stage	Convention
1	5	Above-Stage

Product streams

Name	Stage	Phase	Basis
2	1	Liquid	Mole
3	11	Liquid	Mole

Figure 11-32

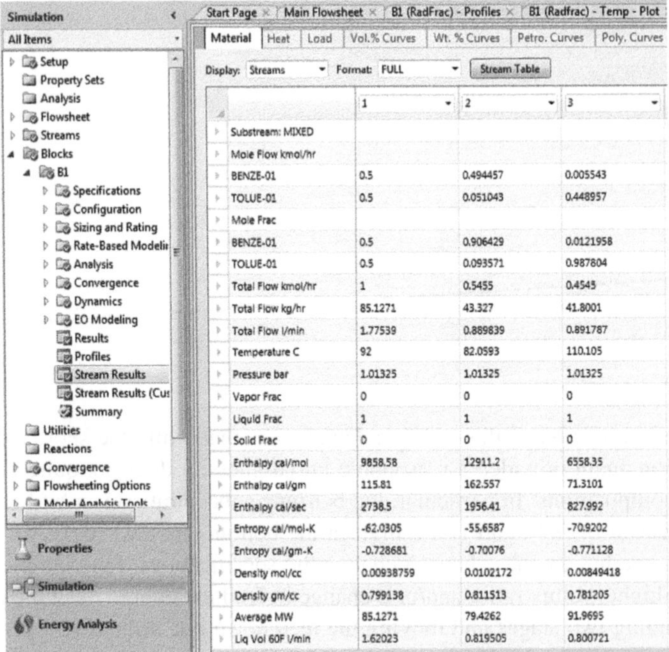

Display: Streams Format: FULL

	1	2	3
Substream: MIXED			
Mole Flow kmol/hr			
BENZE-01	0.5	0.494457	0.005543
TOLUE-01	0.5	0.051043	0.448957
Mole Frac			
BENZE-01	0.5	0.906429	0.0121958
TOLUE-01	0.5	0.093571	0.987804
Total Flow kmol/hr	1	0.5455	0.4545
Total Flow kg/hr	85.1271	43.327	41.8001
Total Flow l/min	1.77539	0.889839	0.891787
Temperature C	92	82.0593	110.105
Pressure bar	1.01325	1.01325	1.01325
Vapor Frac	0	0	0
Liquid Frac	1	1	1
Solid Frac	0	0	0
Enthalpy cal/mol	9858.58	12911.2	6558.35
Enthalpy cal/gm	115.81	162.557	71.3101
Enthalpy cal/sec	2738.5	1956.41	827.992
Entropy cal/mol-K	-62.0305	-55.6587	-70.9202
Entropy cal/gm-K	-0.728681	-0.70076	-0.771128
Density mol/cc	0.00938759	0.0102172	0.00849418
Density gm/cc	0.799138	0.811513	0.781205
Average MW	85.1271	79.4262	91.9695
Liq Vol 60F l/min	1.62023	0.819505	0.800721

Figure 11-33

This set of column parameters exceeds the product composition requirements in that the benzene distillate mole fraction is 0.9064 above the specified 0.9, and its bottom product composition is 0.0122, less than the required 0.02. One could then try other column specifications (e.g., a slightly smaller reflux ratio, or removing one stage from the column, to get closer to the desired compositions.) This is left to the reader to do.

The **Blocks>B1> Summary**, only a part of which is shown in Fig. 11-34, gives information on the reboiler heating and condenser cooling requirements for each 1 kmol/hr of feed for this last column configuration.

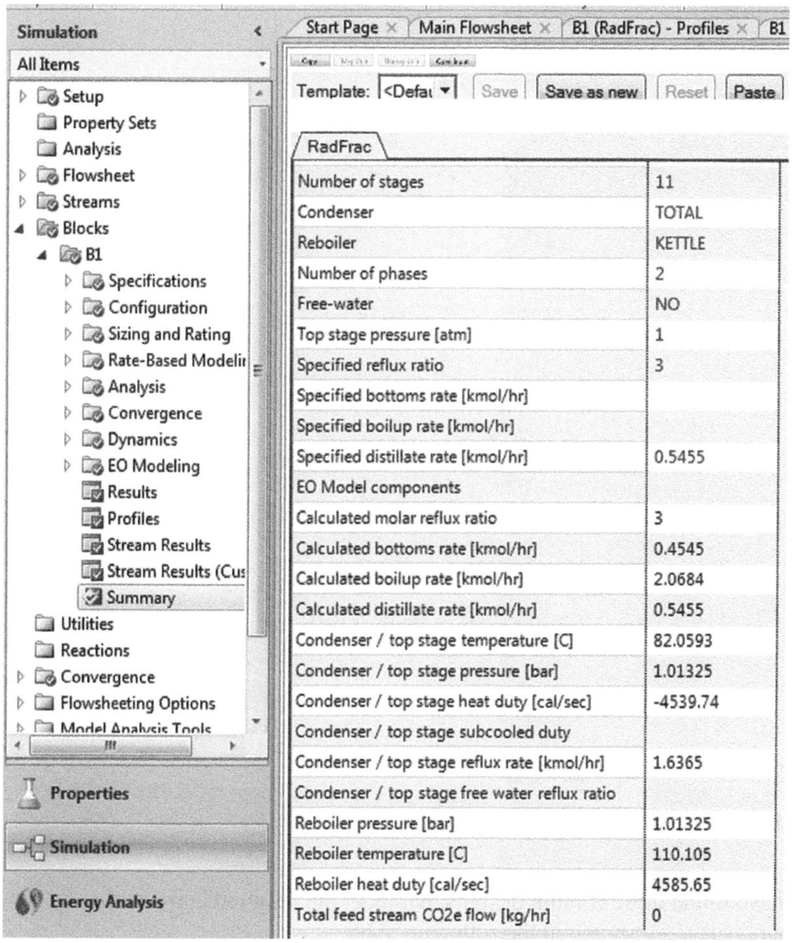
Simulation
All Items
Setup
Property Sets
Analysis
Flowsheet
Streams
Blocks
B1
Specifications
Configuration
Sizing and Rating
Analysis
Convergence
Dynamics
EO Modeling
Results
Profiles
Stream Results
Summary
Utilities
Reactions
Convergence
Flowsheeting Options
Properties
Simulation
Energy Analysis
Start Page
Main Flowsheet
B1 (RadFrac) - Profiles
Template:
Save
Save as new
Reset
Paste
RadFrac
Number of stages 11
Condenser TOTAL
Reboiler KETTLE
Number of phases 2
Free-water NO
Top stage pressure [atm] 1
Specified reflux ratio 3
Specified bottoms rate [kmol/hr]
Specified boilup rate [kmol/hr]
Specified distillate rate [kmol/hr] 0.5455
EO Model components
Calculated molar reflux ratio 3
Calculated bottoms rate [kmol/hr] 0.4545
Calculated boilup rate [kmol/hr] 2.0684
Calculated distillate rate [kmol/hr] 0.5455
Condenser / top stage temperature [C] 82.0593
Condenser / top stage pressure [bar] 1.01325
Condenser / top stage heat duty [cal/sec] -4539.74
Condenser / top stage subcooled duty
Condenser / top stage reflux rate [kmol/hr] 1.6365
Condenser / top stage free water reflux ratio
Reboiler pressure [bar] 1.01325
Reboiler temperature [C] 110.105
Reboiler heat duty [cal/sec] 4585.65
Total feed stream CO2e flow [kg/hr] 0

Figure 11-34

Now plotting the temperature profile (Fig. 11-35) and then for the vapor and liquid compositions at each stage (Fig. 11-36) we get

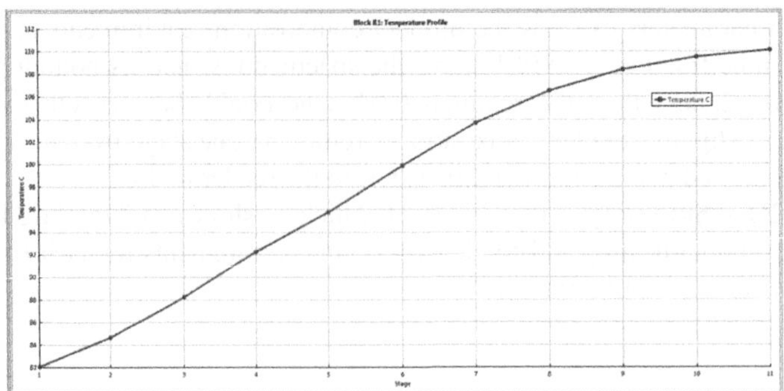

Figure 11-35

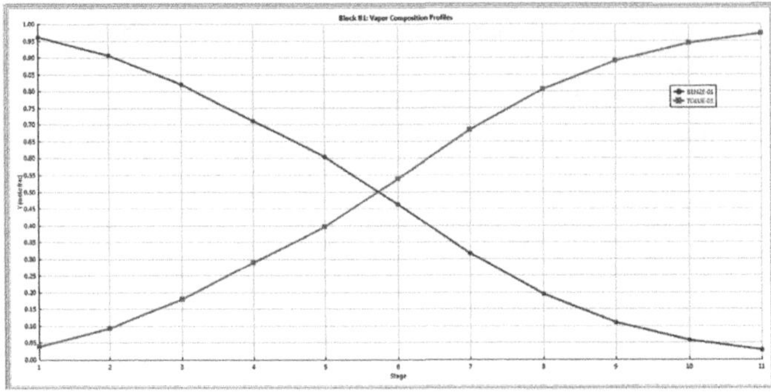

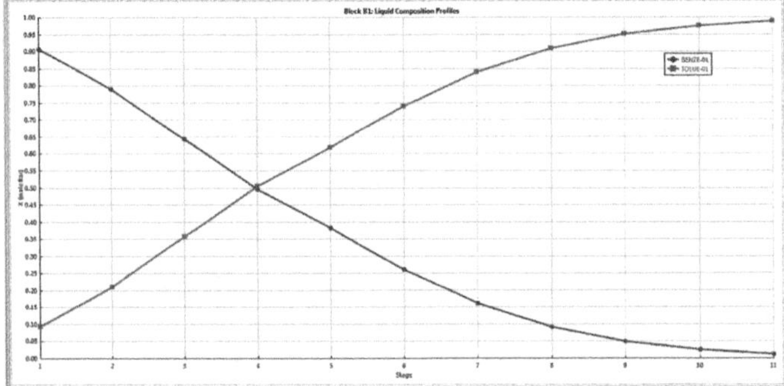

Figure 11-36

So we have found three column designs that meet the required composition specifications. Their parameters are shown in the following table:

	I	II	III
Reflux ratio	1.25	1.115	3.0
Number of stages	15	18	11
Feed stage	7	9	5
Reboiler heat duty (cal/sec)	2601	2447	4586
Condenser heat duty (cal/sec)	2556	2402	4540

In a similar fashion, we can find other choices of column parameters that would meet the composition specifications. So the question is which one of the many possible configurations should be used. To answer that question, one has to bring other factors to bear. If costs were the only consideration, then there would be a trade-off between capital costs (amortized over the life of the project) and operating (heat and cooling) costs. From the table above, we can see that as the reflux ratio increases, the number of stages (and therefore column height) decreases, thereby reducing that part of the capital cost. However, though not shown, the vapor and liquid flow rates increase, so a larger diameter column is needed, which increases the capital cost. Also as the reflux ratio increases, the amount of liquid per unit of product that needs to be boiled in the reboiler increases, as does the amount of vapor that needs to be condensed. These increase the capital costs of these two heat exchangers, and more importantly the ongoing energy costs of both units. [Note that increasing the reflux ratio to 3 from 1.25 almost doubled the energy requirements.]

The old rule-of-thumb, based on many analyses, is that using a reflux ratio between 1.15 and 1.25 times $R_{\min}$ gave a reasonable guess for the economically optimum column design. However, that rule-of-thumb was based on designs of years ago. There have been significant changes in energy and equipment fabrication costs since then, and these two costs have increased at different rates, so this rule-of thumb may no longer be accurate. Therefore, for the optimal design of actual distillation columns, not only are rigorous rating calculations required, but also rigorous economic and environmental analyses must be done.

PROBLEMS

11.1. An equimolar mixture of benzene and 2,2,4-trimethylpentane is to be purified by distillation to produce a distillate containing 0.90 mole fraction of benzene and a bottoms product that contains 0.02 mole fraction of benzene. The distillation column is to operate at 1 atm pressure and the feed to the column is a saturated liquid. Use the RADFRAC program and the NRTL model (with Aspen default parameters) to estimate

(a) the minimum reflux ratio and the minimum number of stages to accomplish the separation;

(b) the actual number of stages and the location of the feed stage if the column is operated at 1.2 times the minimum reflux ratio.

11.2. Repeat the calculation above using the Wilson equation with the Aspen default parameters.

11.3. Repeat the calculation above using the UNIQUAC model with the Aspen default parameters.

11.4. An equimolar mixture of *n*-pentane and *n*-heptane is to be purified by distillation to produce a distillate that contains 98% of the pentane and 1% of the heptane in the initial feed. The column is to operate at 1 bar and the feed is at 20°C and 1 bar. Use the RADFRAC program and the NRTL model (with Aspen default parameters) to estimate

(a) the minimum reflux ratio and the minimum number of stages to accomplish the separation;

(b) the actual number of stages and the location of the feed stage if the column is operated at 1.2 times the minimum reflux ratio.

11.5. Repeat the calculation above using the Wilson equation with the Aspen default parameters.

11.6. Repeat the calculation above using the UNIQUAC model with the Aspen default parameters.

Chapter 12

Liquid–Liquid Extraction

While distillation is the most used separations/purifications process in industry, it is not the only one. Another important process is extraction, which is based on liquid-liquid phase equilibrium. In extraction, when two liquid phases that are only slightly or partially miscible, and therefore exhibit liquid–liquid equilibrium are brought into contact, a solute that initially was in one of the phases is partially extracted into the second phase. For example, if the two liquid phases were water and an organic compound such as chloroform, then a hydrophilic solute in the chloroform would selectively (but not completely) be extracted into the aqueous phase. Conversely, a hydrophobic solute in the aqueous phase would partition between the two phases but selectively into the organic phase.

You may have done extractions in an organic chemistry laboratory by mixing in a separatory funnel organic and aqueous phases, one initially containing a solute, and then separating the two phases to recover the solute (which then may have had to be further concentrated by, for example, evaporation of the solvent). That is a one-stage extraction. In industry it is more common to use multistage extractions, either by repeatedly contacting the initial solute containing liquid with fresh extracting solvent, or more efficiently by using a countercurrent extraction train as will be discussed later.

As an example, we first consider a single stage extraction. An ethyl acetate stream is found to contain 70 wt% ethyl acetate with the remainder being acetone. For the ethyl acetate to be useful in a downstream process, the concentration of acetone has to be decreased to 5 wt% or less.

(a) Determine what the remaining acetone concentration will be in the ethyl acetate phase if 1 kg of this mixture was extracted (brought into contact) with 1 kg of water at 25°C and 1 atm.

(b) Determine what the remaining acetone concentration would be in the ethyl acetate stream from part (a) if it was separated from the water-rich phase and then extracted with another 1 kg of pure water.

(c) Determine what the remaining acetone concentration would be in the ethyl acetate stream from part (b) if it was separated from the water-rich phase and then extracted again with another 1 kg of pure water.

This example will be solved in Aspen Plus® using the **Decanter** block on the **Separators** tab in the **Model Palette** in the **Simulation** mode (Fig. 12-1).

Using Aspen Plus® in Thermodynamics Instruction: A Step-by-Step Guide, First Edition. Stanley I. Sandler.

Figure 12-1

For convenience, in the flow sheet in Fig. 12-2 we have chosen a single decanter (for part (a)), and a train of three decanters (for parts (b) and (c)). Note that the **Decanter** block only allows for a single feed stream. Therefore, **Mixer** blocks have been added as shown below.

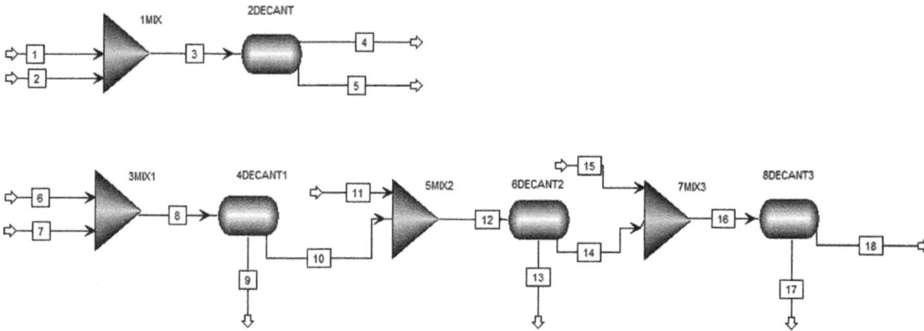

Figure 12-2

The block names have been changed from the defaults (B1, B2, etc.) to recognizable names so that they are easier to identify in the simulation results output.

Next, go through the usual **Setup,** except in **Report Options>Streams** (Fig. 12-3), choose **Mass** as the **Fraction basis** as that is the specification in the product statement.

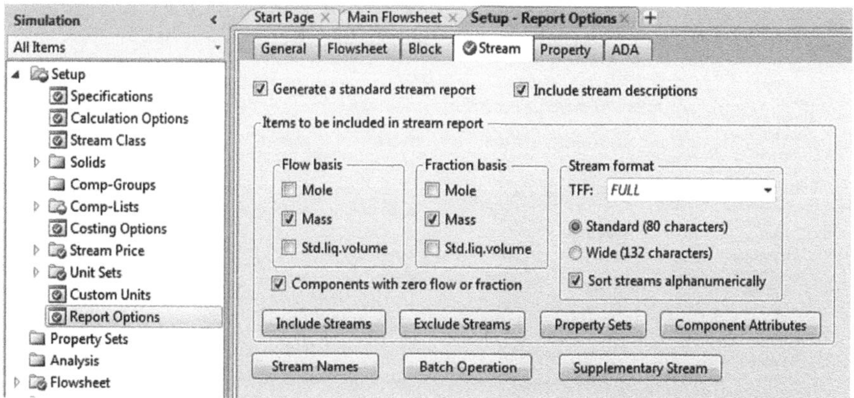

Figure 12-3

In **Properties>Components>Specifications**, add the components water, ethyl acetate, and acetone, resulting in Fig. 12-4.

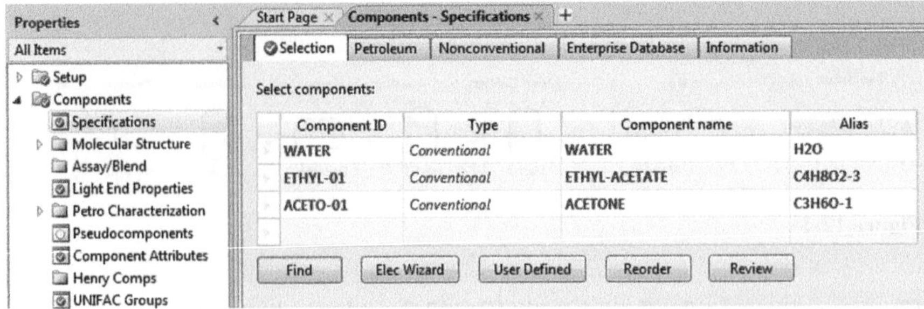

Figure 12-4

The **NRTL** model will then be chosen under **Properties>Methods** in the window of Fig. 12-5. [Though you might also want to try the UNIQUAC model, but not the Wilson model. Why not?]

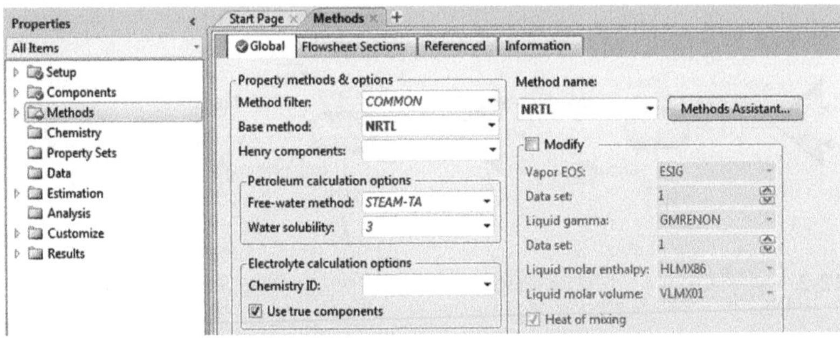

Figure 12-5

Then in **Properties>Methods>Parameters>Binary Interaction** click on **NRTL-1**, which brings up the NRTL parameters in the Aspen Plus database of Fig. 12-6.

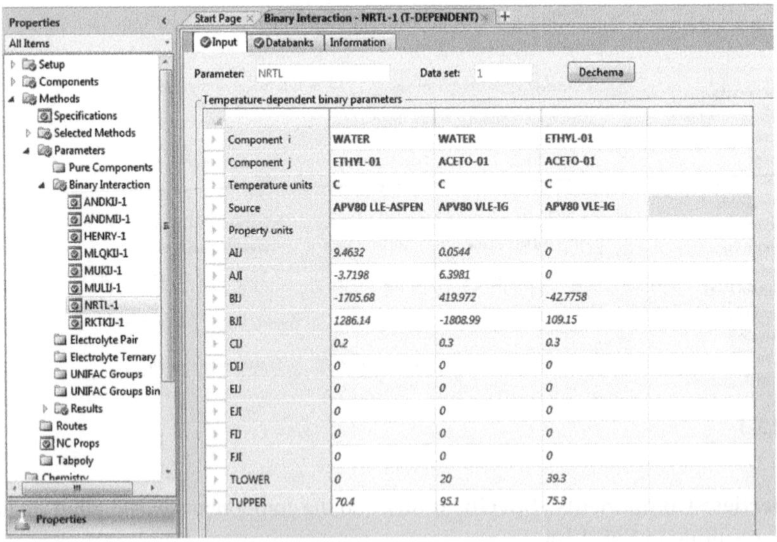

Properties

All Items

Setup
Components
Methods
Specifications
Selected Methods
Parameters
Pure Components
Binary Interaction
ANDKIJ-1
ANDMIJ-1
HENRY-1
MLQKIJ-1
MUKIJ-1
MULIJ-1
NRTL-1
RKTKIJ-1
Electrolyte Pair
Electrolyte Ternary
UNIFAC Groups
UNIFAC Groups Bin
Results
Routes
NC Props
Tabpoly
Chemistry

Properties

Start Page | Binary Interaction - NRTL-1 (T-DEPENDENT)

Input | Databanks | Information

Parameter: NRTL Data set: 1 Dechema

Temperature-dependent binary parameters

Component i	WATER	WATER	ETHYL-01
Component j	ETHYL-01	ACETO-01	ACETO-01
Temperature units	C	C	C
Source	APV80 LLE-ASPEN	APV80 VLE-IG	APV80 VLE-IG
Property units			
AIJ	9.4632	0.0544	0
AJI	-3.7198	6.3981	0
BIJ	-1705.68	419.972	-42.7758
BJI	1286.14	-1808.99	109.15
CIJ	0.2	0.3	0.3
DIJ	0	0	0
EIJ	0	0	0
EJI	0	0	0
FIJ	0	0	0
FJI	0	0	0
TLOWER	0	20	39.3
TUPPER	70.4	95.1	75.3

Figure 12-6

Next going to **Simulation>Streams**, the input streams 1, 6, 11, and 15 are as shown in Fig. 12-7. Note that the problem statement was in terms of a fixed amount of mass, 1 kg. However, Aspen Plus is a simulation program for a continuous process, so we will use as the **Streams** input for each of these streams 1 kg/min (though the Aspen Plus output will be in terms kg/hr, and 1 kg/min = 60 kg/hr).

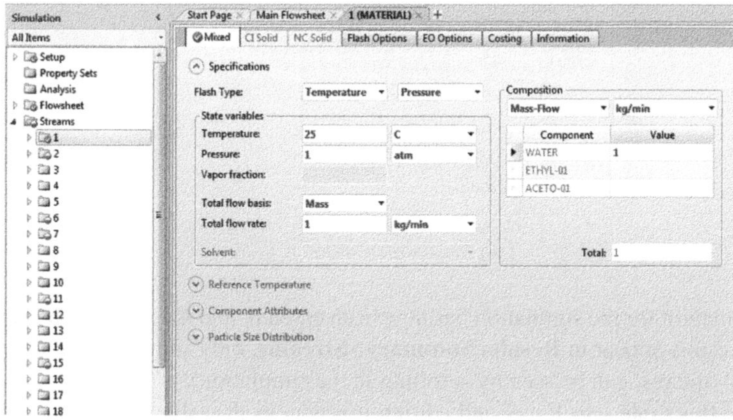

Figure 12-7

While inputs for streams for 2 and 7 are shown in Fig. 12-8.

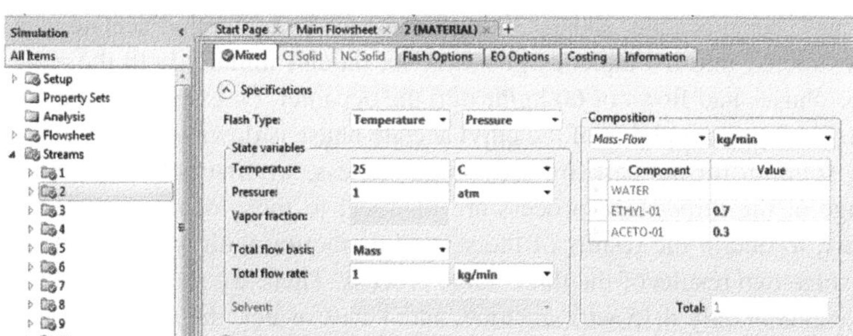

Figure 12-8

Next go to **Blocks** to enter the specifications. For all the decanters, the input used for **Blocks>DECANT** is shown in Fig. 12-9

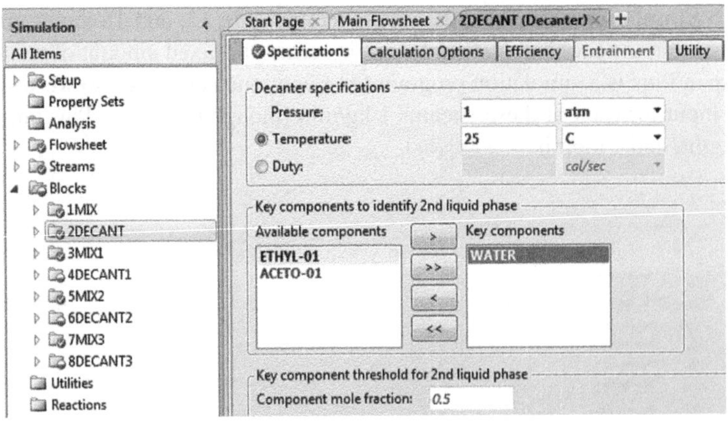

Figure 12-9

All the information for the simulation has now been entered, and the simulation can be run. The stream results appear in **Results Summary>Streams**, only some of which are shown in Fig. 12-10 (the rest can be seen by scrolling in the simulation).

Copying this table into Excel and editing it produces the table in Fig. 12-11. In this edited table, the results for part (a) are on the left and the results for parts (a, b, and c) appear on the right.

With these results, we see for part (a) that in a single stage extraction, the acetone concentration is reduced from 30 wt% in the feed to 16.8 wt% in the extracted ethyl acetate phase. This phase also now contains 5.1 wt% water. The aqueous phase contains 77.65 wt% water, 8.48 wt% ethyl acetate, and 10.30 wt% acetone. So both ethyl acetate and acetone have been extracted into the aqueous phase. Also, though initially both the aqueous and ethyl acetate phases had flows of 60 kg/hr into the decanter, on exiting the aqueous phase has a flow of 74.3 kg/hr and that of the ethyl acetate phase is down to 45.7 kg/hr.

For the solution for the part (b) three-stage process, we first note that the results for the first stage of the three-stage process are identical to those of the one-stage process. Consequently, to obtain the results of the second extraction with water, we only need to look at the stage two results of the three-stage process. There we see that the ethyl acetate phase now contains only 8.15 wt% acetone, and 4 wt% water, the remaining being ethyl acetate. The total flow rate of this stream is 34.2 kg/hr. To obtain the results of the third extraction, part (c) with 1 kg of water, we look at the ethyl acetate stream leaving the third decanter. The acetone content in the product has been reduced to 3.3 wt%, which is less than the 5 wt% maximum specification, the water content is 3.5 wt%, and the ethyl acetate concentration is 93.2 wt%. That is the good news. The bad news is that the total flow rate of this stream is only 26.74 kg/hr, of which 24.92 kg/hr is ethyl acetate. So of the initial 42 kg/hr (0.7 × 60 kg/hr) ethyl acetate that entered into the extraction process, only 24.92 kg/hr or 59.33% is recovered. That is, more than 40% of the initial ethyl acetate has been lost to the aqueous streams that are wastewater.

Material | Heat | Load | Work | Vol.% Curves | Wt. % Curves | Petro. Curves | Poly. Curves | Solids

Display: Allstreams Format: FULL Stream Table Copy All

	1	2	3	4	5	6	7	8	9
Substream: MIXED									
Mole Flow kmol/hr									
WATER	3.33051	0	3.33051	0.129526	3.20098	3.33051	0	3.33051	3.20098
ETHYL-01	0	0.476697	0.476697	0.405201	0.0724956	0	0.476697	0.476697	0.0724956
ACETO-01	0	0.309917	0.309917	0.132611	0.177306	0	0.309917	0.309917	0.177306
Mass Flow kg/hr									
WATER	60	0	60	2.33345	57.6666	60	0	60	57.6666
ETHYL-01	0	42	42	35.7008	6.29922	0	42	42	6.29922
ACETO-01	0	18	18	7.70207	10.2979	0	18	18	10.2979
Mass Frac									
WATER	1	0	0.5	0.0510196	0.776511	1	0	0.5	0.776511
ETHYL-01	0	0.7	0.35	0.780579	0.0848223	0	0.7	0.35	0.0848223
ACETO-01	0	0.3	0.15	0.168402	0.138667	0	0.3	0.15	0.138667
Total Flow kmol/hr	3.33051	0.786614	4.11712	0.667339	3.44978	3.33051	0.786614	4.11712	3.44978
Total Flow kg/hr	60	60	120	45.7363	74.2687	60	60	120	74.2687
Total Flow l/min	1.00608	1.15655	2.2655	0.857052	1.29972	1.00608	1.15655	2.2655	1.29972
Temperature C	25	25	31.3339	25	25	25	25	31.3339	25
Pressure bar	1.01325	1.01325	1.01325	1.01325	1.01325	1.01325	1.01325	1.01325	1.01325
Vapor Frac	0	0	0	0	0	0	0	0	0
Liquid Frac	1	1	1	1	1	1	1	1	1
Solid Frac	0	0	0	0	0	0	0	0	0
Enthalpy cal/mol	-68232.2	-92712.1	-72909.3	-94802.9	-68840.7	-68232.2	-92712.1	-72909.3	-68840.7
Enthalpy cal/gm	-3787.46	-1215.48	-2501.47	-1378.89	-3197.86	-3787.46	-1215.48	-2501.47	-3197.86
Enthalpy cal/sec	-63124.4	-20258	-83382.3	-17518.2	-65968.1	-63124.4	-20258	-83382.3	-65968.1
Entropy cal/mol-K	-38.8501	-99.2728	-50.302	-92.3445	-42.4269	-38.8501	-99.2728	-50.302	-42.4269
Entropy cal/gm-K	-2.15651	-1.30149	-1.72583	-1.3474	-1.97086	-2.15651	-1.30149	-1.72583	-1.97086
Density mol/cc	0.055173	0.0113356	0.0316872	0.0129774	0.0442373	0.055173	0.0113356	0.0316872	0.0442373
Density gm/cc	0.993957	0.864644	0.923575	0.889411	0.952301	0.993957	0.864644	0.923575	0.952301
Average MW	18.0153	76.2763	29.1466	68.5354	21.5271	18.0153	76.2763	29.1466	21.5271
Liq Vol 60F l/min	1.00198	1.16554	2.26746	0.868351	1.29911	1.00198	1.16554	2.26746	1.29911

	10	11	12	13	14	15	16	17	18
Substream: MIXED									
Mole Flow kmol/hr									
WATER	0.129526	3.33051	3.46003	3.38966	0.0763709	3.33051	3.40688	3.35451	0.0523722
ETHYL-01	0.405201	0	0.405201	0.0642389	0.341062	0	0.341062	0.0582658	0.282796
ACETO-01	0.132611	0	0.132611	0.0845893	0.0480219	0	0.0480219	0.0328309	0.015291
Mass Flow kg/hr									
WATER	2.33345	60	62.3335	60.9576	1.37584	60	61.3758	60.4324	0.943501
ETHYL-01	35.7008	0	35.7008	5.65104	30.0497	0	30.0497	5.11359	24.9161
ACETO-01	7.70207	0	7.70207	4.91295	2.78912	0	2.78912	1.90682	0.882295
Mass Frac									
WATER	0.0510196	1	0.589516	0.852296	0.040212	1	0.651447	0.895655	0.0352817
ETHYL-01	0.780579	0	0.33764	0.0790117	0.87827	0	0.31895	0.0760839	0.931725
ACETO-01	0.168402	0	0.0728422	0.0686919	0.0815181	0	0.0296038	0.0282606	0.0329929
Total Flow kmol/hr	0.667339	3.33051	3.99785	3.53299	0.465455	3.33051	3.79596	3.4456	0.35036
Total Flow kg/hr	45.7363	60	105.736	71.5216	34.2147	60	94.2147	67.4728	26.7419
Total Flow l/min	0.857052	1.00608	1.86628	1.22866	0.634896	1.00608	1.64054	1.14661	0.499585
Temperature C	25	25	30.149	25	25	25	29.4493	25	25
Pressure bar	1.01325	1.01325	1.01325	1.01325	1.01325	1.01325	1.01325	1.01325	1.01325
Vapor Frac	0	0	0	0	0	0	0	0	0
Liquid Frac	1	1	1	1	1	1	1	1	1
Solid Frac	0	0	0	0	0	0	0	0	0
Enthalpy cal/mol	-94802.9	-68232.2	-72617.4	-68921.8	-101140	-68232.2	-72267.4	-68968.5	-105110
Enthalpy cal/gm	-1378.89	-3787.46	-2745.63	-3403.99	-1375.91	-3787.46	-2911.69	-3521.98	-1377.15
Enthalpy cal/sec	-17518.2	-63124.4	-80642.5	-67627.4	-13076.8	-63124.4	-76201.2	-66010.5	-10229.9
Entropy cal/mol-K	-92.3445	-38.8501	-47.799	-41.1747	-98.8886	-38.8501	-46.249	-40.5434	-102.798
Entropy cal/gm-K	-1.3474	-2.15651	-1.80726	-2.03858	-1.34528	-2.15651	-1.8634	-2.07041	-1.34681
Density mol/cc	0.0129774	0.055173	0.0357023	0.0479165	0.0122186	0.055173	0.0385641	0.0500838	0.0118304
Density gm/cc	0.889411	0.993957	0.944268	0.970183	0.898171	0.993957	0.957151	0.980757	0.902984
Average MW	68.5354	18.0153	26.4483	20.2474	73.5081	18.0153	24.8197	19.5823	76.3271
Liq Vol 60F l/min	0.868351	1.00198	1.87028	1.22764	0.642643	1.00198	1.64457	1.14538	0.49919

Figure 12-10

	Part a						Part a				Part b				Part c				
	1	2	3	4	5		6	7	8	9	10	11	12	13	14	15	16	17	18
Mass Flow kg/hr																			
WATER	60	0	60	2.333448	57.66655		60	0	60	57.66655	2.333448	60	62.33345	60.9576	1.375844	60	61.37584	60.43235	0.943501
ETHYL-01	0	42	42	35.70078	6.299218		0	42	42	6.299218	35.70078	0	35.70078	5.651044	30.04974	0	30.04974	5.133592	24.91613
ACETO-01	0	18	18	7.702073	10.29793		0	18	18	10.29793	7.702073	0	7.702073	4.912954	2.789119	0	2.789119	1.906824	0.882295
Mass Frac																			
WATER	1	0	0.5	0.05102	0.776511		1	0	0.5	0.776511	0.05102	1	0.589518	0.852296	0.040212	1	0.651447	0.895655	0.035282
ETHYL-01	0	0.7	0.35	0.780579	0.084822		0	0.7	0.35	0.084822	0.780579	0	0.33764	0.079012	0.87827	0	0.31895	0.076084	0.931725
ACETO-01	0	0.3	0.15	0.168402	0.138667		0	0.3	0.15	0.138667	0.168402	0	0.072842	0.068692	0.081518	0	0.029604	0.028261	0.032993
Total Flow kg/hr	60	60	120	45.7363	74.2637		60	60	120	74.2637	45.7363	60	105.7363	71.5216	34.2147	60	94.2147	67.47276	26.74193
Temperature C	25	25	31.33393	25	25		25	25	31.33393	25	25	25	30.149	25	25	25	29.44932	25	25
Pressure bar	1.01325	1.01325	1.01325	1.01325	1.01325		1.01325	1.01325	1.01325	1.01325	1.01325	1.01325	1.01325	1.01325	1.01325	1.01325	1.01325	1.01325	1.01325

Figure 12-11

Clearly this three-stage extraction process can meet the product specifications, but is very inefficient in the use of resources and produces a great deal of waste. So the question that arises is whether there is a more resource-efficient method of doing extractions? The answer is that there is, in fact, a simple method that is in common use. It is countercurrent extraction; one possibility is the three-stage unit shown in Fig. 12-12 (of course, different numbers of stages could be used). In this process the phase to be extracted enters the first stage and the solvent, here fresh water, enters the third stage in this diagram. The aqueous phase from this third extractor is then used as the solvent for the second stage, and the aqueous phase from the second stage is the solvent for the first stage. That is, in the schematic diagram below, the organic phase to be extracted flows from left to right, while the aqueous solvent phase flows from right to left. Such a procedure uses less water (only 60 kg/hr instead of three times that in the process already considered) and, as will be seen, leads to a greater recovery of the ethyl acetate.

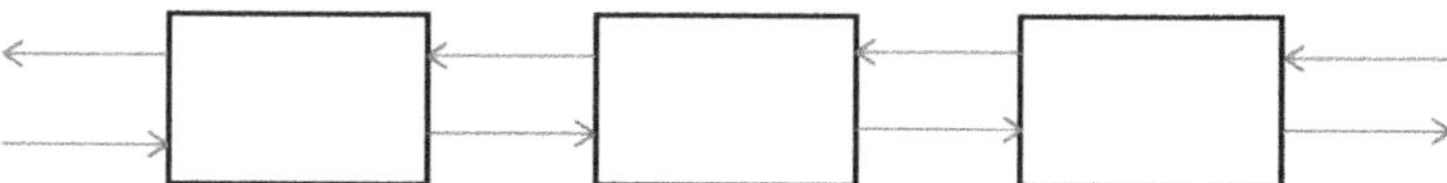

Figure 12-12

This process can be modeled using Aspen Plus. The direct way would be to use decanters as above. However, the problem again is that Aspen Plus only allows a single feed stream into a decanter, so mixers would have to be used as above. A simpler method is to use the **Extract** block found in the **Simulation>Model Palette>Columns>Extract** (Fig. 12-13).

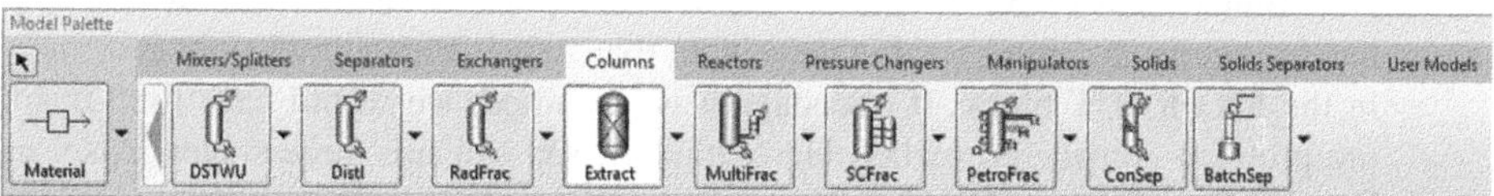

Figure 12-13

So the flow sheet is simply as shown in Fig. 12-14.

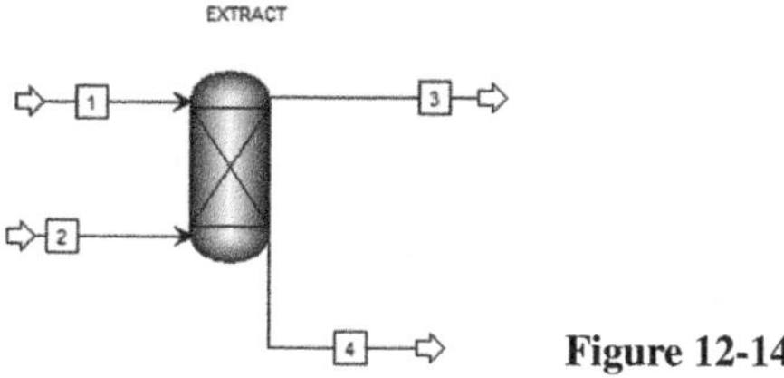

Figure 12-14

The same **Setup**, **Components**, and **Methods** as used previously will be used for **Streams>1** (Fig. 12-15) and for **Streams>2** (Fig. 12-16).

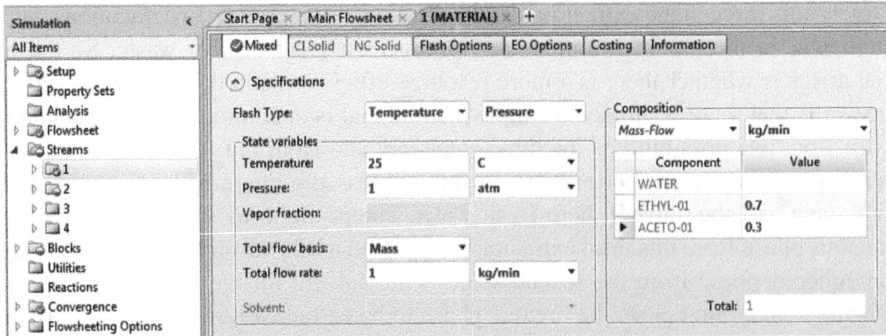

Figure 12-15

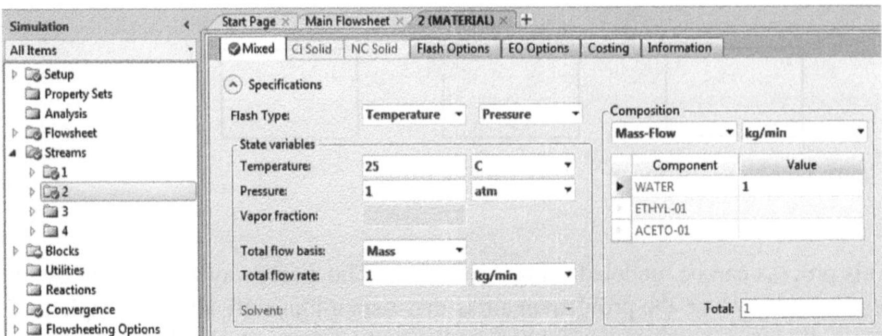

Figure 12-16

Next **Blocks>EXTRACT** requires a number of inputs. In **Blocks>EXTRACT>Specs** one has to specify the number of stages. Note that Aspen Plus allows only 2 or more stages in the **EXTRACT** block. If one stage was permitted, we would have used this block in the problem considered earlier. Here, to start, we will use three stages. Also, using the **Specify Temperature Profile** and by supplying the temperature of the first two stages, the temperature of the third (and additional stages if required in a larger extraction unit) are obtained by extrapolation (or interpolation, if that is appropriate). See Fig. 12-17.

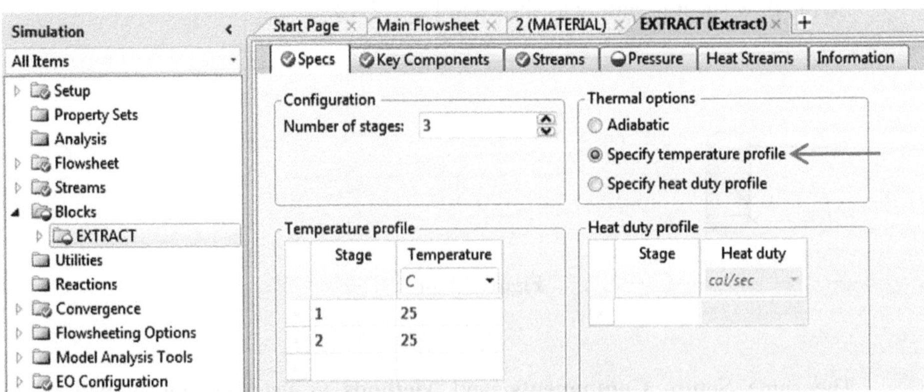

Figure 12-17

The next specification is of the key components in the **Key Components** tab shown in Fig. 12-18.

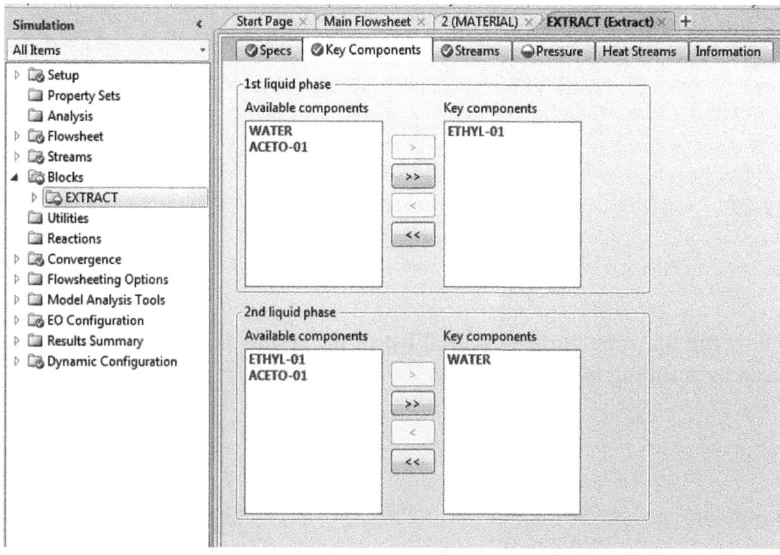

Figure 12-18

Here we have specified ethyl acetate and water as the key components; acetone is the distributed component. The specifications in Fig. 12-19 were then made after clicking on the **Streams** tab, and then after clicking on the **Pressure** tab populate the window as shown in Fig. 12-20.

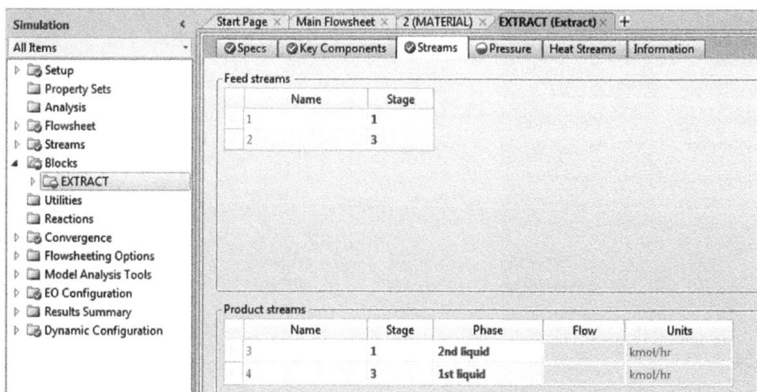

Figure 12-19

Here as in the case of the temperature profile, the pressures for stages other than the ones shown are obtained by extrapolation or interpolation as appropriate.

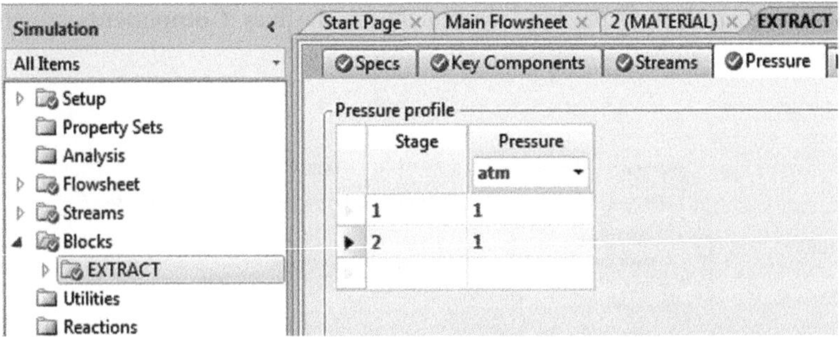

Figure 12-20

We can now run the simulation. A partial list of the results is shown in Fig. 12-21; more can be seen by scrolling in the simulation.

Display: All streams Format: FULL Stream Table Copy All

	1	2	3	4
Substream: MIXED				
Mole Flow kmol/hr				
WATER	0	3.33051	3.24296	0.0875484
ETHYL-01	0.476697	0	0.0816149	0.395082
ACETO-01	0.309917	0	0.256217	0.0536999
Mass Frac				
WATER	0	1	0.725797	0.0399239
ETHYL-01	0.7	0	0.0893324	0.881127
ACETO-01	0.3	0	0.184871	0.0789487
Total Flow kmol/hr	0.786614	3.33051	3.58079	0.53633
Total Flow kg/hr	60	60	80.4947	39.5053
Total Flow l/min	1.15655	1.00608	1.42602	0.732862
Temperature C	25	25	25	25
Pressure bar	1.01325	1.01325	1.01325	1.01325
Vapor Frac	0	0	0	0
Liquid Frac	1	1	1	1
Solid Frac	0	0	0	0
Enthalpy cal/mol	-92712.1	-68232.2	-68788.3	-101350
Enthalpy cal/gm	-1215.48	-3787.46	-3060.03	-1375.91
Enthalpy cal/sec	-20258	-63124.4	-68421.2	-15098.8

Figure 12-21

We see that the product specification of 5 wt % acetone in Stream 4 has been not met with using three stages. Increasing the number of stages to four (Fig. 12-22) and then rerunning the simulation produces the results in Fig. 12-23.

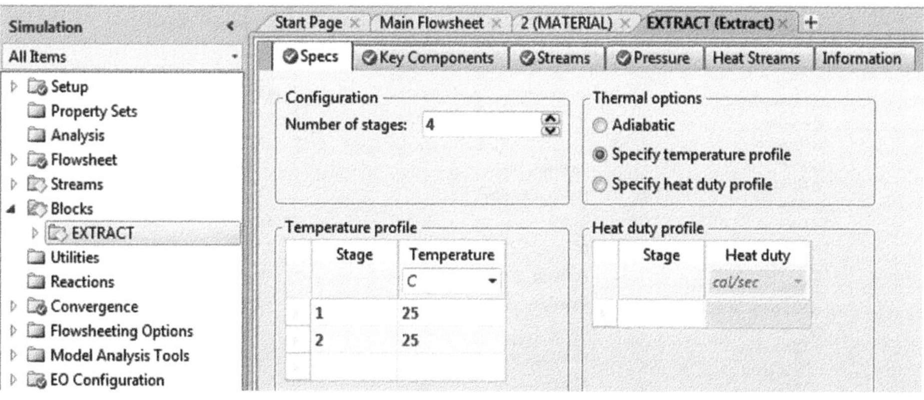
Simulation
All Items
Setup
Property Sets
Analysis
Flowsheet
Streams
Blocks
EXTRACT
Utilities
Reactions
Convergence
Flowsheeting Options
Model Analysis Tools
EO Configuration
Start Page
Main Flowsheet
2 (MATERIAL)
EXTRACT (Extract)
Specs
Key Components
Streams
Pressure
Heat Streams
Information
Configuration
Number of stages: 4
Thermal options
Adiabatic
Specify temperature profile
Specify heat duty profile
Temperature profile
Stage
Temperature
C
1 25
2 25
Heat duty profile
Stage
Heat duty
cal/sec

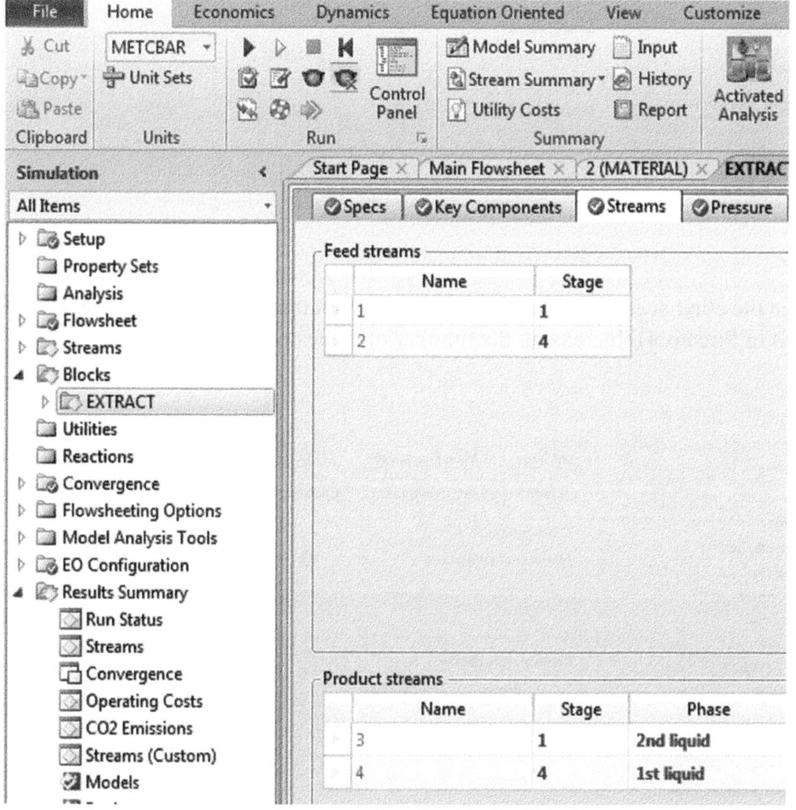
File
Home
Economics
Dynamics
Equation Oriented
View
Customize
Cut
Copy
Paste
Clipboard
METCBAR
Unit Sets
Units
Run
Control Panel
Model Summary
Stream Summary
Utility Costs
Input
History
Report
Summary
Activated Analysis
Simulation
All Items
Setup
Property Sets
Analysis
Flowsheet
Streams
Blocks
EXTRACT
Utilities
Reactions
Convergence
Flowsheeting Options
Model Analysis Tools
EO Configuration
Results Summary
Run Status
Streams
Convergence
Operating Costs
CO2 Emissions
Streams (Custom)
Models
Start Page
Main Flowsheet
2 (MATERIAL)
EXTRAC
Specs
Key Components
Streams
Pressure
Feed streams
Name
Stage
1 1
2 4
Product streams
Name
Stage
Phase
3 1 2nd liquid
4 4 1st liquid

Figure 12-22

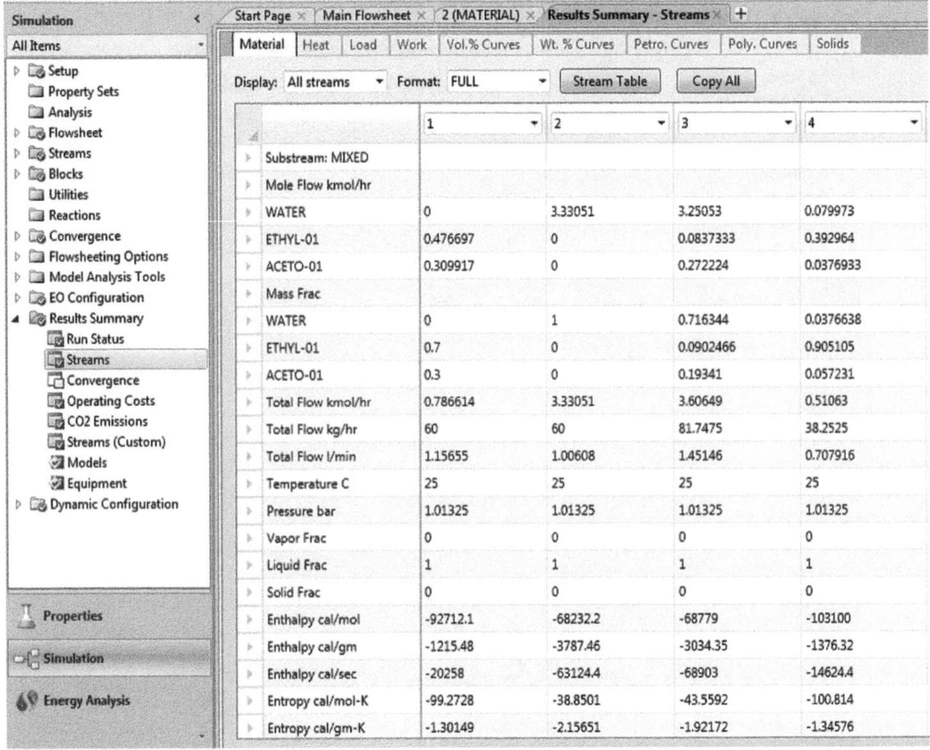

	1	2	3	4
Substream: MIXED				
Mole Flow kmol/hr				
WATER	0	3.33051	3.25053	0.079973
ETHYL-01	0.476697	0	0.0837333	0.392964
ACETO-01	0.309917	0	0.272224	0.0376933
Mass Frac				
WATER	0	1	0.716344	0.0376638
ETHYL-01	0.7	0	0.0902466	0.905105
ACETO-01	0.3	0	0.19341	0.057231
Total Flow kmol/hr	0.786614	3.33051	3.60649	0.51063
Total Flow kg/hr	60	60	81.7475	38.2525
Total Flow l/min	1.15655	1.00608	1.45146	0.707916
Temperature C	25	25	25	25
Pressure bar	1.01325	1.01325	1.01325	1.01325
Vapor Frac	0	0	0	0
Liquid Frac	1	1	1	1
Solid Frac	0	0	0	0
Enthalpy cal/mol	-92712.1	-68232.2	-68779	-103100
Enthalpy cal/gm	-1215.48	-3787.46	-3034.35	-1376.32
Enthalpy cal/sec	-20258	-63124.4	-68903	-14624.4
Entropy cal/mol-K	-99.2728	-38.8501	-43.5592	-100.814
Entropy cal/gm-K	-1.30149	-2.15651	-1.92172	-1.34576

Figure 12-23

So again the ethyl acetate product specification is not met (5.72 wt% acetone vs. the 5 wt% required in Stream 4). Increasing the number of stages to five (Fig. 12-24)

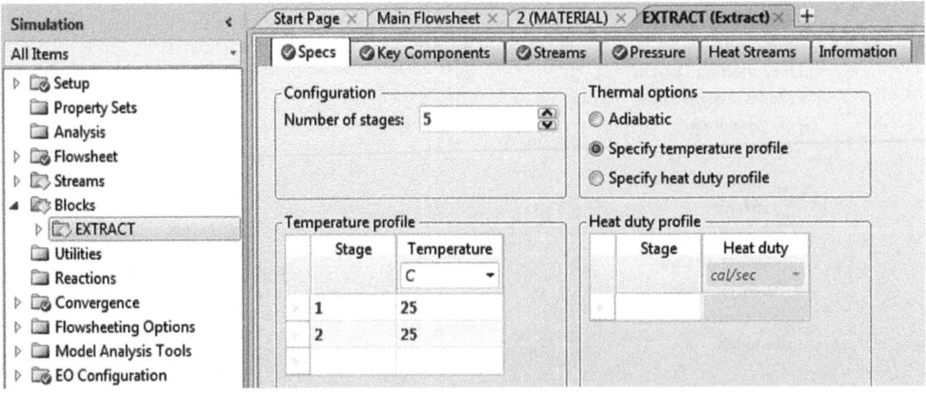

Figure 12-24

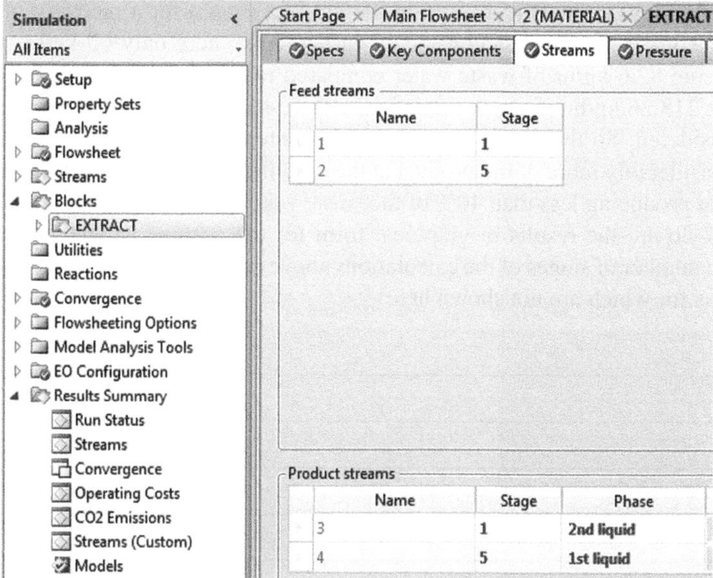

Figure 12-24 *(Continued)*

leads to the results in Fig. 12-25.

	1	2	3	4
Substream: MIXED				
Mole Flow kmol/hr				
WATER	0	3.33051	3.25533	0.0751726
ETHYL-01	0.476697	0	0.0851201	0.391577
ACETO-01	0.309917	0	0.282686	0.0272308
Mass Frac				
WATER	0	1	0.710308	0.036175
ETHYL-01	0.7	0	0.0908342	0.921578
ACETO-01	0.3	0	0.198858	0.0422471
Total Flow kmol/hr	0.786614	3.33051	3.62314	0.49398
Total Flow kg/hr	60	60	82.5638	37.4362
Total Flow l/min	1.15655	1.00608	1.46804	0.69167
Temperature C	25	25	25	25
Pressure bar	1.01325	1.01325	1.01325	1.01325
Vapor Frac	0	0	0	0
Liquid Frac	1	1	1	1
Solid Frac	0	0	0	0
Enthalpy cal/mol	-92712.1	-68232.2	-68773	-104340
Enthalpy cal/gm	-1215.48	-3787.46	-3017.96	-1376.79

Figure 12-25

So we see that with five stages, the product specification of 5 wt% acetone in the ethyl acetate stream is met, and in fact exceeded. Also, of the 42 kg/hr of ethyl acetate in the feed stream, 34.50 kg/hr is recovered in the product stream for a recovery of 82.1% compared to 59.3% in the process considered earlier. Also, here only 60 kg/hr of water is used producing 82.6 kg/hr of waste water compared to using 180 kg/hr of fresh water and producing 213.26 kg/hr of waste water in the three-stage non-countercurrent process originally considered. So the advantages of using countercurrent extraction are obvious: we recover considerably more of the product at the specified purity using only 1/3 as much fresh water and producing less than 40% of the waste water of the original process.

In Fig. 12-26 are the results in graphical form for the acetone weight fraction as a function of the number of stages of the calculations above (as well as for two and six stages, the calculations for which are not shown here).

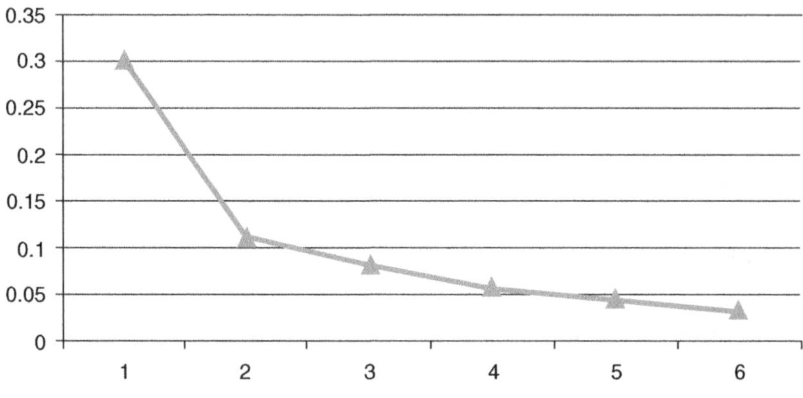

Figure 12-26

As in the earlier comparison of multistage distillation versus repeated flash separations, here we see that multistage countercurrent extraction is more efficient than repeated single-stage operations.

PROBLEMS

12.1. It is desired to remove some of the acetone from a mixture that contains 60 wt% acetone and 40 wt% water by extraction with methyl isobutyl ketone (MIK). If 3 kg of MIK are contacted with 1 kg of the acetone–water mixture, what will be the amounts and compositions of the equilibrium phases? (Illustration 11.2–8 in *Chemical, Biochemical and Engineering Thermodynamics*, 4th ed., S. I. Sandler, John Wiley & Sons, Inc., 2006)

12.2. The acetone–water mixture of the previous problem is to be treated by a two-stage extraction with MIK. In the first stage 1 kg of MIK is contacted with 1 kg of the acetone–water mixture. The water-rich phase goes to a second stage, where it will be contacted with another 1 kg of pure MIK. What will be the amounts and compositions of the equilibrium phases at the exit of each stage? (Illustration 11.2–9 in *Chemical, Biochemical and Engineering Thermodynamics*, 4th ed., S. I. Sandler, John Wiley & Sons, Inc., 2006)

12.3. The acetone–water mixture of Problem 12.1 is to be treated by a countercurrent two-stage extraction with MIK. The feeds to the extraction unit are 1 kg of MIK and 1 kg of the acetone–water mixture. What will be the amounts and compositions of the equilibrium exit stages?

12.4. Repeat the calculation of Problem 12.3 using four countercurrent equilibrium stages.

Chapter 13

Sensitivity Analysis: A Tool for Repetitive Calculations

There are times, particularly, when trying to optimize a process, when one wants to do the same calculations numerous times changing the value of, for example, an input parameter or the conditions in a block. An example of the first case is the countercurrent extraction at the end of the previous chapter; the initial flow rates of the water and ethyl acetate streams were both (arbitrarily) fixed to be equal at 60 kg/hr. When three countercurrent stages were used, it was found that the acetone concentration in the ethyl acetate could not be reduced to the product specification of 5 wt%. The product specification was met by adding two stages. However, another alternative would be to use three stages and increase the water flow rate as we show below. A number of calculations would then be done increasing the water flow rate and examining the exit acetone concentration. As we will show in this chapter, this can be done in a single simulation using the **Model Analysis Tools>Sensitivity** option and instructing Aspen Plus® to do calculations for a number different values of the water flow rate input parameter. The second example is that of systematically varying the value of a process variable, the temperature in a reaction block.

We start with the three-stage **EXTRACT** unit and flow sheet of the previous chapter. The initial **Setup, Components, Methods, Streams**, and **Blocks** are as in the previous chapter, using three stages in the **EXTRACT** block. Then going to **Model Analysis Tools>Sensitivity** brings up the window in Fig. 13-1. Clicking on **New** brings up the small pop-up menu. The default ID of **S-1** will be used (though you can choose another name). Then click on **OK**.

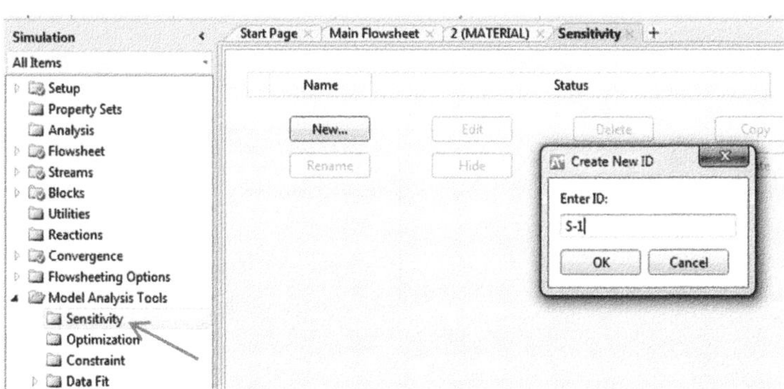

Figure 13-1

Using Aspen Plus® in Thermodynamics Instruction: A Step-by-Step Guide, First Edition. Stanley I. Sandler.

Next click on **Model Analysis Tools>Sensitivity>S-1>Input** to bring up the window in Fig. 13-2 with three tabs **Vary, Define**, and **Tabulate** that will be dealt with in that order.

Start with the **VARY** tab where the variable to be varied and its range are specified. In the example here it is the water mass flow rate (the flow rate of stream 2), and water flow rates of between 60 and 86 kg/hr in increments of 2 kg/hr that will be examined. However, since the base case is 60 kg/hr of water, and the base case always appears in the output, the range here has been set to be from 62 to 86 kg/hr. The other items are as shown in Fig. 13-2.

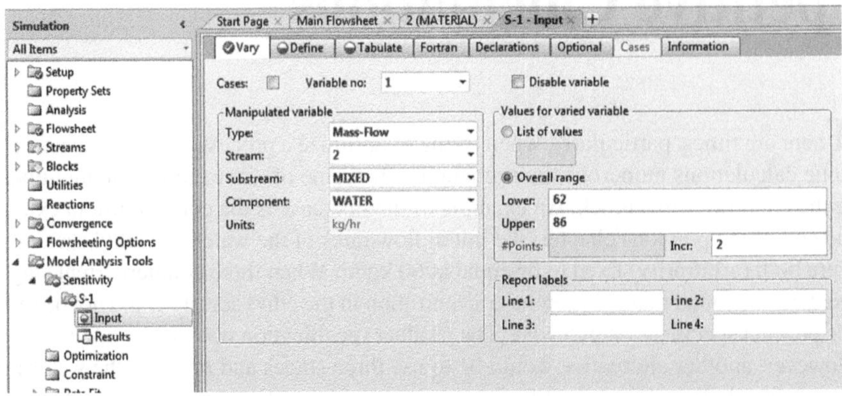

Figure 13-2

Next going to the **Define** tab, clicking on **New** brings up the window in Fig. 13-3, in which you choose a name for one of the variables that you are interested in, here **ACET** for acetone.

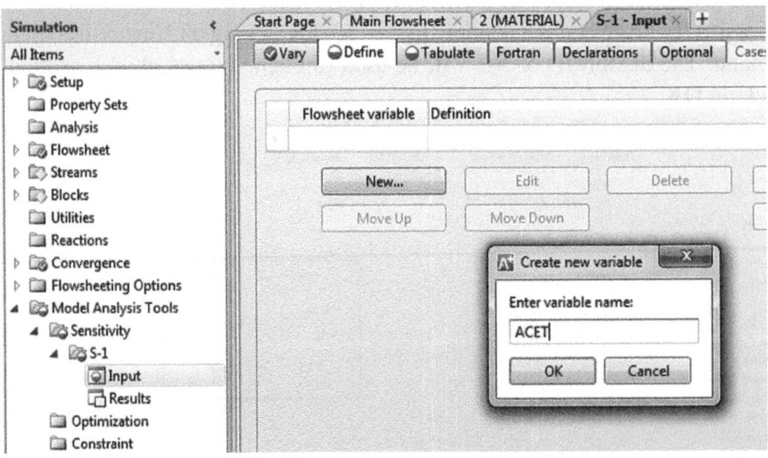

Figure 13-3

Clicking **OK** brings you to the pop-up window of Fig. 13-4, which is then completed as shown in Fig. 13-5 using the drop-down menus that appear.

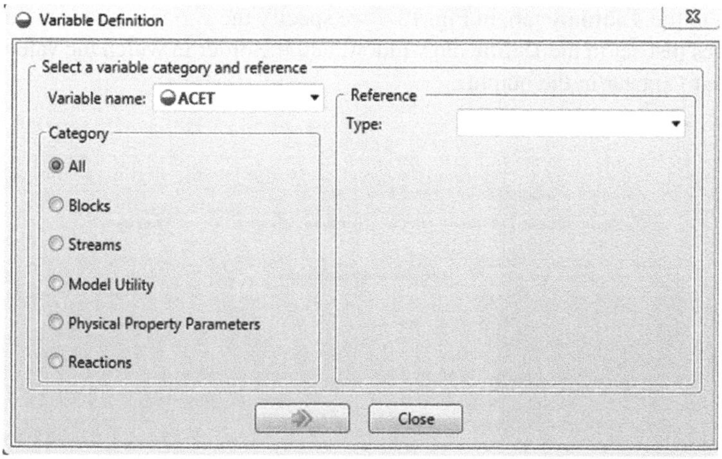

Figure 13-4

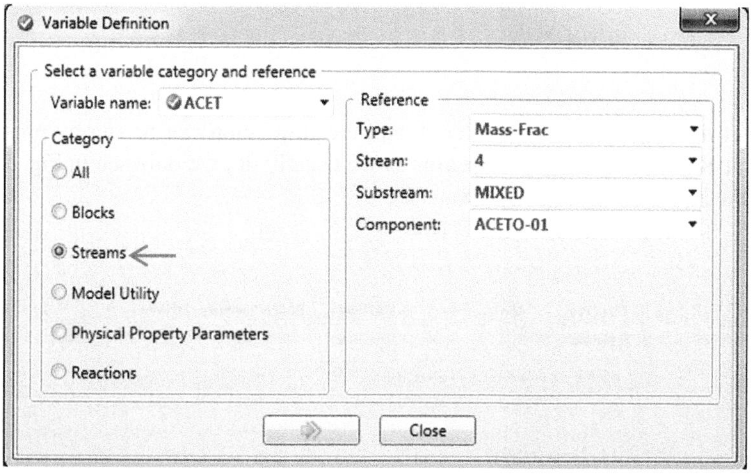

Figure 13-5

Then click **Close**. Repeat this process for water and ethyl acetate (designated as EA) so that the completed **Define** window is as shown in Fig. 13-6.

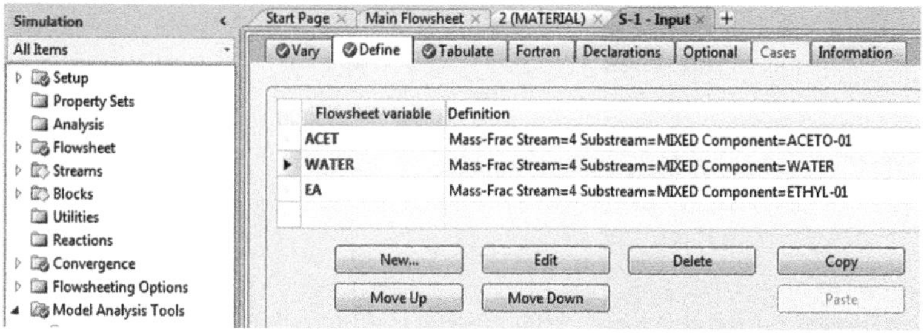

Figure 13-6

Finally, going to the **Tabulate** tab, in Fig. 13-7 we specify the variables we want tabulated using the names just set in the **Define** tab window, and the order in which the values of the parameters are to appear in the output.

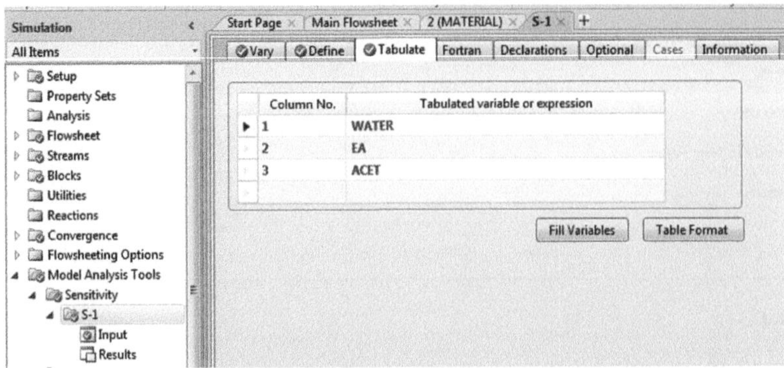

Figure 13-7

All the input data have now been entered, and the simulation can be run by pressing >. If we look at **Results Summary>Streams** as we usually do, we only see in Fig. 13-8 the results for the base case (60 kg/hr of water).

Display: All streams Format: FULL Stream Table Copy All

	1	2	3	4
Substream: MIXED				
Mole Flow kmol/hr				
WATER	0	3.33051	3.24299	0.087519
ETHYL-01	0.476697	0	0.0816894	0.395007
ACETO-01	0.309917	0	0.256281	0.0536366
Mass Frac				
WATER	0	1	0.725706	0.0399214
ETHYL-01	0.7	0	0.089402	0.881201
ACETO-01	0.3	0	0.184892	0.0788772
Total Flow kmol/hr	0.786614	3.33051	3.58096	0.536163
Total Flow kg/hr	60	60	80.5055	39.4945
Total Flow l/min	1.15655	1.00608	1.42622	0.732655
Temperature C	25	25	25	25
Pressure bar	1.01325	1.01325	1.01325	1.01325

Figure 13-8

To see the results of the sensitivity analysis we need to look at **Model Analysis Tools>Sensitivity>S-1>Results** shown in Fig. 13-9.

Row/Case	Status	VARY 1 2 MIXED WATER MASSFLOW KG/HR	WATER	EA	ACET
1	OK	60	0.0399214	0.881201	0.0788772
2	OK	62	0.0394417	0.88611	0.0744481
3	OK	64	0.0390218	0.890699	0.070279
4	OK	66	0.038607	0.895088	0.0663045
5	OK	68	0.0382186	0.899219	0.0625623
6	OK	70	0.0378582	0.90309	0.0590515
7	OK	72	0.0375236	0.906717	0.0557599
8	OK	74	0.0372129	0.910112	0.052675
9	OK	76	0.0369243	0.913291	0.0497847
10	OK	78	0.0366558	0.916268	0.047076
11	OK	80	0.0364064	0.919054	0.04454
12	OK	82	0.0361743	0.921662	0.042164
13	OK	84	0.0359584	0.924103	0.0399382
14	OK	86	0.0357573	0.92639	0.0378525

Figure 13-9

The results show that using a three-stage countercurrent extraction process with a water flow rate of 76 kg/hr or greater with an ethyl acetate stream of 60 kg/hr will produce an acetone concentration of less than 5 wt%. A plot of the acetone concentration in wt% versus water flow in kg/hr can easily be made using **Plot>Results Curve** on the main toolbar (Fig. 13-10).

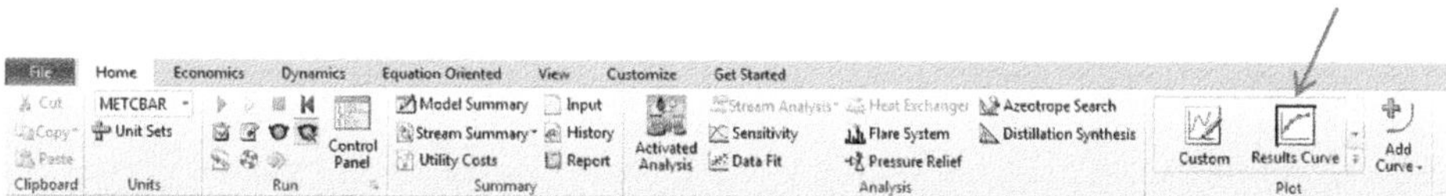

Figure 13-10

That brings up the pop-up window that has been populated as shown in Fig. 13-11.

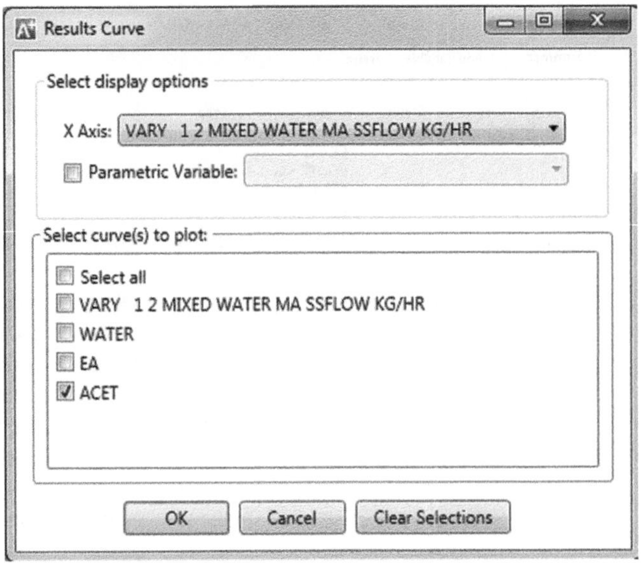

Results Curve

Select display options

X Axis: VARY 1 2 MIXED WATER MA SSFLOW KG/HR

Parametric Variable:

Select curve(s) to plot:

Select all
VARY 1 2 MIXED WATER MA SSFLOW KG/HR
WATER
EA
ACET

OK Cancel Clear Selections

Figure 13-11

Clicking **OK** gives the results in Fig. 13-12.

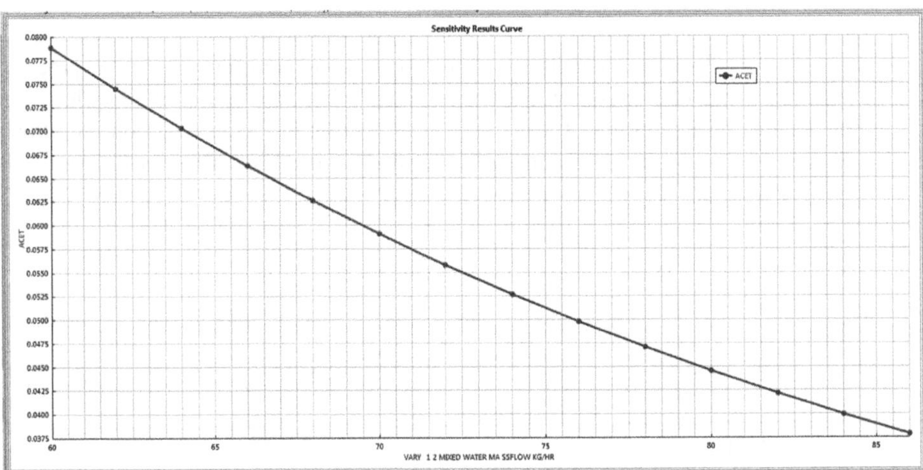

Figure 13-12

So we see that there are there are (at least) two solutions to the extraction problem. We can use a water flow rate of 60 kg/hr and a five-stage extractor as in the previous chapter or a 76 kg/hr or greater water flow rate and a three-stage extractor. In fact, there are other flow rate and number of stage combinations that all are technical possibilities. Which of these possibilities an engineer would choose would depend on additional detailed economic and environmental impact analyses.

Now consider the use of a sensitivity analysis for the equilibrium chemical reaction of Illustrations 13.1-1 to 13.1-3 in the textbook, *Chemical, Biochemical and Engineering Thermodynamics*, 4th ed., S. I. Sandler (John Wiley & Sons, Inc., 2006)

$$N_2O_4 \rightarrow 2NO_2$$

There the reaction was considered at 10 bar and 400 K. Here we want to consider this reaction at 1 bar and over the temperature range of 200 K to 400 K. The **RGIBBS** reactor block will be used in the flow sheet (Fig. 13-13).

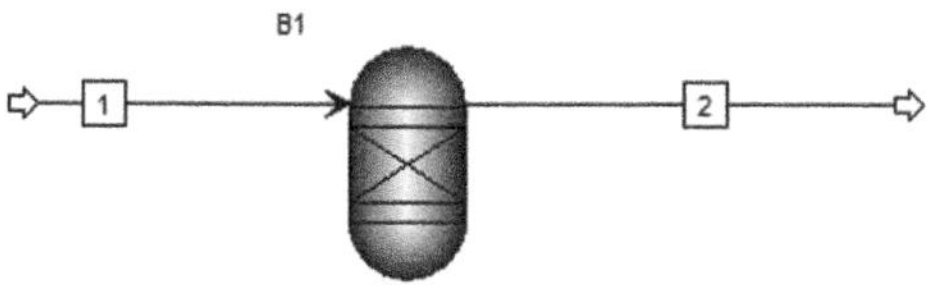

Figure 13-13

The **Setup** is as usual, choosing metric **MET** units (not METCBAR since I want temperature in Kelvin) and in **Report Options>Streams** choosing **Mole** in **Fraction basis**. Then enter nitrogen dioxide and nitrogen tetroxide in **Components** resulting in the populated window shown in Fig. 13-14.

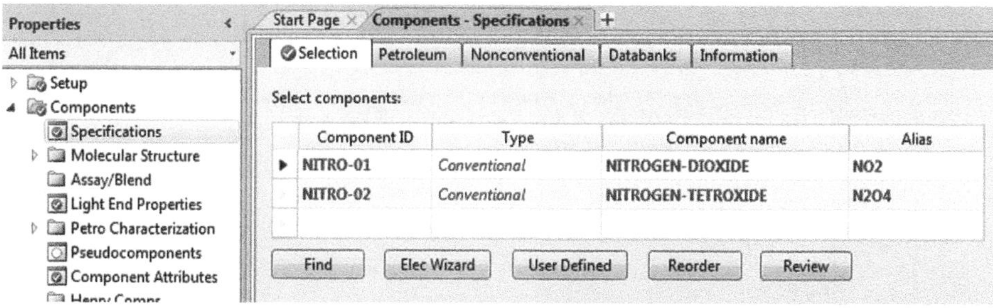

Figure 13-14

Next choose **Ideal** in **Methods** in the window of Fig. 13-15.

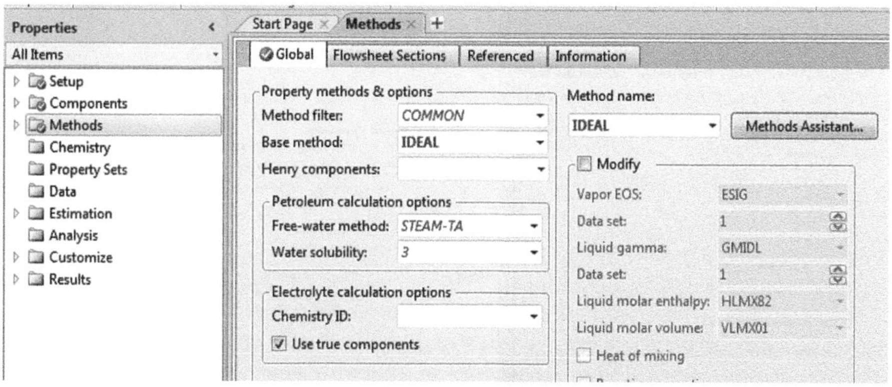

Figure 13-15

Next, go to **Simulation**, then to **Reactions**, select **New**, accept the default name for the reaction, R-1) or choose a name of your own, and finally select **GENERAL** for the reaction type (Fig. 13-16).

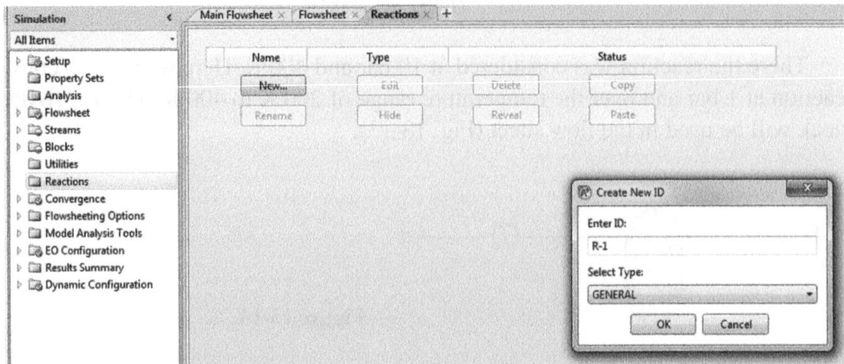

Figure 13-16

Then click OK, which brings up the following window (Fig. 13-17).

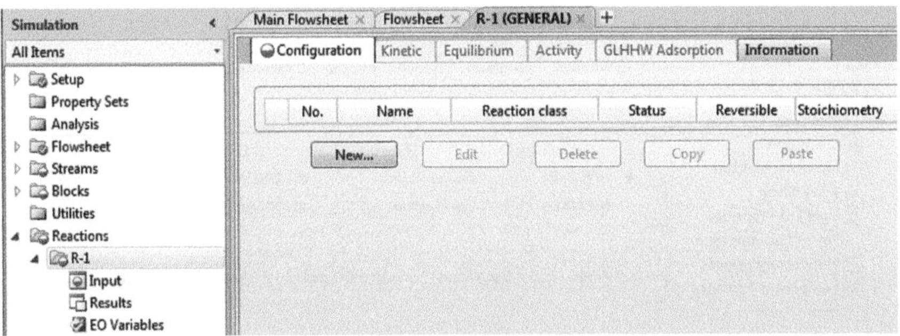

Figure 13-17

Clicking on **New** brings up the stoichiometry window that the user populates as shown in Fig. 13-18.

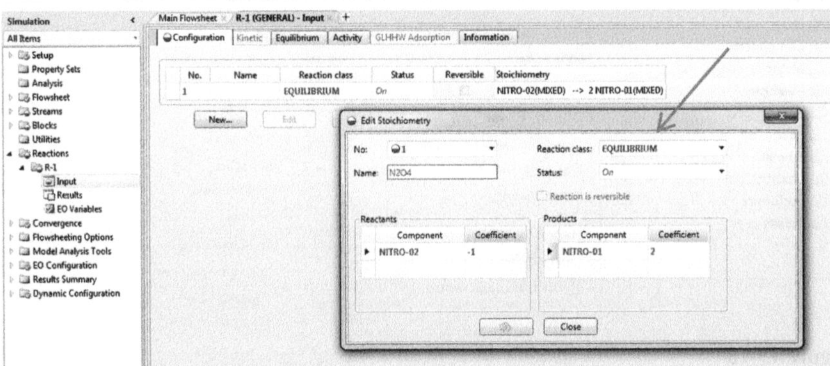

Figure 13-18

Note that ***POWERLAW*** has been changed to **EQUILIBRIUM** for the **Reaction class**. This results in the window of Fig. 13-19.

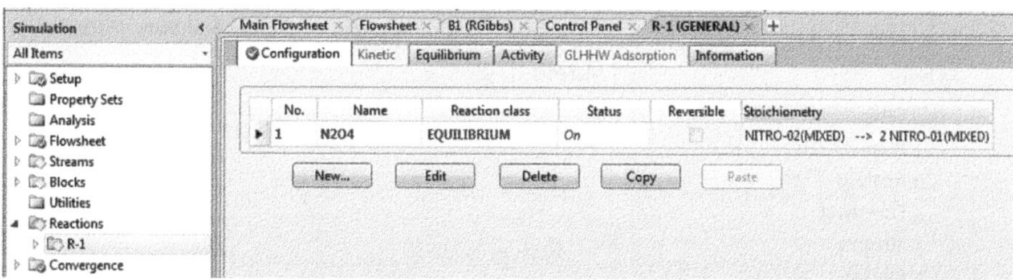

Figure 13-19

Next, go to **Streams>1** and populate this window as in Fig. 13-20.

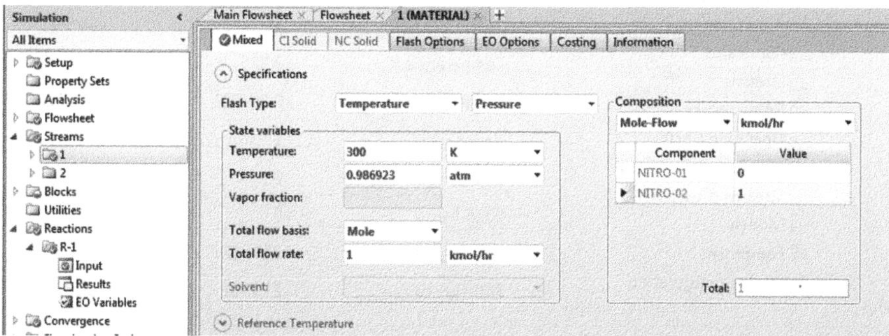

Figure 13-20

Then go to **Blocks>B1** and populate the window that appears as in Fig. 13-21. [Here at 1 bar the operating temperature has been arbitrarily set to 300 K since a range of operating temperatures will be considered.]

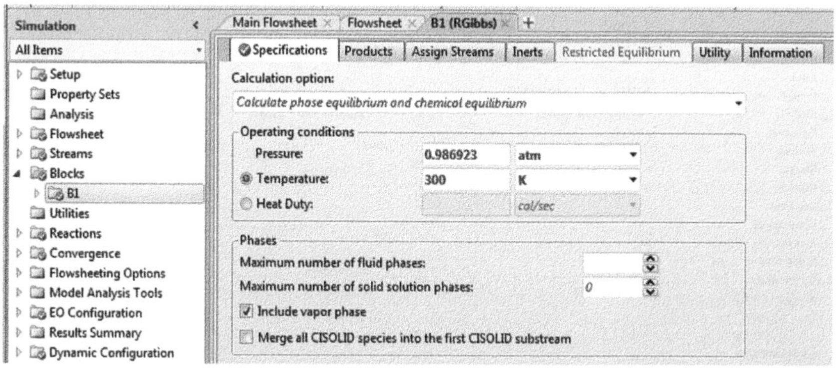

Figure 13-21

Running the simulation then leads to the results seen in Fig. 13-22 under **Results Summary>Streams**.

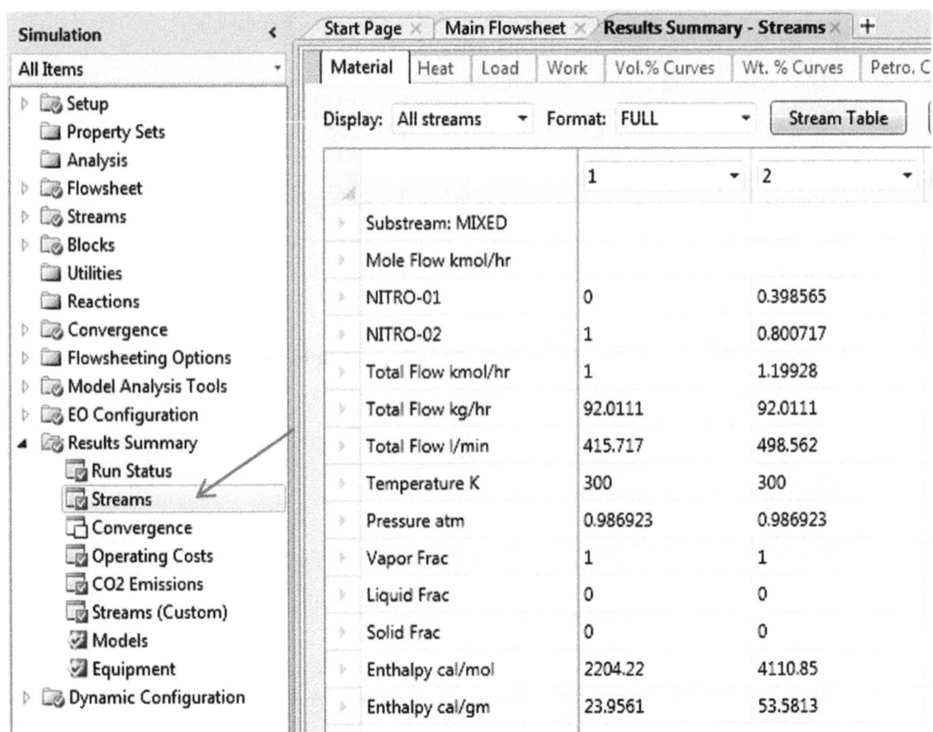

Figure 13-22

Now that we know the simulation is running correctly, we can do a sensitivity analysis by choosing **Model Analysis Tools>Sensitivity**, choosing **New** in the window that appears, and clicking **OK** to use the default ID of **S-1** (or enter your own ID), which brings up Fig. 13-23.

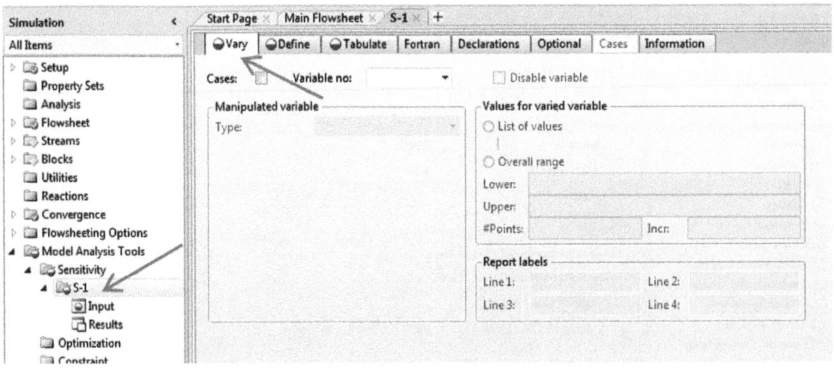

Figure 13-23

Then in the window in Fig. 13-23 click on the **Vary** tab, and for this case populate the window as in Fig. 13-24.

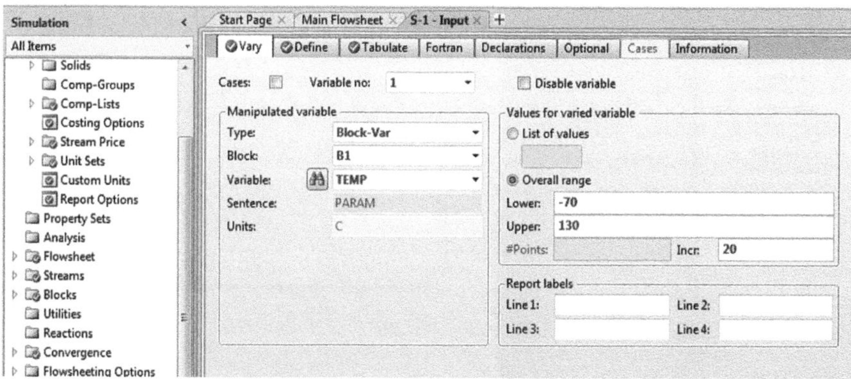

Figure 13-24

Note that Aspen Plus will not allow the temperature to be in Kelvin in a sensitivity analysis. Therefore, centigrade has to be used, and 200 K = −73.25°C and 400 K = 126.85°C. So the temperature range chosen here is −70°C to 130°C.

Next, clicking on the Define tab, the variables (here species mole fractions) of the two components that will be tabulated are added. First, N2O4:

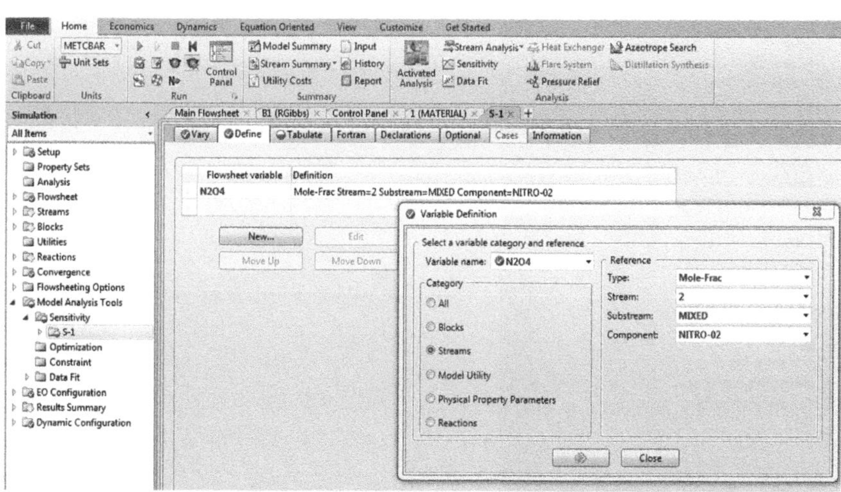

and then after clicking on **Close**, NO2

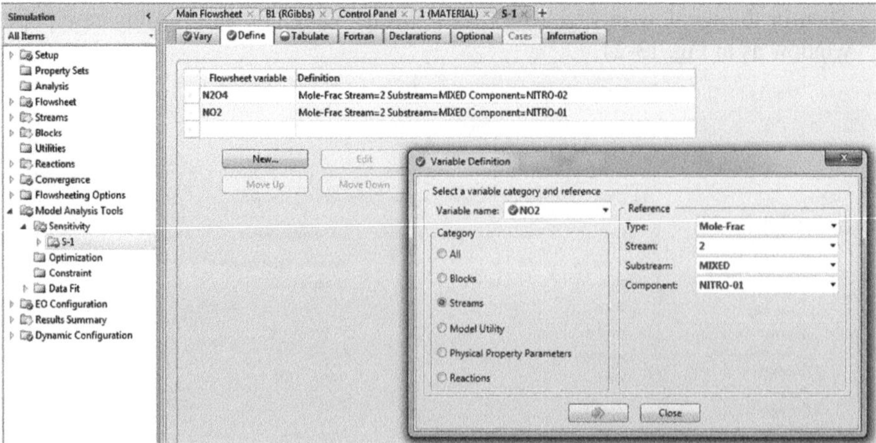

Clicking on **Close** again, and then clicking on the **Tabulate** tab brings up the window in Fig. 13.25 in which you can enter the variables you wish to have tabulated, or click on **Fill Variables**, which will enter the variables identified under the **Define** tab. Note that temperature does not appear in the list. However, by default, it will appear in the results since it is the variable being varied.

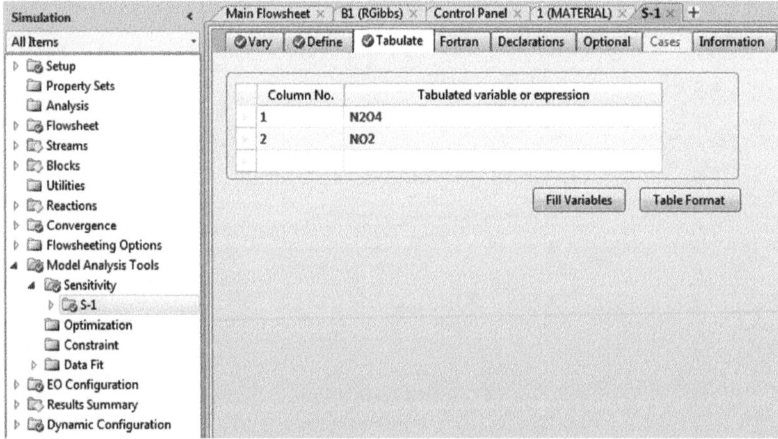

Now run the simulation. Going to **Sensitivity>S-1>Results** in Fig. 13-25 produces the results at a pressure of 1 bar as a function of temperature.

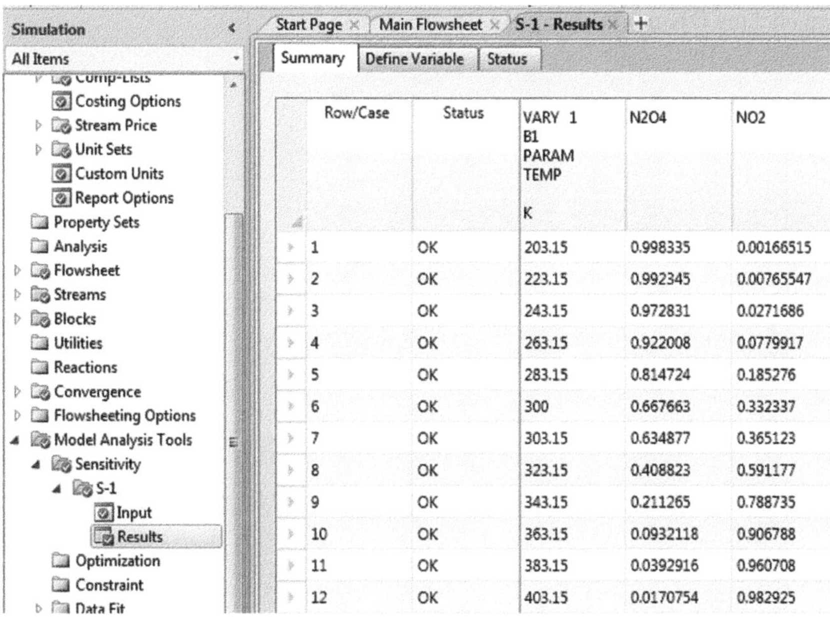

Row/Case	Status	VARY 1 B1 PARAM TEMP K	N2O4	NO2
1	OK	203.15	0.998335	0.00166515
2	OK	223.15	0.992345	0.00765547
3	OK	243.15	0.972831	0.0271686
4	OK	263.15	0.922008	0.0779917
5	OK	283.15	0.814724	0.185276
6	OK	300	0.667663	0.332337
7	OK	303.15	0.634877	0.365123
8	OK	323.15	0.408823	0.591177
9	OK	343.15	0.211265	0.788735
10	OK	363.15	0.0932118	0.906788
11	OK	383.15	0.0392916	0.960708
12	OK	403.15	0.0170754	0.982925

Figure 13-25

Note that the results are in Kelvin even though the sensitivity analysis required the use of centigrade. Now going to the **Plot>Results Curve** on the main toolbar (Fig. 13-26) brings up Fig. 13-27.

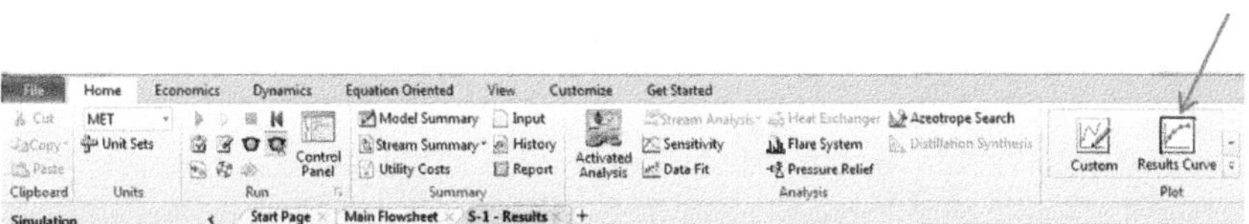

Figure 13-26

Results Curve
Select display options
X Axis: VARY 1 B1 PARAM TEMP K
Parametric Variable:
Select curve(s) to plot:
Select all
VARY 1 B1 PARAM TEMP K
N2O4
NO2
OK
Cancel
Clear Selections

Figure 13-27

Then clicking **OK** produces Fig. 13-28.

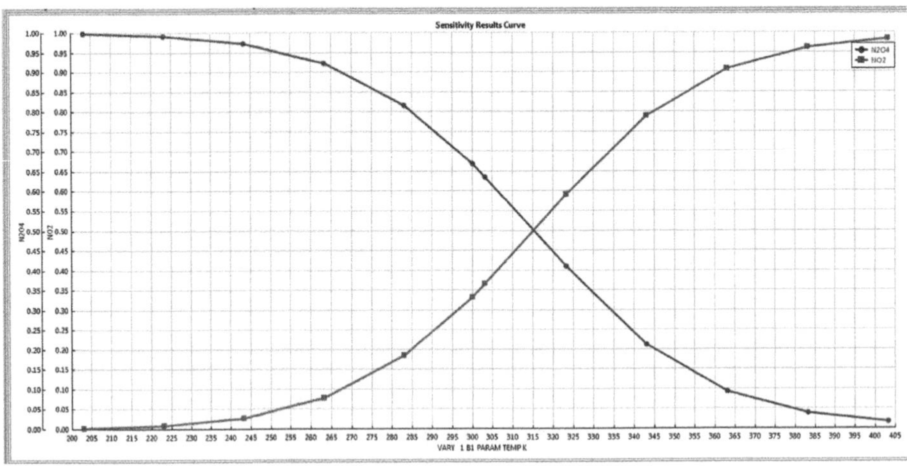

Figure 13-28

So we see that using **Modeling Analysis Tools>Sensitivity** allows the calculation at a number of different temperatures in a single run rather than having to do the calculation at each temperature separately by going to **Blocks>B1**, changing the temperature and rerunning the simulation.

Similarly, we can consider sensitivity of the results to pressure by creating a new sensitivity study, **S-2**, and varying pressure at fixed temperature. Figures 13-29 to 13-31 show that setup without commentary, since the procedure is similar to the case above of varying temperature.

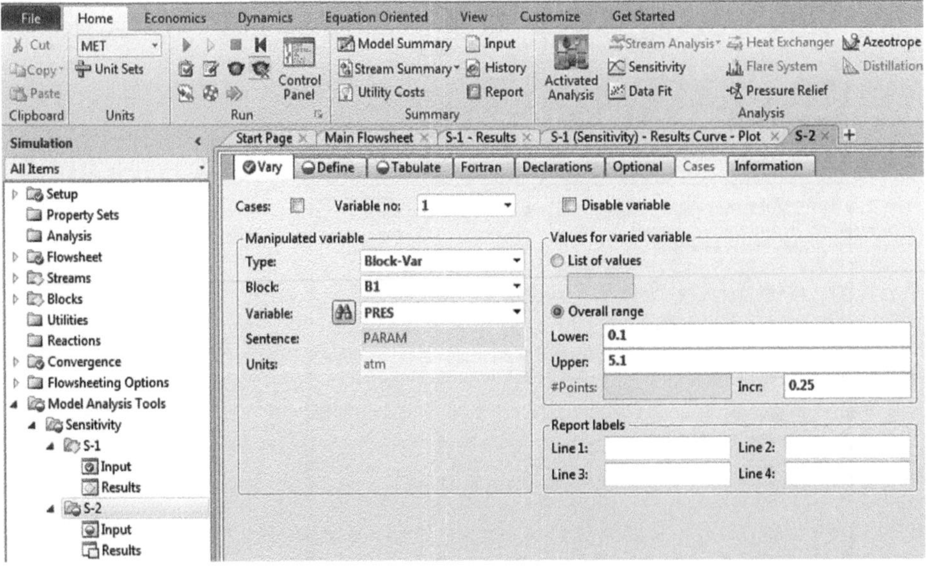

Figure 13-29

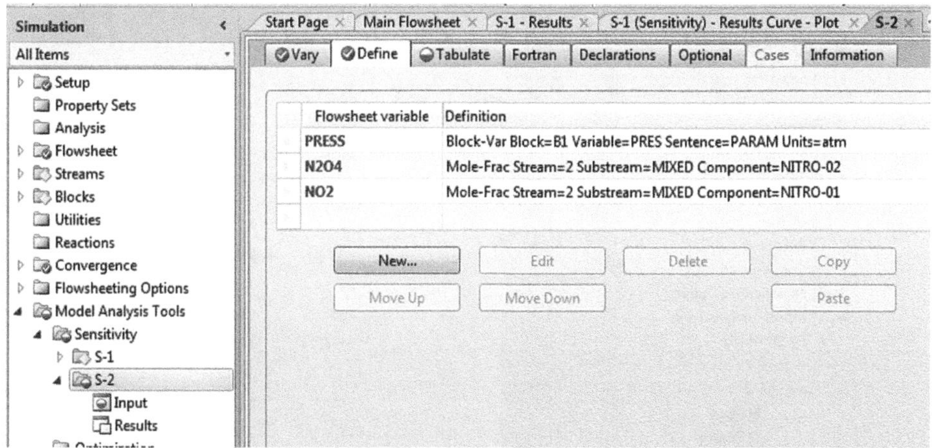

Figure 13-30

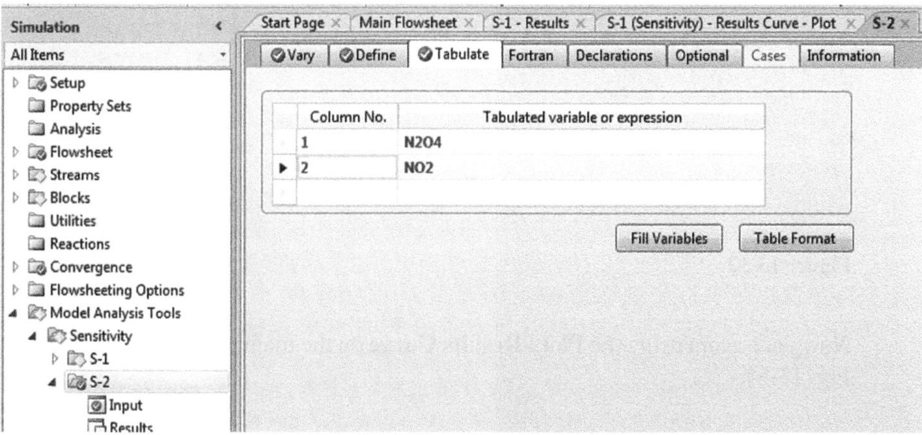

Figure 13-31

Then running the simulation produces the results in Fig. 13-32 (only some of which can be seen, the remainder is visible in the simulation by scrolling).

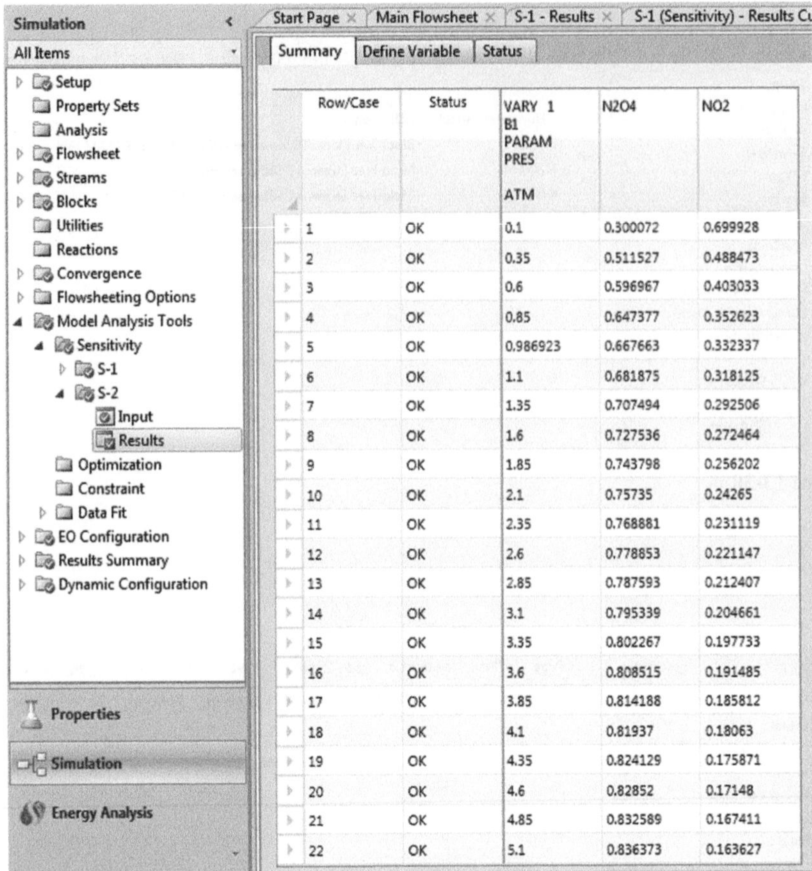

Row/Case	Status	VARY 1 B1 PARAM PRES ATM	N2O4	NO2
1	OK	0.1	0.300072	0.699928
2	OK	0.35	0.511527	0.488473
3	OK	0.6	0.596967	0.403033
4	OK	0.85	0.647377	0.352623
5	OK	0.986923	0.667663	0.332337
6	OK	1.1	0.681875	0.318125
7	OK	1.35	0.707494	0.292506
8	OK	1.6	0.727536	0.272464
9	OK	1.85	0.743798	0.256202
10	OK	2.1	0.75735	0.24265
11	OK	2.35	0.768881	0.231119
12	OK	2.6	0.778853	0.221147
13	OK	2.85	0.787593	0.212407
14	OK	3.1	0.795339	0.204661
15	OK	3.35	0.802267	0.197733
16	OK	3.6	0.808515	0.191485
17	OK	3.85	0.814188	0.185812
18	OK	4.1	0.81937	0.18063
19	OK	4.35	0.824129	0.175871
20	OK	4.6	0.82852	0.17148
21	OK	4.85	0.832589	0.167411
22	OK	5.1	0.836373	0.163627

Figure 13-32

Now once again using the **Plot>Results Curve** on the main toolbar (Fig. 13-33), we obtain Fig. 13-34.

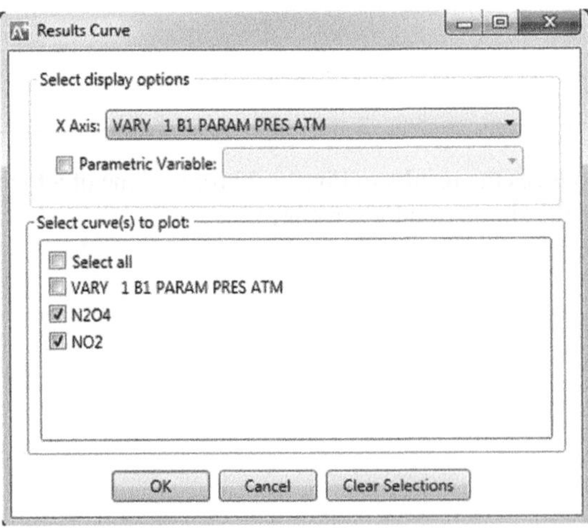

Figure 13-33

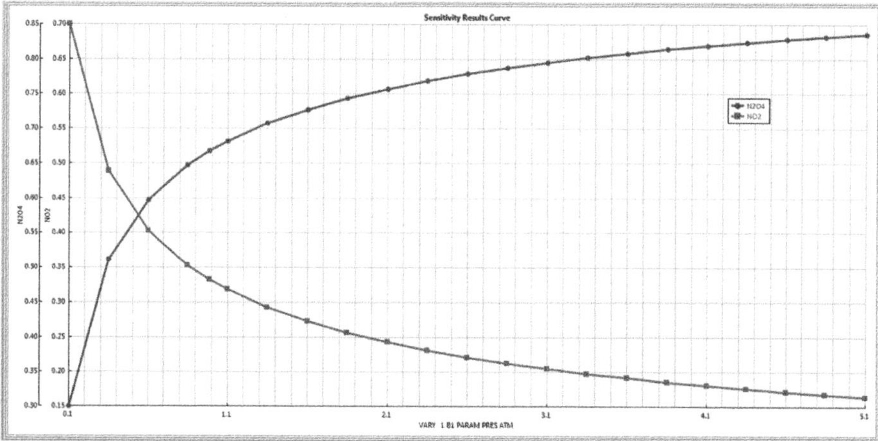

Figure 13-34

Having this simulation set up, it is then easy to consider other cases. For example, to use the **Sensitivity** analysis to do calculations at 10 bar and over the temperature range of 300 to 600 K (26.85 to 326.85°C), one first changes the pressure to 10 bar in **Block B1**, and the temperature range in **Model Analysis Tools>Sensitivity>S-1>Input**, which produces the results in Fig. 13-35.

Simulation
All Items
Setup
Property Sets
Analysis
Flowsheet
Streams
Blocks
B1
Utilities
Reactions
Convergence
Flowsheeting Options
Model Analysis Tools
Sensitivity
S-1
Input
Results
S-2
Optimization
Constraint
Data Fit
EO Configuration
Results Summary
Dynamic Configuration
Properties

Start Page | Main Flowsheet | S-1 - Results

Summary | Define Variable | Status

Row/Case	Status	VARY 1 B1 PARAM TEMP K	N2O4	NO2
1	OK	300	0.879388	0.120612
2	OK	320	0.769062	0.230938
3	OK	340	0.612739	0.387261
4	OK	360	0.431649	0.568351
5	OK	380	0.265841	0.734159
6	OK	400	0.146854	0.853146
7	OK	420	0.0768209	0.923179
8	OK	440	0.0400428	0.959957
9	OK	460	0.0214308	0.978569
10	OK	480	0.0119231	0.988077
11	OK	500	0.0069153	0.993085
12	OK	520	0.0041759	0.995824
13	OK	540	0.00261806	0.997382
14	OK	560	0.00169864	0.998301
15	OK	580	0.00113694	0.998863
16	OK	600	0.000782747	0.999217

Figure 13-35

Then using the **Plot>Results Curve** on the **Main Toolbar** results in the graph in Fig. 13-36.

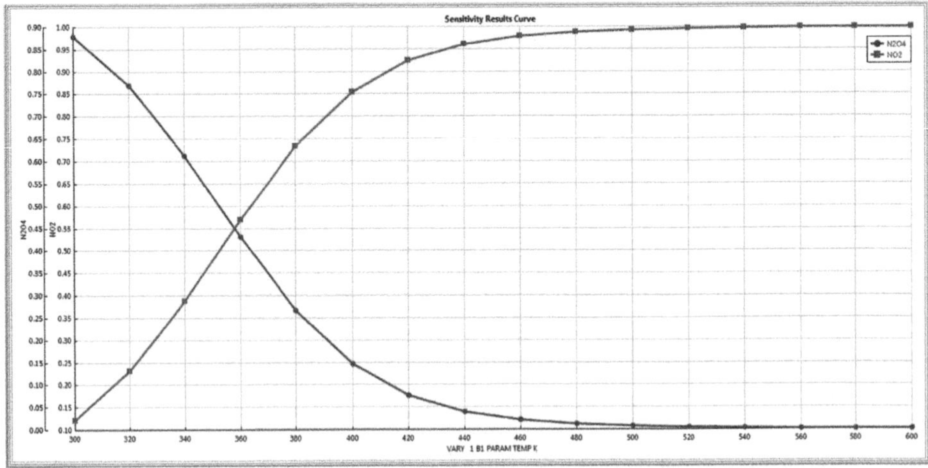

Figure 13-36

It is interesting to note that at 400 K the mole fraction of NO_2 is 0.9806 at 1 bar (indicating almost complete dissociation of N_2O_4), but only is 0.1469 at 10 bar. Is this shift in chemical equilibrium with pressure in accord with the principle of Le Chatelier and Braun?

PROBLEMS

Use the sensitivity analysis tool discussed in this chapter to redo the calculations for the following problems.

13.1. Problem 9.6.

13.2. Problem 9.8.

13.3. Problem 9.15.

13.4. Problem 9.18.

13.5. Problem 9.22.

13.6. Problem 9.23.

Chapter 14

Electrolyte Solutions

Dealing with electrolyte solutions in Aspen Plus® is different than with other mixtures. This is illustrated in this chapter with two examples. The most important differences are (a) that electrolytes ionize so that identifying the species in the mixture is itself a problem, and (b) that because of the long-range Coulombic forces between the ions, different activity coefficient models need to be used than for nonionic systems. Strong electrolytes, such as hydrogen chloride and sodium hydroxide are completely or almost completely ionized in aqueous solution at common conditions

$$\mathrm{HCl} \Leftrightarrow \mathrm{H^+} + \mathrm{Cl^-}$$

$$\mathrm{NaOH} \Leftrightarrow \mathrm{Na^+} + \mathrm{OH^-}$$

so that only the ions are present in solution. However, weak electrolytes such as acetic acid and water are only partially ionized. That is, though acetic acid ionizes

$$\mathrm{CH_3COOH} \Leftrightarrow \mathrm{CH_3COO^-} + \mathrm{H^+}$$

un-ionized acetic acid, the acetate ion and the hydrogen (or hydronium) ion are all present in solution. Fortunately, Aspen Plus provides a wizard, the **Elec Wizard**, discussed below, to retrieve data from the Aspen Plus data bank for the ionization reactions, their equilibrium constants, and the ionic species that are present in solution.

However, before we discuss the use of Aspen Plus with electrolyte solutions, it is useful to review some of the basic ideas and terminology. A general single electrolyte system is schematically represented as

$$\mathrm{A}_{\nu+}B_{\nu-} \Leftrightarrow \nu_+\mathrm{A}^{z+} + \nu_-\mathrm{B}^{z-}$$

where ν is the stoichiometric coefficient of an ion and z is its charge. By electrical neutrality $\nu_+z_+ + \nu_-z_- = 0$. Most thermodynamic models, such as the Debye–Hückel model discussed below, do not lead to predictions of the activity coefficient of the each type of ion. [The ENRTL-RK model in Aspen Plus is an exception in that it does lead to predictions of activity coefficients of the individual ionic species.] Therefore, the most common practice in the literature is to define a mean ionic activity coefficient $\gamma_\pm$ as

$$\gamma_\pm^\nu = \left(\gamma_\mathrm{A}^\square\right)^{\nu_+}\left(\gamma_\mathrm{B}^\square\right)^{\nu_-} \quad \text{where } \nu = \nu_+ + \nu_-$$

and $\gamma_i^{\square}$ is the activity coefficient of species i defined so that it approaches unity as the molality species i, M_i, approaches zero. The molality M_i is defined as the number of moles of species i per 1000 g of solvent. Also, the ionic strength of a solution, I, is defined to be

$$I = \frac{1}{2} \sum_{\text{all ions}} z_i^2 M_i$$

The Debye–Hückel limiting law, based on theory, relating the mean ionic activity coefficient to the ionic strength in very dilute ionic solutions is

$$\ln \gamma_{\pm} = \alpha |z_+ z_-| \sqrt{I}$$

where α is a parameter of order unity in units of $(\text{mol/kg})^{-1/2}$ that depends on the solvent and the temperature. Finally, the pH of a solution is defined as follows

$$\text{pH} = -\log_{10}(a_{\text{H}^+}) = -\log_{10}\left(\frac{\gamma_{\text{H}^+}^{\square} M_{\text{H}^+}}{M = 1}\right) \approx -\log_{10}\left(\frac{M_{\text{H}^+}}{M = 1}\right)$$

The last expression on the right is valid when the value of the hydrogen ion activity coefficient is near unity, which occurs if the solution is very dilute in ions. Many common electrolyte solution models used in thermodynamics start from the Debye–Hückel model and add extra terms for use with solutions that are not highly dilute in ions, that is, for solutions at moderate or high ionic strengths.

To analyze a solution containing electrolytes, Aspen Plus is started and after selecting **New...** in the window in Fig. 14-1, select **Electrolytes,** then **Electrolytes with Metric Units** and then **Create.**

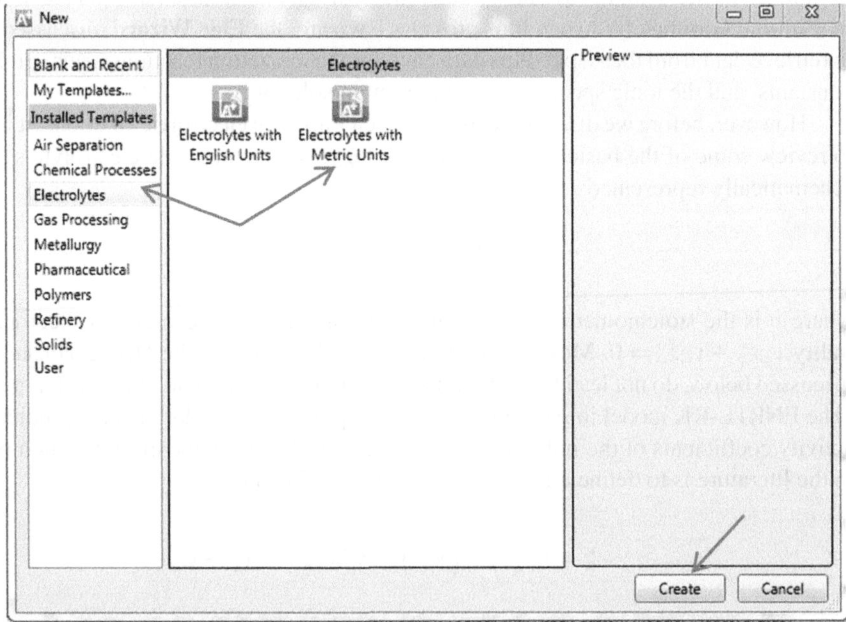

Figure 14-1

Note that with this choice, Aspen Plus opens in **Properties**. Go to **Simulation** (not **Properties**) and in **Setup>Report Options>Stream** both **Mole** and **Mass** have been checked in Fig. 14-2 under **Flow Basis** since there will be interest in having information both on mole fraction and, based on the discussion above, molality.

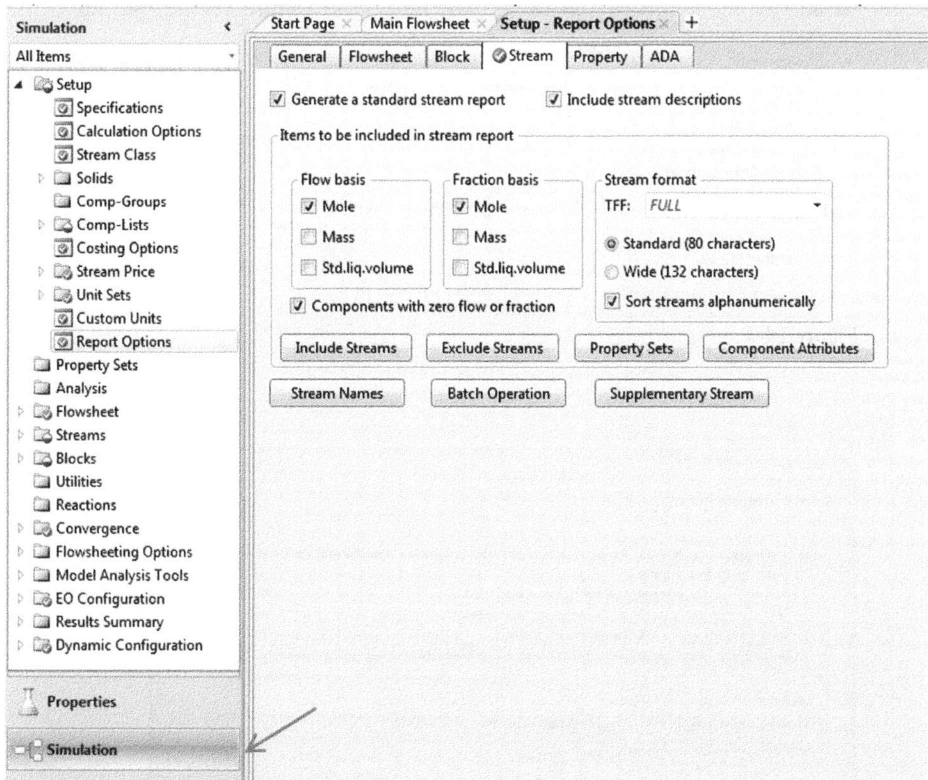

Figure 14-2

For the first example a hydrogen chloride + water mixture at 25°C and 1 atm will be considered. The calculations that follow can be done in several ways. In this chapter, for consistency with a later example, we will use the **Simulation** mode with the **Flash2** separator shown in Fig. 14-3.

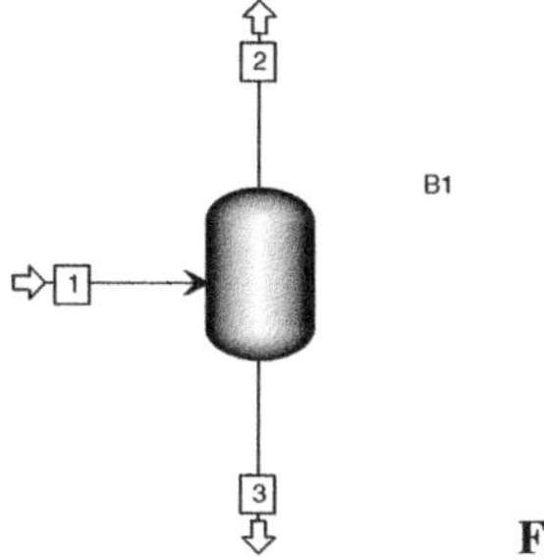

Figure 14-3

Next in the **Properties> Components>Specifications** only HCl has been added as water appears once the **Electrolytes** choice has been made (see Fig. 14-4). Next, click on **Elec Wizard** and the pop-up window in Fig. 14-5 appears:

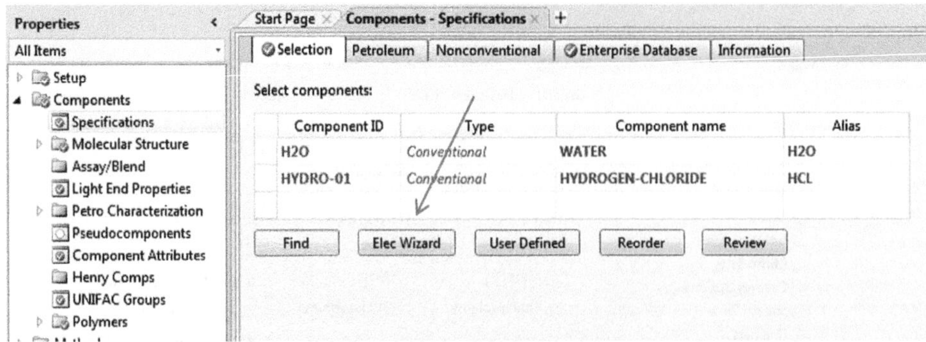

Figure 14-4

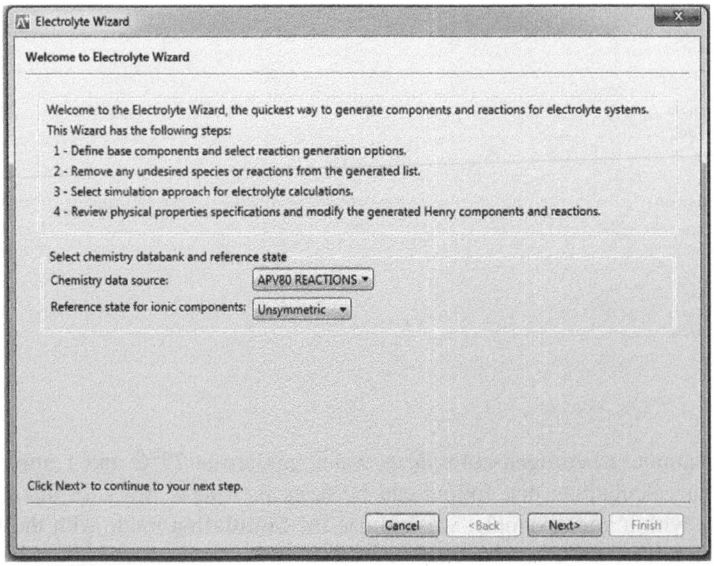

Figure 14-5

Clicking on **Next** brings up the window in Fig. 14-6 with the defaults shown.

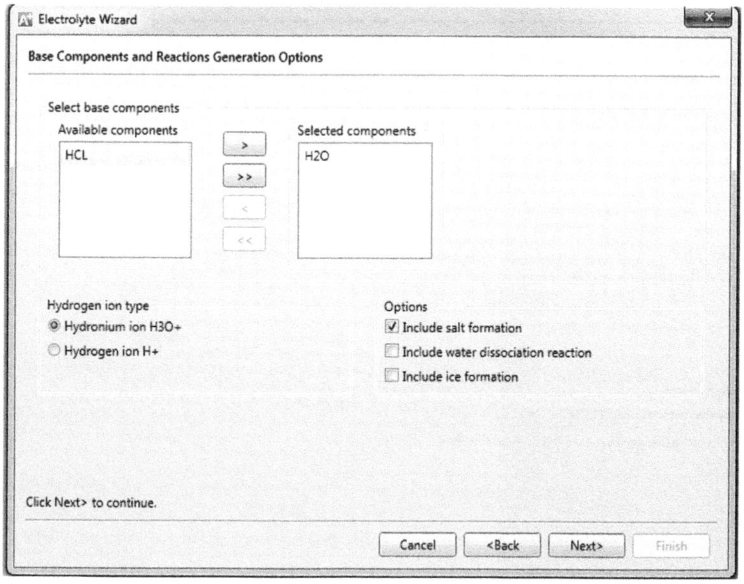

Figure 14-6

Then add **HCl** to the **Selected Components**. Also I chose to add to the default choices to have the results presented in terms of the **Hydrogen ion H+** rather than the **Hydronium ion H3O+**, and **Include water dissociation reaction** and **Include salt formation** under **Options** (Fig. 14-7).

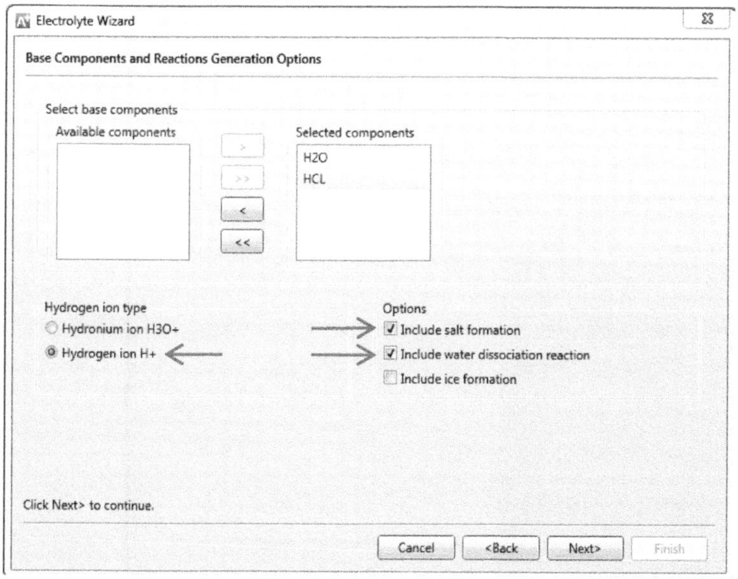

Figure 14-7

For the case here, this brings up the window of Fig. 14-8 in which Aspen has chosen the ionization reactions that will be considered.

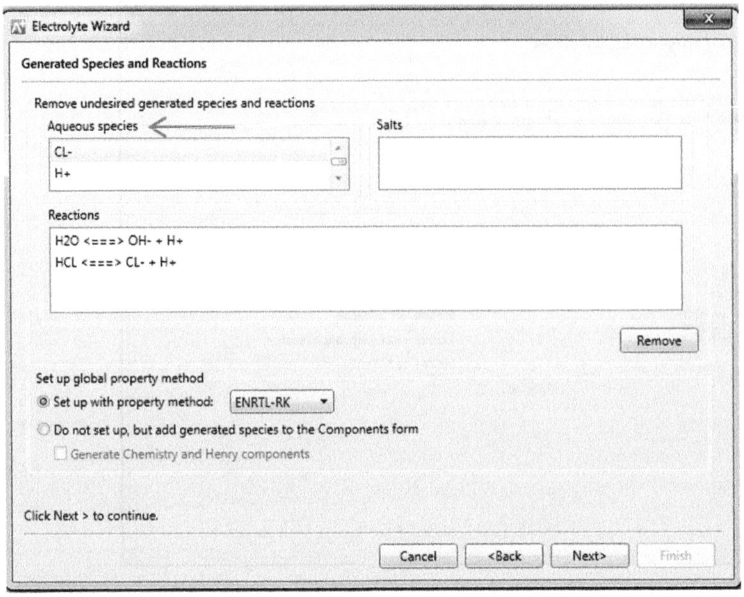

Figure 14-8

Notice that under **Aqueous species** the hydrogen, chlorine, and hydroxide ions are now present (though you have to scroll down to see the OH^- ion) and will be considered, and the reactions that lead to their formation will be added to the simulation. Clicking **Next>** leads to the window in Fig. 14-9.

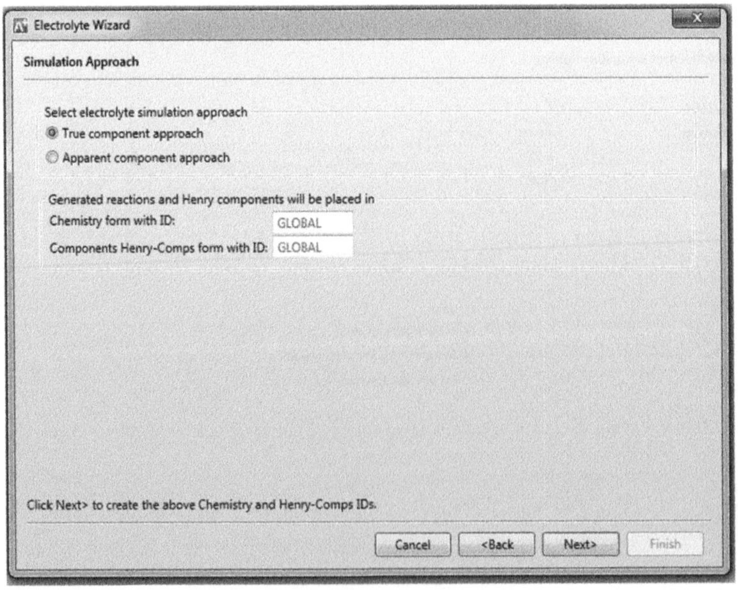

Figure 14-9

Retaining the **True component approach** leads to all ions being explicitly listed in the output. The **Apparent component approach** would lead to compounds rather than ions being listed in the output. In the case of multiple ionization reactions, this difference would be important. Clicking **Next** > gives the window of Fig. 14-10.

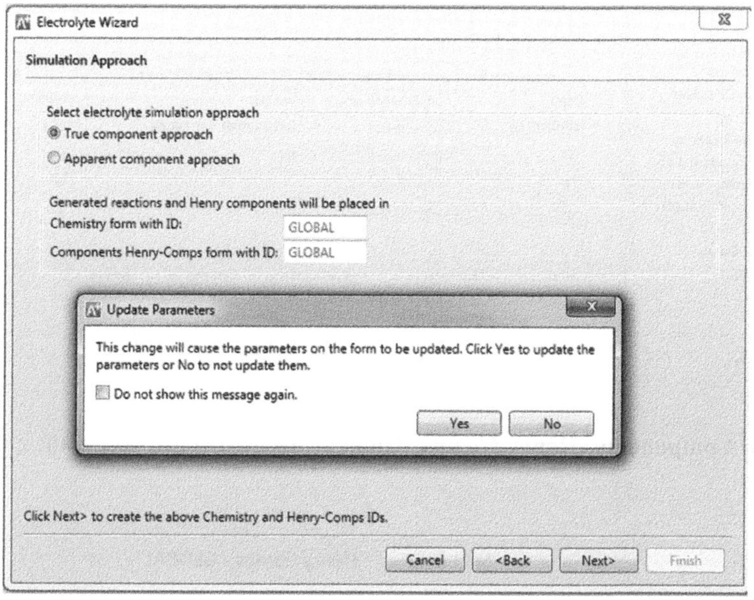

Figure 14-10

Click **Yes** to see the window in Fig. 14-11 and click **Finish**.

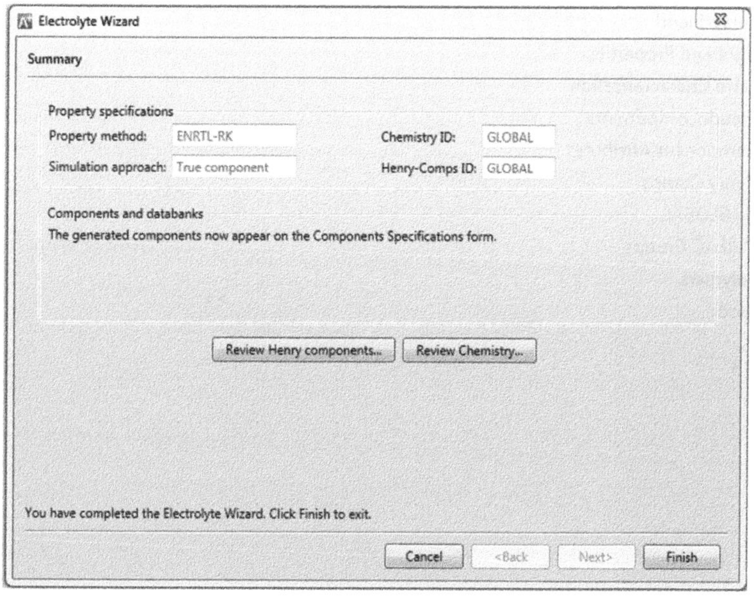

Figure 14-11

The **Components>Specifications>Selection** now has been modified by Aspen Plus to include the ions as shown in Fig. 14-12.

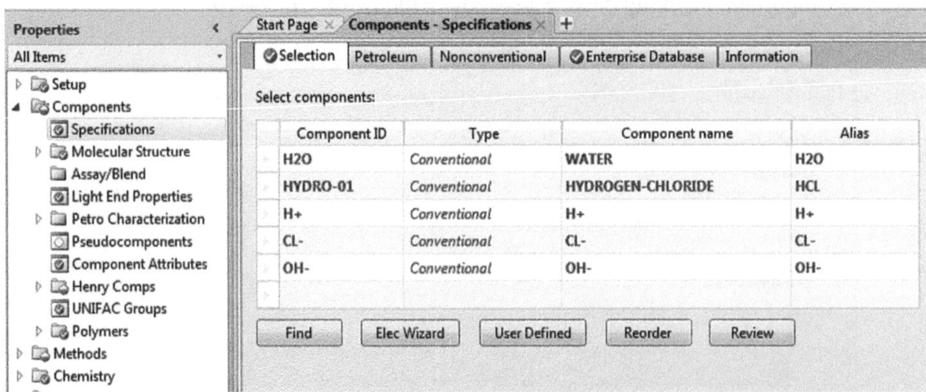

Figure 14-12

Then go to **Components>Henry Comps>Global** (Fig. 14-13) and accept the choice of HCl.

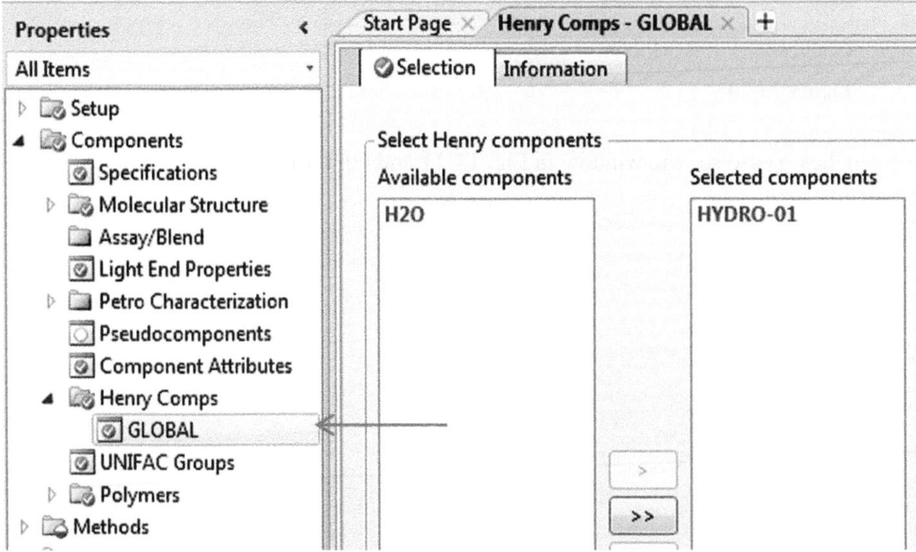

Figure 14-13

Next go to **Methods** (Fig. 14-14) and see the choices that have been made by having selected **Electrolytes with Metric Units** choice in **Setup**.

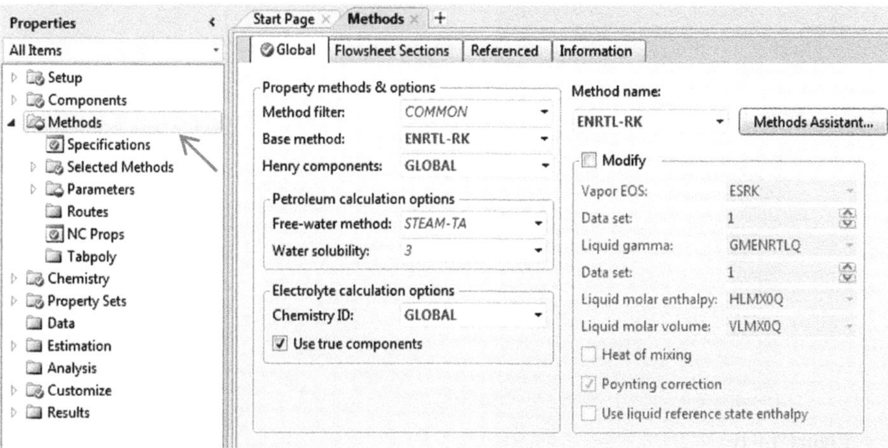

Figure 14-14

Next go to **Methods>Parameters>Binary Interaction** (Fig. 14-15) and accept all the default parameter choices.

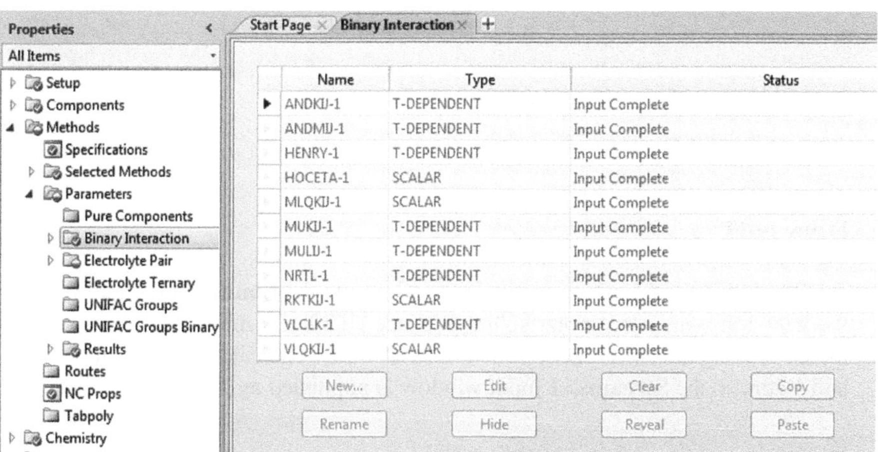

Figure 14-15

Then go to **Methods> Parameters>Electrolyte Pair** (Fig. 14-16) and accept each of the default parameter choices.

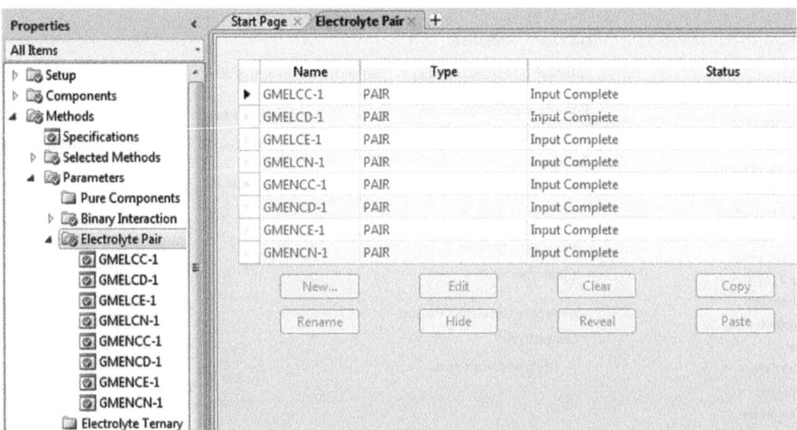

Figure 14-16

As an example, when clicking on **GMELCC-1** the window of Fig. 14-17 appears:

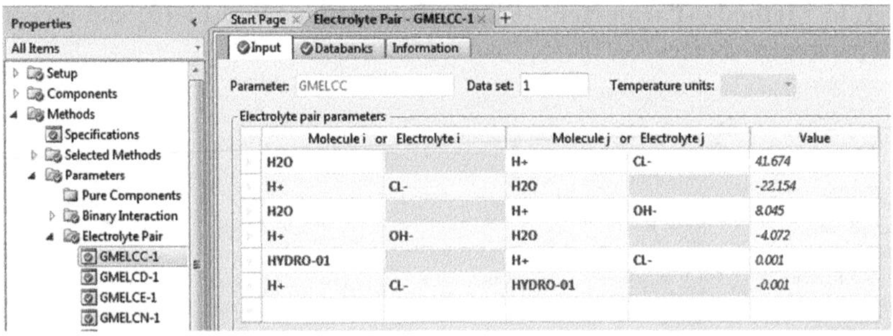

Figure 14-17

The **Properties** window is now complete. Going to the **Simulation** window and **Streams**, we will start with a 0.1 molar solution. Since HCl has a molecular weight of 36.46, that means 3.646 grams of HCl per 1000 grams of water. The calculations will be done at 25°C and 1 atm, so the **Streams>1** input window is populated as shown in Fig. 14-18.

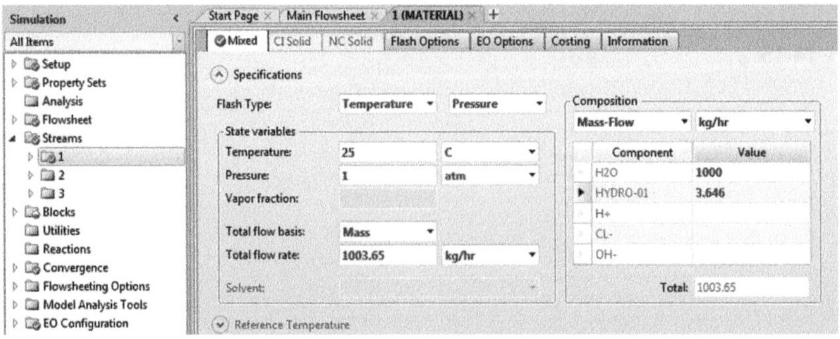

Figure 14-18

[Note that 1003.646 appears as 1003.65 since only six significant figures are displayed.] Next go to **Blocks>B1>Specification** and specify 25°C and 1 atm (Fig. 14-19), where **Blocks>B1** has been previously set in the **Flowsheet** to be the **Flash2** separator block.

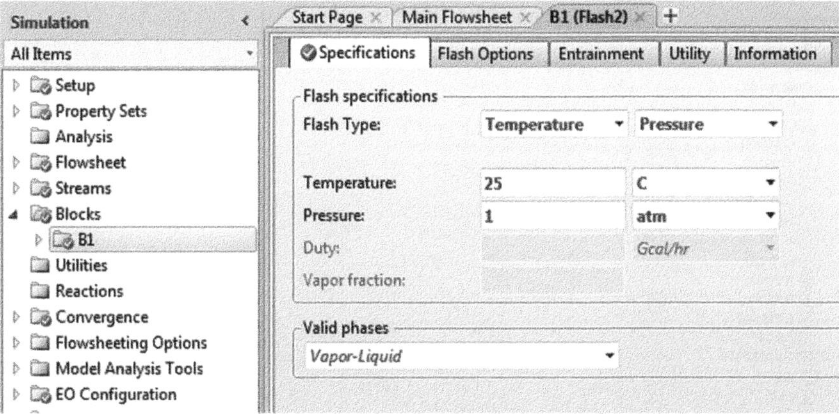

Figure 14-19

This completes the input for the simulation. Running the simulation gives the results in **Results Summary>Streams** shown in Fig. 14-20.

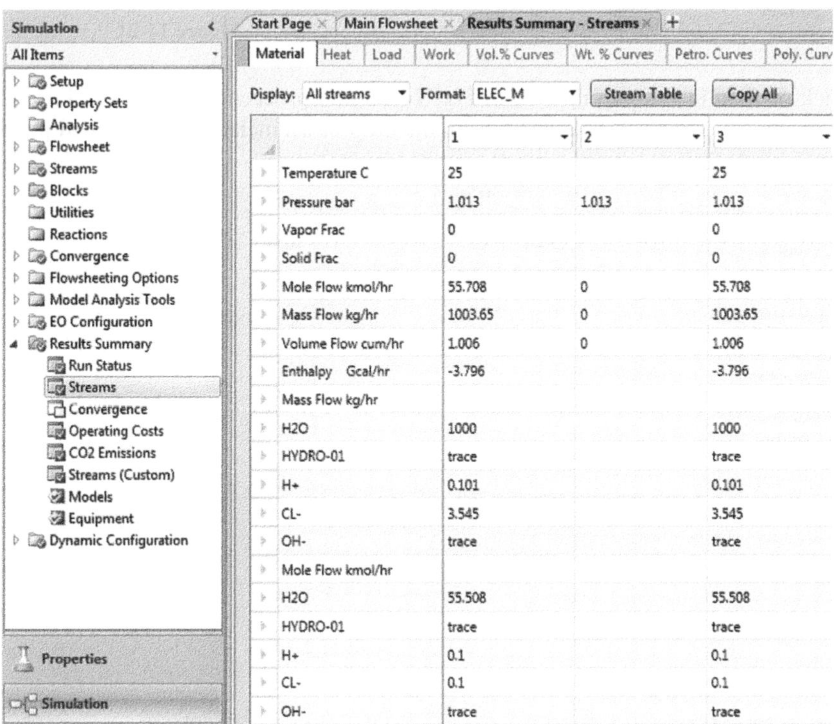

	1	2	3
Temperature C	25		25
Pressure bar	1.013	1.013	1.013
Vapor Frac	0		0
Solid Frac	0		0
Mole Flow kmol/hr	55.708	0	55.708
Mass Flow kg/hr	1003.65	0	1003.65
Volume Flow cum/hr	1.006	0	1.006
Enthalpy Gcal/hr	-3.796		-3.796
Mass Flow kg/hr			
H2O	1000		1000
HYDRO-01	trace		trace
H+	0.101		0.101
CL-	3.545		3.545
OH-	trace		trace
Mole Flow kmol/hr			
H2O	55.508		55.508
HYDRO-01	trace		trace
H+	0.1		0.1
CL-	0.1		0.1
OH-	trace		trace

Figure 14-20

The results in Fig. 14-20 are incomplete and not very interesting since they do not provide information on the solution pH or the species activity coefficients.

Now go back to **Properties** and click on **Property Sets**, which brings up a list of Aspen Plus templates for a number of properties, for example **PH** for pH as shown in Fig. 14-21. Click on **New**.

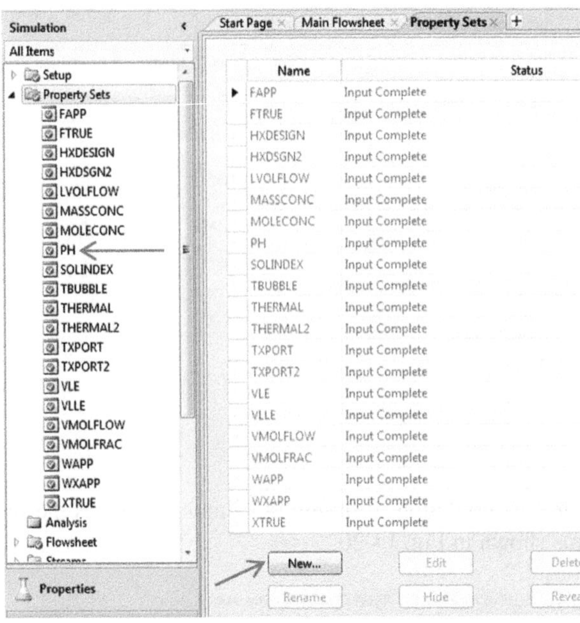

Figure 14-21

Click on **OK** in the pop-up box in Fig. 14-22 and proceed, as shown in Fig. 14-23, to **Properties** by choosing **GAMMA** from the drop-down menu under **Physical properties.**

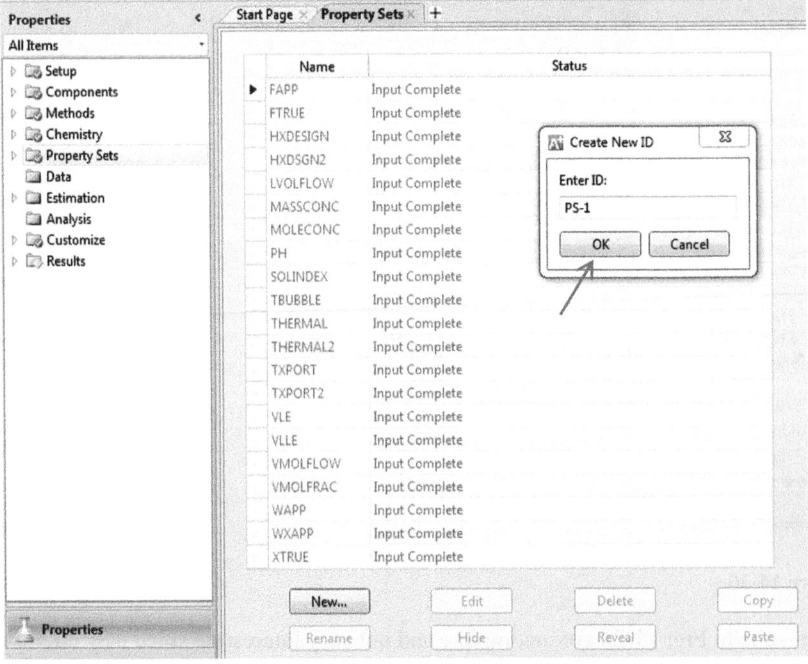

Figure 14-22

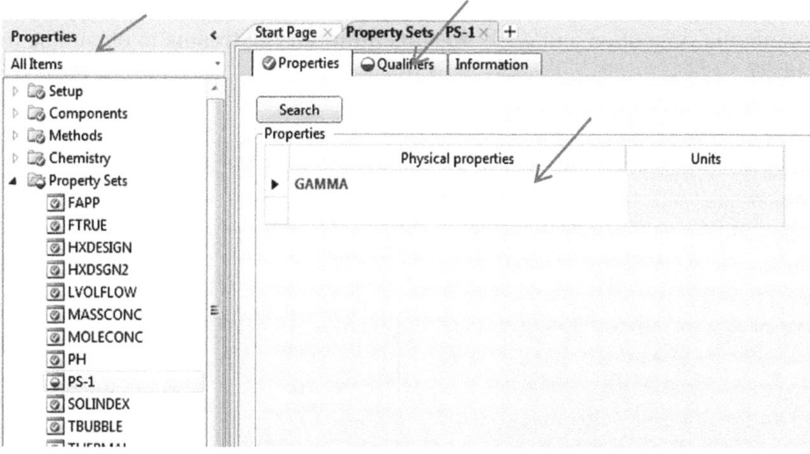

Figure 14-23

Then after clicking on the **Qualifiers** tab (Fig. 14-24), populate the window as shown with each of the ions.

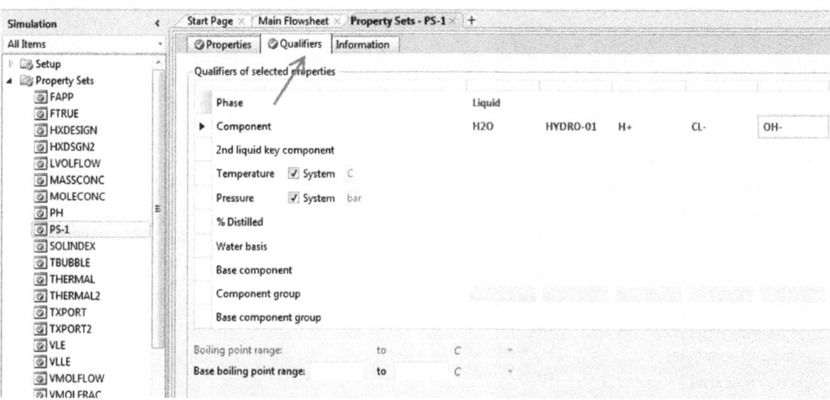

Figure 14-24

Now go back to **Setup>Report Options>Stream** and click on **Property Sets** (Fig. 14-25).

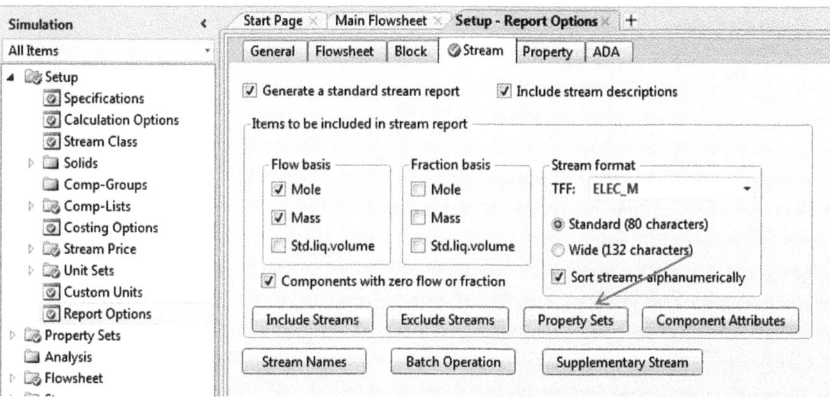

Figure 14-25

Then scroll down to **PS-1** and **PH** (Fig. 14-26) and move them to **Selected property sets**. Now rerun the simulation and select **Results Summary>Streams** to obtain the results in Fig. 14-27.

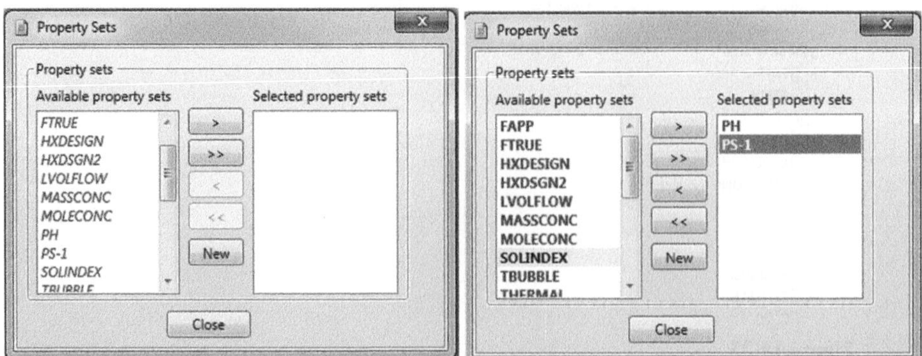

Figure 14-26

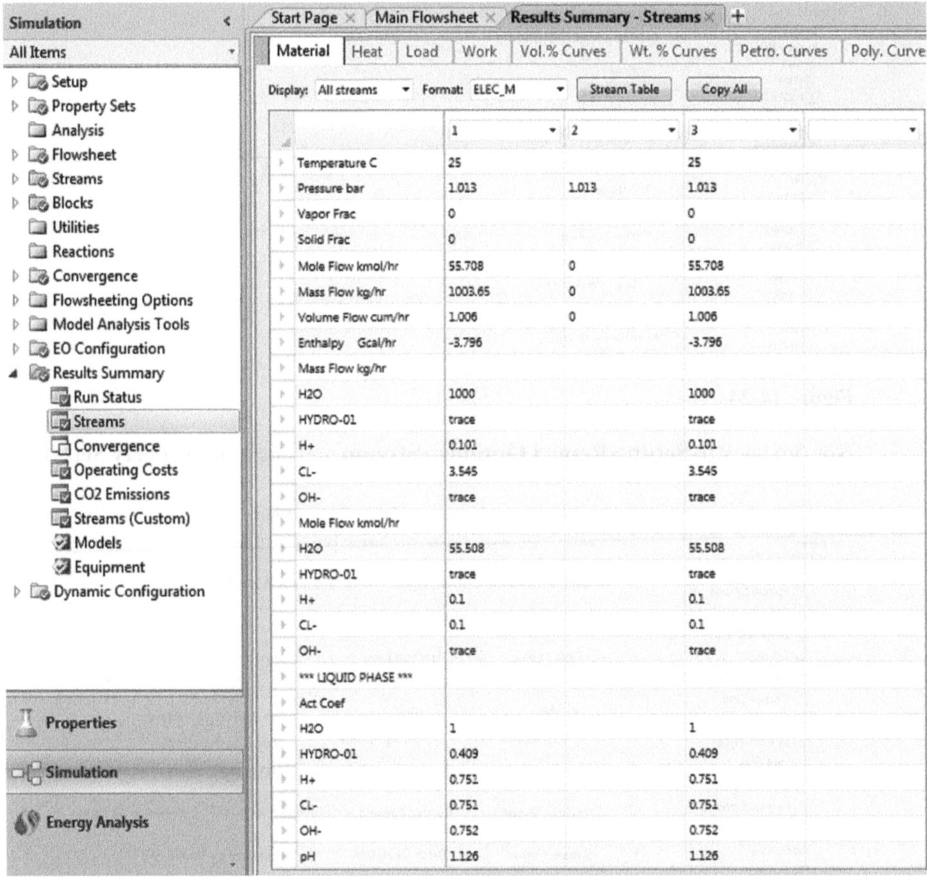

	1	2	3
Temperature C	25		25
Pressure bar	1.013	1.013	1.013
Vapor Frac	0		0
Solid Frac	0		0
Mole Flow kmol/hr	55.708	0	55.708
Mass Flow kg/hr	1003.65	0	1003.65
Volume Flow cum/hr	1.006	0	1.006
Enthalpy Gcal/hr	-3.796		-3.796
Mass Flow kg/hr			
H2O	1000		1000
HYDRO-01	trace		trace
H+	0.101		0.101
CL-	3.545		3.545
OH-	trace		trace
Mole Flow kmol/hr			
H2O	55.508		55.508
HYDRO-01	trace		trace
H+	0.1		0.1
CL-	0.1		0.1
OH-	trace		trace
*** LIQUID PHASE ***			
Act Coef			
H2O	1		1
HYDRO-01	0.409		0.409
H+	0.751		0.751
CL-	0.751		0.751
OH-	0.752		0.752
pH	1.126		1.126

Figure 14-27

So that now we have the activity coefficients of each of the species (ions and water) and the pH of the solution supplied by Aspen Plus.

To verify how the pH was computed we can go back to

$$\mathrm{pH} = -\log_{10}(a_{\mathrm{H}^+}) = -\log_{10}\left(\frac{\gamma_{\mathrm{H}^+}^{\square} M_{\mathrm{H}^+}}{M = 1\mathrm{M}}\right) \approx -\log_{10}\left(\frac{M_{\mathrm{H}^+}}{M = 1\mathrm{M}}\right)$$

so that

$$\mathrm{pH} \approx -\log_{10}\left(\frac{M_{\mathrm{H}^+}}{M = 1\mathrm{M}}\right) = -\log_{10}\left(\frac{0.1}{M = 1\mathrm{M}}\right) = 1$$

which is only approximately correct, while

$$\mathrm{pH} = -\log_{10}\left(\frac{\gamma_{\mathrm{H}^+}^{\square} M_{\mathrm{H}^+}}{M = 1\mathrm{M}}\right) = -\log_{10}\left(\frac{0.751 \times 0.1}{M = 1\mathrm{M}}\right) = 1.124$$

is, to within roundoff error, in agreement with the result in the table above.

It is of interest to repeat the calculation at a collection of HCl molalities. This is most easily accomplished using the sensitivity analysis introduced in the previous chapter. This is done as follows. Start at **Model Analysis Tools>Sensitivity>New** (Fig. 14-28).

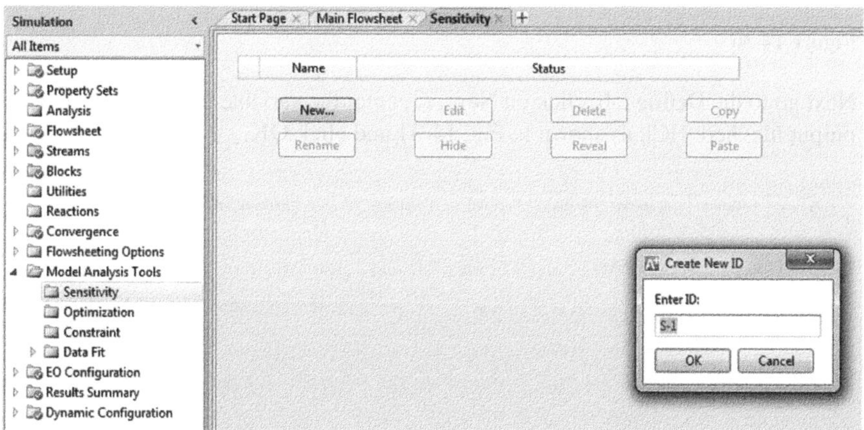

Figure 14-28

Click on **OK** to accept the default name of the sensitivity study, **S-1**, (or choose a name of your own), which brings up the window in Fig. 14-29. Start at the **Vary** tab and populate the window it brings up as shown in Fig. 14-30. Note that the HCl flows chosen cover the range of 0 to 2.0 molar.

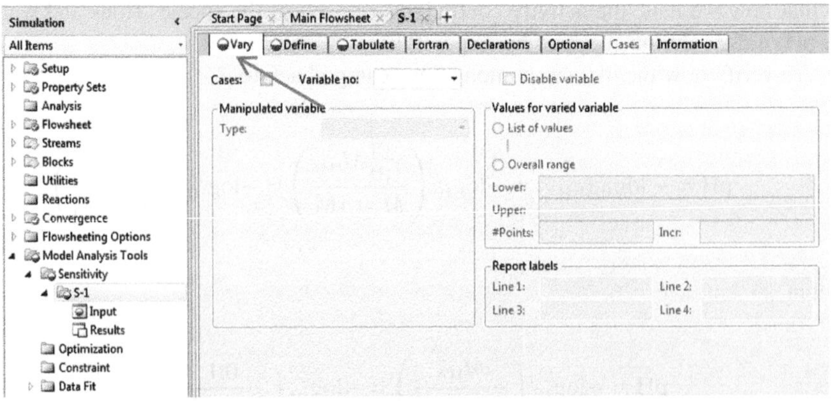

Figure 14-29

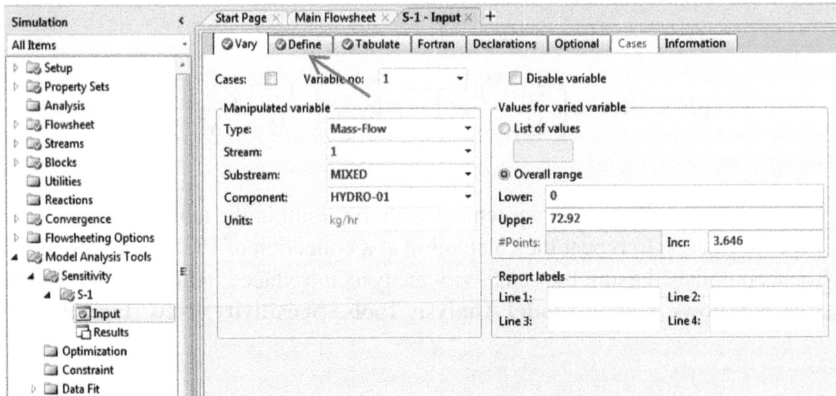

Figure 14-30

Next go to the **Define** tab, click on **New . . .**, enter the variable name you wish to use in the output file, here HCl, as shown in Fig. 14-31 and click **OK**.

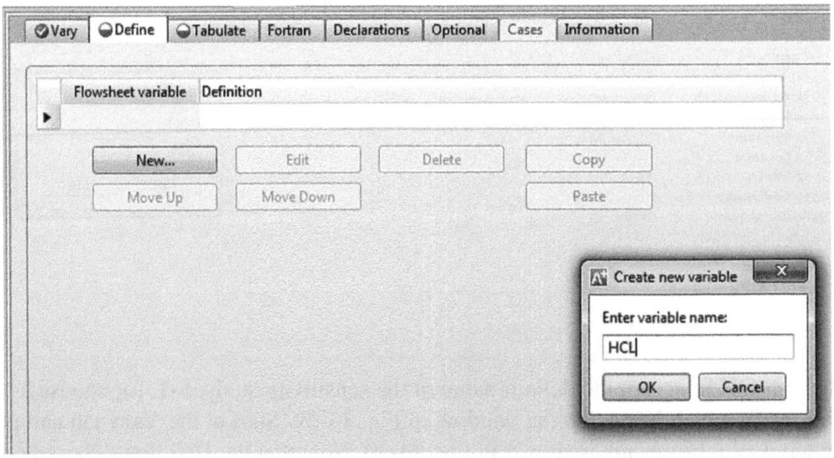

Figure 14-31

Then populate the window as shown in Fig. 14-32 to define that variable.

Simulation
All Items
Setup
Property Sets
Analysis
Flowsheet
Streams
Blocks
Utilities
Reactions
Convergence
Flowsheeting Options
Model Analysis Tools
Sensitivity
S-1
Input
Results
Optimization
Constraint
Data Fit
EO Configuration
Results Summary
Dynamic Configuration
Properties
Simulation
Start Page
Main Flowsheet
S-1
Vary
Define
Tabulate
Fortran
Declarations
Optional
Cases
Information
Flowsheet variable
Definition
HCL
Mass-Flow Stream=1 Substream=MIXED Component=HYDRO-01 Units=kg/hr
New...
Edit
Delete
Copy
Move Up
Move Down
Paste
Variable Definition
Select a variable category and reference
Variable name: HCL
Category
All
Blocks
Streams
Model Utility
Physical Property Parameters
Reactions
Reference
Type: Mass-Flow
Stream: 1
Substream: MIXED
Component: HYDRO-01
Units: kg/hr
Close

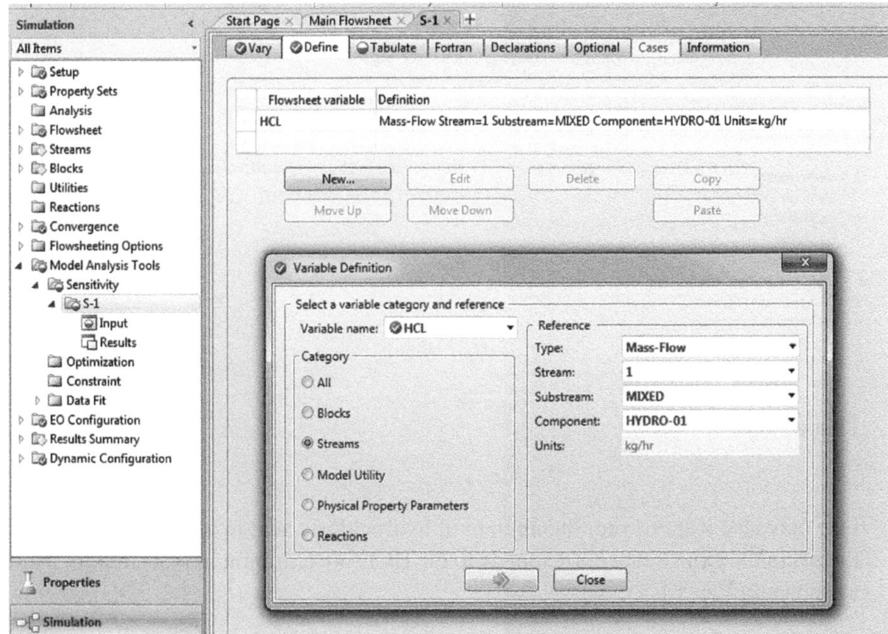

Figure 14-32

Next click on **New...** and define a second variable **PH** (Fig. 14-33).

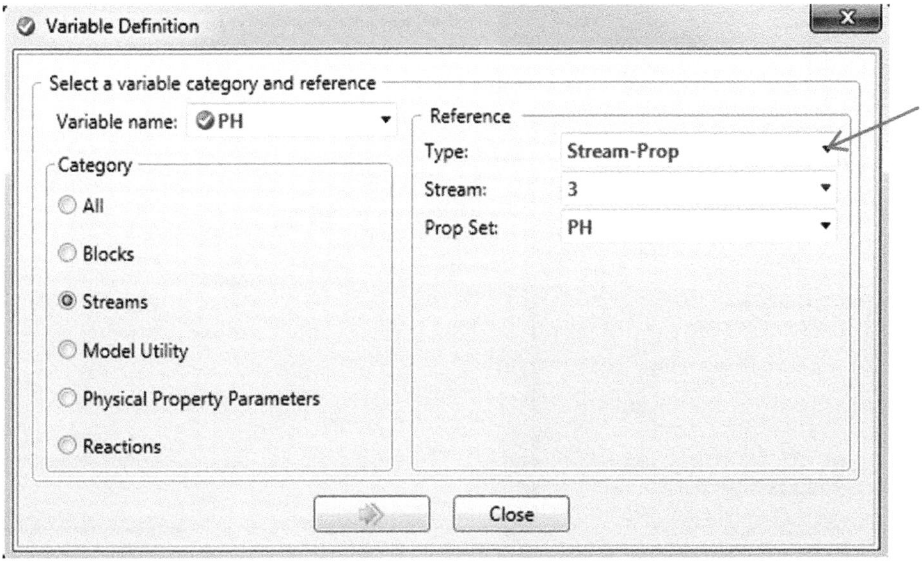

Variable Definition
Select a variable category and reference
Variable name: PH
Category
All
Blocks
Streams
Model Utility
Physical Property Parameters
Reactions
Reference
Type: Stream-Prop
Stream: 3
Prop Set: PH
Close

Figure 14-33

Note that **pH** is chosen to be a stream property (**Stream-Prop**), which provides the possibility of choosing pH as a **Prop Set** (Fig. 14-34). So now we have (Fig. 14-34):

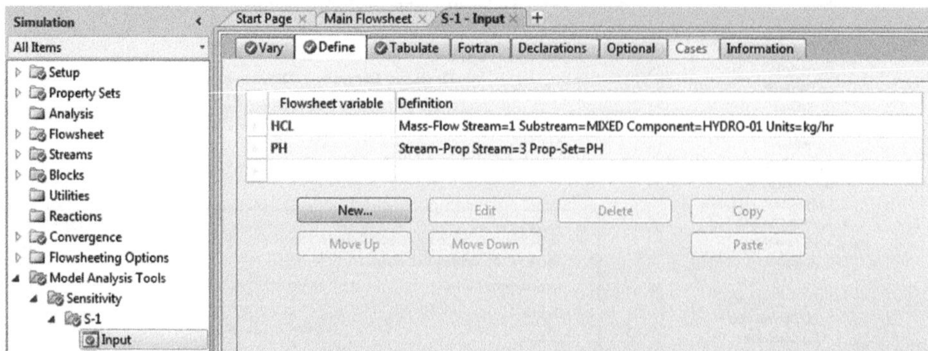

Figure 14-34

One may also want to vary the amount of hydrochloric acid in the exit liquid stream (since the pH in the exit stream corresponds to the HCl concentration in that stream). This is done as shown in Fig. 14-35.

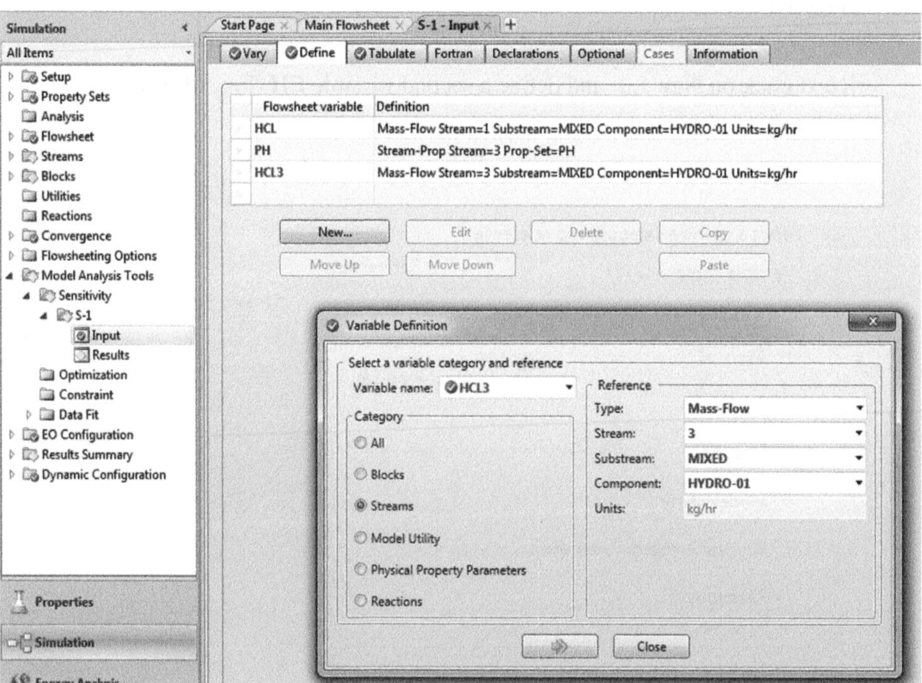

Figure 14-35

Then move on to the **Tabulate** tab, which brings up a window that has been populated as shown in Fig. 14-36.

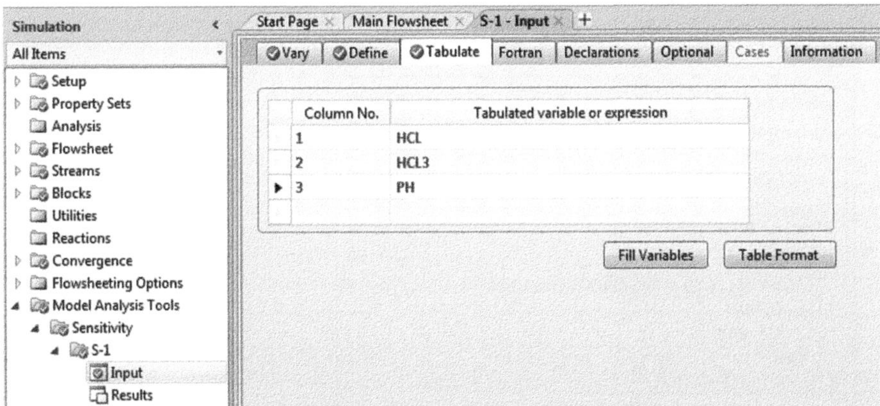

Figure 14-36

Finally, running the simulation and going to **Model Analysis Tools>Sensitivity>S-1>Results**, we see the results in Fig. 14-37. In particular, note that there is no change in HCl concentration from the feed to the exiting liquid; that is because essentially no HCl is lost to the vapor stream or to ionization. Also note that the pH is low (except at 0 molar) and is even negative at high HCl molarities.

Row/Case	Status	VARY 1 1 MIXED HYDRO-01 MASSFLOW KG/HR	HCL KG/HR	HCL3 KG/HR	PH
1	OK	0	0	0	6.99918
2	OK	3.646	1.21189e-14	1.21189e-14	1.12599
3	OK	7.292	9.55708e-14	9.55708e-14	0.865
4	OK	10.938	4.3542e-13	4.3542e-13	0.715794
5	OK	14.584	1.56827e-12	1.56827e-12	0.61059
6	OK	18.23	4.92303e-12	4.92303e-12	0.528508
7	OK	21.876	1.40679e-11	1.40679e-11	0.460394
8	OK	25.522	3.74594e-11	3.74594e-11	0.401447
9	OK	29.168	9.4269e-11	9.4269e-11	0.348855
10	OK	32.814	2.26304e-10	2.26304e-10	0.300845
11	OK	36.46	5.21633e-10	5.21633e-10	0.256238
12	OK	40.106	1.16002e-09	1.16002e-09	0.214215
13	OK	43.752	2.498e-09	2.498e-09	0.174192
14	OK	47.398	5.22401e-09	5.22401e-09	0.135742
15	OK	51.044	1.06347e-08	1.06347e-08	0.0985457
16	OK	54.69	2.11157e-08	2.11157e-08	0.0623619
17	OK	58.336	4.09607e-08	4.09607e-08	0.0270055
18	OK	61.982	7.7738e-08	7.7738e-08	-0.00766694
19	OK	65.628	1.44526e-07	1.44526e-07	-0.0417671
20	OK	69.274	2.63503e-07	2.63503e-07	-0.0753825
21	OK	72.92	4.71608e-07	4.71608e-07	-0.108582

Figure 14-37

Going to the main toolbar, and clicking on **Plot>Results Curve** brings up the window in Fig. 14-38, where the choice that has been made is to plot pH versus HCl flow (into a stream of 1000 kg water/hr), resulting in Fig. 14-39.

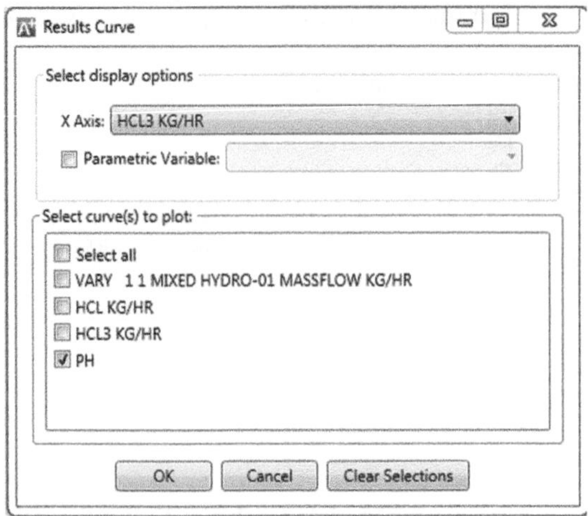

Figure 14-38

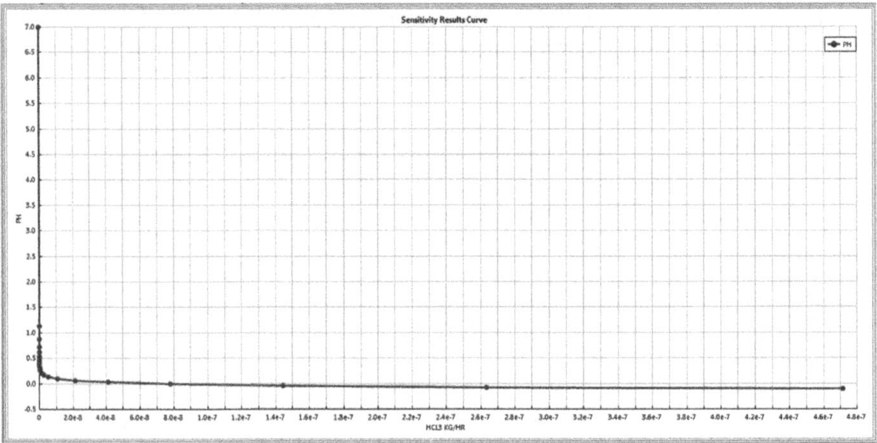

Figure 14-39

For the second example a water + acetic acid + sodium chloride solution at 25°C and 1 bar will be used, and in **Components>Specifications**, water, sodium chloride, and acetic acid are added as shown in Fig. 14-40.

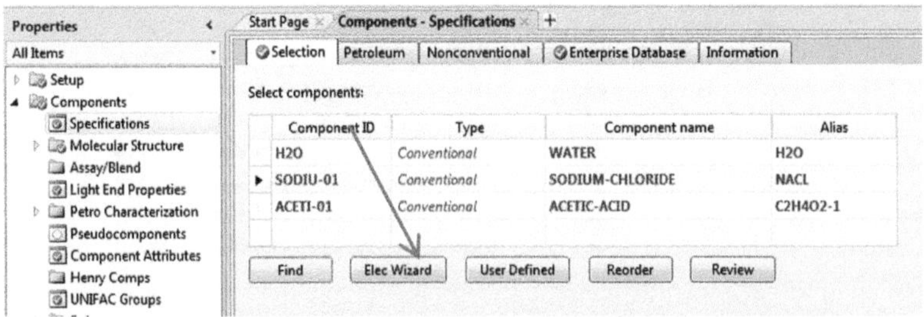

Figure 14-40

Then clicking on **Elec Wizard** results in the window in Fig. 14-41.

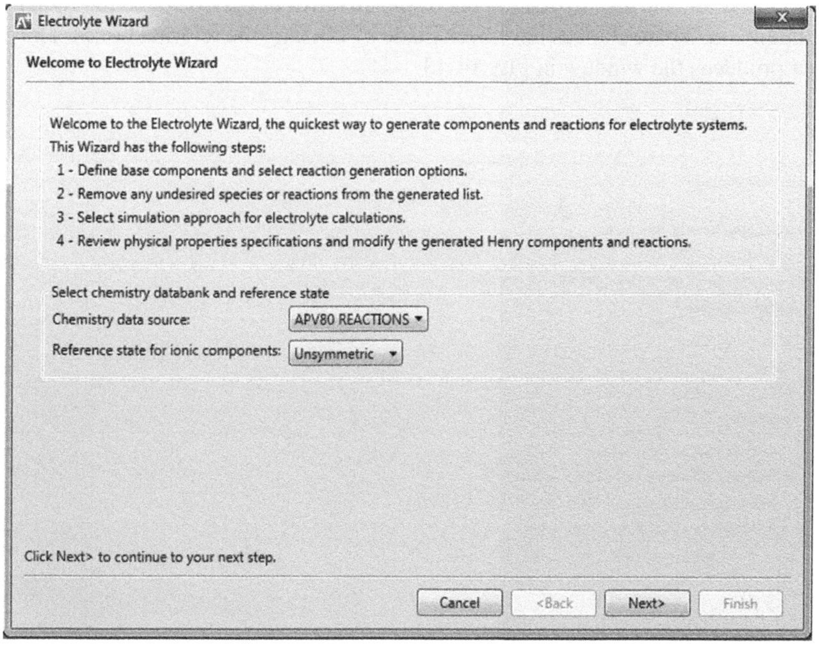

Figure 14-41

Here the Aspen Plus default choices will be accepted. Then press **Next>** to see the window in Fig. 14-42.

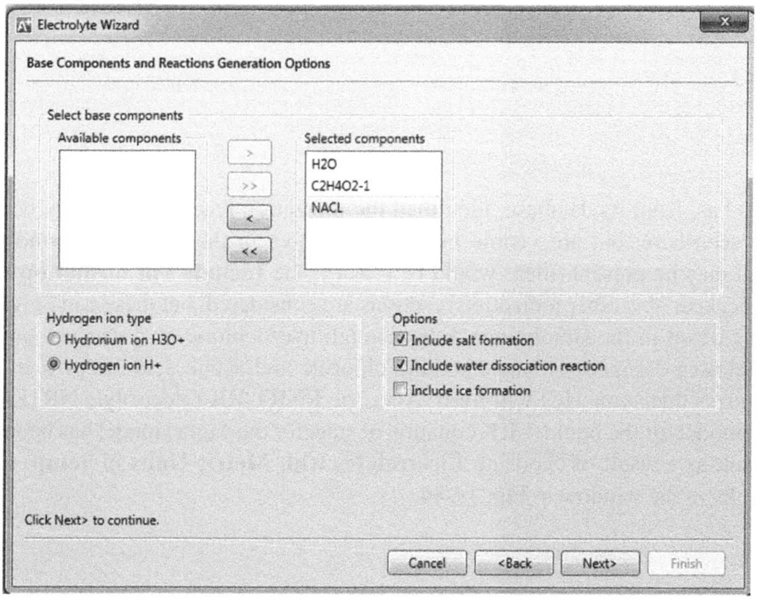

Figure 14-42

In this window the water, acetic acid, and sodium chloride have all been chosen for inclusion. Also, the water dissociation reaction has been checked for inclusion, as has the possibility of salt formation, and also to express the results in terms of the hydrogen ion rather than the hydronium ion. These choices have been made by clicking the relevant boxes. The pressing **Next>** produces the window in Fig. 14-43.

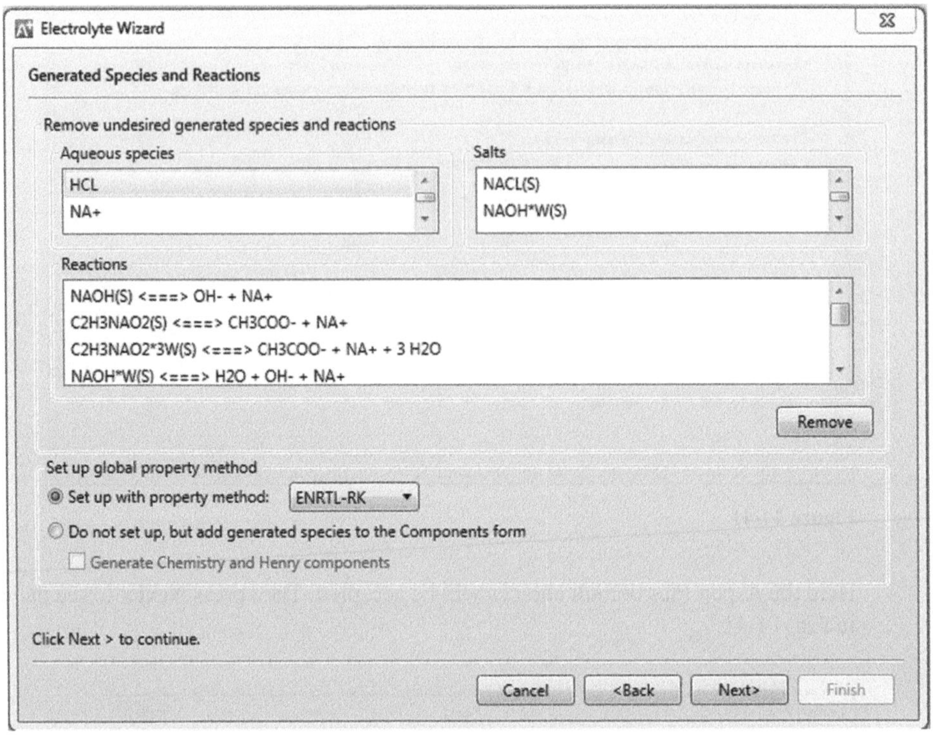

Electrolyte Wizard

Generated Species and Reactions

Remove undesired generated species and reactions

Aqueous species

HCL

NA+

Salts

NACL(S)

NAOH*W(S)

Reactions

NAOH(S) <===> OH- + NA+

C2H3NAO2(S) <===> CH3COO- + NA+

C2H3NAO2*3W(S) <===> CH3COO- + NA+ + 3 H2O

NAOH*W(S) <===> H2O + OH- + NA+

Remove

Set up global property method

Set up with property method: ENRTL-RK

Do not set up, but add generated species to the Components form

Generate Chemistry and Henry components

Click Next > to continue.

Cancel <Back Next> Finish

Figure 14-43

Aspen Plus has, from its database, identified the ionization reactions that may occur (not all can be seen here, but are visible by scrolling down in the simulation window) and species that may be present (there would be fewer if the **Include salt formation** box had not been checked; also other ionization reactions are considered, but these can only be seen by scrolling down in the simulation). Even though hydrochloric acid was not one of the originally chosen compounds, since sodium chloride and acetic acid are present, Aspen Plus recognizes that some HCl will form. Also, the **ENRT-RK** (electrolyte NRTL activity coefficient model for the liquid + RK equation of state for the vapor) model has been chosen as the default as a result of choosing **Electrolytes with Metric Units** in **Setup**. Clicking **Next>** results in the window in Fig. 14-44.

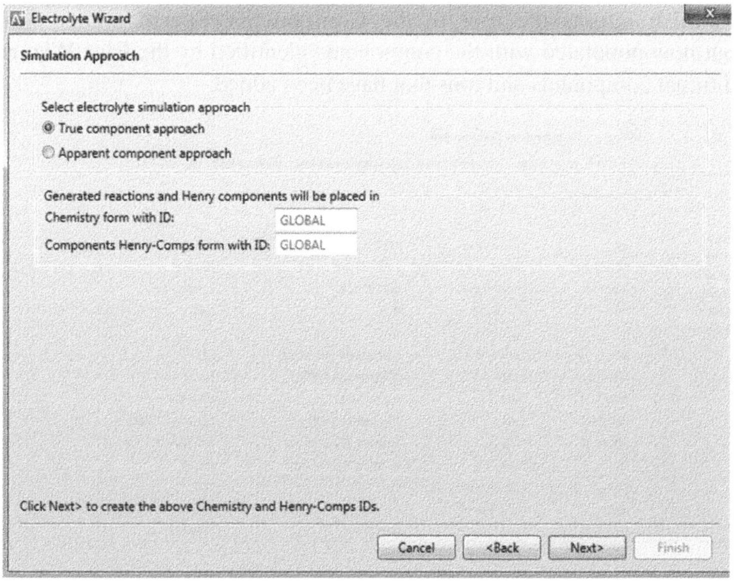

Figure 14-44

The choice here is again between the **True component approach**, in which each ionic species appears in the output, or the **Apparent component approach** in which only compounds, not ions, appear in the output. The former will be used here. Clicking on **Next>** produces the warning in Fig. 14-45.

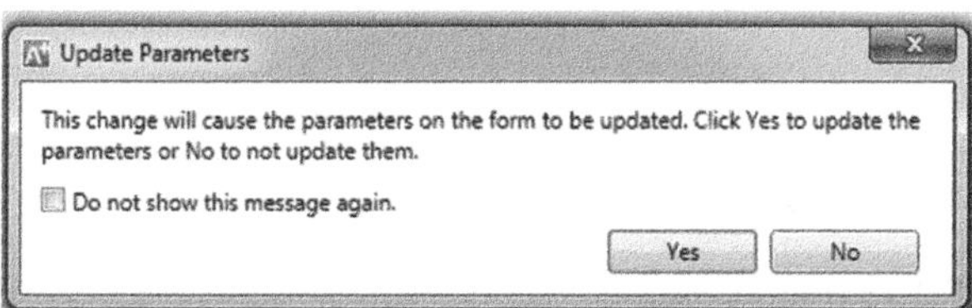

Figure 14-45

Choose **Yes** and then **Next>** to come to the **Summary** window in Fig. 14-46.

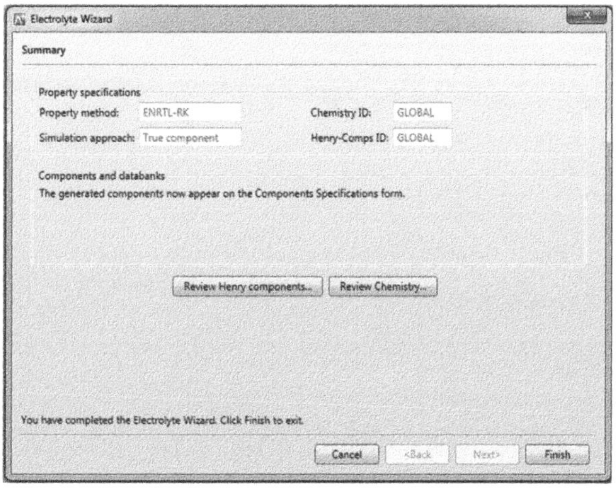

Figure 14-46

Clicking on **Finish** returns the user to the **Components>Specifications** window of Fig. 14-47, but now populated with the components identified by the **Elec Wizard**. Note all of the additional compounds and ions that have been added.

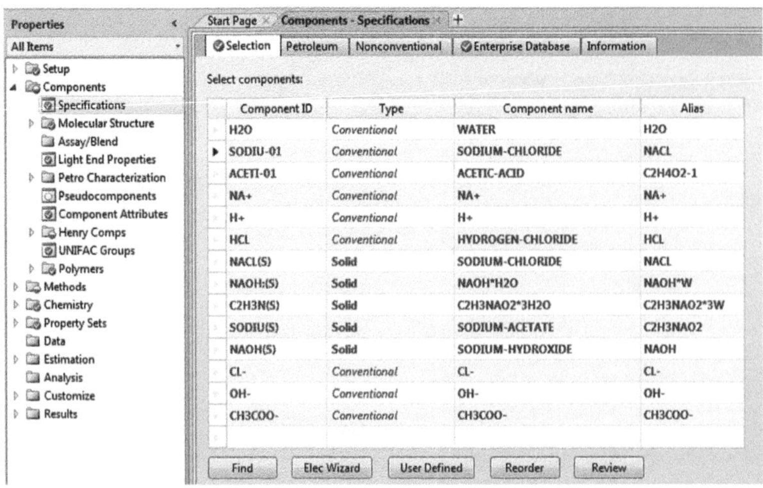

Figure 14-47

Next going to **Henry Comps>Global** a decision has to be made as to the components that will be described by Henry's Law. The Aspen Plus default is hydrogen chloride, HCl, though the user can add others to the list. The default choice will be accepted here as shown in Fig. 14-48 as the other components are not sufficiently volatile to appear in the vapor phase.

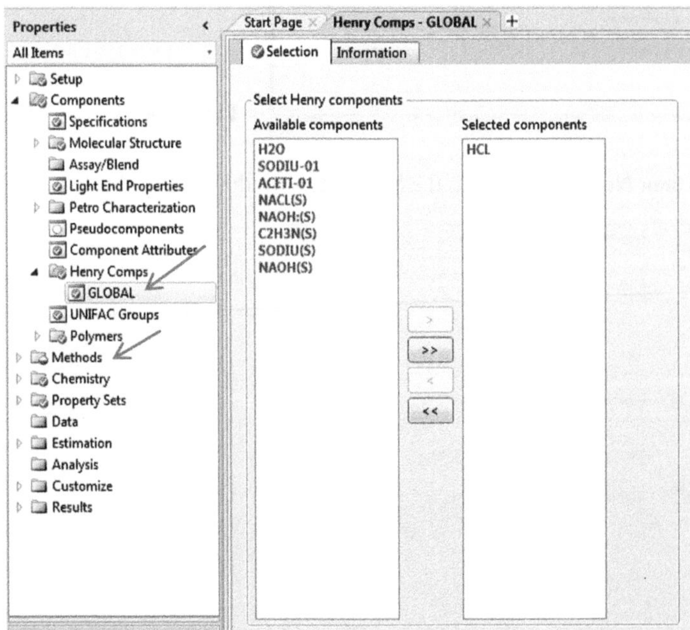

Figure 14-48

The next step is to go to **Methods** and populate the resulting window as shown in Fig. 14-49.

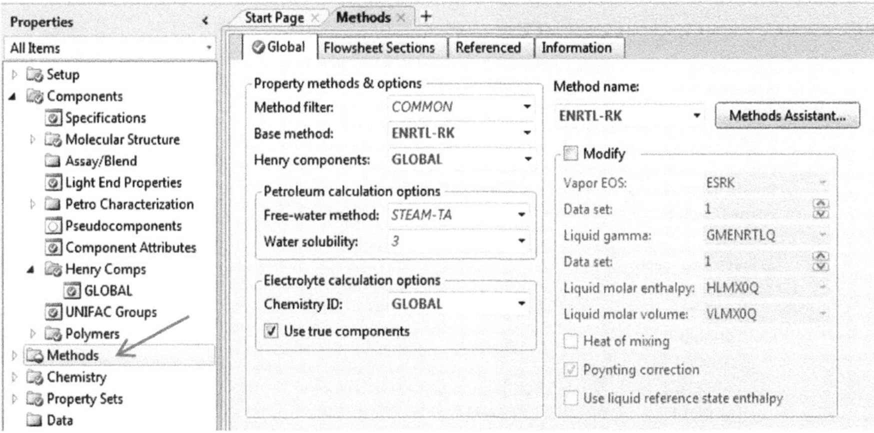

Figure 14-49

Next go to **Parameters>Binary Interaction** (Fig. 14-50), and then to **HENRY-1** to show the values of the water–HCl and HCl–acetic acid parameters in the Aspen Plus data bank (Fig. 14-51). We will accept these parameters here though the user can specify their own.

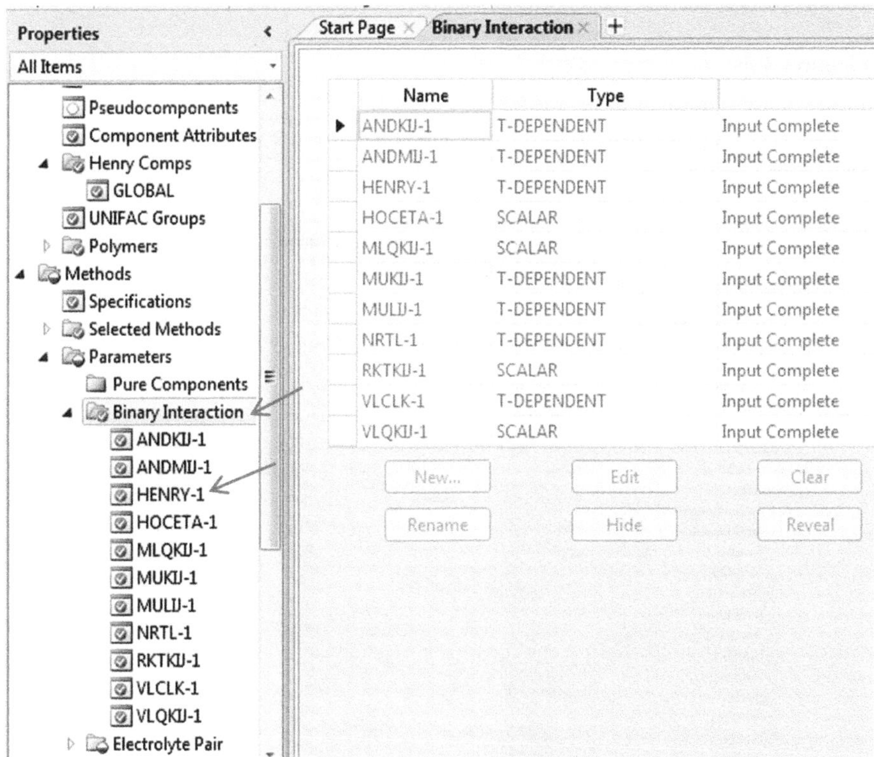

Figure 14-50

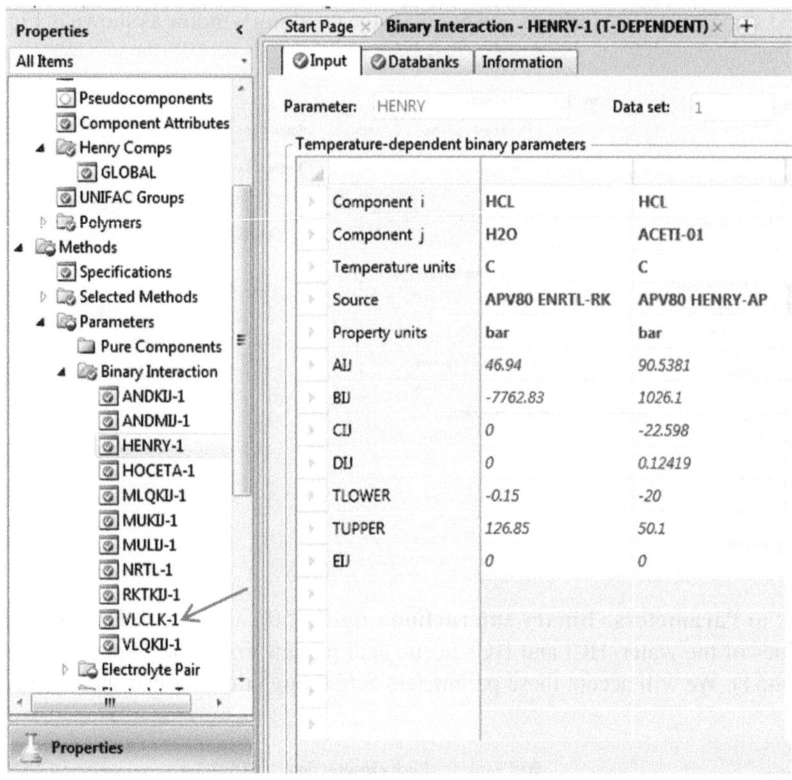

Figure 14-51

Next choose **Binary Interaction>VLCLK-1** to show the values of the ion–ion interactions (Fig. 14-52).

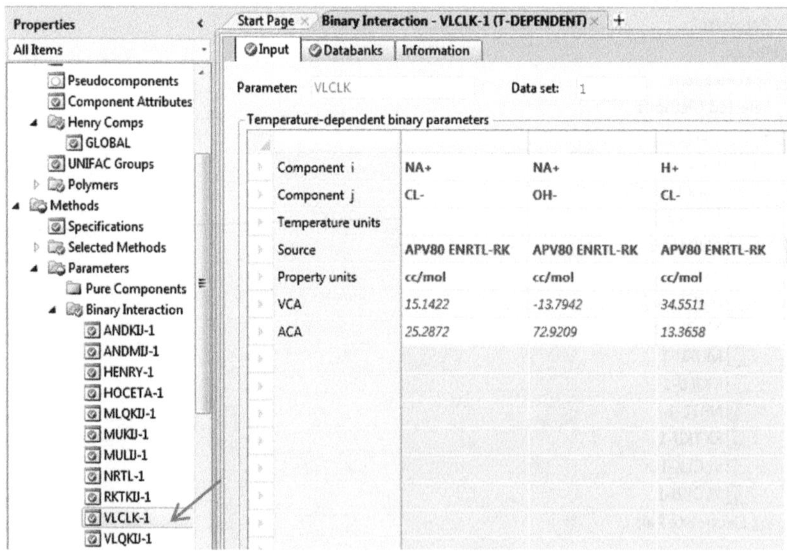

Figure 14-52

Again, the default parameter values will be accepted for this example. Finally, going to **Parameters>Electrolyte Pair** brings up a whole collection of ion–ion parameters that must either be specified, or one has to accept the choices presented by clicking on them. Two example sets of parameter values, **GMELCC-1** and **GMELCN-1**, are shown in Figs. 14-53 and 14-54. Note that because of the complexity of electrolyte systems and the models used, all these **Electrolyte Pair** parameter sets must be specified, either by accepting the Aspen Plus default values or by the user specifying their own values.

Start Page | Electrolyte Pair - GMELCC-1

Input | Databanks | Information

Parameter: GMELCC Data set: 1 Temperature units:

Electrolyte pair parameters

Molecule i or Electrolyte i		Molecule j or Electrolyte j		Value
H2O		NA+	CL-	5.9802
NA+	CL-	H2O		-3.78917
H2O		NA+	OH-	6.738
NA+	OH-	H2O		-3.77122
H2O		NA+	CH3COO-	8.40108
NA+	CH3COO-	H2O		-4.49111
H2O		H+	CL-	41.674
H+	CL-	H2O		-22.154
H2O		H+	OH-	8.045
H+	OH-	H2O		-4.072
HCL		NA+	CL-	15
NA+	CL-	HCL		-8
HCL		NA+	OH-	15
NA+	OH-	HCL		-8
HCL		H+	CL-	0.001
H+	CL-	HCL		-0.001
NA+	CL-	NA+	OH-	1.95044
NA+	OH-	NA+	CL-	8.40768

Figure 14-53

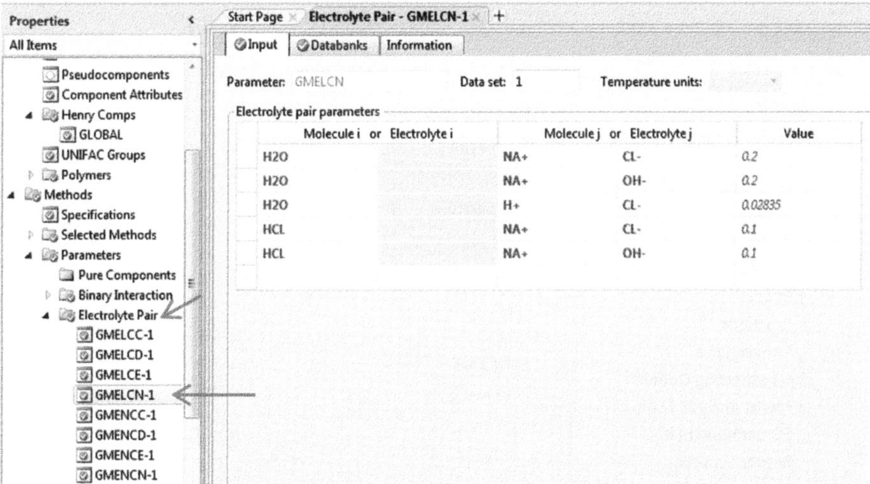

Start Page | Electrolyte Pair - GMELCN-1

Input | Databanks | Information

Parameter: GMELCN Data set: 1 Temperature units:

Electrolyte pair parameters

Molecule i or Electrolyte i		Molecule j or Electrolyte j		Value
H2O		NA+	CL-	0.2
H2O		NA+	OH-	0.2
H2O		H+	CL-	0.02835
HCL		NA+	CL-	0.1
HCL		NA+	OH-	0.1

Figure 14-54

Once all these parameters have been specified and the **Properties** box is checked, one can proceed to do a simulation. A simple two-phase flash, **Flash2**, will be used here as in the previous illustration.

Now going to **Streams>1**, the input for this example, we will choose the feed stream to be 1.3 mols of water, 0.2 mols of dissolved NaCl (not solid salt, which would be NaCl(s)), and 0.5 mols of acetic acid. Note that this stream data input window (Fig. 14-55) also provides the opportunity to include the amounts of each of the ions, which are unknown unless one has already solved the ionization equilibrium equations.

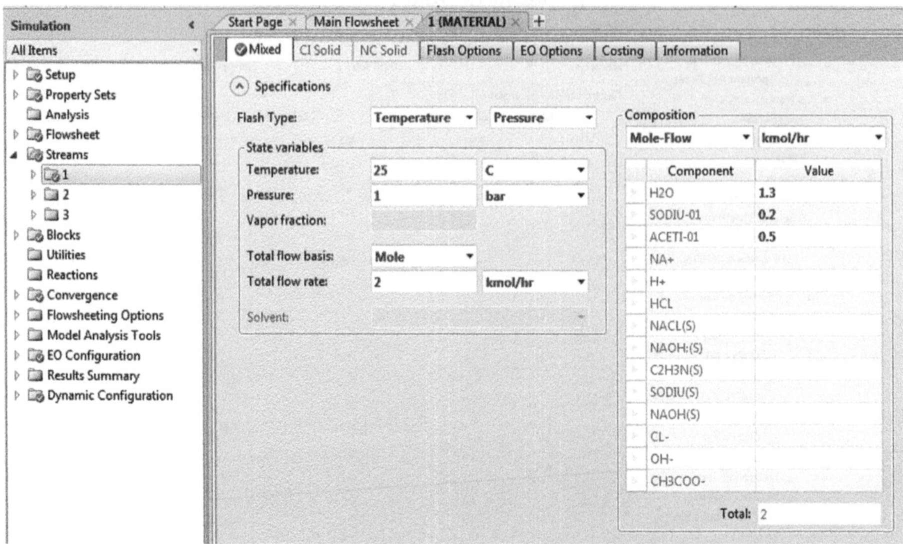

Figure 14-55

Next going to **Blocks>B1**, the two-phase flash operating conditions are set, here 25°C and 1 bar, in the window of Fig. 14-56.

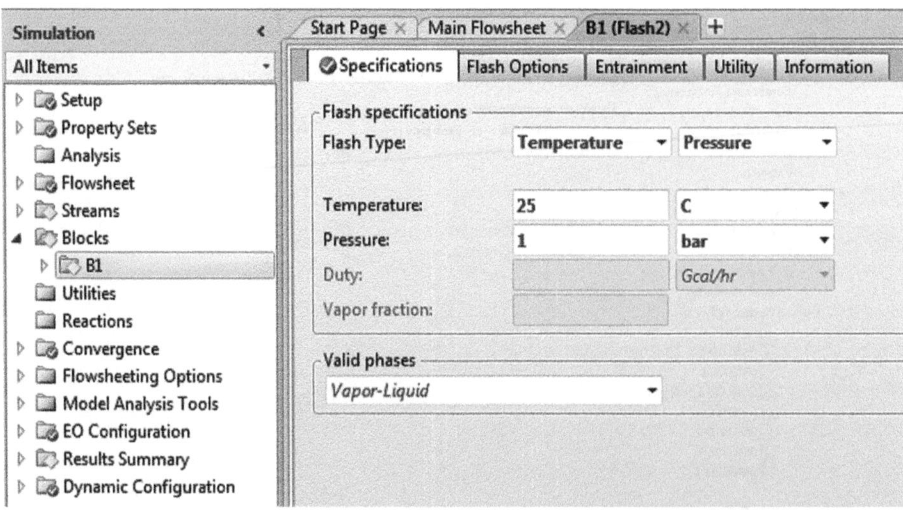

Figure 14-56

The simulation can now be run. The following results are obtained (presented in Fig. 14-57 in abbreviated and reformatted form by copying and pasting into EXCEL and keeping only the items shown in that figure).

	1	2	3				
Temperature C	25		25				
Pressure bar	1	1	1				
Vapor Frac	0		0				
Solid Frac	0		0				
Mole Flow kmol/hr	2.204	0	2.204				
Mole Flow kmol/hr				Mole Frac			
H2O	1.3	0	1.3	H2O	0.59	0	0.59
NACL	0	0	0	NACL	0	0	0
C2H4O-01	0.496	0	0.496	C2H4O-01	0.225	0	0.225
H+	0.004	0	0.004	H+	0.002	0	0.002
NA+	0.2	0	0.2	NA+	0.091	0	0.091
HCL	0	0	0	HCL	0	0	0
C2H3N(S)	0	0	0	C2H3N(S)	0	0	0
NACL(S)	0	0	0	NACL(S)	0	0	0
SODIU(S)	0	0	0	SODIU(S)	0	0	0
NAOH(S)	0	0	0	NAOH(S)	0	0	0
NAOH:(S)	0	0	0	NAOH:(S)	0	0	0
OH-	0	0	0	OH-	0	0	0
CH3COO-	0.004	0	0.004	CH3COO-	0.002	0	0.002
CL-	0.2	0	0.2	CL-	0.091	0	0.091

Figure 14-57

There are several interesting observations about these results. First, even though we specified the feed composition as water:NaCl:acetic acid in the ratio of 1.3:0.2:0.5, Aspen Plus first solved the ionization equations to find that in the feed the NaCl is completely ionized (as expected for a strong electrolyte) and that acetic acid was only very slightly ionized. These calculated results are what now appear in the molar flow rates and mole fractions. Second, for the conditions chosen, there is no vapor phase and no solid salts present. Finally, given the small amount of acetic acid that is ionized, the hydrogen ion concentration (from the acetic acid) is low, and there is essentially no water ionization as indicated by the hydroxide ion (OH^-) concentration being zero to the accuracy of the calculations.

It is also of interest to know the values of the activity coefficients of the species in the liquid phases. To get this information, we again need to use the **Properties>Prop-sets** (Fig. 14-58).

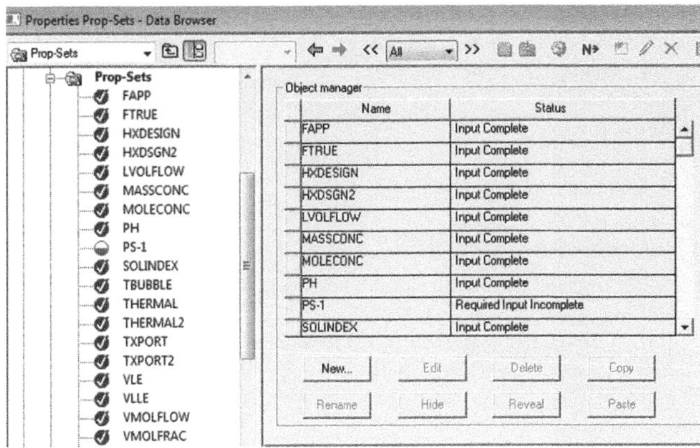

Figure 14-58

Then click on **New...** that brings up the pop-up menu in the window of Fig. 14-59 in which the default name of **PS-1** has been accepted.

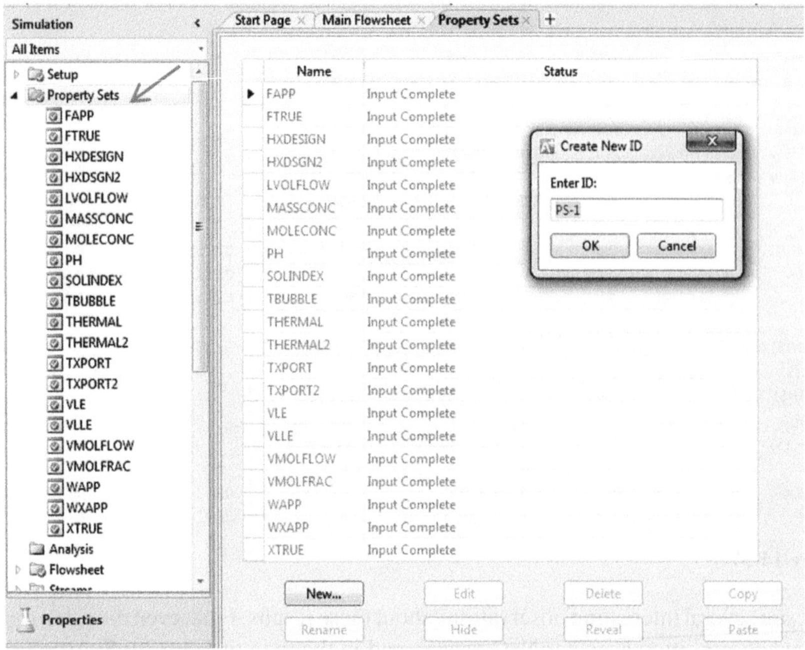

Figure 14-59

As before, **GAMMA** has been chosen from the drop-down list of properties as seen in Fig. 14-60.

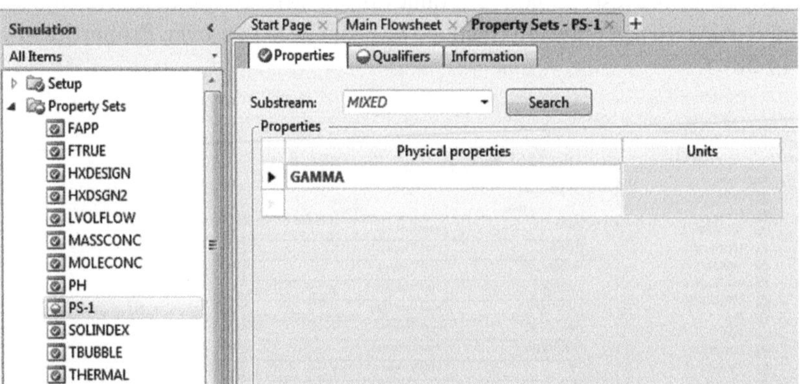

Figure 14-60

Then after clicking on the **Qualifiers** tab, the choices in Fig. 14-61 were made.

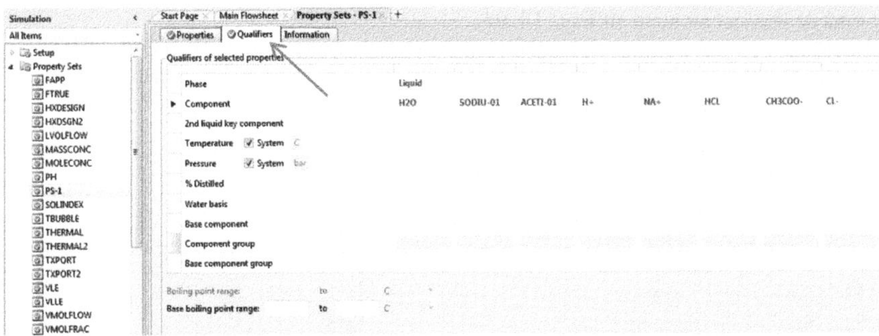

Figure 14-61

Now returning to **Setup**, click on **Report Options>Stream>Property Sets** (Fig. 14-62),

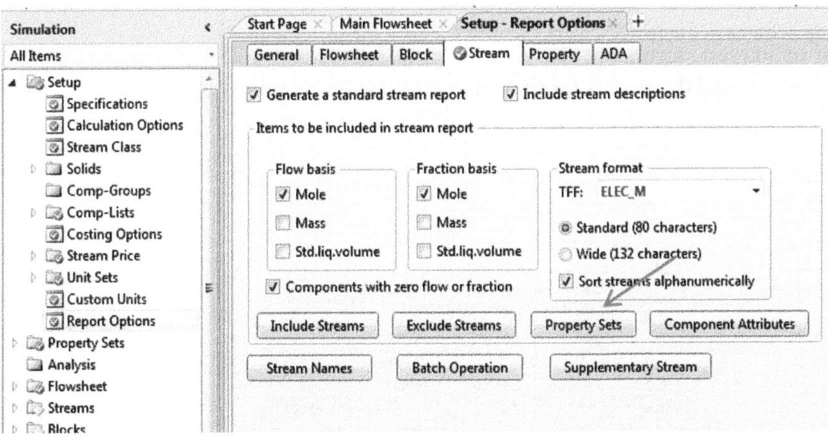

Figure 14-62

then select and move **PS-1** and **PH** from the drop-down menu on the left of the window in Fig. 14-63, and then click on **Close.**

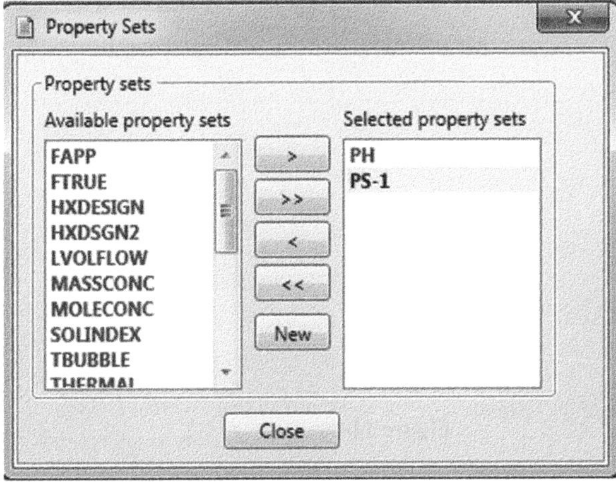

Figure 14-63

Now rerunning the simulation at 25°C and 1 bar gives the same results as above, with the exception that the values for the activity coefficients and the solution pH are appended at the end of the list as seen in Fig. 14-64.

*** LIQUID PHASE ***		
Act Coef		
H2O	0.738	0.738
SODIU-01		
ACETI-01	0.875	0.875
H+	0.011	0.011
NA+	1.171	1.171
HCL	0.875	0.875
CH3COO-	2.042	2.042
CL-	1.1	1.1
pH	3.011	3.011

Figure 14-64

Out of interest, rerunning the simulation at 125°C and 1.25 bar produces the results in Fig. 14-65, and the results for the activity coefficients and pH in Fig. 14-66.

	1	2	3
Temperature C	25	125	125
Pressure bar	1	1.25	1.25
Vapor Frac	0	1	0
Solid Frac	0	0	0
Mole Flow kmol/hr	2.204	0.167	2.035
Mass Flow kg/hr	65.135	4.41	60.725
H2O	1.3	0.133	1.167
SODIU-01			
ACETI-01	0.496	0.033	0.465
NA+	0.2		0.2
H+	0.004		0.002
HCL	trace	trace	trace
NACL(S)			
NAOH:(S)			
C2H3N(S)			
SODIU(S)			
NAOH(S)			
CL-	0.2		0.2
OH-	trace		trace
CH3COO-	0.004		0.002

Mole Frac			
H2O	0.59	0.8	0.573
SODIU-01			
ACETI-01	0.225	0.2	0.228
NA+	0.091		0.098
H+	0.002		947 PPM
HCL	trace	trace	trace
NACL(S)			
NAOH:(S)			
C2H3N(S)			
SODIU(S)			
NAOH(S)			
CL-	0.091		0.098
OH-	trace		trace
CH3COO-	0.002		947 PPM

Figure 14-65

*** LIQUID PHASE ***		
Act Coef		
H2O	0.738	0.758
SODIU-01		
ACETI-01	0.875	0.874
H+	0.011	0.004
NA+	1.171	1
HCL	0.875	0.874
CH3COO-	2.042	1.523
CL-	1.1	0.967
pH	3.011	3.637

Figure 14-66

Note that the activity coefficients and pH values are different in the feed and the liquid product stream since the feed was at 25°C while the flash unit was operated at 125°C, and that there is a change in composition of the liquid stream since a vapor stream containing water and acetic acid has been produced. Also, note that the activity coefficients and pH are only calculated for the liquid streams as they have no meaning for a vapor stream.

PROBLEMS

14.1. Calculate the pH and species activity coefficients of aqueous 0.25 M hydrogen chloride solutions at 5 atm over the temperature range of 25°C to 100°C.

14.2. Calculate the pH and mean ionic activity coefficients of aqueous hydrogen chloride solutions at 25°C and 1 atm over the composition range of 0 to 6 M using Aspen Plus and the ENRT-RK model, and compare the activity coefficient values with the following experimental data.

Molality	$\gamma\pm$
0.0	1.00
0.1	0.778
0.25	0.720
0.5	0.681
0.75	0.665
1.0	0.657
2.0	0.669
3.0	0.714
4.0	0.782
5.0	0.873
6.0	0.987

14.3. Calculate the pH and species activity coefficients of aqueous acetic acid solutions at 25°C and 1 atm over the range of 0 to 2 M acetic acid.

14.4. Calculate the pH and species activity coefficients of aqueous sodium hydroxide solutions at 25°C and 1 atm over the range of 0 to 2 M sodium hydroxide.

14.5. Use Aspen Plus and the ENRT-RK model to predict for the mean ionic activity coefficients of lithium bromide in aqueous solution at 25°C and compare with the following experimental data.

M_{LiBr}	$\gamma\pm$	M_{LiBr}	$\gamma\pm$	M_{LiBr}	$\gamma\pm$
0.001	0.967	0.5	0.739	10	19.92
0.005	0.934	1.0	0.774	12	46.3
0.01	0.891	3	1.156	14	104.7
0.05	0.847	5	2.74	16	198.
0.1	0.790	8	8.61	20	485.

14.6. Use Aspen Plus and the ENRT-RK model to predict for the mean ionic activity coefficients of the following salts in aqueous solution at 25°C and compare with the experimental data below.

		$\gamma\pm$	
Molality	KCl	$CrCl_3$	$Cr_2(SO_4)_3$
0.1	0.770	0.331	0.0458
0.2	0.718	0.298	0.0300
0.3	0.688	0.294	0.0238
0.5	0.649	0.314	0.0190
0.6	0.637	0.335	0.0182
0.8	0.618	0.397	0.0185
1.0	0.604	0.481	0.0208

14.7. Use Aspen Plus and the ENRT-RK model to predict the mean ionic activity coefficients of the following compounds in aqueous solution at 25°C and compare with the experimental data below for the mean ionic activity coefficients.

M	HCl	$CaCl_2$	$ZnSO_4$
0.001	0.966	0.888	0.734
0.005	0.928	0.789	0.477
0.01	0.905	0.732	0.387
0.05	0.830	0.584	0.202
0.1	0.796	0.531	0.148
0.5	0.757	0.457	0.063
1.0	0.809	0.509	0.044
2.0	1.009	0.807	0.035
3.0	1.316	1.055	0.041

Index

Note: Since this is a monograph about thermodynamics and Aspen Plus®, some terms are used very many times. Citing each such occurrence in this index would not be very useful. Therefore, this index includes only the first and other significant occurrences of each item. Aspen Plus commands or terms specific to this software are shown in **bold** font.

Using Aspen Plus® in Thermodynamics Instruction: A Step-by-Step Guide, First Edition. Stanley I. Sandler.

Printed and bound by CPI Group (UK) Ltd, Croydon, CR0 4YY

27/10/2024

14580280-0003